SECOND EDITION

Practical Handbook on
IMAGE PROCESSING for
SCIENTIFIC and
TECHNICAL APPLICATIONS

Bernd Jähne
University of Heidelberg

CRC PRESS

Boca Raton London New York Washington, D.C.

Library of Congress Cataloging-in-Publication Data

Jèahne, Bernd, 1953-
Practical handbook on image processing for scientific and technical applications / Berne Jèahne.— 2nd ed.
     p. cm.
    Includes bibliographical references and index.
    ISBN 0-8493-1900-5 (alk. paper)
    1. Image processing—Digital techniques—Handbooks, manuals, etc. I. Title.

TA1637.J347 2004
621.36′7—dc22
                                                           2004043570

Visit the CRC Press Web site at www.crcpress.com

No claim to original U.S. Government works
International Standard Book Number 0-8493-1900-5
Library of Congress Card Number 2004043570
Printed in the United States of America 1 2 3 4 5 6 7 8 9 0
Printed on acid-free paper

Preface

What This Handbook Is About

Digital image processing is a fascinating subject in several aspects. Human beings perceive most of the information about their environment through their visual sense. While for a long time images could only be captured by photography, we are now at the edge of another technological revolution that allows image data to be captured, manipulated, and evaluated electronically with computers.

With breathtaking pace, computers are becoming more powerful and at the same time less expensive. Thus, the hardware required for digital image processing is readily available. In this way, image processing is becoming a common tool to analyze multidimensional scientific data in all areas of natural science. For more and more scientists, digital image processing will be the key to studying complex scientific problems they could not have dreamed of tackling only a few years ago. A door is opening for new interdisciplinary cooperation merging computer science with corresponding research areas. Thus, there is a need for many students, engineers, and researchers in natural and technical disciplines to learn more about digital image processing.

Since original image processing literature is spread over many disciplines, it is hard to gather this information. Furthermore, it is important to realize that image processing has matured in many areas from ad hoc, empirical approaches to a sound science based on well-established principles in mathematics and physical sciences.

This handbook tries to close this gap by providing the reader with a sound basic knowledge of image processing, an up-to-date overview of advanced concepts, and a critically evaluated collection of the best algorithms, demonstrating with real-world applications. Furthermore, the handbook is augmented with usually hard-to-find practical tips that will help to avoid common errors and save valuable research time. The wealth of well-organized knowledge collected in this handbook will inspire the reader to discover the power of image processing and to apply it adequately and successfully to his or her research area. However, the reader will not be overwhelmed by a mere collection of all available methods and techniques. Only a carefully and critically evaluated selection of techniques that have been proven to solve real-world problems is presented.

Many concepts and mathematical tools, which find widespread application in natural sciences, are also applied to digital image processing. Such analogies are pointed out because they provide an easy access to many complex problems in digital image processing for readers with a general background in natural sciences. The author — himself educated in physics and computer science — merges basic research in digital image processing with key applications in various disciplines.

This handbook covers all aspects of image processing from image formation to image analysis. Volumetric images and image sequences are treated as a natural extension of image processing techniques from two to higher dimensions.

Prerequisites

It is assumed that the reader is familiar with elementary matrix algebra as well as the Fourier transform. Wherever possible, mathematical topics are described intuitively, making use of the fact that image processing is an ideal subject to illustrate even complex mathematical relations. Appendix B outlines linear algebra and the Fourier transform to the extent required to understand this handbook. This appendix serves also as a convenient reference to these mathematical topics.

How to Use This Handbook

This handbook is organized by the tasks required to acquire images and to analyze them. Thus, the reader is guided in an intuitive way and step by step through the chain of tasks. The structure of most chapters is as follows:

1. A summary page highlighting the major topics discussed in the chapter.
2. Description of the tasks from the perspective of the application, specifying and detailing what functions the specific image processing task performs.
3. Outline of concepts and theoretical background to the extent that is required to fully understand the task.
4. Collection of carefully evaluated procedures including illustration of the theoretical performance with test images, annotated algorithms, and demonstration with real-world applications.
5. Ready-to-use reference data, practical tips, references to advanced topics, emerging new developments, and additional reference material. This reference material is parted into small units, consecutively numbered within one chapter with boxed numbers, e. g., 3.1 . The reference item is referred to by this number in the following style: ≻ 3.1 and 3.3.

Exceptions from this organization are only the two introductory Chapters 1 and 2. The individual chapters are written as much as possible in an internally consistent way.

Another key to the usage of the handbook is the detailed indices and the glossary. The glossary is unique in the sense that it covers not only image processing in a narrow sense but all important associated topics: optics, photonics, some important general terms in computer science, photogrammetry, mathematical terms of relevance, and terms from important applications of image processing. The glossary contains a brief definition of terms used in image processing with cross-references to find further information in the main text of the handbook. Thus, you can take the glossary as a starting point for a search on a specific item. All terms contained in the indices are emphasized by typesetting in *italic* style.

Acknowledgments

Many of the examples shown in this handbook are taken from my research at Scripps Institution of Oceanography (University of California, San Diego) and at the Institute for Environmental Physics and the Interdisciplinary Center for Scientific Computing (University of Heidelberg). I gratefully acknowledge financial support for this research from the US National Science Foundation (OCE91-15944-02, OCE92-17002, and OCE94-09182), the US Office of Naval Research (N00014-93-J-0093, N00014-94-1-0050), and the German Science Foundation, especially through the interdisciplinary research unit

FOR240 "Image Sequence Analysis to Study Dynamical Processes". I cordially thank my colleague F. Hamprecht. He contributed the last chapter about classification (Chapter 17) to this handbook.

I would also express my sincere thanks to the staff of CRC Press for their constant interest in this handbook and their professional advice. I am most grateful for the invaluable help of my friends at AEON Verlag & Studio in proofreading, maintaining the databases, and in designing most of the drawings.

I am also grateful to the many individuals, organizations, and companies that provided valuable material for this handbook:

- Many of my colleagues — too many to be named individually here — who worked together with me during the past seven years within the research unit "Image Sequence Analysis to Study Dynamical Processes" at Heidelberg University
- Dr. M. Bock, DKFZ Heidelberg
- Dr. J. Klinke, PORD, Scripps Institution of Oceanography, University of California, San Diego
- Prof. H.-G. Maas, Institute of Photogrammetry and Remote Sensing, University of Dresden
- Prof. J. Ohser, FH Darmstadt
- Dr. T. Scheuermann, Fraunhofer Institute for Chemical Technology, Pfinztal, Germany
- Prof. Trümper, Max-Planck-Institute for Extraterrestric Physics, Munich
- ELTEC Elektronik GmbH, Mainz, Germany
- Dr. Klee, Hoechst AG, Frankfurt, Germany
- Optische Werke G. Rodenstock, Precision Optics Division, D-80469 Munich
- Prof. J. Weickert, University of Saarbrücken, Germany
- Zeiss Jena GmbH, Jena, Germany
- Dr. G. Zinser, Heidelberg Engineering, Heidelberg, Germany

The detailed description on imaging sensors in Chapter 5 is based on an extensive camera test program. I am grateful to the manufacturers and distributors who provided cameras at no cost: Adimec, Allied Vision, Basler Vision Technologies, IDS, PCO, Pulnix, and Stemmer Imaging (Dalsa, Jai).

Most examples contained in this handbook have been processed using **heurisko®**, a versatile and powerful image processing package. **heurisko®** has been developed by AEON[1] in cooperation with the author.

In a rapid progressing field such as digital image processing, a major work like this handbook is never finished or completed. Therefore, any comments on further improvements or additions to the handbook are very welcome. I am also grateful for hints on errors, omissions, or typing errors, which despite all the care taken may have slipped my attention.

Heidelberg, Germany, January 2004   Bernd Jähne

---

[1] AEON Verlag & Studio, Hanau, Germany, http://www.heurisko.de

Contents

I From Objects to Images

II Handling and Enhancing Images

III From Images to Features

V Appendices

1 Introduction

1.1 Highlights

Electronic imaging and digital image processing constitute — after the invention of photography — the second revolution in the use of images in science and engineering (Section 1.2). Because of its inherently interdisciplinary nature, image processing has become a major integrating factor stimulating communication throughout engineering and natural sciences.

For technical and scientific applications, a wide range of quantities can be imaged and become accessible for spatial measurements. Examples show how appropriate optical setups combined with image processing techniques provide novel measuring techniques including:

Geometric measurements (Section 1.3):

- size distribution of pigment particles and bubbles (Sections 1.3.1 and 1.3.2)
- counting and gauging of cells in bioreactors (Section 1.3.3)

Radiometric measurements (Section 1.4)

- spatio-temporal concentration fields of chemical specimen (Section 1.4.1)
- surface temperature of plant leaves and tumors (Section 1.4.2)
- slope measurements of short ocean wind waves (Section 1.4.3)
- radar imaging in Earth Sciences (Section 1.4.4)
- X-ray satellite astronomy (Section 1.4.5)
- spectroscopic imaging in atmospheric sciences (Section 1.4.6)

Three-dimensional measurements from volumetric images (Section 1.5)

- surface topography measurements of press forms and the human retina (Sections 1.5.1 and 1.5.2)
- 3-D microscopy of cell nuclei (Section 1.5.3)
- X-ray and magnetic resonance 3-D imaging (Section 1.5.4)

Velocity measurements from image sequences (Section 1.6)

- particle tracking velocimetry for 2-D flow measurements (Section 1.6.1)
- flow tomography for 3-D flow measurements (Section 1.6.2)
- study of motor proteins (Section 1.6.3)

a   *b*

Figure 1.1: *From the beginning of science, researchers tried to capture their observations by drawings.* **a** *With this famous sketch, Leonardo da Vinci [1452-1519] described turbulent flow of water.* **b** *In 1613, Galileo Galilei — at the same time as others — discovered the sun spots. His careful observations of the motion of the spots over an extended period led him to the conclusion that the sun is rotating around its axis.*

1.2 From Drawings to Electronic Images

From the beginning, scientists tried to record their observations in pictures. In the early times, they could do this only in the form of *drawings*. Especially remarkable examples are from *Leonardo da Vinci* (Fig. 1.1a). He was the primary empiricist of visual observation for his time. *Saper vedere* (knowing how to see) became the great theme of his many-sided scientific studies. Leonardo da Vinci gave absolute precedence to the illustration over the written word. He didn't use the drawing to illustrate the text, rather, he used the text to explain the picture.

Even now, illustrations are widely used to explain complex scientific phenomena and they still play an important role in descriptive sciences. However, any visual observation is limited to the capabilities of the human eye.

The invention of *photography* triggered the first revolution in the use of images for science. The daguerretype process invented by the French painter *Jaque Daguerre* in 1839 became the first commercially utilized photographic process. Now, it was possible to record images in an objective way. Photography tremendously extended the possibilities of visual observation. Using flash light for illumination, phenomena could be captured that were too fast to be recognized by the eye. It was soon observed that photographic plates are also sensitive to non-visible radiation such as ultraviolet light and electrons. Photography played an important role in the discovery of *X-rays* by *Wilhelm Konrad Röntgen* in 1895 and led to its widespread application in medicine and engineering.

However, the cumbersome manual evaluation of photographic plates restricted the quantitative analysis of images to a few special areas. In astronomy, e. g., the position

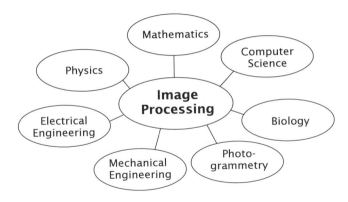

Figure 1.2: *Sciences related to image processing.*

and brightness of stars is measured from photographic plates. Photogrammetrists generate maps from aerial photographs. Beyond such special applications, images have been mostly used in science for qualitative observations and documentation of experimental results.

Now we are experiencing the second revolution of *scientific imaging*. Images can be converted to electronic form, i. e., digital images, that are analyzed quantitatively using computers to perform exact measurements and to visualize complex new phenomena.

This second revolution is more than a mere improvement in techniques. Techniques analogous to the most powerful human sense, the visual system, are used to get an insight into complex scientific phenomena. The successful application of image processing techniques requires an interdisciplinary approach. All kinds of radiation and interactions of radiation with matter can be used to make certain properties of objects visible. Techniques from computer science and mathematics are required to perform quantitative analyses of the images. Therefore, image processing merges a surprisingly wide range of sciences (Fig. 1.2): mathematics, computer science, electrical and mechanical engineering, physics, photogrammetry and geodesy, and biological sciences. As actually no natural science and no engineering field is excluded from applications in image processing, it has become a major integrating factor. In this way, it triggers new connections between research areas. Therefore, image processing serves as an integrating factor and helps to reverse the increasing specialization of science.

This section introduces typical scientific and technical applications of image processing. The idea is to make the reader aware of the enormous possibilities of modern visualization techniques and digital image processing. We will show how new insight is gained into scientific or technical problems by using advanced visualization techniques and digital image processing techniques to extract and quantify relevant parameters. We briefly outline the scientific issues, show the optical setup, give sample images, describe the image processing techniques, and show results obtained by them.

1.3 Geometric Measurements: Gauging and Counting

Counting of *particles* and measuring their *size distribution* is an ubiquitous image processing task. We will illustrate this type of technique with three examples that also demonstrate that a detailed knowledge of the image formation process is required to make accurate measurements.

a

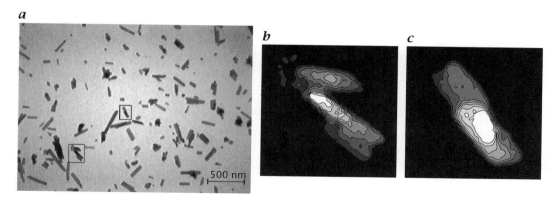

*Figure 1.3: Electron microscopy image of color pigment particles. The crystalline particles tend to cluster; **b** and **c** are contour plots from the areas marked in **a** . Images courtesy of Dr. Klee, Hoechst AG, Frankfurt.*

1.3.1 Size Distribution of Pigment Particles

The quality and properties of paint are largely influenced by the size distribution of the coloring pigment particles. Therefore, a careful control of the size distribution is required. The particles are too small to be imaged with standard light microscopy. Two transmission electron microscopy images are shown in Fig. 1.3. The images clearly demonstrate the difficulty in counting and gauging these particles. While they separate quite well from the background, they tend to form clusters (demonstrated especially by Fig. 1.3b). Two clusters with overlaying particles are shown in Fig. 1.3b and c as contour plots. Processing of these images therefore requires several steps:

1. Identify non-separable clusters
2. Identify overlaying particles and separate them
3. Count the remaining particles and determine their size
4. Compute the size distribution from several images
5. Estimate the influence of the clusters and non-separable particles on size distribution

1.3.2 Gas Bubble Size Distributions

Bubbles are submerged into the ocean by breaking waves and play an important role in various small-scale air-sea interaction processes. They form an additional surface for the exchange of climate-relevant trace gases between the atmosphere and the ocean, are a main source for marine aerosols and acoustic noise, and are significant for the dynamics of wave breaking. Bubbles also play an important role in chemical engineering. They form the surface for gas-liquid reactions in bubble columns where a gas is bubbling through a column through which a liquid is also flowing.

Figure 1.4a shows an underwater photograph of the *light blocking* techniques used to visualize air bubbles. The principle is illustrated in Fig. 1.4b. The bubbles are observed as more or less blurred black ellipses (Fig. 1.5a).

Although this sounds like a very simple application, only careful consideration of details led to a successful application and measurement of bubble size distributions. It is obvious that with this technique the sampling volume is not well defined. This problem could be resolved by measuring the degree of blurring, which is proportional

a

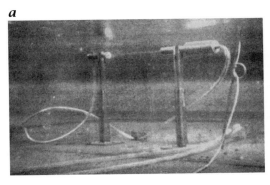

b

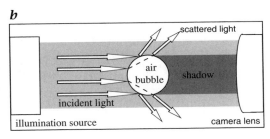

Figure 1.4: a *Underwater photography of the optical setup for the measurements of bubble size distributions. The light source (on the left) and the optical receiver with the CCD camera (on the right) are mounted 5 to 20 cm below the water's surface. The measuring volume is located in the middle of the free optical pass between source and receiver and has a cross-section of about $6 \times 8\ mm^2$.* **b** *Visualization principle: bubbles scatter light away from the receiver for almost the entire cross section and thus appear as black circles (Fig. 1.5).*

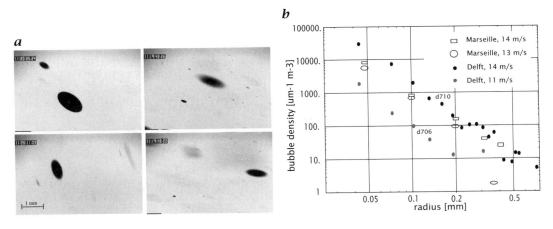

Figure 1.5: a *Four example images of bubbles that can be observed as more or less blurred black ellipses;* **b** *size distribution of bubbles measured with the instrument shown in Fig. 1.4 at wind speeds of $14\ ms^{-1}$ (black circles) and $11\ ms^{-1}$ (gray circles) and 5 cm below the mean water surface in the wind/wave flume of Delft Hydraulics, The Netherlands. The results are compared with earlier measurements in the Marseille wind/wave flume using a laser light scattering technique.*

to the distance of the bubble from the focal plane [69, 70]. Such a type of technique is called *depth from focus* and only one image is required to measure the distance from the focal plane. This approach has the additional benefit that the depth of the sampling volume is proportional to the radius of the bubbles. In this way, larger bubbles that occur much more infrequently can be measured with significantly better counting statistics. Figure 1.5b shows a sample of a size distribution measured in the large wind/wave flume of Delft Hydraulics in The Netherlands.

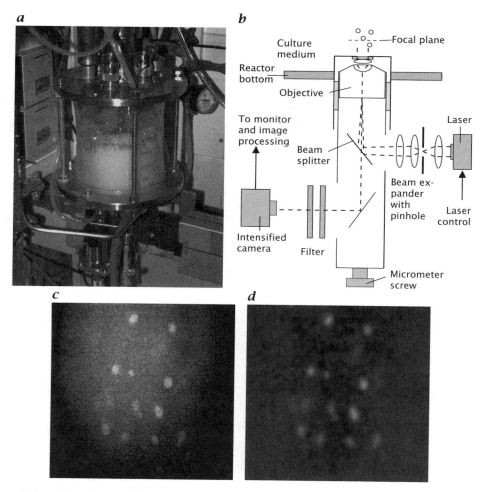

Figure 1.6: *a Experimental bioreactor at the university of Hannover used to test the new in situ microscopy technique for cell counting. The laser can be seen in the lower right of the image, the intensified camera at the lower left of the camera below the bioreactor. b Sketch of the optical setup for the in situ microscopy. The illumination with a pulsed nitrogen laser is applied via a beam splitter through the objective of the microscope. c Example cell images, made visible by excitation of the metabolism-related NADH/NADPH fluorescence with a nitrogen laser; d improved by adaptive filtering. From Scholz [125].*

1.3.3   In Situ Microscopy of Cells in Bioreactors

A similar technique as for the measurement of bubble size distributions can also be used to measure the concentration of cells in *bioreactors* in situ. Conventional off-line measuring techniques require taking samples to measure cell concentrations by standard *counting* techniques in the microscope. Flow *cytometry* has the disadvantage that probes must be pumped out of the bioreactor, which makes the sterility of the whole setup a much more difficult task. A new in-situ microscopy technique uses a standard flange on the bioreactor to insert the microscope so that contamination is excluded. This technique that is suitable for an in situ control of fermentation processes has been developed in cooperation between the ABB Research Center in Heidelberg, the

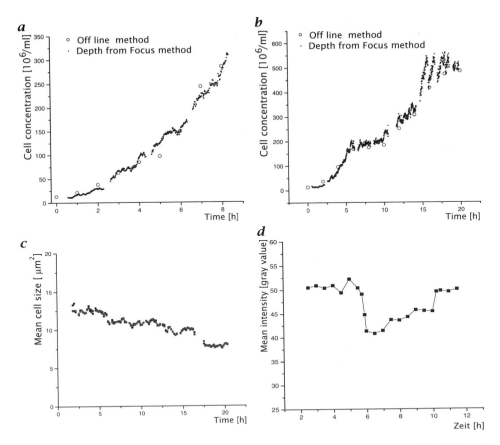

Figure 1.7: *Fermentation processes in a bioreactor as they can be observed with in situ microscopy.* ***a*** *Fed-batch culture (continuous support of nutrition after the culture reaches the stationary phase).* ***b*** *Batch culture fermentation (no additional nutrition during the fermentation).* ***c*** *Decrease of the mean cell size as observed in the fermentation shown in a.* ***d*** *Temporal change of the mean intensity of the NADH/NADPH fluorescence as observed during the fermentation shown in* ***b***. *From Scholz [125].*

Institute for Technical Chemistry at the University of Hannover, and the Institute for Environmental Physics at Heidelberg University [141].

Figure 1.6a, b shows the setup. The cells are made visible by stimulation of the NADH/ NADPH *fluorescence* using a *nitrogen laser*. In this way, only living cells are measured and can easily be distinguished from dead ones and other particles. Unfortunately, the NADH/NADPH fluorescence is very weak. Thus, an intensified CCD camera is required and the cell images show a high noise level (Fig. 1.6c).

With this high noise level it is impossible to analyze the blurring of the cells for a precise concentration determination. Therefore, first an adaptive smoothing of the images is applied that significantly reduces the noise level but doesn't change the steepness of the edges (Fig. 1.6d). This image material is now suitable for determining the degree of blurring by using a multigrid algorithm and the distance of the cell to the focal plane on a Laplacian pyramid (Section 14.3.5).

The *cell concentrations* determined by the in situ microscopy compare well with off-line cell counting techniques (Fig. 1.7a and b). The technique delivers further information about the cells. The mean *cell size* can be determined and shows a decrease

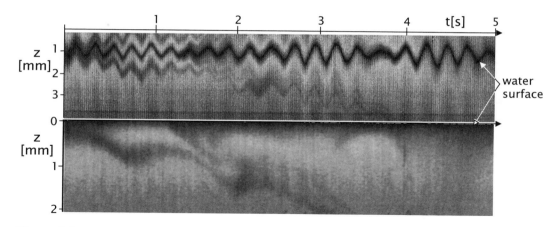

Figure 1.8:
A 5 second time series of vertical concentration profiles of gases dissolved in water right at the interface as measured in the circular wind wave flume at the Institute for Environmental Physics at Heidelberg University. Traces of HCl gas were injected into the air space of the facility. The absorption of HCl at the water surface induces the transport of a fluorescent pH indicator across the aqueous boundary layer giving rise to lower concentration levels of the fluorescent dye towards the water surface. The wind speed was 2.4 m/s. The lower image shows the same sequence of vertical profiles but in a coordinate system moving with the water surface.

during the course of the fermentation process (Fig. 1.7c). One also gets a direct insight into the metabolism. After an initial exponential increase of the cell concentration during the first six hours of the fermentation (Fig. 1.7a), the growth stagnates because all the glucose has been used up. At this point, the NADP fluorescence decreases suddenly (Fig. 1.7d) as it is an indicator of the metabolic activity. After a while, the cells become adapted to the new environment and start burning up the alcohol. Growth starts again and the fluorescence intensity comes back to the original high level.

1.4 Radiometric Measurements: Revealing the Invisible

The radiation received by an imaging sensor at the image plane from an object reveals some of its features. While visual observation by the human eye is limited to the portion of the electromagnetic spectrum called light, imaging sensors are now available for almost any type of radiation. This opens up countless possibilities for measuring object features of interest.

1.4.1 Fluorescence Measurements of Concentration Fields

Here we describe two imaging techniques where concentration fields of chemical species are measured remotely by using fluorescence. The first example concerns the *exchange of gases* between the atmosphere and the ocean, the second the *metabolism* of endolithic cells, which live within the skeleton of *corals*.

The exchange of climate relevant gases such as CO_2, methane, or fluorocarbons between the atmosphere and the ocean are controlled by a thin layer at the water surface, the so-called aqueous viscous boundary layer. This layer is only 30 to 300 μm thick and dominated by molecular diffusion. Its thickness is determined by the degree of turbulence at the ocean interface.

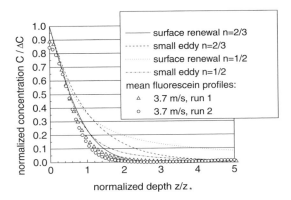

Figure 1.9: *Comparison of the mean concentration profiles computed from sequences of profiles as shown in Fig. 1.8. The measured profile is compared with various theoretical predictions. From [105].*

A new visualization technique was developed to measure the *concentration of gases* dissolved in this layer with high spatial and temporal resolution. The technique uses a chemically reactive gas (such as HCl or NH_3) and *laser-induced fluorescence (LIF)* [67]. Although the technique is quite complex in detail, its result is simple. The measured fluorescence intensity is proportional to the concentration of a gas dissolved in water. Figure 1.8 shows time series of vertical concentration profiles. The thin either bright (NH_3) or dark (HCl) layer indicates the water surface undulated by waves. Because of the total reflection at the water surface, the concentration profiles are seen twice. First directly below the water surface and second as a distorted mirror image apparently above the water surface. Image processing techniques are used to search the extreme values in order to detect the water surface. Then, the concentration profiles can directly be drawn as a function of the distance to the water surface (Fig. 1.8). Now, it can clearly be seen that the boundary layer thickness shows considerable fluctuations and that part of it is detached and transported down into the water bulk. For the first time, a technique is available that gives direct insight into the mechanisms of air-sea gas transfer. Mean profiles obtained by averaging (Fig. 1.9) can directly be compared with profiles computed with various models.

The optical measurement of oxygen is based on the ability of certain dyes or luminophores to change their optical properties corresponding to a change of concentration of the analyte - oxygen. These indicators can be incorporated in polymers and easily spread on transparent support foils allowing the 2D measurement of oxygen and the "look through" (Fig. 1.10c and d). With a special measuring system, called modular luminescence lifetime imaging system (MOLLI), it is possible to use the "delayed" luminescence for the oxygen measurement and white light illumination for structural images. The use of the decaying luminescence (light emitted when the excitation light source is switched off) is possible with a special CCD camera with a fast electronical shutter and additional modulation input (sensicam sensimod, PCO AG), if the luminescence decay times are in the range of μs or larger.

In the presented application, the light dependent metabolism of endolithic cells, which live within the skeleton of massive corals, was investigated. These cells usually see only minimum amounts of light, since most of the light is absorbed in the surface layer of the coral by the coral symbionts. Therefore the oxygen production within the skeleton was measured in relation to various illumination intensities. One result is

Figure 1.10: *Optical measurement of oxygen in corals: a Sample place for the coral, the lagoon of Heron Island, Capricorn Islands, Great Barrier Reef, Australia. b Set-up with the coral in the glass container placed on top of the LED frame while the measuring camera was looking at the cut coral surface from below. c Close up of the set-up with a cut coral porite placed on top of a transparent planar oxygen optode, which was fixed at the bottom of a glass container, filled with lagoon sea water. The blue light is the excitation light, coming from blue LEDs arranged in a frame below the glass, the orange-red light corresponds to the excited luminescence of the optode; The fixed white tube (upper right part) served for flushing the water surface with a constant air stream to aerate the water. d Measured 2D oxygen distribution (view to cut coral surface) given in % air saturation. The oxygen image is blended into a grayscale image from the coral structure. The high oxygen values were generated by endolithic cells, which live within the coral skeleton. The production was triggered by a weak illumination through the oxygen sensor, which simulated the normal light situation of these cells at daylight in the reef. Images courtesy of Dr. Holst, PCO AG, Kelheim, Germany. (See also Plate 1.)*

shown in the Fig. 1.10d. Cleary the oxygen production by the ring of endolithic cells can be seen. The coral was sampled in the lagoon of Heron Island (Capricorn Islands, Great Barrier Reef, Australia) and the experiments were made at Heron Island Research Station in a cooperation between the Max Planck Institute for Marine Microbiology, Microsensor Research Group, Bremen, Germany http://www.mpi-bremen.de and Dr. Peter Ralph, Department of Environmental Sciences, University of Technology, Sydney, Australia (http://www.science.uts.edu.au/des/prresum.html).

a *b*

Figure 1.11: *Application of thermography in botanical research. **a** Infrared image of a ricinus leaf. The patchy leaf surface indicates that evaporation and, therefore, opening of the stomata is not equally distributed over the surface of the leaf. **b** The fast growth of a plant tumor destroys the outer waxen cuticle which saves the plant from uncontrolled evaporation and loss of water. In the temperature image, therefore, the tumor appears significantly cooler (darker) than the trunk of the plant. It has about the same low temperature as the wet blurred sponge in the background of the image. The images are from unpublished data of a cooperation between U. Schurr and M. Stitt from the Institute of Botany of the University of Heidelberg and the author.*

1.4.2 Thermography for Botany

This section shows two interesting interdisciplinary applications of *thermography* in botanical research. With thermography, the temperature of objects at environmental temperatures can be measured by taking images with infrared radiation in the 3-5 μm or 8-14 μm wavelength range.

1.4.2a Patchiness of Photosynthesis. Figure 1.11a shows the surface temperature of a ricinus leaf as measured by an infrared camera. These cameras are very sensitive. The one used here has a temperature resolution of 0.03 °C. The temperature of the leaf is not constant but surprisingly shows significant patchiness. This patchiness is caused by the fact that the *evaporation* is not equally distributed over the surface of the leaf. Evaporation of water is controlled by the stomata. Therefore, the thermographic findings indicate that the width of the stomata is not regular. As photosynthesis is controlled by the same mechanism via the CO_2 exchange rate, *photosynthesis* must have a patchy distribution over the surface of the leaf. This technique is the first direct proof of the patchiness of evaporation. So far it could only be concluded indirectly by investigating the chlorophyll fluorescence.

1.4.2b Uncontrolled Evaporation at Tumor Surfaces. Another interesting phenomenon yielded by infrared thermography can be seen in Fig. 1.11b. The tumor at the trunk of the ricinus plant is significantly cooler than the trunk itself. The tumor has about the same low temperature as a wet sponge, which is visible blurry in the background of the image. From this observation it can be concluded that the evaporation at the tumor surface is unhindered with a rate similar to that of a wet sponge. As a signifi-

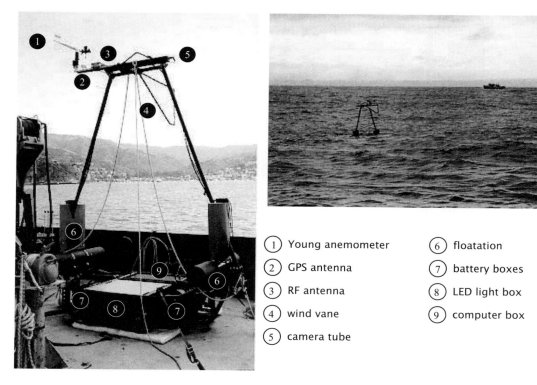

(1) Young anemometer		(6) floatation	
(2) GPS antenna		(7) battery boxes	
(3) RF antenna		(8) LED light box	
(4) wind vane		(9) computer box	
(5) camera tube			

Figure 1.12: *Instrumentation to take image sequences of the slope of short wind waves at the ocean interface. Left: The wave-riding buoy with a submerged light source and a camera at the top of the buoy which observes an image sector of about* 15×20 *cm^2. Right: Buoy deployed from the research vessel New Horizon. Almost all parts of the buoy including the light source are submerged. From Klinke and Jähne [79].*

cant negative effect of the tumor, the plant loses substantial amounts of water through the tumor surface. Normally, a plant is protected from uncontrolled evaporation by a waxen layer (cuticle) and the evaporation mainly occurs at the stomata of the leaves.

1.4.3 Imaging of Short Ocean Wind Waves

Optical measuring techniques often must be used in hostile environments. One of the most hostile environments is the ocean, especially for measurements close to the ocean surface. Nevertheless, it is possible to use imaging optical techniques there.

Recently, short ocean wind waves ("ripples") have become a focus of interest for scientific research. The reason is their importance for modern remote sensing and significant influence on small-scale exchange processes across the ocean/atmosphere interface. Short ocean *wind waves* are essentially the scatters that are seen by modern remote sensing techniques from satellites and airplanes using micro waves. In understanding the role of these waves on the electromagnetic backscatter in the microwave range, it is required to know the shape and spatio-temporal properties of these small-scale features at the ocean surface.

Measurements of this kind have only recently become available but were limited to measurements in simulation facilities, so-called wind-wave flumes. Figure 1.12 shows a sketch of a new type of wave-riding buoy that was specifically designed to measure

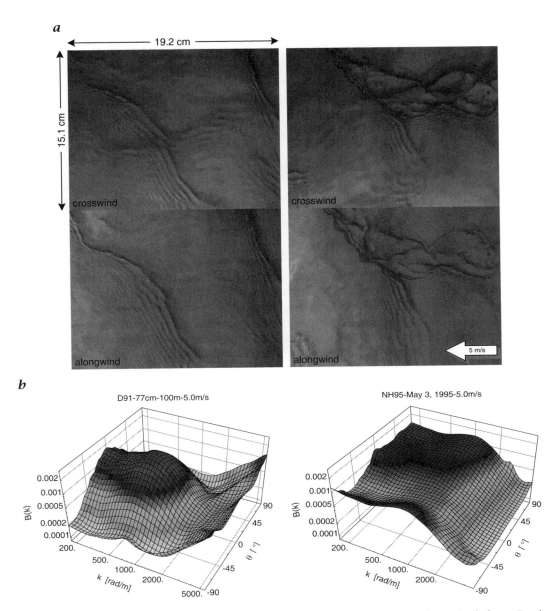

Figure 1.13: *a Sample images taken from the wave-riding buoy at a wind speed of about 5 m/s. The left image shows the slope in horizontal direction and the right image the slope in vertical direction. The image sector is about 15×20 cm². b 2-D wave number spectra computed from about 700 images as in a (right) compared with spectra obtained under similar conditions in a laboratory facility (Delft wind/wave flume, left). The spectra are shown in logarithmic polar coordinates. One axis is the logarithm of the wave number, the other the direction of wave propagation, where zero means propagation in wind direction. From Klinke and Jähne [79].*

the slope of short ocean wind waves [79]. The light source of the instrument consists of 11,000 light-emitting diodes (LEDs) which are arranged in two arrays so that two perpendicular intensity wedges are generated. By pulsing the two perpendicular intensity wedges shortly after each other, quasi-simultaneous measurements of both the

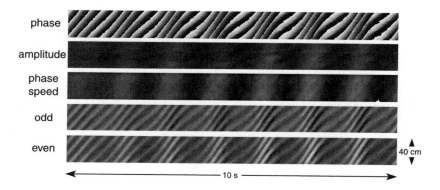

Figure 1.14: *Analysis of the interaction of mechanically generated water waves using Hilbert filtering and local orientation.*

along-wind and cross-wind slope of the waves can be performed with a single camera. While most of the buoy, including the light sources, resides below the water surface, a small CCD camera and other instruments to measure the wind speed are sticking out of the water to take the images of the water surface at a sector of about 15×20 cm^2. Figure 1.12 shows the buoy during deployment from the research ship New Horizon.

Sample images are shown in Fig. 1.13a. One of the simplest evaluations is the computation of the two-dimensional wave number spectrum by using two-dimensional Fourier transform techniques. The resulting spectra (Fig. 1.13b) essentially show how the wave slope is distributed over different wavelengths and directions.

While this type of evaluation gives very valuable results for understanding the micro-wave backscatter from the ocean surface, it does not give a direct insight into the dynamics of the waves, i. e., how fast they grow when the wind is blowing, how strongly they interact with each other, and how fast they decay by viscous and turbulent dissipation. Principally, all of the information about the dynamics of the waves is contained in these wave slope image sequences. In order to illustrate how advanced image processing techniques can be used to study the dynamics and interaction of waves, the simple example of the interaction of two mechanically generated water waves in a linear glass tunnel is shown in Fig. 1.14. The images show a stripe of 40 cm in the direction of wave propagation observed for 10 s. The waves enter the stripe at the lower part and can be seen exiting it a little time later at the upper part. First, only 2.5 Hz waves occur which are later superimposed and modulated by a longer wave with a period of 0.7 Hz. Already in the original image (marked with **even**) one can see the modulation or the wavelength, wave frequency, amplitude, and also — by the change of the inclination — the modulation of the phase speed.

A basic tool for the analysis of the interaction is the so-called Hilbert filter (Section 14.4.2b) which leaves the amplitude constant but shifts the waves by 90° (see stripe marked with **odd** in Fig. 1.14). From the even and odd signals, both the amplitude of the short wave and the phase of the short wave can be computed (see corresponding stripes in Fig. 1.14). The phase speed can be computed more directly. It results from the inclination of the constant gray values in the original image and has been computed by a technique called *local orientation*.

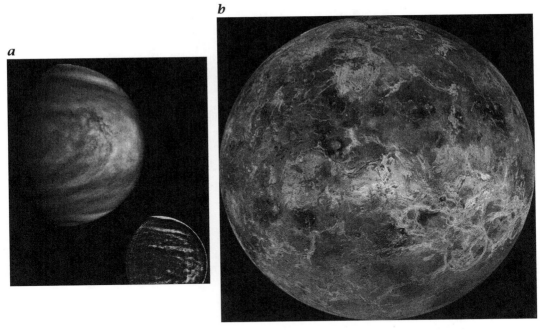

Figure 1.15: *Images of the planet Venus: **a** Images in the ultraviolet (top left) show patterns at the very top of Venus' main sulfuric acid haze layer while images in the near infrared (bottom right) show the cloud patterns several km below the visible cloud tops. **b** Topographic map of the planet Venus shows the elevation in a color scheme as it is usually applied for maps (from blue over green to brownish colors). This image was computed from a mosaic of Magellan radar images that have been taken in the years 1990 to 1994. Source: http//www.jpl.nasa.gov. (See also Plate 2.)*

1.4.4 SAR Imaging for Planetology and Earth Sciences

In this section, we introduce the first non-optical imaging technique. It is an active technique, since an airborne or spaceborne instrument emits micro waves (electromagnetic waves with wavelengths of a few cm, see Section 3.3.1, Fig. 3.3) from an antenna. Depending on the nature of the reflecting surface, especially its roughness, part of the radiation is reflected back and received by the satellite antenna. From the time elapsed between the emission of the microwave signal and its reception, the distance of the reflecting surface can also be measured.

SAR means synthetic aperture radar. Antennas of only a few meters in diameter result in a very poor image resolution for wavelengths of several centimeters. SAR imaging uses a clever trick. During the flight of the satellite, a certain point of the ocean surface is illuminated for a certain time span, and the returned signal contains information about the point not only from the diameter of the satellite but spanning the whole path during the illumination period. In this way, a much larger synthetic aperture can be constructed resulting in a correspondingly higher image resolution.

A remarkable feature of radar imaging is its capability to penetrate even thick clouds. This can convincingly be demonstrated by the planet Venus. Optical images in the visible electromagnetic range show the planet Venus as a uniform white object since the light is reflected at the top of the dense cloud cover. Ultraviolet and near infrared radiation can only penetrate a certain range of the cloud cover (Fig. 1.15a). In contrast, Venus' clouds are transparent to microwaves. More than a decade of radar in-

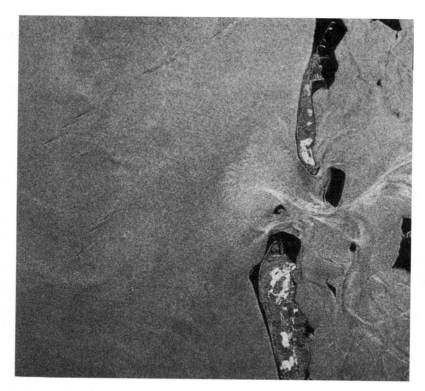

Figure 1.16: *Radar image of the Dutch coast including the islands of Fleeland and Terschelling taken with a synthetic aperture radar of the SEASAT satellite on October 9, 1978. In the mud-flats between the two islands, strong variations in the radar backscatter can be observed which first puzzled scientists considerably. Later, it turned out that they were caused by a complex chain of interactions. Because of the low water depth, there are strong tidal currents in the region which are modulated by the varying water depth. The changing currents in turn influence the small-scale water surface waves. These small waves form the roughness of the sea surface and influence the backscatter of the microwaves. Image courtesy of D. van Halsema, TNO, the Netherlands.*

vestigations culminating in the 1990 to 1994 Magellan mission revealed the topography with a resolution of about 100 m. The hemispheric view of Venus shown in Fig. 1.15b was computed from a mosaic of Magellan images.

A historic oceanographic example of an SAR image is shown in Fig. 1.16. It was taken in 1978 during the three-month life span of the first synthetic aperture radar on board the *SEASAT satellite.* To the big surprise of the involved scientists, the bottom topography in the mud-flats became visible in these images because of a long chain of complex interactions finally leading to the modulation of small-scale water surface waves by current gradients.

The variety of parameters that can be retrieved from radar imagery is nicely demonstrated with four images from the *TOPEX*/Poseidon mission where *radar altimetry* and passive microwave radiometry were used to study global atmospheric and oceanographic circulation. Figure 1.17a shows the dynamic topography, i. e., the "highs" and "lows" of the ocean currents as a deviation from a surface of constant gravitational energy. Variations in the dynamic topography of about two meters have been measured with an accuracy of about 3 cm by measuring the time of flight of short radar pulses. The significant wave height (a measure for the mean height of wind waves) has been determined from the shape of the returned radar pulses (Fig. 1.17b). Figure 1.17c

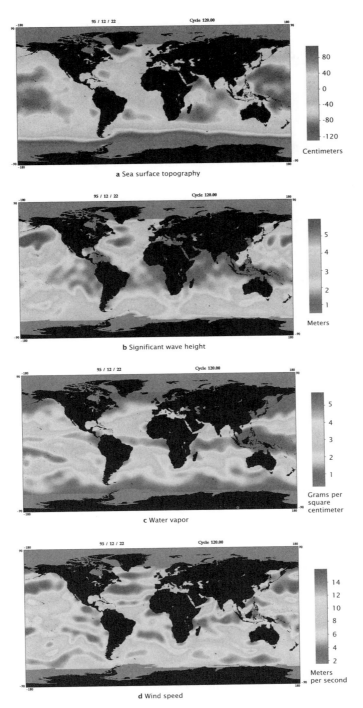

a Sea surface topography

b Significant wave height

c Water vapor

d Wind speed

Figure 1.17: *Images of the TOPEX/Poseidon mission derived from radar altimetry and passive micro wave radiometry (Source: http://www.jpl.nasa.gov). **a** dynamic topography (highs and lows of the ocean currents) shown as a deviation from an area of constant gravitational energy. **b** significant wave height, **c** water vapor content of the atmosphere in g/cm² measured by passive micro wave radiometry, **d** wind speed determined from the strength of the backscatter. All images have been averaged over a period of 10 days around December 22, 1995. (See also Plate 4.)*

Figure 1.18: *Synthetic color image of a* 30.2 × 21.3 *km sector of the tropical rain forest in west Brazil (Source: image p-46575 published in http://www.jplinfo.jpl.nasa.gov). Three SAR images taken with different wavelengths (lower three images: Left: X-band, Middle: C-band, Right: L-band) have been composed into a color image.*
Pristine rain forest appears in pink colors while clear areas for agricultural usage are greenish and bluish. A heavy rain storm appears in red and yellow colors since it scatters the shorter wavelength micro waves. Image taken with the imaging radar-C/X-band aperture radar (SIR-C/X-SAR) on April 10, 1994 on board the space shuttle Endeavor.   (See also Plate 3.)

shows the total water vapor content of the atmosphere (in g/cm^2) as it has been measured by passive micro wave radiometry. This quantity is an important parameter for global climatology, but it is also required for the precise measurement of the time of flight. An unknown water vapor content in the atmosphere would result in a measurement uncertainty of the dynamic topography of the ocean of up to 30 cm. Finally, the wind speed over the sea surface can be determined from the strength of the radar backscatter. A calm sea is a better reflector than a rough sea at high wind speeds. All the images shown in Fig. 1.17 have been averaged over 10 days around the 22nd of December 1995.

Multi-frequency SAR images can be displayed as color images as shown in Fig. 1.18. The color image was created by combining images taken with three separate radar frequencies into a composite image. The three black and white images in Fig. 1.18 represent the individual frequencies. Left: X-band, vertically transmitted and received, blue component; middle: C-band, horizontally transmitted and vertically received, green component; right: L-band, horizontally transmitted and vertically received, red component. A heavy rain storm with large droplets scatters the short wavelength in the X-band range and thus appears as a black cloud in the expanded image. The same area shows up only faintly in the C-band image and is invisible in the L-band image. Reflection of radar wavelength depends on the roughness of the reflecting surfaces in the centimeter to meter range. Therefore, pristine tropical rain forest (pink areas) can clearly be distinguished from agriculturally used (blue and green) patches.

1.4.5 X-Ray Astronomy with ROSAT

Since their discovery in 1895 by Konrad W. Röntgen, *X-ray* images have found widespread application in medicine, science, and technology. As it is easy to have a point X-ray source, optics are not required to take an absorption image of an object from X-ray examination.

To take an image of an X-ray source, however, X-ray optics are required. This is a very difficult task as the index of refraction for all materials is very low in the X-ray region of the electromagnetic spectrum. Therefore, the only way to build a telescope is to use mirrors with grazing incident rays. The configuration for such a telescope, the Wolters telescope, is shown in Fig. 1.19a. This telescope is the main part of the German X-ray satellite *ROSAT* which has been exploring the sky since July 1990 in the X-ray region. The most spectacular objects in the X-ray sky are the clouds of exploding stars (supernovae). In 1995, fragments of an exploding star could be observed for the first time. Figure 1.19c, d shows six fragments marked from A to F that are the remains of the supernova explosion of a star in the Vela constellation. This object is only about 1,500 light years away from the earth and the almost circular explosion cloud has a diameter of 200 light years. If the atmosphere were transparent to X-rays and if we could image X-rays, we would see a bright object with a diameter of 16 times that of the moon (Fig. 1.19b). The explosion clouds of supernovae become visible in the X-ray region because they move with supersonic speed heating the interstellar gas and dust to several million degrees centigrade. This is the first time that explosion fragments have been observed giving new insight into the physics of dying stars.

1.4.6 Spectroscopic Imaging for Atmospheric Sciences

The measurement of the concentration of trace gases in the atmosphere is a typical example where imaging with a single wavelength of radiation is not sufficient. As illustrated in Fig. 1.20, the absorption spectra of various gases such as sulfur dioxide (SO_2) and ozone (O_3) overlap each other and can thus only be separated by measuring at many wavelengths simultaneously. Such a technique is called *hyperspectral* or *spectroscopic imaging*.

An example of an instrument that can take such images is the GOME instrument of the ERS2 satellite. The instrument is designed to take a complete image of the earth every three days. At each pixel a complete spectrum with 4000 channels in the ultraviolet and visible range is taken. The total atmospheric column density of a gas can be determined by the characteristic absorption spectrum using a complex nonlinear regression analysis. A whole further chain of image processing steps is required to separate the stratospheric column density from the tropospheric column density.

As a result, global maps of trace gas concentrations, such as the example images of the tropospheric column density of NO_2 shown in Fig. 1.21, are obtained. NO_2 is one of the most important trace gases for the atmospheric ozone chemistry. The main sources for tropospheric NO_2 are industry and traffic, forest and bush fires (biomass burning), microbiological soil emissions, and lighting. Satellite imaging allows for the first time to study the regional distribution of NO_2 and to better identify the various sources. Meanwhile years of such data are available so that the weekly and annual cycles and temporal trends can be studied as well [154]. The three example images in Fig. 1.21 give already an idea of the annual cycle with significantly higher NO_2 concentration on the northern hemisphere during winter time.

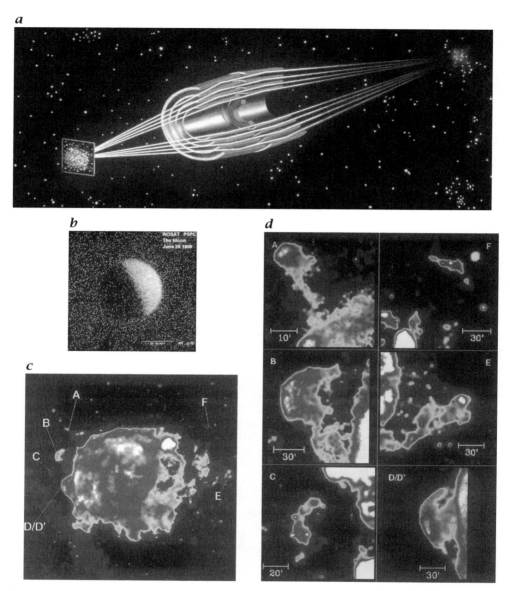

Figure 1.19: *a Artist's impression of an X-ray telescope, the so-called Wolters telescope, the primary X-ray instrument on the German ROSAT satellite. Because of the low index of refraction, only telescopes with grazing incident rays can be built. Four double-reflecting pairs of concentric mirrors are used to increase the aperture and thus the brightness of the image. b X-ray image of the moon, with 15' (1/4°) diameter a small object in the sky as compared to c explosion cloud of the supernova Vela with a diameter of about 4° as observed with the German X-ray satellite ROSAT. Scientists from the Max Planck Institute for Extraterrestrial Physics (MPE) discovered six fragments of the explosion, marked from A to F. The intensity of the X-ray radiation in the range from 0.1 to 2.4 keV is shown in pseudo colors from light blue over yellow, red to white and covers a relative intensity range of 500. The bright white circular object at the upper right edge of the explosion cloud is the remains of another supernova explosion which lies behind the VELA explosion cloud and has no connection to it. d Enlarged parts of figure c of the fragments A through F at the edge of the explosion cloud. Images courtesy of the Max Planck Institute for Extraterrestrial Physics, Munich, Germany. (See also Plate 5.)*

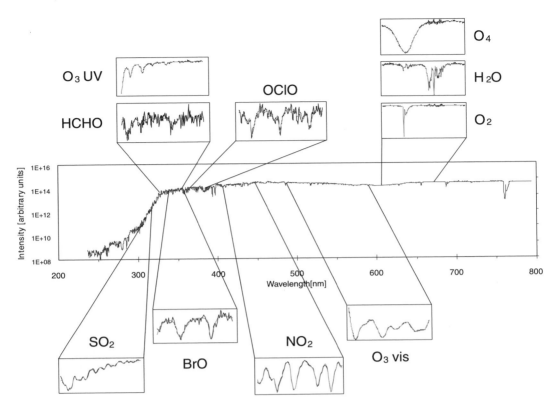

Figure 1.20: *Absorption ranges of various trace gases in the sun light spectrum that has passed through the earth's atmosphere. The thick lines in the various inset figures indicate the least squares fit curve of the absorption spectra to the measured absorption spectra (thin lines) by the GOME instrument on the ERS2 satellite. From Kraus et al. [84].*

1.5 Depth Measurements: Exploring 3-D Space

Classical imaging techniques project the 3-D space onto a planar image plane. Therefore the depth information is lost. Modern imaging techniques, in combination with image processing, now offer a wide range of possibilities to either reconstruct the depth of opaque objects or to take 3-D images. Here, examples are given both for depth imaging as well as for 3-D imaging.

1.5.1 Optical Surface Profiling

Accurate measurements of the surface topography is a ubiquitous task in technical and research applications. Here, we show just one of the many techniques for optical surface profiling as an example. It is a new type of confocal microscopy that has recently been introduced by Scheuermann et al. [122]. The advantage of this technique lies in the fact that no laser scanning is required; instead, a statistically distributed pattern is projected through the microscope optics onto the focal plane. Then, this pattern only appears sharp on parts that lie in the focal plane. On all other parts, this pattern gets more blurred the larger the distance from the focal plane is. A significant advantage of this technique is that it does not require objects with texture. It also

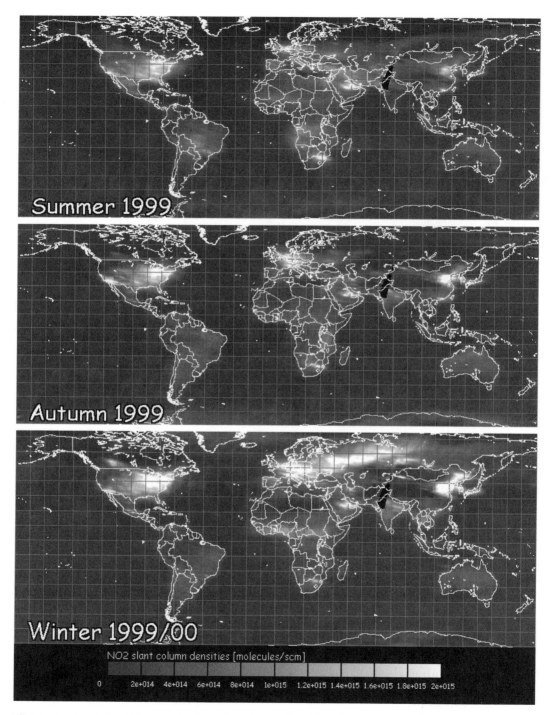

Figure 1.21: *Maps of tropospheric NO₂ column densities showing three three-month averages from summer 1999, fall 1999, and winter 1999/2000 (courtesy of Mark Wenig, Institute for Environmental Physics, University of Heidelberg).*

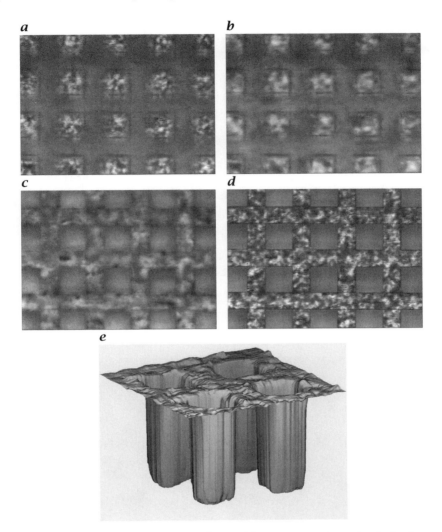

Figure 1.22: *Focus series of a PMMA press form with narrow rectangular holes imaged with a confocal technique using statistically distributed intensity patterns. The images are focused on the following depths measured from the bottom of the holes: **a** 16 μm, **b** 160 μm, **c** 480 μm, and **d** 620 μm. **a** is focused on the bottom of the holes while **d** is focused on the surface of the form. **e** 3-D reconstruction from the focus series. From Scheuermann et al. [122].*

works well with completely unstructured objects from which the surface depth cannot be measured with other optical techniques.

This technique is demonstrated with the measurement of the height of a press form for micro structures. The form is made out of PMMA, a semi-transparent plastic material with a smooth surface. The form has 500 μm deep narrow and rectangular holes which are very difficult to measure with any other technique. In the focus series shown in Fig. 1.22a–d, it can be seen that first the patterns of the material in the bottom of the holes become sharp while, after moving towards the optics, the final image focuses at the surface of the form. The depth of the surface can be reconstructed by searching the position of maximum contrast for each pixel in the focus series (Fig. 1.22e).

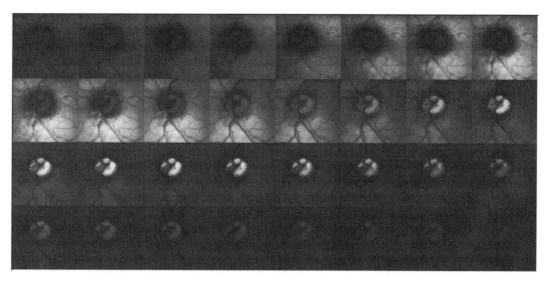

Figure 1.23: *A series of 32 confocal images of the retina. The depth of the scan increases from left to right and from top to bottom. Images courtesy of R. Zinser, Heidelberg Engineering. (See also Plate 6.)*

a   *b*

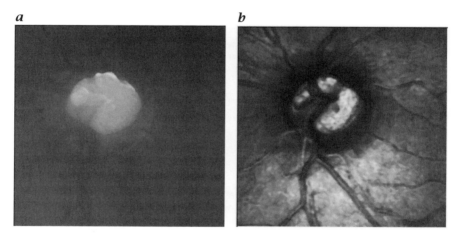

Figure 1.24: *Reconstruction of the topography of the retina from the focus series in Fig. 1.23. a Depth map: deeper lying structures at the exit of the optical nerve are coded brighter. b Reconstructed reflection image showing all parts of the image sharply despite significant depth changes. Images courtesy of R. Zinser, Heidelberg Engineering. (See also Plate 7.)*

1.5.2   3-D Retina Imaging

Conventional 2-D microscopy has significant disadvantages when 3-D structures are to be observed. Because of the very low depth of focus, essentially only planar objects could be observed. New techniques, especially confocal laser scanning microscopy (Section 4.4.3), are currently revolutionizing microscopy and leading the way from 2-D to 3-D microscopy. Simply spoken, in confocal microscopy only the plane of focus is illuminated so that one can easily scan through the objects to obtain a true 3-D image.

Confocal laser scanning techniques can also be used to take 3-D images of the retina in the eye. Figure 1.23 shows a series of 32 confocal images of the retina. The depth of the image increases from left to right and from top to bottom. It can clearly be recognized that the exit of the optical nerve lies deeper than the surrounding retina. As the focused plane is characterized by a maximum reflectivity, a depth image can be reconstructed from the focus series (Fig. 1.24a). The depth in this image is coded in such a way that deeper parts of the surface appear brighter. Besides the depth map, an integration of over-the-focus series results in a type of standard reflection image; however, with the big advantage in contrast to conventional microscopy, all parts of the surface are sharp despite significant depth differences (Fig. 1.24b).

3-D imaging of the retina has gained significant importance in ophthalmology as it opens new diagnostic possibilities. One example is early diagnosis of illnesses such as glaucoma.

1.5.3 Distribution of Chromosomes in Cell Nuclei

Confocal microscopy (Section 4.4.3) can not only be used to extract the depth of surfaces but also makes 3-D imaging a reality. 3-D microscopy will be demonstrated with the significant example of the study of the distribution of chromosomes in cell nuclei.

Until recently, chromosomes, the carriers of the genes, could only be individually made visible during the metaphase of the cell cycle. Basic scientific questions such as the distribution of chromosomes in the cell nucleus could hardly be investigated experimentally but remained a field of speculative models.

With the progress in biochemistry, it is now possible to selectively stain individual chromosomes or even specific parts of them. Figure 1.25 shows a focus series of a female human nucleus. Chromosomes X and 7 are stained by a greenish fluorescent dye. In order to distinguish both chromosomes, a substructure of chromosome 7 is stained red; the whole nucleus is counter-stained with a deep blue fluorescent dye. As chromosomes in human cells appear always in pairs, two of each of the chromosomes are seen in Fig. 1.25.

The very noisy and still not completely resolved structures in Fig. 1.25 indicate that a 3-D segmentation of chromosomes in the nucleus is not a trivial image processing task. The segmentation problem was solved by an iterative partition of the 3-D image using Voronoi diagrams [25]. The final result of the segmentation represented as a 3-D reconstruction is shown in Fig. 1.25i. Volume measurements of the active and inactive X chromosome proved that, in contrast to previous assumptions, the volume of the active and inactive chromosome does not differ significantly. The territories of the active X chromosome, however, are characterized by a flatter form and a larger surface.

1.5.4 X-Ray and Magnetic Resonance 3-D Imaging

As we have seen in Sections 1.5.2 and 1.5.3, light is suitable for 3-D imaging of cell structures and the retina. However, light does not penetrate many other objects and therefore other, more penetrating radiation is required for 3-D imaging.

The oldest example are *X-rays*. Nowadays, X-ray imaging is not only an invaluable diagnostic tool in medicine but also used in many technical and scientific applications. It is especially useful in material sciences, where the 3-D properties of complex composite or inhomogeneous materials must be analyzed. Fig. 1.26a shows a lightweight but stiff metal foam with open pores as an example.

Such 3-D X-ray images are created by a complex procedure, known as *computer tomography* using projections from many directions. A single projection taken with

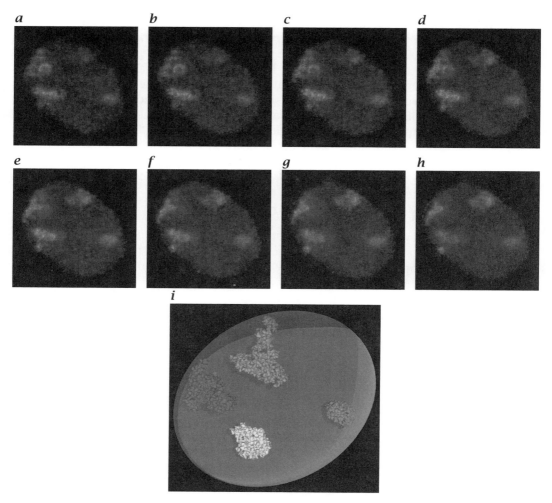

Figure 1.25: *a - h Part of a depth scan using confocal microscopy across a female human nucleus. Visible are chromosomes X and 7 (green). For differentiation between chromosomes X and 7, a substructure of chromosome 7 has been additionally colored with a red dye. The depth increment between the individual 2-D images is 0.3 μm; the image sector is 30 × 30 μm. i 3-D reconstruction. The image shows the inactive chromosome X in red, the active chromosome X in yellow, chromosome 7 in blue, and its centromeres in magenta. The shape of the whole nucleus has been modeled as an ellipsoid. From Eils et al. [25]. (See also Plate 8.)*

penetrating radiation is not of much use, because it integrates the absorption from all penetrated layers. Only a complex reconstruction process forms a 3-D image.

An even more complex image formation process using radio waves (Fig. 3.3) in conjunction with strong magnetic fields is known as *magnetic resonance imaging (MRI)* [126]. It turned out that this imaging technique is much more versatile than X-ray imaging because it visualizes the interaction of the spin (magnetic moment) of atom nuclei with their surrounding. In contrast to X-ray tomography, MRI can, for example, distinguish gray and white brain tissues (Fig. 1.26b). It is also possible to make images of certain chemical species and to measure velocities, e.g., through blood vessels or the stem of plants [49].

a *b*

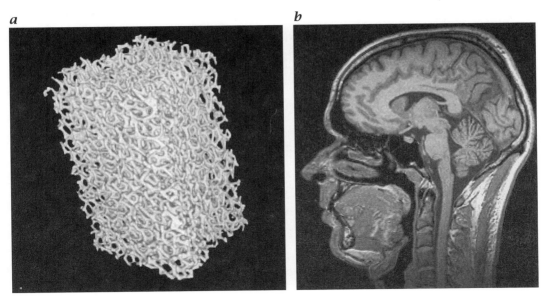

*Figure 1.26: Examples of 3-D images taken with penetrating electromagnetic radiation: **a** High resolution 3-D X-ray image of nickel foam (courtesy of Joachim Ohser, FH Darmstadt, Germany); **b** slice from a 3-D magnetic resonance image of a human head in T1 mode (courtesy of Michael Bock, DKFZ Heidelberg).*

1.6 Velocity Measurements: Exploring Dynamic Processes

Change is a common phenomenon through all sciences. Actual research topics include — to name only a few — the growth of organisms and their genetic control, heterogeneous catalytic reactions, and the dynamics of non-linear and chaotic systems. Image sequences open a way to explore the kinematics and dynamics of scientific phenomena or processes. Principally, image sequences of volumetric images capture the complete information of a dynamic process. As examples of applications of velocity measurements we discuss two quantitative visualization techniques for turbulent flows and the study of motor proteins in motility assays.

1.6.1 Particle Tracking Velocimetry

While techniques of flow visualization have been used from the beginnings of hydrodynamics, quantitative analysis of image sequences to determine flow parameters has only become available recently. The standard approach is the so-called *particle image velocimetry* (PIV). With this technique, however, only a snapshot of the flow field can be obtained. Methods that track particles go one step further. They cannot only determine the instantaneous flow field but by tracking a particle over a long image sequence, it is also possible to obtain the Lagrangian flow field and to get an insight into the temporal variation of flow fields. This technique is called *particle tracking velocimetry* (PTV).

The flow is made visible by injecting small neutrally buoyant particles and by making them visible using a suitable illumination technique. A long tracking requires, in contrast to the thin light sheets used in particle imaging velocimetry, a rather thick light sheet so that the particles stay for a sufficiently long time in this light sheet. Figure 1.27 shows two example images of particle traces as they have been used to measure the flow beneath a water surface undulated by waves. It can be seen that the

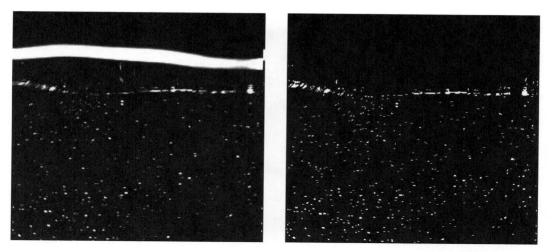

Figure 1.27: *Example images of particle traces for the visualization of flows to below the water surface undulated by waves.* **a** *Original image.* **b** *Segmented image, reflections from the water surface removed. From these images, particles are extracted and tracked until they leave the illuminated light sheet (Fig. 1.28). From Hering et al. [55].*

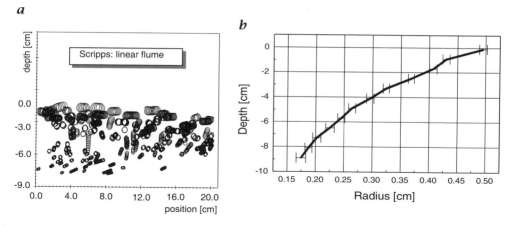

Figure 1.28: **a** *Particle traces as extracted by particle tracking velocimetry over long image sequences.* **b** *Radius of the orbital motion computed from the traces in Fig. 1.28a as a function of the water depth. From Hering et al. [55].*

particles trace the flow below mechanically generated surface waves over several wave periods (Fig. 1.28a). From the particle traces, further parameters can be extracted. Figure 1.28b, for example, shows the radius of the orbital motions as a function of the distance to the water surface.

1.6.2   3-D Flow Tomography

The flow of liquids can also be marked by fluorescent dyes (Fig. 1.29a). Such techniques are standard now in experimental fluid dynamics. Three-dimensional measurements of flow fields, however, are a very demanding task as they require the acquisition of 4-D imagery with three spatial and one temporal coordinate. Therefore, such an approach is currently only possible in rather slow flows. Then, it is possible to take a depth scan

a *b*

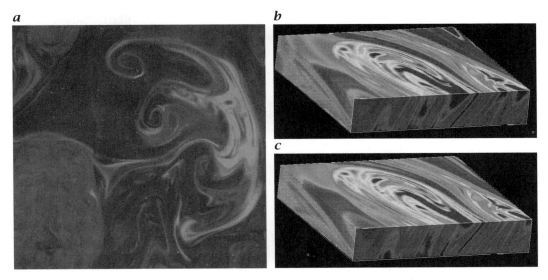

c

Figure 1.29: *Fluorescent dyes can also be used to make flows visible: **a** visualization of a mixing process; **b** and **c** two consecutive $15 \times 15 \times 3\,mm^3$ volumetric images consisting of 50 layers of 256×256 pixels taken 100 ms apart. The voxel size is $60 \times 60 \times 60\,\mu m^3$. From Maas [94]. (See also Plate 9.)*

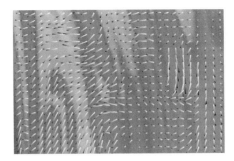

Figure 1.30: *Flow field determined by the least-squares matching technique from volumetric image sequences. The flow vectors are superimposed to the concentration field. From Maas [94]. (See also Plate 10.)*

rapidly enough so that the images are acquired quasi-simultaneously. Here, we show an example that did not use particles but a fluorescent dye to make the flow visible. Each volumetric image consists of 50 scans with 256×256 pixels. Five hundred scans could be taken per second resulting in a temporal resolution of 10 volumetric images per second. Two consecutive volume images are shown in Fig. 1.29b, c.

The volumetric image sequences were evaluated with a least-squares matching technique to determine the displacement from image to image. This proven technique is intensively used in photogrammetry to solve, for instance, the stereo correspondence problem. In contrast to simple correlation techniques, this method has the advantage that not only the velocity itself but also first derivatives of the velocity field such as divergence or rotation can be determined directly.

A result of this technique is shown in Fig. 1.30. In the figure, the determined flow vectors are superimposed to the concentration field of the dye.

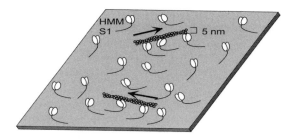

Figure 1.31: *Principle of the in vitro motility assay. Isolated actin filaments fluorescently labeled with rhodamine-phalloidin move over a surface of immobilized myosin molecules. The actin movement is recorded with epi-fluorescence imaging using high sensitivity intensified CCD-cameras; from Uttenweiler et al. [147].*

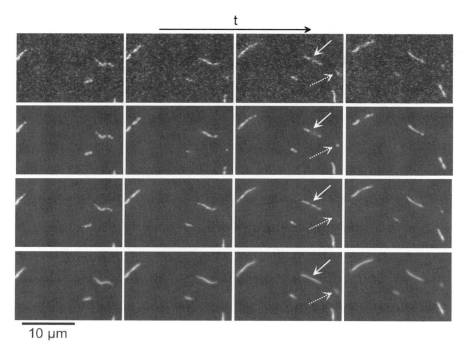

Figure 1.32: *Comparison of various filtering methods for enhancement of the motility assay image sequences; top row: unprocessed microscopic fluorescence images; second row: isotropic nonlinear spatial smoothing; third row: isotropic nonlinear spatial and temporal smoothing; bottom row: 3D-spatio-temporal anisotropic diffusion filtering. From Uttenweiler et al. [148].*

1.6.3 Motor Proteins

The advances in new microscopic techniques allow studying cellular processes on the basis of the underlying molecular interactions. Especially, fluorescence microscopy has become a universal tool for studying physiological systems and the underlying biophysical principles down to the molecular level. The analysis of fluorescence image sequences is still a very challenging and demanding task. When recording high spatially and temporally resolved dynamic processes, the signal-to-noise ratio is generally very low due to the limited amount of fluorescence photons available for detection.

As an example, the study of motor proteins is discussed. They convert chemical energy into mechanical energy, which is used for cellular movement. The most abundant of the motor proteins is myosin, which in conjunction with actin filaments plays the major role in the contraction of heart and skeletal muscle as well as in intracellular motility.

The interaction of actin filaments with the motor protein myosin can be studied in the in vitro motility assay as shown in Fig. 1.31. The original image sequences taken by epi-fluorescence imaging using high sensitivity intensified CCD-cameras are very noisy making it hardly possible to segment the filaments from the background or to make reliable estimates of their velocity. Nonlinear smoothing techniques such as inhomogeneous and anisotropic diffusion filters proved to be very successful in improving the signal-to-noise ratio [147].

2 Tasks and Tools

2.1 Highlights

Task

The human visual system is essential in understanding the basic functionality of machine vision and the principal properties that can be measured by image processing and analysis (Section 2.2.1). Before starting to apply image processing techniques, it is essential to understand that

- there are basic differences in the purpose and functionality of human and machine vision (Section 2.2.2),
- any measurement shows two types of uncertainties (Section 2.2.3),
- an adequate representation of information yields compact data and fast algorithms (Section 2.2.4),
- efficient and accurate measurements require a correct but as simple as possible model describing the essential properties of the image formation and processing steps.

We gain access to the complex image processing task from image formation to object recognition and classification by parting it into hierarchically cascaded tasks (Section 2.2.6). We will learn to master these tasks — one by one — in the following chapters.

Tools

The basic tools for digital image processing are (Section 2.3):

Tool	Usage
Electronic imaging system (camera) consisting of	
Image formation system (lens)	Collects radiation to form an image of the object features of interest
Radiation converter (sensor)	Converts irradiance at the image plane into an electric signal
Frame grabber	Converts electric signal into a digital image and stores it in the computer
Personal computer or workstation	Provides platform for digital processing of image
Image processing software	Provides algorithms to process and analyze contents of digital image

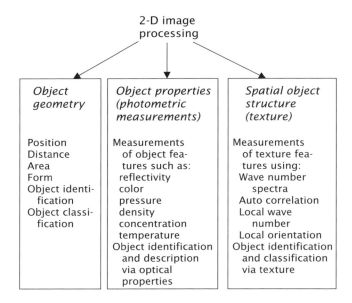

Figure 2.1: *Features of interest for scientific and technical applications that can be extracted from 2-D image data.*

Rapid progress in computer technology and photonics has reached a critical level of performance. Standard computer systems with only a little extra hardware can be used even for quite complex image processing tasks. Equally important, image processing is about to change from a heuristic approach to a sound science.

Mathematical foundation of image processing identifies soluble tasks, yields predictable, reliable, and — in some cases — *optimal* results, enables novel approaches, and delivers fast and efficient algorithms.

2.2   Basic Concepts

2.2.1   Goals for Applications of Image Processing

The general goal of image processing for scientific and technical applications is to use radiation emitted by objects. An imaging system collects the radiation to form an image. Then, image processing techniques are used to perform an area-extended measurement of the object features of interest. For scientific and technical applications, area-extended measurements constitute a significant advantage over point measurements, since also the spatial and not only the temporal structure of the signals can be acquired and analyzed (see the examples in Sections 1.3–1.6). Figures 2.1 and 2.2 give a detailed and systematic classification of the features that can be extracted from image data. First, it has to be distinguished whether the imaging system results in a 2-D image (Fig. 2.1) or whether a 3-D reconstruction of the imaged object is required (Fig. 2.2). Another important characteristic is the dynamics of the objects to be observed. Is it sufficient to take only one snapshot of the scene or are the motion and dynamics of objects of interest?

The goals for 2-D image processing (Fig. 2.1) can be divided into three principle categories: determine the *geometry*, *photometry*, and *spatial structure* of objects. In

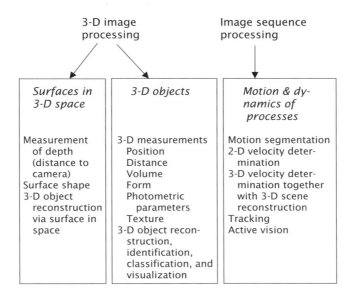

Figure 2.2: *Features of interest for scientific and technical applications that can be extracted from 3-D image data.*

the simplest case, the measuring task requires only measurement of position, size, and form of the objects. Then, the requirements for the illumination and setup of the imaging system seem to be rather simple. The objects of interest must clearly be distinguished from the background.

Photometric applications are more demanding. Then, the irradiance at the image plane should directly reflect the object feature to be measured and should not be influenced by other parameters. Photometric techniques show a wide variety. This starts with the selection of the imaging technique. Figure 2.3 gives a systematic overview of imaging systems using the following classification schemes: a) type of radiation, b) properties used for imaging, and c) techniques used to form an image.

With respect to image analysis, purely geometric or photometric techniques are rather simple and straightforward. The real difficulty of these techniques is the selection and adequate setup of the imaging techniques. Image analysis gets complex when combined geometric and photometric measurements must be carried out. Then, objects cannot be recognized by their gray value but only after analyzing the spatial structure.

The first step into 3-D image processing (Fig. 2.2) is the determination of the depth or the distance of the objects from the camera. This approach corresponds to the functionality of our human visual system. By a number of different techniques, e. g., stereo vision, we are able to recognize the distance of objects in 3-D space. We do not need true 3-D information. It is sufficient to determine the depth of the surface of opaque objects.

True 3-D images cannot be captured with our visual system. Here, imaging techniques in computer vision step beyond the visual capabilities of human beings. Since our world is three dimensional, 3-D imaging techniques have an enormous potential. They allow us to look inside of objects. While medical applications had a pioneering role in 3-D imaging, it can now be found in all branches of natural and engineering sciences. As we cannot directly visualize 3-D images, we always need some special tools to grasp their contents. We could either reduce the 3-D image again to surfaces

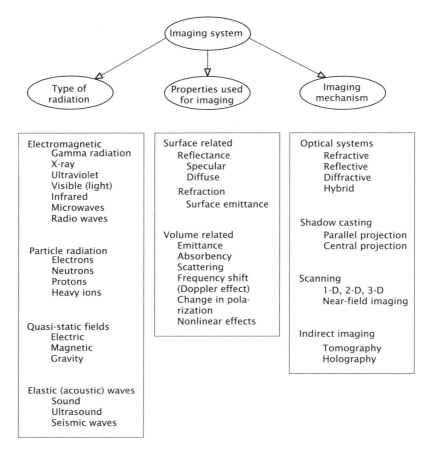

Figure 2.3: *Classification of an imaging system according to the type of radiation, the properties used for imaging, and the techniques used to form an image.*

which then can be observed under different points of view or we can slice through a 3-D image plane by plane.

Finally, the analysis of motion gives access to the dynamics of processes. Motion analysis is generally a very complex problem. The true motion of objects can only be inferred from their motion at the image plane after the 3-D reconstruction of the scene. Another important aspect is the response of the imaging system to the dynamics of the scene. Generally, this is known as *active vision*. In its simplest form, the so-called *tracking*, an imaging system, follows a moving object to keep it in the center of the image.

2.2.2   Measuring versus Recognizing

The basic difference between human and machine vision may be briefly summarized as "humans recognize, machines measure". Of course, this statement is a crude oversimplification but it points to the major difference between *human vision* and scientific and technical applications of image processing with respect to the task. It is often overlooked that the task of human vision is quite different from that of *machine vision*. The most important goal for human vision is to recognize the environment. Put yourself in an absolutely dark room or environment and you will quickly realize how much

Table 2.1: *Comparison of the function modules of human and machine vision.*

Task	Human vision	Machine vision
Visualization	Passive, mainly by reflection of light by opaque surfaces	Passive and active (controlled illumination) using electromagnetic, particulate, and acoustic radiation
Image formation	Refractive optical system	Various systems (see Chapter 4)
Control of irradiance	Muscle-controlled pupil (range $\sim 1\!:\!10^6$)	Motorized apertures, filter wheels, tunable filters
Focusing	Muscle-controlled change of focal length	Auto focus systems based on various principles of distance measurements
Irradiance resolution	Logarithmic sensitivity $\sim 2\%$ resolution	Linear sensitivity, quantization between 8 and 16 bits (256 to 64 k gray levels), resolution > 1 ‰
Tracking	Highly mobile eye ball	Scanner and robot-mounted cameras
Image processing and analysis	Hierarchically organized massively parallel processing	Parallel processing still in its infancy

we depend on our visual sense. The focus of scientific and technical applications in machine vision is more on quantitative than on qualitative information. Although considerable progress has been made in introducing qualitative information and reasoning into machine vision systems (*fuzzy logic*), the main focus is still on quantitative information. Moreover, a human observer will never be capable of performing certain image processing tasks with the accuracy and endurance of a machine vision system.

A detailed comparison of the function modules used for certain image processing tasks is given in Table 2.1. In strictly technical aspects, the function modules of machine systems are often superior to those in human and biological vision. However, machine vision systems by far do not match up the capability of human vision to adapt to different image processing tasks, to solve difficult recognition and classification problems, and to build up knowledge about the observed scenes.

Although much has been learned from biological vision systems for image processing and analysis, the markedly different tasks and capabilities of the function modules have to be kept in mind. It is not possible to simply transfer or copy the human biological vision to machine vision systems. However, the basic functionality of the human vision system can guide us in the task of how these principles can be transferred to technical vision systems.

Another aspect in the relation between human and computer vision is of importance. We cannot think of image processing without human vision. We observe and evaluate the images that we are processing. This key fact forces us to acquire at least a basic knowledge about the functionality of human vision. Otherwise, we are easily misled in the evaluation of images and image processing operations. We will point out this elementary knowledge about the human vision system in the appropriate sections of this handbook. A detailed comparison of human and computer vision can be found in Levine [89].

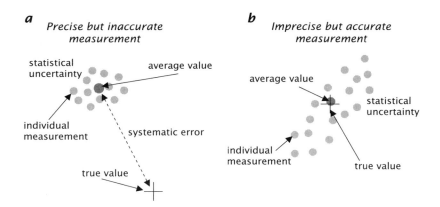

Figure 2.4: *Illustration of systematic and statistical errors distinguishing precision and accuracy for the measurement of position in 2-D images. The statistical error is given by the distribution of the individual measurements, while the systematic error is the difference between the true value and the average of the measured value.*

2.2.3 Signals and Uncertainty

Image processing can be regarded as a subarea of signal processing or measuring technology. This generally means that any measurement we take from images such as the size or the position of an object or its mean gray value only makes sense if we also know the uncertainty of our measurement. This basic fact, which is well known to any scientist and engineer, had often been neglected in the initial days of image processing since it was hard to obtain any reliable measurements at all. Using empirical and not well-founded techniques made reliable error estimates impossible. Fortunately, knowledge in image processing has advanced considerably. Nowadays, many sound techniques are available that include error estimates.

It is important to distinguish two important classes of errors, statistical and systematic errors (Fig. 2.4). *Statistical errors* reflect the fact that if one and the same measurement is repeated over and over again, the results will not be exactly the same but rather scatter and form a cloud of points as illustrated in Fig. 2.4. A suitable measure for the width of the cloud gives the statistical error and its center of gravity the mean measured value. This mean value may differ from the true value. This deviation is called the *systematic error*. A precise but inaccurate measurement is encountered when the statistical error is low but the systematic error is high (Fig. 2.4a). If the reverse is true, i. e., the statistical error is large and the systematic error is low, the individual measurements scatter widely but their mean value is close to the true value (Fig. 2.4b).

While it is easy to get — at least in principle — an estimate of the statistical error by repeating the same measurement again and again, it is much harder to control systematic errors. They are often related to a lack in understanding of the measuring procedure. Unknown parameters influencing the measuring procedure may easily lead to systematic errors. A simple example for a systematic error is a *calibration error*.

With respect to image processing, we can conclude that it is important to understand all the processes that form the image and to understand the functionality of all operations performed on the images.

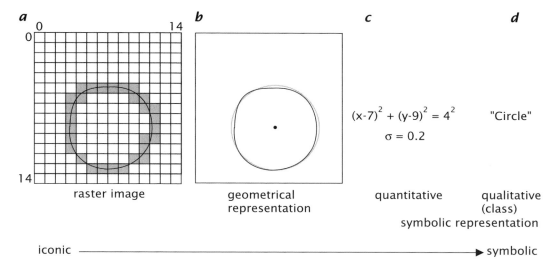

Figure 2.5: *Steps from an iconic towards symbolic representation of a simple geometric object, a circle. The quantitative symbolic description includes a statistical error, the variance, that shows how well the equation for a circle describes the shape of the object.*

2.2.4 Representation and Algorithms

In the first place, an image is stored in a computer as a so-called *raster image*. The area of the image is parted into small cells and in each cell, a digital value, normally between 0 and 255 (8 bits), represents the brightness of the image in the corresponding cell. If the rasterization is fine enough, we do not become aware of it, but rather perceive a continuous image, as, for instance, on a monitor.

A raster image is only one way to present the information contained in an image. A raster image representation is often also called *iconic representation*. Actually, if we perform image processing and analysis, our goal is to retrieve other representations that describe the object adequately and consume less storage space. Figure 2.5 illustrates the change of representation that goes with the analysis of a raster image. The figure contains a very simple object, a circle, which is shown in Fig. 2.5a in its iconic representation. It is obvious that the circle is not very well represented on a raster image and that the first image processing tasks try to go to a vectorial representation in which it is shown as a smooth curve (Fig. 2.5b).

In the next processing step, we could try to fit a circle to this curve (Fig. 2.5b) and then yield an equation describing this circle together with the standard deviation which is the mean deviation from the fitted curve of the real one (Fig. 2.5c). In our case, the circle has a center of 7 in x direction, 9 in y direction, and a radius of 4. In this way, we arrive at quantitative *symbolic representation* of the object given by a mathematical formula. In a more qualitative symbolic representation, we could just say that this figure contains a circle in contrast to an ellipse or square. The qualitative symbolic representation puts the focus on the class of geometric objects while the quantitative representation actually describes the precise size and shape of this object.

It is obvious that along with the processing of the image from iconic to symbolic representation, there is a considerable data reduction. However, it is not clear which representations to choose and also which representations allow the processing in the fastest and easiest way. In this respect, there is a very close relation between the chosen

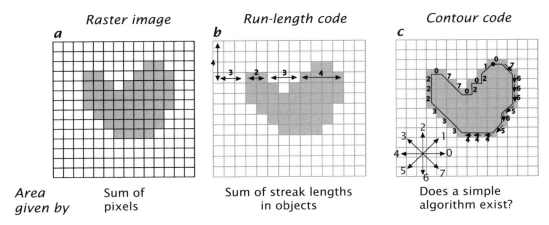

Figure 2.6: *Demonstration of various possibilities to describe a binary 2-D object without holes and to determine its area.*

representation and the methods with which we can perform the process. The rules or prescriptions used to do the processing are called *algorithms.*

We introduce this relation with a simple example of describing a binary 2-D object without holes and the task to compute its area (Fig. 2.6). The most simple and straightforward representation is pixel-based in that we put a 1 in each element that belongs to the object and a 0 for the pixels which do not belong to the object. Computation of the area is very simple. We just have to count the pixels whose value is 1. This representation, however, is very data-intensive since we need to store each pixel of the object. As an object is a connected area, there is actually no need to store each pixel so that we can easily derive a more compact representation. A simple representation of this kind is the so-called *run-length coding* which basically parts the image into horizontal stripes. We scan the image row by row and denote at which offset we find the first pixel of the object. Then, we only need to store the length of the continuous stripe contained in the object. If the shape of the object is complex, it could, of course, be that we have several separated stripes in one line but this can easily be handled by this code. In essence, the algorithm to compute the area is still very simple. We just have to add up the lengths of the stripes containing object pixels. Such an algorithm is much faster than the simple counting of pixels as there are much less stripes than pixels.

Yet another way to represent a binary object starts with the observation that it is sufficient to know the edge of the object. From this information, we can always fill in its interior. A fast way to represent the contour of an object is the so-called *contour* or *chain code*. We start this code by denoting the position of one pixel at the edge of the object and then move along the edge of the object. On a discrete rectangular grid, we have only very few options to move to the next pixel of the contour. If we choose to move along axes and diagonals, eight possible directions exist (8-neighborhood). These directions can be denoted by numbers between 0 and 7 (Fig. 2.6c). We then follow the contour of the object as long as we reach the starting position again. This representation is even more compact than the run-length code since only 3 bits are required to code the movement to the next pixel on the contour. It is less obvious, though, whether a fast and simple algorithm exists to determine the area from the contour code representation. The answer is that a fast and even simple algorithm exists (Section 16.4.3a).

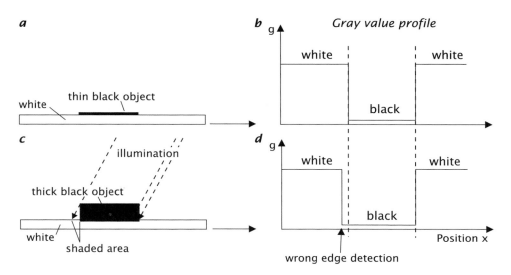

Figure 2.7: *Demonstration of a systematic error which cannot be inferred from the perceived image. **a**, **c** sketch of the object and illumination conditions. **b** and **d** resulting gray value profiles for **a** and **c**, respectively.*

This example nicely shows that one of the important tasks in image processing is to choose the right representation and to find one which is not only compact but also allows fast processing.

2.2.5 Models

The term *model* reflects the fact that any natural phenomenon can only be described to a certain degree of accuracy and correctness. Therefore, it is important to seek the simplest and most general description that still describes the observations with minimum deviations. It is the power and beauty of the basic laws of physics that even complex phenomena can be understood and quantitatively be described on the base of a few simple and general principles.

In the same way, it is the correct approach for a certain image processing task to seek the simplest model that describes the task in the most accurate way. Actually, an analysis of statistical and systematic errors as performed in Section 2.2.3 completely relies on certain model assumptions and is only valid as long as the model agrees to the experimental setup.

Even if the results seem to be in perfect agreement with the model assumption, there is no guarantee that the model assumptions are correct. This is due to the fact that in general very different model assumptions may lead to the same result. Figure 2.7 shows a simple case. We assume that the object to be observed is a flat object lying on a white background and that it is illuminated homogeneously (Fig. 2.7a). Then, the object can clearly be identified by low gray values in the image, and the discontinuities between the high and low values mark the edges of the object. If, however, the black object is elevated and the scene is illuminated by a parallel oblique light (Fig. 2.7c), we receive exactly the same profile in an image. However, in this case only the right edge is detected correctly. In the image, the left edge is shifted to the left because of the shadowed region resulting in a too large size of the object. This example clearly

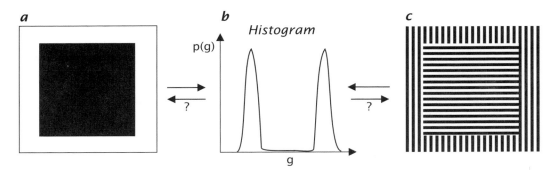

Figure 2.8: *Demonstration of a systematic deviation from a model assumption (object is black, background white) that cannot be inferred from the image histogram.*

demonstrates that even in very simple cases we can run into situations where the model assumptions appear to be met as judged by the results but actually are not.

Another case is shown in Fig. 2.8. A black flat object fills half of the image on a white background. The histogram (the distribution of the gray values) clearly shows a bimodal shape. This tells us that basically only two gray values occur in the image, the lower being identified as the black object and the higher as the white background. This does not mean, however, that any bimodal histogram stems from such an image and that we can safely conclude that we have black objects on a white background or vice versa. The same bimodal histogram is also gained from an image in which both the object and the background have black and white stripes. The object can still be identified easily since the stripes are oriented in different directions. However, the low gray values no longer belong to the object and the high ones to the background. Thus, we misinterpreted the bimodal histogram. A simple segmentation procedure in which we identify all pixels below a certain threshold to the object and the others to the background or vice versa would not extract the desired object but the black stripes. This simple procedure only works if the model assumption is met that the objects and the background are of uniform brightness. While it is quite easy to see the failure of the model assumption in this simple case, this may be much more difficult in more complex cases.

2.2.6 Hierarchy of Image Processing Tasks

In this handbook, we take a task-oriented approach. We ask the question, which steps are required to solve certain image processing tasks? In this respect, the functionality of the human visual system can serve as a reliable guide for breaking up the complex image processing task. First, an optical system forms an image of the observed objects. A sensor converts this image into a form that is usable for digital processing with a computer system. The first processing step, denoted as *low-level image processing*, enhances, restores, or reconstructs the image formed. Further processing extracts features from the images that finally lead to the identification and classification of the objects in the images. In this way, the circle is closed, leading us from objects that are imaged and processed back to their recognition and description.

The four major parts of this handbook, Part I: From Objects to Images, Part II: Enhancing, Restoring, and Reconstructing Images, Part III: From Images to Features, and Part IV: From Features to Objects, reflect these four major steps in the hierarchy of image processing tasks. Figure 2.9 further details them and identifies the chapters

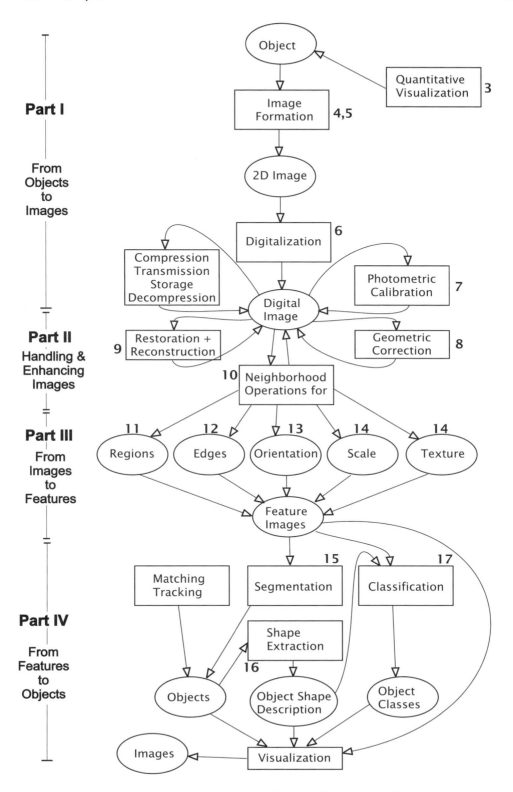

Figure 2.9: *Hierarchy of image processing tasks. The bold figures at the boxes indicate chapter numbers.*

that discuss the various processes. In the remainder of this chapter, a brief overview of the four parts of this handbook is given.

2.2.6a   From Objects to Images (Part I).   In a strict sense, the formation of images does not belong to image processing. It must be, however, regarded as an integral part since only an accurate knowledge of the *image formation* process allows an adequate processing of the images. For geometric measurements, an object or an object feature must be made visible in such a way that it can easily be distinguished from the background by its irradiance. For photometric measurements, the object feature of interest should best be directly proportional to the irradiance received at the image plane. Chapter 3 discusses the properties of electromagnetic, particulate, and acoustic radiation with respect to imaging, and the basic types of interaction between radiation and matter that can be used for quantitative visualization.

Tasks such as position, size, and depth measurements, measurement of the surface orientation or special quantitative visualization of a certain feature require different illumination techniques which are described in Section 3.4. The reference section of Chapter 3 contains additional valuable material about radiation properties, light sources, and optical properties of various materials.

Chapter 4 describes the formation of images. Both the geometry and radiometry of image formation are discussed. Chapter 5 focuses on the *imaging sensors* that convert the irradiance at the image plane into an electronic image and includes a reference section on the available types of CCD imagers. In the final step of the image formation process, the image is digitized and quantized. Chapter 6 discusses the basic principles, especially the sampling theorem, describes procedures to measure the overall performance of the image formation process, and gives an overview of *frame grabbers*, the hardware required to capture an image and to convert it into a *digital image*.

2.2.6b   Handling and Enhancing Images (Part II).   The second part of the handbook deals with all the low-level processing techniques required to remove various types of distortions caused by the image formation process. Chapter 7 describes the point operations that can be used for *photometric calibration*, *color conversion*, and *windowing* of images. The topic of Chapter 8 is *geometric correction* procedures while Chapter 9 discusses *restoration* procedures for distorted images (such as blurring) and the *reconstruction* of images from projections (such as in tomographic imaging).

2.2.6c   From Images to Features (Part III).   The third part of the handbook deals with operations that can extract *features* from images which will later be used to identify and classify objects. Common to all these operators is that they do not operate on the whole image but only in small neighborhoods to analyze the spatial structure. Therefore, a whole chapter (Chapter 10) is devoted to the discussion of the basic principles of neighborhood operations, the central type of image operators. The other chapters detail operators that can extract specific features that identify and classify objects:

- Regions of constant gray values (Chapter 11)
- Detection of discontinuities (edges), a complementary approach to the identification of regions (Chapter 12)
- Orientation of gray value structures of constant features (Chapter 13)
- Scale of gray value structures (Chapter 14)

Chapter 14 also takes all the information collected in this part together to describe the spatial structure of patterns (texture).

All neighborhood operations convert the original intensity-based image into one single or more feature images. Many features (even simple ones as edges) cannot sufficiently be described by scalar quantities but require a vectorial or even a tensorial description.

2.2.6d   From Features to Objects (Part IV).   The fourth part of the handbook finally describes all the steps that lead from feature images to the *identification* and *classification* of objects using measurements of their *shape* (geometry) and features related to photometric quantities and texture. The first step in this chain of processing steps is called *segmentation* (Chapter 15). A decision is made whether a pixel belongs to an object or not. Chapter 16 describes the procedures to measure the shape of objects and Chapter 17 deals with the classification of objects.

2.3   Tools

2.3.1   Overview

Widespread application of image processing is possible as the tools are readily available and are becoming more and more inexpensive. Basically, four components are required:

- a camera to capture an image,
- a frame grabber to convert it into a digital image readable by a computer (Section 2.3.2),
- a personal computer or a workstation (Section 2.3.3), and
- image processing software to process the digital image (Section 2.3.4).

2.3.2   Camera and Frame Grabber

In the simplest case, the *camera* consists of two parts (Fig. 2.10a): a lens that collects the appropriate type of radiation emitted from the object of interest and that forms an image of the real object and a semiconductor device — either a *charged coupled device* (*CCD*) or a *CMOS sensor* — which converts the irradiance at the image plane into an electric signal. The task of the *frame grabber* is to convert the analog electrical signal from the camera into a digital image that can be processed by a computer.

Image display has become an integral part of a personal computer or workstation. Consequently, a modern frame grabber no longer requires its own image display unit. It only needs circuits to digitize the electrical signal from the imaging sensor and to store the image in the memory of the computer. Figure 2.10c shows an example of such a modern frame grabber which only consists of a few components and can be programmed to send the image data as a continuous data stream without intervention of the central processing unit (CPU) to the memory (RAM) of the PC, where the images are processed.

Modern digital cameras include meanwhile most of the functionality of a frame grabber. CMOS sensors make it possible to integrate the whole functionality of a camera onto a single chip including analog/digital converter and digital output (Fig. 2.10b). All it takes then to get the digital image data to the computer is to connect the camera via a standard high-speed serial interface such as *Firewire* (*IEEE 1394*) or *USB2*.

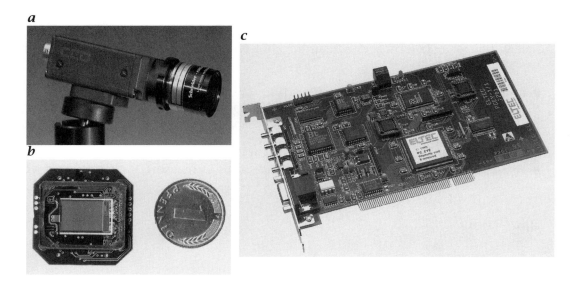

Figure 2.10: *Examples of hardware required to convert a personal computer into an image processing workstation: **a** A miniature monochrome 1/3" CCD camera (shown here KP-M3, Hitachi) with a lens. **b** The VICHI CMOS sensor: the complete functionality of a camera including digital and analog output is integrated onto a single CMOS chip (image courtesy of K. Meier, Kirchhoff-Institute for Physics, University of Heidelberg), [92]). **c** A frame grabber (shown here PC_EYE1, Eltec, Mainz, Germany). This PCI bus board contains an analog/digital converter, a lookup table, and a DMA PCI bus controller to transfer images directly to the main RAM.*

2.3.3 Computer

The tremendous progress of micro computer technology in the past 25 years has brought digital image processing to the desk of any scientist and engineer. For a general purpose computer to be useful for image processing, four key demands must be met: high-resolution image display, sufficient memory transfer bandwidth, sufficient storage space, and sufficient computing power.

In all four areas, a critical level of performance has been reached that makes it possible to process images on standard hardware. In the near future, it can be expected that general-purpose computers can also handle volumetric images and/or image sequences. In the following, we will outline the four key areas of image processing on a general-purpose computer.

2.3.3a Image Display. As already mentioned, *image display* has become an integral part of modern PCs and workstations. Thus, expensive extra hardware is no longer required for image display. Frame grabbers to digitize images (see above) have dropped in price to a fraction of the costs of a general-purpose computer or are no longer required any more with digital cameras that connect to the computer via a standard bus system.

Integral image display means a number of additional advantages. Original and processed images can be displayed simultaneously in multiple windows and graphical tools can be used to inspect their contents (Fig. 2.11). Besides the display of gray-scale images with up to 256 shadings (8 bits), also true-color images with up to 16.7 million colors (3 channels with 8 bits each) can be displayed on inexpensive high-end graphic boards with a resolution of up to 1600×1200 pixels.

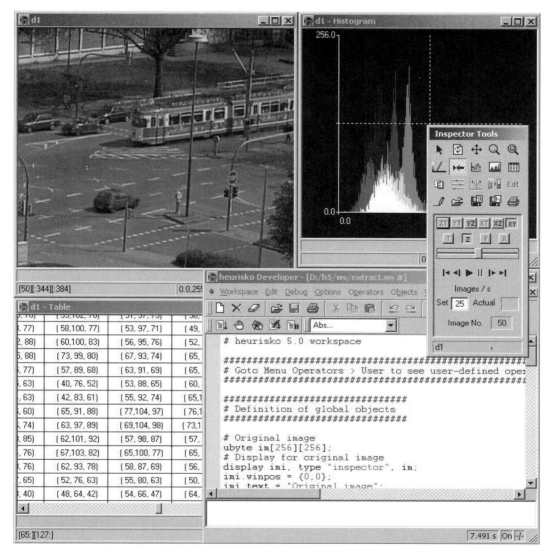

Figure 2.11: *Screenshot of the graphical interface of a modern image processing software (**heurisko®**) featuring multiwindow image display, flexible image "inspectors" for interactive graphical evaluation of images, and a powerful set of image operators, which can easily be combined to user-defined operators adapted to specific image processing tasks. This software package is not limited to 2-D image processing with 8-bit depth (256 gray scales) but can also handle multichannel images of various data types, multiscale images (pyramids), volumetric images, and image sequences. (See also Plate 11.)*

2.3.3b Memory. General-purpose computers now include sufficient *random access memory* (*RAM*) to store multiple images. Theoretically, a 32-bit computer can address up to 4 GB of memory. The design of the motherboards is often more restrictive. Nevertheless, even complex image processing operations that require the storage of multiple intermediate images can be handled without the need to store intermediate results on external storage devices. While in the early days of personal computers hard disks had a capacity of 5 to 10 MB, now disk systems with 10 to 100 GB are standard.

Thus, a large database of images can be stored on disks, which is an important requirement for scientific image processing. For permanent data storage and PC exchange,
the CD-ROM, and more recently the DVD-R, is playing an increasingly important role.
One medium can contain up to 600 MB (DVD-R: 4.7 GB) of image data that can be read
independent of the operating system on MS Windows, MacIntosh, and UNIX platforms.
Cheap CD and DVD writers are available allowing everybody to produce and distribute
his own CDs/DVDs in a cost effective way.

2.3.3c Data Transfer Bandwidth. The standard AT bus system on IBM compatible
PCs used to be one of the most serious bottlenecks for image processing on personal
computers. With the advent of new fast bus systems such as the PCI bus, memory
transfer rates to peripheral devices can reach up to 100 MB per second. Frame grabbers
that digitize standard video signals produce — depending on resolution and color — a
data stream between 6 and 40 MB per second that can be stored in real time in the RAM
of the PC. The transfer rates to external storage devices have also considerably been
improved. Uncompressed real-time image transfer from and to hard disks is currently
at the edge of the technology of standard hard disks. Fast real-time recording with
sustained transfer rates between 15 and 50 MB per second is, however, possible with
standard *RAID disk arrays*.

Also, for processing of images on general-purpose computers, high transfer rates
between the memory and the CPU are of importance. From the figures in Table 2.2,
it can be concluded that these transfer rates have increased from about 100 MB/s in
1996 to 400 MB in 2003 on a standard PC.

2.3.3d Computing Power. Within the short history of *microprocessors* and personal
computers, the computing power has tremendously increased. Figure 2.12 illustrates
exemplarily with the Intel X86 processor family that within the time frame from 1978
to 1996 the clock rate has increased by a factor of almost 50, and another factor of
15 to 2003. The number of elementary operations such as integer addition and multiplication or floating point addition and multiplication has increased even more. While
simple operations such as integer addition have increased "only" by a factor of about
4000, complex instructions such as floating point multiplication have increased by a
factor of almost 200 000. On a 2003 state-of-the-art processor such as a Pentium IV
with 3 GHz clock rate, complex instructions such as floating point multiplication have
a peak performance of 6 000 million floating point operations per second (*MFLOPS*).
Figure 2.12 also illustrates that nowadays floating point arithmetic seems to be even
faster than the much simpler integer arithmetic.

More important than peak performance values are the results from benchmarks
with real-world image processing applications. Table 2.2 summarizes for comparison the results with a 1996 personal computer, a 133 MHz Pentium processor, with a
2003 computer, a 1.4 GHz Pentium IV processor. Both systems run the *heurisko* image
processing software[1].

First of all, it can be observed that the peak rates of Fig. 2.12 are by far not reached.
This is not surprising because the peak rates indicate the maximum performance of
single operations under optimum conditions when the computer is doing nothing else
than this operation. Thus, these times do not include the overhead associated with
loading the data into the registers and storing them back to memory. The increase
in speed with the ten times faster clocked Pentium IV, however, roughly matches the
increase in peak performance.

---

[1]If you want to play around with your own benchmark test, you can download a demo version of the
software from http://www.heurisko.de.

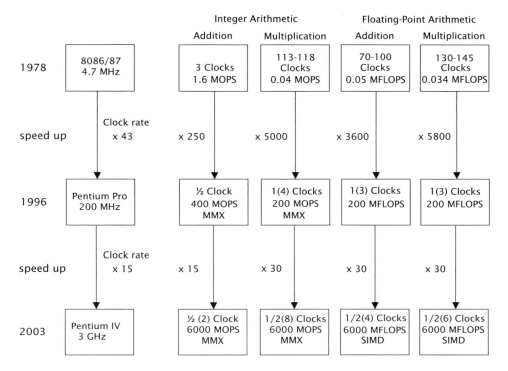

Figure 2.12: *Development of the computing power within the last 25 years exemplarily demonstrated with the Intel X86 processor family. Shown are the clock rates of elementary instructions for addition and multiplication for integer and floating point arithmetic operating on processor registers, i. e., without any overhead for data transfer between the processor and the memory. The figure lists the clock cycles for the instructions and the corresponding number of operations that can be performed in a second. Fractional numbers as 1/2 means that two operations can be executed in parallel within one clock cycle; numbers in parenthesis give the latency, i. e., the number of clocks until the result is available. (MOPS = million operations per second, MFLOPS = million floating-point operations per second).*

With computational intensive operations such as convolutions, a maximum floating point performance of close to 600 and 950 MFLOPS has been measured with optimized C-code (Table 2.2). The maximum integer performance using assembler code with MMX2 instructions reaches up to 1100 MOPS. This is due to the fact that with these instructions four 16-bit operations can be performed in parallel.

Binary operations are significantly faster because in a 32-bit word 32 binary pixels can be computed in parallel. A binary erosion or dilation of a 512×512 binary image takes less than 1.0 ms. Even complex operations on floating point or integer images require computing times that are faster than the standard video frame rate of 30 frames per second or 33.3 ms period. Even the fast Fourier transform on a 512×512 image, for example, takes only about 30 ms.

At any rate, these figures prove that many image processing tasks can be performed nowadays on a standard personal computer in real time. Even complex tasks can be performed within seconds, so that systematic studies with thousands of images come within the reach of every engineer and scientist. This is why image processing has such a high potential for scientific and technical applications.

Table 2.2: *Computing times required for typical image processing tasks on a modern personal computer. The reported benchmarks were measured on a 133 MHz Pentium and a 1.4 GHz Pentium IV with the heurisko image processing software running under Windows '95. The table includes the time in seconds to perform the specified operation with 512×512 images, the effective number of arithmetic operations in a million operations per second (MOPS for **a** and **b**, MFLOPS for **c**), and the effective memory transfer rates for **a** binary, **b** byte (8-bit), and **c** float images.*

Type of Operation	Time [ms]	MOPS/ MFLOPS	Transfer MB/s	Time [ms]	MOPS/ MFLOPS	Transfer MB/s
	133 MHz Pentium			**1.4 GHz Pentium IV**		
a Binary image arithmetics						
Clear image	0.7	—	46	0.072	—	450
Copy image	1.2	—	50	0.109	—	550
Logical And	1.9	133	50	0.156	1600	600
3×3 binary erosion or dilation	15	281	4.1	0.75	5600	82
5×5 binary erosion or dilation	37	330	1.7	1.81	6700	35
b Integer arithmetics with byte (8-bit) images						
Clear image	3.1	—	80	0.25	—	1000
Copy image	6.6	—	76	0.50	—	1000
Add two images	24	10.6	32	0.76	330	1000
General 3×3 convolution (9 multiplications and 8 additions)	331	12.8	0.76	3.8	1100	66
General 5×5 convolution (25 multiplications and 24 additions)	886	13.8	0.28	11.3	1100	22
c Floating point arithmetic						
Clear image	12.1	—	83	2.0	—	500
Copy image	24.4	—	82	3.5	—	570
Add two images	54	4.65	37	4.2	60	480
General 3×3 convolution	267	15.9	7.5	7.0	600	290
General 5×5 convolution	667	18.4	3.0	13.0	950	150
Fast Fourier Transform	635	9.9	32	29.8	210	700

2.3.4 Software and Algorithms

The rapid progress of computer hardware may distract us from the importance of software and the mathematical foundation of the basic concepts for image processing. In the early days, image processing may have been characterized more as an "art" than as a science. It was like tapping in the dark, empirically searching for a solution. Once an algorithm worked for a certain type of image, you could be sure that it did not work with other images and you did not even know why.

Fortunately, this is gradually changing. Image processing is about to mature to a well-developed science. The deeper understanding has also led to a more realistic assessment of today's capabilities of image processing and analysis which in many respects is still worlds away from the capability of human vision. It is a widespread misconception that a better mathematical foundation of image processing is of interest only to the theoreticians but has no real consequences for the applications. The contrary is true. The advantages are tremendous. In the first place, mathematical analysis allows a distinction between image processing problems that can and those that cannot be solved. This is already very helpful. Image processing algorithms become predictable and accurate, and in some cases optimal results are known. New mathematical methods often result in novel approaches which could solve problems that were thought to be very difficult especially in areas where the general belief was that no further improvements were possible. Finally, an important issue is also that a detailed mathematical analysis also leads to faster and more efficient algorithms, i. e., it becomes possible to perform the same image processing task with less operations.

Image processing software has matured from a collection of subroutines that could be handled only by experts to software packages with a user-friendly graphical interface (see Fig. 2.11 for an example). Images can be displayed in multiple windows of user-selectable size. "Inspectors" are available that allow for an interactive quantitative evaluation of original and processed images. The user can take the built-in operators as building blocks to create operators optimally adapted to specific tasks in order to facilitate routine analysis. We will illustrate this concept with many examples throughout this handbook.

Part I

From Objects to Images

3   Quantitative Visualization

3.1   Highlights

Three types of radiation can be used to image objects

- electromagnetic radiation including gamma rays, X-rays, ultraviolet, light, infrared, microwaves, and radio waves,
- particulate radiation (Section 3.3.2), e. g., electrons and neutrons,
- acoustic (elastic) waves in gases, liquids, and solids (Section 3.3.3).

Electromagnetic waves (Section 3.3.1) show a dual nature. They can either be described as transversal waves or as a stream of particles (photons; Section 3.3.2). In vacuum, they travel with light speed, in matter with reduced, wavelength-dependent speed.

Acoustic waves need matter as a carrier. In gases and liquids, only longitudinal acoustic waves can propagate while in solids longitudinal (compression) and transversal (shear) waves are possible. The propagation speed of the various types of acoustic waves directly reflects the elastic properties of the medium.

Radiation interacts with matter. These interactions can either be surface-related (Section 3.3.6) or volume-related (Section 3.3.7). Radiation can be emitted from surfaces because of thermal vibration (Section 3.3.6a) or various types of external stimulation. The bulk optical properties of matter cause absorption and scattering of radiation. Optically active material rotates the plane of polarization.

Radiation carries several pieces of information along that can be used to identify the imaged objects and measure specific properties of them:

1. Frequency or, equivalently, wavelength and (for electromagnetic particulate radiation) particle energy
2. Amplitude (intensity)
3. Polarization mode (for transversal waves only)
4. Phase (This quantity is only assessable with coherent imaging techniques such as interferometric and holographic techniques.)

Radiometry (Section 3.3.4) is important to estimate the radiant energy received by the imaging sensor. Radiometric terms describe the flow of energy in radiation with respect to energy or number of particles: the total emission of a radiation source, the directional flow of radiation emitted from, passing through, or incident at a surface.

The corresponding photometric terms set these terms in relation to the spectral sensitivity of the human eye.

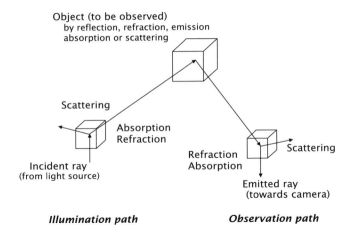

Figure 3.1: *Schematic illustration of the interaction between radiation and matter for the purpose of object visualization. The relation between the emitted radiation towards the camera and the object feature can be disturbed by scattering, absorption, and refraction of the incident and the emitted ray.*

The practical aspects for a successful and optimal setup for quantitative visualization of object properties include:

1. Proper selection of a radiation source (Section 3.5.2)
2. Knowledge about optical properties (Section 3.5.4)

3.2 Task

An imaging system collects radiation emitted by objects to make them visible. Radiation consists of a flow of atomic and subatomic particles, electromagnetic waves or acoustic waves. While in classical computer vision scenes and illumination are taken and analyzed as they are given, in scientific imaging a different approach is required. There, the first task is to establish the quantitative relation between the object feature of interest and the emitted radiation. In order to optimize the imaging of a certain object feature, it is required to carefully select the best type of radiation and to set up an appropriate illumination. Figure 3.1 illustrates that both the incident ray as well as the ray emitted by the object towards the camera may be influenced by additional processes. The position of the object can be shifted by refraction of the emitted ray. Scattering and absorption of the incident and emitted rays lead to an attenuation of the radiant flux that is not caused by the observed object itself but by the environment and thus falsifies its observation. It is the task of a proper setup to insure that these additional influences are minimized and that the received radiation is directly related to the object feature of interest. In cases where we do not have any influence on the illumination or setup, we can still choose the most appropriate type and wavelength range of the radiation.

A wealth of phenomena is available for imaging of objects and object features including self-emission, induced emission (luminescence), reflection, refraction, absorption, and scattering of radiation. These effects depend on the optical properties of the object material and on the surface structure of the object. Basically, we can distinguish

Task List 1: Quantitative Visualization

Task	Procedures
Measurement of geometric object features: position, distance, length, perimeter, area, surface, volume	Setup of adequate illumination to distinguish the objects from the background; object radiance should be homogeneous and not depend on viewing angle
Measurement of surface orientation	Setup of an adequate illumination where emitted radiation should depend on surface orientation only and not on optical properties of object (such as reflectivity)
Measurement of object feature such as density, temperature, concentration of a chemical species, pressure, pH, redox potential	Choice of type of radiation, choice of type of interaction of radiation with matter that relates feature of interest to emitted radiation with minimal influence of other parameters. Use of visualization techniques such as staining, coating, etc.
Object identification and classification	Select type of radiation and interaction that best identifies and distinguishes the objects to be classified.

between surface-related interactions caused by discontinuities of optical properties at the surface of objects and volume-related interactions.

It is obvious that the complexity of the procedures for quantitative visualization strongly depends on the image processing task. If our goal is only to perform a precise geometrical measurement of the objects, it is sufficient to set up an illumination in which the objects are uniformly illuminated and clearly distinguished from the background. In this case, it is not required that we really establish quantitative relations between the object features and the radiation emitted towards the camera.

If we want, however, to measure certain object features such as density, temperature, orientation of the surface, or the concentration of a chemical species, we need to know the exact relation between the selected feature and the emitted radiation. A simple example is the detection of an object by its color, i. e., the spectral dependency of the reflection coefficient.

In most applications, however, the relationship between the parameters of interest and the emitted radiation is much less evident. In satellite images, for example, it is easy to recognize urban areas, forests, rivers, lakes, and agricultural regions. But by which features do we recognize them? And, an even more important question, why do they appear in this way in the images?

Likewise, in medical research one very general question of image-based diagnosis is to detect pathological aberrations. A reliable decision requires a good understanding of the relation between the biological parameters that define the aberration and their appearance in the images.

Because of the complex nature of image formation, it often happens that new visualization techniques promptly discover unexpected phenomena. Some examples are shown in Chapter 1 (e. g., Fig. 1.11, 1.16, 1.19, and 1.25).

The setup of quantitative visualization techniques often includes techniques that manipulate the objects themselves. Typical examples are the use of dyes and tracer particles for flow visualization, the use of stains to identify certain parts in biological tissue, and coating of objects with thin metallic films for electron microscopy. Task list 1 summarizes the tasks involved in quantitative visualization classified by the goal to be achieved.

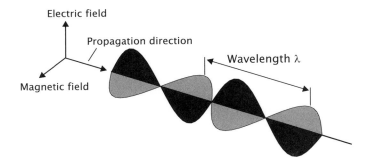

Figure 3.2: *Electromagnetic radiation is a transversal wave. The electrical field E and the magnetic field B are perpendicular to each other and to the direction of propagation. The wavelength λ is the distance of one cycle of the field oscillation and the frequency ν the number of cycles per unit time.*

3.3 Concepts

For the purpose of imaging, three types of radiation are to be considered. The first type is electromagnetic radiation traveling with the velocity of light in free space. The second type consists of a flow of atomic and subatomic particles traveling at less than the speed of light in free space. Finally, a third class of radiation, acoustic or elastic waves, requires matter as a carrier.

3.3.1 Electromagnetic Waves

Electromagnetic radiation consists of alternating *electric* and *magnetic fields*. In an *electromagnetic wave*, these fields are directed perpendicular to each other and to the direction of propagation (Fig. 3.2). They are classified by the *frequency* and the *wavelength* of the fluctuation. Electromagnetic waves span an enormous 24 decades of frequencies. Only a very small fraction, about one octave, falls in the visible region, the part to which the human eye is sensitive.

Figure 3.3 indicates the classification usually used for electromagnetic waves of different frequency or wavelengths. This partition is somewhat artificial and has mainly historical reasons given by the way these waves are generated or detected. Especially confusing is the further subdivision of the *ultraviolet* and *infrared* part of the electromagnetic spectrum (≻ 3.1 and 3.3).

Regardless of the frequency, all electromagnetic radiation obeys the same equations. The only difference is the distance or scale over which the phenomena operate. If all dimensions are scaled by the wavelengths of the radiation, then light cannot be distinguished from radio waves or gamma radiation. In free space, all electromagnetic waves travel with the *speed of light*, $c \approx 3 \cdot 10^8$ ms^{-1} (Table 3.1). Thus, the relation between wavelength and frequency of an electromagnetic wave is given by

$$\lambda \nu = c \quad \text{or} \quad \omega = ck, \tag{3.1}$$

where $\omega = 2\pi\nu$ is the *circular frequency* and $k = 2\pi/\lambda$ the *wave number*.

The wave equation for electromagnetic waves can directly be derived from the fundamental Maxwell equations describing all electromagnetic phenomena. In free space, i. e., without any electric currents and charges, the wave equations for the electric field

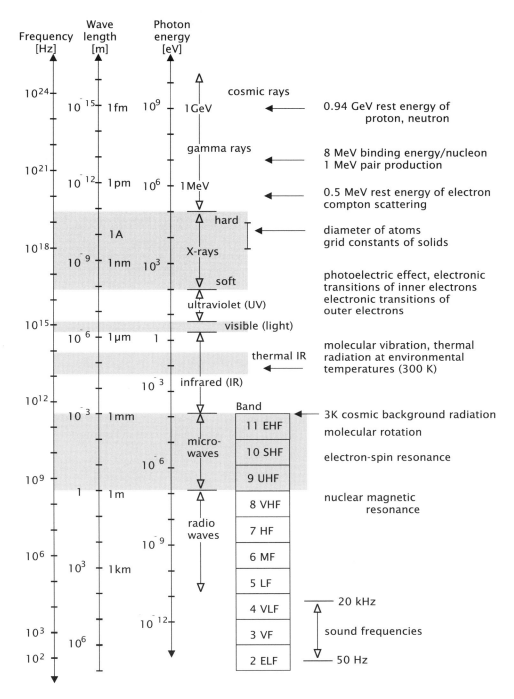

Figure 3.3: *The electromagnetic spectrum covers an enormous 24 decades of frequencies. The graph contains wavelength, frequency, and photon energy scales and indicates the ranges for the various types of electromagnetic radiation from cosmic rays to radio waves. Also indicated are the most prominent types of interaction between radiation and matter at different wavelengths.*

Table 3.1: *Important physical constants related to radiation and interaction between radiation and matter.*

Constant	Value and Units
Velocity of light	$c = 2.9979 \cdot 10^8 \, \text{ms}^{-1}$
Planck's constant	$h = 6.6262 \cdot 10^{-34} \, \text{Js}$
	$\hbar = h/2\pi = 1.0546 \cdot 10^{-34} \, \text{Js}$
Elementary electric charge	$e = 1.6022 \cdot 10^{-19} \, \text{As}$
Boltzmann's constant	$k_B = 1.3806 \cdot 10^{-23} \, \text{J/K} = 0.8167 \cdot 10^{-4} \text{eV/K}$
Stefan-Boltzmann constant	$\sigma = 5.6696 \cdot 10^{-8} \, \text{Wm}^{-2}\text{K}^{-4}$
Wien's constant	$\lambda_m T = 2.8978 \cdot 10^{-3} \, \text{Km}$
Rest energy/mass electron	$m_e c^2 = 0.51100 \, \text{MeV}, \; m_e = 9.1096 \cdot 10^{-31} \, \text{kg}$
Rest energy/mass proton	$m_p c^2 = 938.26 \, \text{MeV}, \; m_p = 1.6726 \cdot 10^{-27} \, \text{kg}$

E and the magnetic field B in SI units are given by

$$\Delta E - \frac{1}{c^2}\frac{\partial^2 E}{\partial t^2} = 0, \quad \Delta B - \frac{1}{c^2}\frac{\partial^2 B}{\partial t^2} = 0, \quad \text{and} \quad \nabla \times E + \frac{\partial B}{\partial t} = 0. \tag{3.2}$$

Without loss in generality, we can assume that the wave is propagating in z direction and the electrical field vector is oriented in x direction. Then, any general function of the form

$$E_x = cB_y = f(z - ct) \tag{3.3}$$

solves the homogeneous wave equation. This general solution simply indicates that the electrical and magnetic fields of a form given by this function in z direction is traveling without any change in form with the light speed c in z direction. A special solution of the wave equation is the *harmonic planar wave* with the circular frequency ω and the wave number k:

$$E_x = a_0 \exp[-\mathrm{i}(\omega/c)(z - ct)] = a_0 \exp[-\mathrm{i}(kz - \omega t)]. \tag{3.4}$$

In free space all waves are traveling with the same speed, the velocity of light (Table 3.1). Such waves are called nondispersive. They do not change their form while propagating through the medium.

Things get much more complex if an electromagnetic wave is propagating in matter. Now, the electric and magnetic fields of the waves are interacting with the electric charges, electric currents, electric fields, and magnetic fields in the medium. Nonetheless, the basic solutions remain the same, except for two modifications: the propagation of the wave is slowed down and the wave is attenuated.

The simplest case is given when the medium reacts in a linear way on the disturbance of the electric and magnetic fields caused by the electromagnetic wave and when the medium is isotropic. Then we can write the following solution:

$$E_x = a_0 \exp(-\omega \chi z/c) \exp[-\mathrm{i}(\omega n z/c - \omega t)]. \tag{3.5}$$

The influence of the medium is expressed in the complex *index of refraction*, $\eta = n + \mathrm{i}\chi$. The real part, n, or ordinary index of refraction, is the ratio of the velocity of light in a vacuum, c, to the propagation velocity u in the medium, $n = c/u$. The imaginary component of η, χ is related to the attenuation of the wave amplitude. The penetration depth of the wave is given according to Eq. (3.5) as $c/(\omega\chi)$.

3.3.1a Dispersion. Generally, the index of refraction depends on the frequency or wavelength of the electromagnetic wave. Therefore, the travel speed of a wave is no longer independent of the wavelength. This effect is called *dispersion* and the wave is called a dispersive wave.

The index of refraction and, thus, dispersion is a primary parameter characterizing the optical properties of a medium. In the context of imaging it can be used to identify a chemical species or any other physical parameter influencing it. The downside of the effect of dispersion is that it is difficult to built refractive optical imaging systems that work not only for monochromatic light but for a certain wavelength range.

3.3.1b Superposition Principle; Nonlinear Effects. The wave equation Eq. (3.2) or, more generally, the Maxwell equations are linear equations. This means that any superposition of solutions of the equation is also a solution. This important principle makes the solution of equations easy. In the context of electromagnetic waves this means that we can decompose any complex wave pattern into some basic ones such as plane harmonic waves. Or, the other way round, that we can superimpose any two or more electromagnetic waves and are sure that they are still electromagnetic waves.

This principle only breaks down for waves with very high field strengths. Then, the material no longer acts in a linear way on the electromagnetic wave and gives rise to *nonlinear optical phenomena*. These phenomena have become obvious only quite recently with the availability of very intense light sources such as lasers. A prominent nonlinear phenomenon is the *frequency doubling* of light. This effect is now widely used in lasers to produce output beams of the double frequency (half wavelength). From the perspective of quantitative visualization, a whole new world is opening since these nonlinear effects can be used to visualize very specific phenomena and material properties.

3.3.1c Polarization. The superposition principle can be used to explain the polarization of electromagnetic waves. Polarization is defined by the orientation of the electric field vector. If this vector is confined to a plane as we have used in the previous examples of a plane harmonic wave, the radiation is called *linearly polarized*. In general, electromagnetic waves are not polarized. To discuss the general case, we consider two waves traveling in z direction, one with the electric field component in x direction and the other with the electric field component in y direction. The amplitudes E_1 and E_2 are constant and ϕ is the phase difference between the two waves. If $\phi = 0$, the electromagnetic field vector is confined to a plane (Fig. 3.4a). The angle α of this plane with respect to the x axis is given by

$$\alpha = \arctan \frac{E_2}{E_1}. \tag{3.6}$$

Another special case arises if the phase difference $\phi = \pm 90°$ and $E_1 = E_2$, and the wave is called *circularly polarized* (Fig. 3.4b). In this case, the electric field vector rotates around the propagation direction with one turn per period of the wave (Fig. 3.4b). The general case where both the phase difference is not $\pm 90°$ and the amplitudes of both components are not equal is called *elliptically polarized*. In this case, the E vector is rotating around the propagation direction on an ellipse. Any type of polarization can also be composed of a right and left circularly polarized beam. A left circular and right circular beam of the same amplitude, for instance, result in a linearly polarized beam. The direction of the polarization plane depends on the phase shift between the two circularly polarized beams.

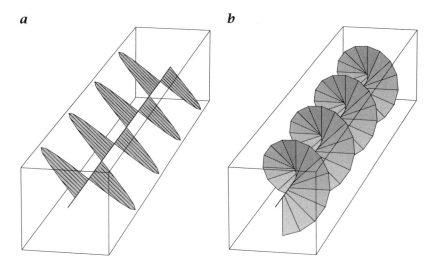

Figure 3.4: a Linearly and **b** circularly polarized electromagnetic radiation.

3.3.1d   Coherence.   An important concept for electromagnetic waves is the term *coherence*. Two beams of radiation are said to be coherent if there exists a systematic relationship between the phases of the electromagnetic field vectors. If this relationship is random, the radiation is *incoherent*. It is obvious that incoherent radiation superposes in a different way than coherent radiation. In case of coherent radiation, destructive inference is possible in the sense that two wave trains with equal amplitude erase each other if their phase shift is 180°.

Normal light sources are incoherent and they do not send out one continuous planar wave but rather short-duration wave packages without definitive phase relationship. Even incoherent light can show interference effects, when the beam is split by partial reflection into two that superimpose with a path difference shorter than the coherence length. This can happen with thin films, when partially reflected beams bounce between the front and rear surface of the films. The commonly available interference filter that transmits all incident radiation in one spectral band and reflects all other radiation is based on this effect. Interference effects can also reduce the quality of image sensors because they can be covered by thin films and microlens arrays.

3.3.1e   Energy Flux Density.   The *energy flux density* carried by an electromagnetic wave is equally distributed between the electric and magnetic field and proportional to the square of the field strength. It is given by the time averaged *Poynting vector S*

$$S = \overline{E \times H} = \frac{1}{2}\frac{n}{Z_0}E_0^2, \tag{3.7}$$

where $Z_0 \approx 377 V/A$ is the wave impedance of the vacuum and n the index of refraction. The second part of the equation is only valid for nonmagnetic media. The units for the energy flux density are W/m^2.

3.3.1f   Particulate Nature of Electromagnetic Waves: Photons.   Electromagnetic radiation has also a particle-like property in addition to those characterized by waves. The electromagnetic energy Q is quantized in that for a given frequency its energy can only occur in multiples of the quantity $h\nu$ in which h is *Planck's constant*, the *action* *quantum* (Table 3.1):

$$Q = h\nu = \hbar\omega. \qquad (3.8)$$

The quantum of electromagnetic energy is called the *photon*.

In any interaction of radiation with matter, be it absorption of radiation or emission of radiation, energy can only be exchanged in multiples of these quanta. The energy of the photon is often given in the energy unit electron Volts (eV) which is the kinetic energy an electron would acquire in being accelerated through a potential difference of one Volt. A photon of yellow light, as an example, has an energy of approximately 2 eV. Figure 3.3 contains also the photon energy scale in eV. The higher the frequency of electromagnetic radiation, the more its particulate nature becomes apparent because the quantization of its energy gets larger. The energy of photons can become larger than the energy associated with the rest mass of elementary particles. In this case it is possible that electromagnetic energy is spontaneously converted into mass in the form of a pair of particles. Although a photon has no rest mass, a momentum is associated with it since it moves with the speed of light and thus has a finite energy. The momentum, p, is given by

$$p = h/\lambda = hk. \qquad (3.9)$$

The quantization of the energy of electromagnetic waves is important for imaging since sensitive radiation detectors (Chapter 5) can measure the absorption of a *single* photon ("photon counting" devices). Thus, the lowest energy amount that can be detected is $h\nu$. The random nature of arrival of photons at the detector gives rise to an uncertainty ("noise") in the measurement of radiation energy. An estimate for the standard deviation is: $\sigma_n = \sqrt{N}$, where N is the number of counted photons (Section 5.3.2g).

3.3.2 Particle Radiation

Unlike electromagnetic waves, most *particulate radiation* moves at a speed less than the speed of light because the particles have a nonzero rest mass. With respect to imaging, the most important particulate radiation is *electrons*, also known as *beta radiation* emitted by radioactive elements. Other important particulate radiation includes the positively charged nucleus of the hydrogen atom or the *proton*, the nucleus of the helium atom or *alpha radiation* which has a double positive charge, and the neutron.

Particulate radiation also shows a wave-like character. The wavelength λ and the frequency ω is directly related to the energy and momentum of the particle:

$$\begin{aligned} \nu &= 1/T &= E/h \quad &\text{Bohr frequency condition,} \\ k &= 1/\lambda &= p/h \quad &\text{de Broglie wave number relation.} \end{aligned} \qquad (3.10)$$

These are the same relations as for the photon, Eqs. (3.8) and (3.9). Their significance for imaging purposes lies in the fact that particles typically have much shorter wavelength radiation. Electrons, for instance, with an energy of about 30 keV have a wavelength of $7 \cdot 10^{-12}$ m or 7 pm. This is about 100 000 times less than the wavelength of light. Since the resolving power of any (far field) imaging system is limited to scales in the order of a wavelength of the radiation, imaging systems based on electrons (*electron microscopes*) have a much higher potential resolving power than any light microscope.

3.3.3 Acoustic Waves

In contrast to electromagnetic waves, *acoustic* or *elastic waves* need a carrier. Acoustic waves propagate elastic deformations. So-called *longitudinal acoustic waves* are generated by isotropic pressure, causing a uniform compression and thus a deformation in the direction of propagation. The local density ρ, the local pressure p, and the local velocity v are governed by the same wave equation

$$\frac{\partial^2 \rho}{\partial t^2} = u^2 \Delta \rho, \quad \frac{\partial^2 p}{\partial t^2} = u^2 \Delta p, \quad \frac{\partial^2 v}{\partial t^2} = u^2 \Delta v. \tag{3.11}$$

The velocity of sound is denoted by u and given by

$$u = \frac{1}{\sqrt{\rho_0 \beta_{ad}}}, \tag{3.12}$$

where ρ_0 is the static density and β_{ad} the adiabatic *compressibility*. The adiabatic compressibility is given as the ratio of the relative volume change caused by a uniform pressure (force/unit area) under the condition that no heat exchange takes place:

$$\beta_{ad} = -\frac{1}{V}\frac{dV}{dP}. \tag{3.13}$$

Equation (3.12) relates the *speed of sound* in a universal way to the elastic properties of the medium. The lower the density and the compressibility, the higher is the speed of sound. Acoustic waves travel much slower than electromagnetic waves. Their speed in air, water, and iron at 20°C is 344 m/s, 1485 m/s, and 5100 m/s, respectively.

In solids, propagation of sound is much more complex.

1. Solids are generally not isotropic. In this case, the elasticity of a solid can no longer be described by a scalar compressibility. Instead, a tensor description is required.

2. Shear forces in contrast to pressure forces give rise to transversal acoustic waves, where the deformation is perpendicular to the direction of propagation as with electromagnetic waves.

Thus, sound waves travel with different velocities in a solid.

Despite all these complexities, the velocity of sound depends only on the density and the elastic properties of the medium as indicated in Eq. (3.12). Therefore, acoustic waves show no dispersion, i. e., waves of different frequencies travel with the same speed. This is an important basic fact for *acoustic imaging* techniques.

3.3.4 Radiometric Terms

Radiometry is the topic in optics describing and measuring radiation and its interaction with matter. Because of the dual nature of radiation, the radiometric terms refer either to energy or to particles; in case of electromagnetic radiation, the particles are photons (Section 3.3.1f). If it is required to distinguish between the two types of terms, the indices e and p are used for energy-based and photon-based radiometric terms, respectively.

Radiometry is not a complex subject. It has only generated confusion by the different, inaccurate, and often even wrong usage of the terms. Moreover, radiometry is taught less frequently and in less detail as other subjects in optics. Thus, knowledge about radiometry is less widespread. It is, however, a very important subject for imaging. Geometrical optics only tells us where the image of an object is located, whereas radiometry says how much radiant energy has been collected from an object.

Table 3.2: *Important radiometric terms describing radiation in terms of energy and photons, dA_0 is an element of area in the surface, θ the angle of incidence, Ω the solid angle. For energy- and photon-based terms, often the indices e and p, respectively, are used.*

Term	Energy-based	Units	Photon-based	Units
Energy	Radiant energy	Ws	Number of photons	1
Energy flux (power)	Radiant flux $\Phi = \dfrac{\partial Q}{\partial t}$	W	Photon flux	s^{-1}
Energy density	Irradiation $H = \dfrac{\partial Q}{\partial A_0}$	Wm^{-2}	Photon irradiation	m^{-2}
Incident energy flux density	Irradiance $E = \dfrac{\partial \Phi}{\partial A_0}$	Wm^{-2}	Photon irradiance	$m^{-2}s^{-1}$
Excitant energy flux density	Excitance $M = \dfrac{\partial \Phi}{\partial A_0}$	Wm^{-2}	Photon excitance	$m^{-2}s^{-1}$
Energy flux per solid angle	Intensity $I = \dfrac{\partial \Phi}{\partial \Omega}$	Wsr^{-1}	Photon intensity	$s^{-1}sr^{-1}$
Energy flux density per solid angle	Radiance $L = \dfrac{\partial^2 \Phi}{\partial \Omega \partial A_0 \cos \theta}$	$Wm^{-2}\ sr^{-1}$	Photon radiance	$m^{-2}s^{-1}sr^{-1}$

3.3.4a Radiant Energy and Radiant Flux. Since radiation is a form of energy, it can do work. A body absorbing the radiation is heated up. Radiation can set free electric charges in a suitable material designed to detect radiation. *Radiant energy* is denoted by Q and given in units of Ws (Joule) or number of particles (photons). Using the quantization of the energy of electromagnetic waves as expressed in Eq. (3.8), the relation between the photon based energy Q_p and Q_e in energy units is

$$Q_p = \frac{Q_e}{h\nu}. \tag{3.14}$$

The power of radiation, i. e., the energy per unit time, is known as *radiant flux* and denoted by Φ:

$$\Phi = \frac{\partial Q}{\partial t}. \tag{3.15}$$

This term is important to describe the total energy emitted by a light source per unit time. It has units of Joules/s (J/s), Watts (W), or photons per s (s^{-1}).

3.3.4b Areal Densities of Radiant Quantities. Areal densities of radiant quantities are of special interest to quantify the "brightness" of light sources and images at a sensor plane. The radiant flux per unit area, the flux density, is known by two names:

$$\text{irradiance} \quad E = \frac{\partial \Phi}{\partial A_0}, \qquad \text{excitance} \quad M = \frac{\partial \Phi}{\partial A_0}. \tag{3.16}$$

The *irradiance*, E, is the radiant flux incident upon a surface per unit area, for instance a sensor that converts the radiant energy into an electric signal. The units for irradiance are Wm^{-2} or $m^{-2}s^{-1}$. If the radiation is emitted from a surface, the radiant flux density is sometimes also called *excitance* or *emittance* and denoted by M.

The areal density of radiant energy, the

$$\text{irradiation} \quad H = \frac{\partial Q}{\partial A_0}, \tag{3.17}$$

is not often used in radiometry, but is of importance for imaging sensors. Typically a sensor element is illuminated by an irradiance $E(t)$ for a certain *exposure time*. Then the received *irradiation H* is

$$H = \int_{t_1}^{t_2} E(t)\mathrm{d}t. \tag{3.18}$$

If the sensor element has a linear response, the generated electrical signal is proportional to the irradiation H and the area of photosensitive part of the sensor element. The irradiation has units of J/m^{-2} or photons per area (m^{-2}).

3.3.4c   Intensities and Solid Angle.   The concept of the *solid angle* is eminent for an understanding of the angular distribution of radiation. Consider a source of small extent at the center of a sphere of radius R beaming radiation outwards in a cone of directions (Fig. 3.5a). The boundaries of the cone outline an area A on the sphere. The solid angle (Ω) measured in steradians (sr) is the area A divided by the square of the radius ($\Omega = A/R^2$). Although the steradian is a dimensionless quantity, it is advisable to use it explicitly when a radiometric term referring to a solid angle can be confused with the corresponding nondirectional term. The solid angle of a whole sphere and a hemisphere are 4π and 2π, respectively.

The (total) radiant flux per solid angle emitted by a source is called the *radiant intensity I* (units W/sr or $s^{-1}sr^{-1}$):

$$I = \frac{\partial \Phi}{\partial \Omega}. \tag{3.19}$$

It is obvious that this term makes only sense to describe point sources, i.e., when the distance from the source is much larger than its extent. This region is also often called the far-field of a radiator. The intensity is also useful to describe light beams.

3.3.4d   Radiance.   For an extended source, the radiation per unit area and per solid angle is more important. The *radiance L* is the area and solid angle density of the radiant flux, i.e., the radiant flux in a specific direction at a specified point of the surface per area projected and per solid angle (Fig. 3.5b):

$$L = \frac{\partial^2 \Phi}{\partial A \partial \Omega} = \frac{\partial^2 \Phi}{\partial A_0 \cos \theta \partial \Omega}. \tag{3.20}$$

The radiation can either be emitted from, pass through, or be incident on the surface. The radiance L depends on the angle of incidence to the surface, θ (Fig. 3.5b), and the azimuth angle ϕ. For a planar surface, θ and ϕ are contained in the interval $[0, \pi/2]$ and $[0, 2\pi]$, respectively. It is important to realize that the radiance is defined per *projected* area, $dA = dA_0 \cdot \cos \theta$. Thus, the effective area from which the radiation is emitted increases with the angle of incidence. The units for energy-based and photon-based radiance are $Wm^{-2}sr^{-1}$ and $s^{-1}m^{-2}sr^{-1}$, respectively.

Often — especially incident — radiance is called brightness. It is better not to use this term at all since it has contributed much to the confusion between radiance and irradiance. Although both quantities have the same dimension, they are quite different. The radiance L describes the angular distribution of radiation while the

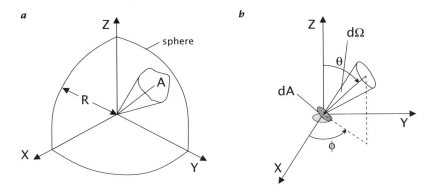

Figure 3.5: a *Definition of the solid angle.* **b** *Definition of radiance, the radiant power emitted per surface area, d A, projected into the direction of propagation and per solid angle.*

irradiance E integrates the radiance incident to a surface element over a solid angle range corresponding to all directions under which it can receive radiation:

$$E = \int_\Omega L(\theta, \phi) \cos\theta \, d\Omega = \int_0^{\pi/2} \int_0^{2\pi} L(\theta, \phi) \cos\theta \sin\theta \, d\theta d\phi. \tag{3.21}$$

The factor $\cos\theta$ arises from the fact that the radiance is defined per unit projected area (Fig. 3.5b), while the irradiance is defined per actual unit area at the surface.

3.3.4e Spectroradiometry. Since any interaction between matter and radiation depends on the wavelength or frequency of the radiation, it is necessary to treat all radiometric quantities as a function of the wavelength. Therefore, we refer to all these quantities as per unit interval of wavelength. Alternatively, it is also possible to use unit intervals of frequencies or wave numbers. The wave number denotes the number of wavelength per unit length interval. To keep the various spectral quantities distinct, we specify the dependency, e. g., $L(\lambda)$, $L(\nu)$, and $L(k)$, when the bandwidth is measured in units of wavelengths, frequencies, and wave numbers, respectively.

Note that different definitions for the term *wave number* exist:

$$k = \frac{2\pi}{\lambda} \quad \text{or} \quad k = \frac{1}{\lambda}. \tag{3.22}$$

Physicists usually include the factor 2π into the definition of the wave number: $k = 2\pi/\lambda$, similar to the definition of the circular frequency $\omega = 2\pi/T = 2\pi\nu$. In optics and spectroscopy, however, it is defined as the inverse of the wavelength without the factor 2π (i. e., number of wavelength per unit length) and usually denoted by $\tilde{\nu} = \lambda^{-1}$. Throughout this handbook the definition of the wave number $k = 1/\lambda$ without the factor 2π is used, because this definition is more intuitive and results in simpler equations, especially for the Fourier transform (Section B.3).

3.3.5 Photometric Terms

The radiometric terms discussed in the previous section measure the properties of radiation in terms of energy or number of photons. *Photometry* relates the same quantities to the human eyes' response to them. Photometry is of importance to scientific imaging in two respects: (1) Photometry gives a quantitative approach to radiometric

Table 3.3: *Basic photometric terms and their radiometric equivalent (compare Table 3.2).*

Term	Photometric quantity	Symbol	Units	Radiometric quantity
Flux	Luminous flux	Φ_v	lumen (lm)	Radiant flux
Energy	Luminous energy	Q_v	lm s	Radiant energy
Flux density	Illuminance	E_v	$lm\ m^{-2} = lux\ (lx)$	Irradiance
Energy density	Exposure	H_v	$lm\ s\ m^{-2} = lx\ s$	Irradiation
Intensity	Luminous intensity	I_v	$lm\ sr^{-1} = candela\ (cd)$	Radiant intensity
Flux density per solid angle	Luminance	L_v	$lm\ m^{-2}sr^{-1} = cd\ m^{-2}$	Radiance

terms as they are sensed by the human eye. (2) Photometry serves as a model as how to describe the response of any type of radiation sensor used to convert irradiance into an electric signal. The key in understanding photometry is to look at the spectral response of the human eye. Otherwise, there is nothing new to photometry. The same symbols are used for photometric terms as for the corresponding radiometric terms (Table 3.3). If required, photometric terms receive the index v.

3.3.5a Spectral Response of the Human Eye. The *human visual system* responds only to electromagnetic radiation having wavelengths between about 360 and 800 nm. It is very insensitive at wavelengths between 360 and about 410 nm and between 720 and 830 nm. Even for individuals without vision defects, there is some variation in the spectral response. Thus, the visible range in the electromagnetic spectrum (*light*, Fig. 3.3) is somewhat uncertain.

The retina of the eye onto which the image is projected contains two general classes of receptors, rods, and cones. Photopigments in the outer segments of the receptors absorb radiation. The absorbed energy is then converted into neural electrochemical signals which are transmitted to subsequent neurons, the optic nerve, and to the brain. Three different types of photopigments in the cones make them sensitive to different spectral ranges and, thus, enable color vision. Vision with cones is only active at high and medium illumination levels and is also called *photopic vision*. At low illumination levels, vision is taken over by the rods. This type of vision is called *scotopic vision*.

At first glance it might seem impossible to measure the spectral response of the eye in a quantitative way since we can only rely on the subjective impression of how the human eye senses "radiance". However, the spectral response of the human eye can be measured by making use of the fact that it can sense brightness differences very sensitively. Based on extensive studies with many individuals, in 1924 the International Lighting Commission (CIE) set a standard for the spectral response of the human observer under photopic conditions that was slightly revised several times later on. Figure 3.6 and Table 3.3 show the 1980 values. The relative spectral response curve for scotopic vision, $V'(\lambda)$, is similar in shape but the peak is shifted from about 555 nm to 510 nm (Fig. 3.6).

Physiological measurements can only give a relative spectral luminous efficiency function. Therefore, it is required to set a new unit for luminous quantities. This new unit is the *candela* and one of the seven fundamental units of the metric system

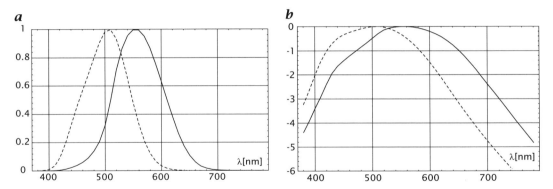

Figure 3.6: *Relative spectral response of the "standard" human eye as set by the CIE in 1980 under medium to high irradiance levels (photopic vision, $V(\lambda)$, solid line), and low radiance levels (scotopic vision, $V'(\lambda)$, dashed line): **a** linear plot, **b** logarithmic plot; data from Photonics Handbook [88].*

(Système Internationale, or SI). The candela is defined to be the luminous intensity of a source that emits monochromatic radiation of a frequency of $5.4 \cdot 10^{14}$ Hz and that has a radiant intensity of $1/683$ W/sr. The odd factor $1/683$ has historical reasons because the candela was previously defined independently from radiant quantities.

With this definition of the luminous intensity and the capability of the eye to detect small changes in brightness, the luminous intensity of any light source can be measured by comparing it to a standard light source. This approach, however, would refer the luminous quantities to an individual observer. Therefore, it is much better to use the standard spectral luminous efficacy function. Then, any luminous quantity can be computed from its corresponding radiometric quantity by:

$$
\begin{aligned}
Q_v &= 683\frac{\text{lm}}{\text{W}} \cdot \int_{380\,\text{nm}}^{780\,\text{nm}} Q(\lambda)V(\lambda)\mathrm{d}\lambda \qquad \text{photopic vision} \\
Q_{v'} &= 1754\frac{\text{lm}}{\text{W}} \cdot \int_{380\,\text{nm}}^{780\,\text{nm}} Q(\lambda)V'(\lambda)\mathrm{d}\lambda \quad \text{scotopic vision,}
\end{aligned}
\tag{3.23}
$$

where $V(\lambda)$ and $V'(\lambda)$ are the spectral luminous efficacy for photopic and scotopic vision, respectively. Table 3.3 lists all photometric quantities and their radiant equivalent.

3.3.5b Photometric Quantities. The *luminous flux* is the photometric equivalent of the radiant flux in radiometry. It describes the total "light energy" emitted by a light source per unit time as perceived by the human eye. The units for the luminous flux are lumen (lm).

Illuminance is the photometric equivalent of *irradiance*. The illuminance, E_v, is the luminous flux incident upon a surface per unit area. It has units of $\text{lm}\,\text{m}^{-2}$ or lux (lx). Illuminance leaving a surface is often called *luminous excitance*.

The photometric quantity *exposure* corresponds to the radiometric term *irradiation*. It describes the total "light energy" per area emitted by a light source or received by the human eye. The units for the exposure are $\text{lx}\,\text{s}$ or $\text{lm}\,\text{s}\,\text{m}^{-2}$.

Luminous intensity is the photometric equivalent of radiant *intensity*. It is the total luminous flux emitted by a source per solid angle in a specified direction. Therefore, it has the units of lumen/steradian (lm/sr) or candela (cd). The luminous intensity is useful to describe point sources and light beams.

Finally, the quantity *luminance* corresponds to the radiometric quantity *radiance*. It describes quantitatively the subjective perception of "brightness". Luminance is the luminant flux per unit projected area at a point in the surface and per unit solid angle in the given direction. The light flux can either be emitted from, pass through, or be incident on the surface. Generally, it depends on the angle of incidence, θ, to the surface and the angle. The units for luminance are $\mathrm{lm\,m^{-2}sr^{-1}}$ or $\mathrm{cd/m^{-2}}$. It is important to realize that the luminance — just as the radiance — is defined per projected area (Section 3.3.4d).

3.3.5c   Radiation Luminous Efficacy.   *Radiation luminous efficacy*, K_r, is defined as the ratio of luminous flux (light in lumens) to radiant flux (in W) in a beam of radiation. Luminous efficiency is a measure of the effectiveness of a beam of radiation in stimulating the perception of light in the human eye.

$$K_r = \frac{\Phi_v}{\Phi_e}\left[\frac{\mathrm{lm}}{\mathrm{W}}\right]. \tag{3.24}$$

Therefore, the luminous efficacy of a beam of infrared or ultraviolet radiation is zero, as it is not perceived by the human eye. From the definition of the luminous intensity (candela), it can be inferred that the maximum luminous efficacy is 683 lm/W. This maximum value can only be reached by a monochromatic light source radiating at a wave length of 555 nm. Any other light source that includes other wavelengths must necessarily have a lower luminous efficacy ($\succ$ 3.6).

3.3.5d   Lighting System Luminous Efficacy.   This is a different kind of luminous efficiency than radiation luminous efficacy. It relates the luminous flux to the total electrical power in W supplied to the light source:

$$K_s = \frac{\Phi_v}{P_e}\left[\frac{\mathrm{lm}}{\mathrm{W}}\right]. \tag{3.25}$$

K_s is a direct measure of how efficiently electrical energy is converted into luminous flux.

The higher K_s is the less energy is required to get the same luminous flux. The *lighting system luminous efficacy* is always smaller than the radiation luminous efficacy since it also includes the efficiency with which electrical energy is converted into radiant energy. For a summary of the electric lighting system luminous efficacy of some common light sources and an illustration of the difference between the two types of efficacy, see $\succ$ 3.7 and 3.8.

3.3.6   Surface-Related Interactions of Radiation with Matter

Basically, two classes of interactions of radiation with matter can be distinguished. The first class is related to the discontinuities of optical properties at the interface between two different materials (Fig. 3.7). The second class is volume-related and depends on the bulk properties of the material. A sharp interface between two materials is only a model. For radiation of a certain wavelength, the optical properties appear only as a discontinuity if they change at the interface over a distance which is much shorter than the wavelength of the radiation.

The principal possibilities for interaction of radiation and matter at an interface are summarized in Fig. 3.7. The simplest case of interaction is the self-emission of

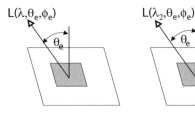 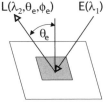 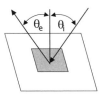 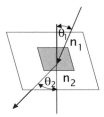

Surface emission	Stimulated emission	Reflection	Refraction
passive thermal imaging	laser induced fluorescence	specular (direct)	refractive optical imaging
passive microwave imaging	active thermal imaging	diffuse	shape-from-refraction
	emission of radiation stimulated by irradiation with electrons, etc.	mixed radar scatterometry SAR imaging shape-from-shading	

Figure 3.7: *Principal possibilities for interaction of radiation and matter at the surface of an object, i. e., at the discontinuity of optical properties. Some important applications are also listed.*

radiation. Emission of radiation may also be stimulated by other effects including the irradiance of radiation at a different wavelength.

As the optical properties change abruptly at the interface, part of the radiation is reflected and the other part is penetrating into the material. The direction of the penetrating beam is generally different from the direction of the incident beam. This effect is called *refraction*.

3.3.6a   Thermal Emission.   *Emission* of electromagnetic radiation occurs at any temperature and is thus an ubiquitous form of interaction between matter and electromagnetic radiation. It ranges from the cosmic background radiation in the microwave range to X-rays emitted from the shock waves in explosion fragments of supernovae heated to several million degrees centigrade. The cause for the spontaneous emission of electromagnetic radiation is the thermal molecular motion which increases with temperature. By emission of radiation, thermal energy of matter is transferred to electromagnetic radiation and according to the universal law of energy conservation, the matter is cooling down.

An upper level for thermal emission exists. According to the laws of thermodynamics, the fraction of radiation at a certain wavelength that is absorbed must also be re-emitted; thus, there is an upper limit for the emission, when the absorptivity is one. A perfect absorber — and thus a maximal emitter — is called a *blackbody*.

The correct theoretical description of the radiation of a blackbody by *Planck* in 1900 required the assumption of emission and absorption of radiation in discrete energy quantities $E = h\nu$ and initiated the field of quantum physics. The spectral radiance of a blackbody with the absolute temperature T is (Fig. 3.8a):

$$L_e(\lambda, T) = \frac{2hc^2}{\lambda^5} \frac{1}{\exp\left(\frac{hc}{k_B T \lambda}\right) - 1}, \tag{3.26}$$

with

$$
\begin{array}{ll}
h & \text{Planck's constant,} \\
k_B & \text{Boltzmann's constant, and} \\
c & \text{speed of light in vacuum.}
\end{array}
\tag{3.27}
$$

a **b**

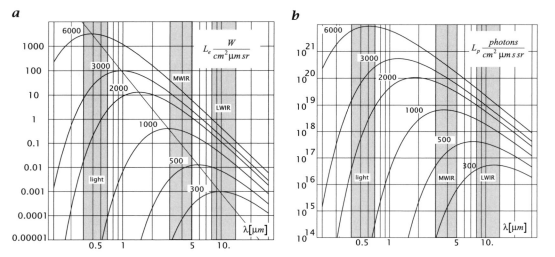

Figure 3.8: *Spectral radiance of a blackbody at different absolute temperatures T in a double logarithmic plot: **a** energy-based radiance, **b** photon radiance (number of photons emitted per second, steradian, unit wavelength of 1 μm, and square meter). The thin line in **a** marks the wavelength of maximum emission as a function of the absolute temperature.*

By using the relations $E = h\nu$ and $c = \nu\lambda$, the photon radiance is (Fig. 3.8b):

$$L_p(\lambda, T) = \frac{2c}{\lambda^4} \frac{1}{\exp\left(\frac{hc}{k_B T \lambda}\right) - 1}. \tag{3.28}$$

Blackbody radiation has the important feature that the emitted radiation does not depend on the viewing angle. Such a radiator is called a *Lambertian radiator*. Therefore the spectral emittance (constant radiance integrated over a hemisphere) is π times higher than the radiance:

$$M_e(\lambda, T) = \frac{2\pi hc^2}{\lambda^5} \frac{1}{\exp\left(\frac{hc}{k_B T \lambda}\right) - 1}, \quad M_p(\lambda, T) = \frac{2\pi c}{\lambda^4} \frac{1}{\exp\left(\frac{hc}{k_B T \lambda}\right) - 1}. \tag{3.29}$$

The total emittance of a blackbody integrated over all wavelengths is proportional to T^4 according to the law of Stefan and Boltzmann

$$M_e = \int_0^\infty M_e(\lambda)\mathrm{d}\lambda = \frac{2}{15}\frac{k_B^4 \pi^5}{c^2 h^3} T^4 = \sigma T^4, \tag{3.30}$$

where $\sigma \approx 5.67 \cdot 10^{-8} \mathrm{Wm^{-2}K^{-4}}$ is the *Stefan-Boltzmann constant* (Table 3.1). The wavelength of maximum emittance of a blackbody is given by *Wien's law* (Table 3.1):

$$\lambda_m \approx \frac{2.898 \cdot 10^{-3} \mathrm{K\,m}}{T}. \tag{3.31}$$

The maximum emittance at room temperature (300 K) is in the infrared at about 10 μm and at 3000 K (incandescent lamp) in the near infrared at 1 μm.

Real objects emit less radiation than a blackbody. The ratio of the emission of a real body to the emission of the blackbody is called (specific) *emissivity* ϵ and depends on the wavelength. As already stated above, the emissivity $\epsilon(\lambda)$ must be equal to the

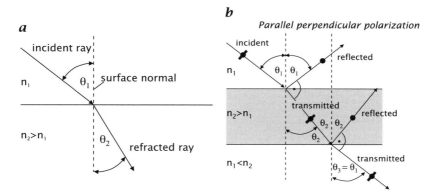

Figure 3.9: *a A ray changes direction at the interface between two optical media with a different index of refraction. b Parallel polarized light is entirely transmitted and not reflected when the angle between the reflected and transmitted beam is 90°. This condition occurs both at the transition from the optically thinner and thicker medium.*

absorptivity $\alpha(\lambda)$, the fraction of energy of the incident radiation absorbed by the material:

$$\epsilon(\lambda) = \alpha(\lambda) \quad \textit{Kirchhoff's law}. \tag{3.32}$$

Because of energy conservation, the fraction of the incident radiation that is not absorbed can either be transmitted or reflected. Therefore,

$$\alpha(\lambda) + \tau(\lambda) + \varrho(\lambda) = 1, \tag{3.33}$$

where $\tau(\lambda)$ and $\varrho(\lambda)$ are the *transmissivity* and the *reflectivity* of the material, respectively.

3.3.6b Refraction. At the interface between two optical media, the transmitted ray is *refracted*, i. e., changes direction according to *Snell's law* (Fig. 3.9):

$$\frac{\sin \theta_1}{\sin \theta_2} = \frac{n_2}{n_1}, \tag{3.34}$$

where θ_1 and θ_2 are the angles of incidence and refraction, respectively. Refraction is the basis for transparent optical elements (lenses) that can form an image of an object. This means that all rays emitted from a point of the object and passing through the optical element converge at another point at the image plane (Fig. 3.10).

A general difficulty for refractive optical elements — in contrast to reflective elements — is related to *dispersion*, the dependence of the refractive index of each material on the wavelength (Section 3.3.1a).

3.3.6c Reflection and Transmission at Specular Surfaces. A *specular surface* behaves like a mirror. The angles of incidence and reflection are equal. The incident beam, the reflected beam, and the surface normal lie in one plane. The ratio of the reflected radiant flux to the incident flux at the surface is called the *reflectivity* ρ.

Specular reflectance only occurs when all parallel incident rays are reflected as parallel rays. A surface needs not be perfectly smooth for specular reflectance because of the wave-like nature of electromagnetic radiation. It is sufficient that the residual roughness elements are significantly smaller than the wavelength.

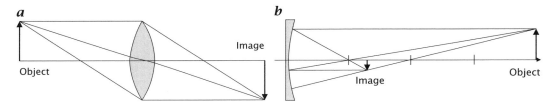

Figure 3.10: a *Refraction and* **b** *reflection are the base for imaging optical elements. All rays emitted from a point of the object and passing through a lens are collected at a corresponding point in the image plane.*

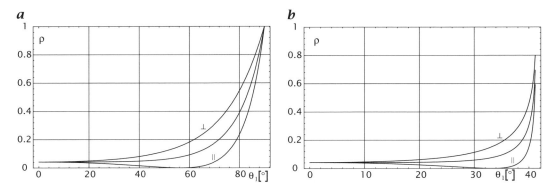

Figure 3.11: *Interface reflectivities for parallel (∥) and perpendicular (⊥) polarized light and unpolarized light incident from **a** air (n_1 = 1.00) to BK7 glass (n_2 = 1.517), **b** BK7 glass to air.*

The reflectivity ρ depends on the angle of incidence, the refractive indices, n_1 and n_2, of the two media meeting at the interface, and the polarization state of the radiation (Section 3.3.1c). Linearly polarized light is called parallel and perpendicular, if the electric field vector is parallel and perpendicular to the plane of incidence, the plane containing the directions of incidence, reflection, and the surface normal. The reflectivity for parallel polarized light is given by *Fresnel's equations*:

$$\rho_\| = \frac{\tan^2(\theta_1 - \theta_2)}{\tan^2(\theta_1 + \theta_2)}, \tag{3.35}$$

for perpendicularly polarized light

$$\rho_\perp = \frac{\sin^2(\theta_1 - \theta_2)}{\sin^2(\theta_1 + \theta_2)}, \tag{3.36}$$

and for unpolarized light (see Fig. 3.11)

$$\rho = \frac{\rho_\| + \rho_\perp}{2}, \tag{3.37}$$

where θ_1 and θ_2 are the angles of the incident and refracted rays related by Snell's law, respectively.

At normal incidence ($\theta = 0$), the reflectivity does not depend on the polarization state:

$$\rho = \frac{(n_1 - n_2)^2}{(n_1 + n_2)^2} = \frac{(n-1)^2}{(n+1)^2} \quad \text{with} \quad n = n_1/n_2. \tag{3.38}$$

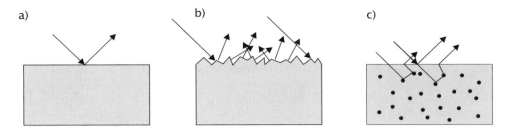

Figure 3.12: *Various types of reflecting surfaces:* **a** *specular reflection at a smooth surface;* **b** *diffuse reflection at a rough surface with micro facets;* **c** *mixed specular and diffuse reflection at a partially transparent surface with volume reflectance.*

As illustrated in Fig. 3.11, parallel polarized light is not reflected at all at a certain angle, the polarizing or *Brewster angle* θ_b. This condition occurs when the refracted and reflected rays are perpendicular to each other (Fig. 3.9b):

$$\theta_b = \arcsin \frac{1}{\sqrt{1 + n_1^2/n_2^2}}. \tag{3.39}$$

When a ray enters into a medium with lower refractive index, there is a critical angle, θ_c

$$\theta_c = \arcsin \frac{n_1}{n_2} \quad \text{with} \quad n_1 < n_2 \tag{3.40}$$

beyond which all light is reflected and none enters the optically thinner medium. This phenomenon is called *total reflection*.

3.3.6d Reflection at Rough Surfaces. Most materials do not have smooth specular reflecting surfaces (Fig. 3.12a). Indeed, surfaces useful for reflective or refractive optical elements require significant technological effort so that they can be used for imaging elements. Most natural and also technical objects do not directly reflect light but show a diffuse reflectance since surface micro roughness causes reflection in various directions depending on the slope distribution of the reflecting facets (Fig. 3.12b).

There is a great variety in how these rays are distributed over the emerging solid angle. Some materials produce strong forward scattering effects while others scatter almost equally in all directions. Other materials show a kind of mixed reflectivity which is partly specular due to reflection at the smooth surface and partly diffuse caused by body reflection (Fig. 3.12c). In this case, light is partly penetrating into the object where it is scattered at optical inhomogeneities. Part of this scattered light is leaving the object again causing a diffuse reflection. For imaging of objects that do not emit radiation by themselves but passively reflect incident light, it is essential to know how the light is reflected.

Generally, the relation between the incident and emitted radiance can be expressed as the ratio of the radiance emitted under the polar angle θ_e and the azimuth angle φ_e and the irradiance received under the incidence angle θ_i. This ratio is called the *bidirectional reflectance distribution function* (BRDF) or *reflectivity distribution*, since it generally depends on the angles of both the incident and excitant radiance:

$$f(\theta_i, \phi_i, \theta_e, \phi_e) = \frac{L_e(\theta_e, \phi_e)}{E_i(\theta_i, \phi_i)}. \tag{3.41}$$

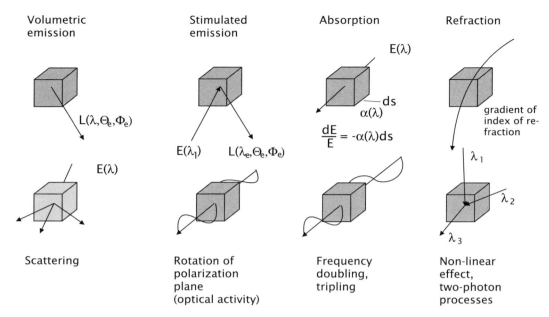

Figure 3.13: *Schematic illustration of the principal volume-related interactions between matter and radiation.*

For a perfect mirror (specular reflection), f is zero everywhere, except for $\theta_i = \theta_e$ and $\phi_e = \pi + \phi_i$, hence

$$f(\theta_i, \phi_i, \theta_e, \phi_e) = \delta(\theta_i - \theta_e) \cdot \delta(\phi_i + \pi - \phi_e). \qquad (3.42)$$

The other extreme is a perfect diffuser, reflecting incident radiation equally into all directions independently of the angle of incidence. Such a reflector is known as *Lambertian* surface or reflector. The radiance of such a surface is independent of the viewing direction:

$$L_e = \frac{1}{\pi} E_i \quad \text{or} \quad f(\theta_i, \phi_i, \theta_e, \phi_e) = \frac{1}{\pi}. \qquad (3.43)$$

3.3.7 Volume-Related Interactions of Radiation with Matter

This section deals with the optical properties governing radiation propagation within an optical medium. Four basic effects can be distinguished. Radiation can be absorbed or scattered and its plane of polarization can be rotated. Additionally, the scattered radiation can show a frequency shift, known as the Doppler effect.

3.3.7a Absorptance and Transmittance. Radiation traveling in matter is more or less absorbed and converted into different energy forms, especially heat. The radiant intensity I diminishes in a thin layer dx always by the same fraction. Therefore, the ratio

$$\alpha(\lambda, x) = -\frac{1}{I} \frac{dI(\lambda)}{dx} \qquad (3.44)$$

is defined as the *absorption coefficient* α. It is a property of the medium and depends on the wavelength of the radiation. It is a reciprocal length and has the units of m^{-1}. By integration of Eq. (3.44), we can compute the attenuation of the radiant intensity over the distance from 0 to x:

$$I(x) = I(0) \cdot \exp \left(- \int_0^x \alpha(\lambda, x) dx' \right),$$ (3.45)

or, if the medium is homogeneous, i. e., α does not depend on the position x',

$$I(x) = I(0) \exp(-\alpha(\lambda)x).$$ (3.46)

The exponential attenuation of radiation in a homogeneous medium, as expressed by Eq. (3.46), is often referred to as *Lambert Beer's* or Bouger's law. After a distance of $1/\alpha$, the radiation is attenuated to $1/e$ of its initial value.

The path integral over the absorption coefficient

$$\tau(x_1, x_2) = \int_{x_1}^{x_2} \alpha(x') dx'$$ (3.47)

results in a dimensionless quantity that is known as the *optical thickness* or *optical depth*. The optical depth is a logarithmic expression of radiation attenuation and means that along the path from the point x_1 to point x_2 the radiation has been attenuated to $e^{-\tau}$.

If radiation travels in a composite medium, often only one chemical species — at least at certain wavelengths — is responsible for the attenuation of the radiation. Therefore, it makes sense to relate the absorption coefficient to the concentration of that species,

$$\alpha = \varepsilon \cdot c \quad [\varepsilon] = \left[\frac{1}{\text{mol m}^{-1}} \right],$$ (3.48)

where c is the concentration in mole/l. Then, ϵ is known as the *molar absorption coefficient*. For some applications, e. g., absorption of radiation in the atmosphere, it is useful to relate the absorption coefficient to the mass of the absorbed material along the path. This simple linear relation Eq. (3.48) holds for a very wide range of radiant intensities but breaks down at very high intensities, e. g., the absorption of high-intensity laser beams. At that point, the world of nonlinear optical phenomena is entered.

Since the absorption coefficient is a distinct optical feature of chemical species, it can be used in imaging applications to identify chemical species and to measure their concentrations.

Finally, the term *transmittance* means the fraction of radiation that remains after the radiation has traveled a certain path in the medium. Often, transmittance and *transmissivity* are confused. In contrast to transmittance, the term transmissivity is related to a single surface. It means the fraction of radiation that is not reflected but entering the medium.

3.3.7b Scattering. The attenuation of radiation by scattering can be described with the same concepts as for loss of radiation by absorption. The *scattering coefficient* is defined by

$$\beta(\lambda) = -\frac{1}{I} \frac{dI(\lambda)}{dx}.$$ (3.49)

It is a reciprocal length with the units m^{-1}. If radiation is attenuated both by absorption and scattering, the two effects can be combined in the *extinction coefficient* $\kappa(\lambda)$:

$$\kappa(\lambda) = \alpha(\lambda) + \beta(\lambda).$$ (3.50)

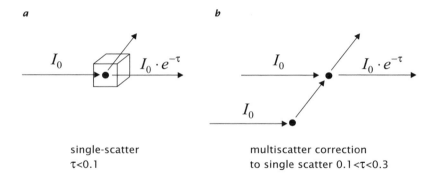

single-scatter
$\tau < 0.1$

multiscatter correction
to single scatter $0.1 < \tau < 0.3$

Figure 3.14: a For optical depths less than 0.1, the probability for scattering is so low that only single scattering events must be considered. b For optical depths between 0.1 and 0.3, multiscatter corrections to single scatter must be considered since multiply scattered rays can reenter the primary beam.

Unfortunately, there is no unified terminology and symbols for these various coefficients. The different communities use different symbols and slightly different definitions.

Although scattering appears to be similar to absorption, it is a much more difficult phenomenon. The above formula can only be used if the radiation from the individual scatters adds up incoherently at a distance far from the particles. Only then, the total scattered radiation is superposed by the scattered radiation from each individual particle into the same direction. The complexity of scattering is related to the fact that the scattered radiation (without additional absorption) is never lost. Scattered light is scattered again. Therefore, a fraction of the beam can reenter and amplify the original beam.

The probability that radiance will be scattered in a certain path length more than once is directly related to the total attenuation by scattering along the path of the beam or the optical depth τ. If τ is smaller than 0.1, less than 10% of the radiance is scattered. The probability for a scattered ray to be scattered a second time is therefore less than 10% or in total less than 1%. This means that single scattering is a dominant process. With increasing optical depth more and more of the radiance is scattered and, thus, the probability that multiple scattered rays enter the primary beam again is no longer negligible (Fig. 3.14b). For optical depths less than 0.3, multi-scatter corrections to single-scatter theory are sufficient. At higher optical depths, however, multiple scattering dominates.

The scattering coefficient only describes the loss of radiation in the incident beam, but not the properties of the scattered light. This is the more complex but also the more interesting part of light scattering. The total amount of scattered light and the analysis of the angular distribution is related to the optical properties of the scattered medium. Consequently, scattering is caused by the optical inhomogeneity of the medium. In further discussions we assume that small spherical particles with radius a and index of refraction n are imbedded in a homogeneous optical medium.

Scattering by a particle is described by the *cross section*. It is defined in terms of the ratio of the flux Φ_s removed by the particle in relation to the flux Φ incident on the particle:

$$\sigma_s = \frac{\Phi_s}{\Phi} \pi r^2. \tag{3.51}$$

The cross section has the units of an area. It can be regarded as the effective area of the particle for scattering that completely scatters all incident radiative flux. Therefore, the *efficiency factor* for scattering Q_s is defined as

$$Q_s = \sigma_s / (\pi r^2). \tag{3.52}$$

The angular distribution of the scattered radiation is given by the *differential cross section*, $d\sigma_s / d\Omega$, the flux density scattered per unit solid angle. The total cross section is given as the integral over the sphere of the differential cross section

$$\sigma_s = \int \frac{d\sigma_s}{d\Omega} d\Omega. \tag{3.53}$$

The relation between the scattering coefficient β Eq. (3.49) is given by taking the number of particles per unit volume N. Then,

$$\beta = N\sigma. \tag{3.54}$$

In a unit volume, there are N particles. Thus, the total effective scattering cross section covers the area $N \cdot \sigma$. This area compared to the unit area gives the fraction of area that removes the incident flux and is thus equal to the scattering coefficient β.

The scatter by small particles is most significantly influenced by the ratio of the particle size to the wavelength of the radiation expressed in the dimensionless particle size $q = 2\pi r/\lambda = rk$. If $q \ll 1$ (Rayleigh scatter), the scatter is very weak and proportional to λ^{-4}:

$$\sigma_s / (\pi r^2) = \frac{8}{3} (q)^4 \left| \frac{n^2 - 1}{n^2 + 2} \right|. \tag{3.55}$$

For $q \gg 1$, the scatter can be described by geometrical optics. If the particle completely reflects the incident radiation, the scattering cross section is equal to the geometrical cross section $(\sigma_s / (\pi r^2) = 1)$ and the differential cross section is constant (isotropic scattering, $d\sigma / d\Omega = r^2/2$).

The scatter for particles with sizes of about the wavelength of the radiation (*Mie scatter*) is very complex due to diffraction and interference effects of the light scattered from different portions of the surface of the particle [149]. The differential cross section shows strong variations with the scattering angle and is directed mostly in forward direction, while Rayleigh scatter is rather isotropic.

3.3.7c Optical Activity. An optically active material rotates the plane of polarization of electromagnetic radiation. The rotation is proportional to the concentration of the optical active material and the path length x,

$$\varphi = [\alpha](\lambda) c x. \tag{3.56}$$

The constant $[\alpha]$ is known as the specific rotation and has the units $[m^2 \, mol]$ or $[cm^2 \, g^{-1}]$; it depends strongly on the wavelength of the radiation. Generally, the specific rotation is significantly larger at shorter wavelengths.

Two well-known optically active materials are quartz crystals and sugar solution. Optically active materials — including the measurement of the wavelength dependency — can be used to identify chemical species and to measure their concentration. With respect to visualization, optical activity is significant since it can be induced by various external forces, among others, electrical fields (*Kerr effect*) and magnetic fields (*Faraday effect*).

Table 3.4: *Various forms of luminescence classified by the lifetime of the excited state and the process generating the excited state.*

Type of luminescence	Description
Fluorescence	Short lifetime luminescence, occurring within 1-200 ns of excitation; typically a spin-allowed transition
Phosphorescence	Delayed luminescence, occurring milliseconds to minutes after excitation; frequently a spin-forbidden transition
Photoluminescence	Luminescence arising from the absorption of photons
Electroluminescence (cathodoluminescence)	Luminescence arising from an electric current in solids or solution or in gases during an electrical discharge
Thermoluminescence	Thermally stimulated luminescence of materials previously activated by radiation
Radioluminescence	Luminescence arising from the passage of ionizing radiation or particle radiation in matter
Chemiluminescence	Luminescence arising during chemical reactions
Bioluminescence	Chemiluminescence in living organisms

3.3.7d Luminescence. *Luminescence* is the emission of radiation from materials that arises from a radiative transition between an excited state and a lower state. *Fluorescence* is luminescence characterized by short lifetimes of the excited state (in the order of nano seconds), while the term *phosphorescence* is used for longer lifetimes (milliseconds to minutes). Luminescence is an enormously versatile process since it can be triggered by various processes (Table 3.4). On the one hand, the emitted radiation is characteristic for the material and thus can be used to identify chemical species and to measure their concentration. On the other hand, the luminescence can be used to measure the process that generated the luminescence. Luminescence in crystals (e.g., thallium-activated sodium iodide) or solutions of highly fluorescent organic scintillators are widely used to measure γ and β radiation (*radioluminescence*). In *chemiluminescence*, the energy required to generate the excited state is derived from the energy released by a chemical reaction. Chemiluminescence normally has only low efficiencies (i. e., number of photons emitted per reacting molecule) in the order of 1 % or less. Flames are the classical example of a low-efficient chemiluminescent process.

Bioluminescence is a chemiluminescence in living organisms. Fireflies and the glow of marine microorganisms are well-known examples. The firefly reaction involves the enzymatic oxidation of luciferin. In contrast to most chemiluminescent processes, this reaction has an efficiency close to 100 %. Low-level bioluminescent processes are common to many essential biological processes. Imaging of these processes is becoming an increasingly important tool to study biochemical processes. Marking biomolecules with fluorescent dyes is becoming another more and more sophisticated tool in biochemistry. It has become even possible to mark individual chromosomes or gene sequences in chromosomes with fluorescent dyes (see Section 1.5.3).

Luminescence always has to compete with other processes that deactivate the excited state without radiation emission. A prominent radiationless deactivation process is the energy transfer during the collision of molecules. Some types of molecules, especially electronegative molecules such as *oxygen*, are very efficient in deactivating

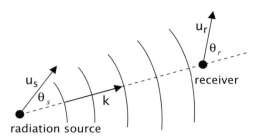

Figure 3.15: *Geometry of the classical Doppler effect: Source and receiver are moving relative to the medium in which the radiation is propagating with velocities $\mathbf{u}_s$ and $\mathbf{u}_r$, respectively.*

excited states during collisions. This process is referred to by the term *quenching*. The presence of a quenching molecule, often in surprisingly low concentration, causes the fluorescence to decrease. Therefore, the measurement of the fluorescent irradiance can be used to measure the concentration of the quenching molecule. The dependence of the fluorescent intensity on the concentration of the quencher is given by the Stern-Vollmer equation

$$\frac{L}{L_0} = \frac{1}{1 + K_q c_q},$$
(3.57)

where L is the fluorescent radiance, L_0 the fluorescent radiance when no quencher is present, c_q the quencher concentration, and K_q the quenching constant. Efficient quenching requires that the excited state has a sufficiently long lifetime.

3.3.7e Doppler Effect. A velocity difference between a radiating source and a receiver causes the receiver to measure a different frequency than emitted by the source. This phenomenon is known as the *Doppler effect*. The frequency shift is directly proportional to the velocity difference according to

$$v_r - v_s = \Delta v = \frac{(\mathbf{u}_s - \mathbf{u}_r)\mathbf{k}}{1 - \mathbf{u}_s \bar{\mathbf{k}}/c},$$
(3.58)

where v_s is the frequency of the source, v_r the frequency measured at the receiver, $\bar{\mathbf{k}} = \mathbf{k}/|\mathbf{k}|$ is a unit vector in the direction of wave propagation, c the propagation speed of the radiation, and $\mathbf{u}_s$ and $\mathbf{u}_r$ the velocities of the source and receiver relative to the medium in which the wave is propagating (Fig. 3.15). Only the velocity component in the direction to the receiver causes a frequency shift. If the source is moving towards the receiver ($\mathbf{u}_s \bar{\mathbf{k}} > 0$), the frequency is increasing since the wave fronts come faster after each other. A critical limit is reached when the source moves with the propagation speed of the radiation. Then, the radiation is left behind the source.

For small velocities relative to the wave propagation speed, the frequency shift is directly proportional to the relative velocity between source and receiver.

$$\Delta v \approx (\mathbf{u}_s - \mathbf{u}_r)\mathbf{k} \quad \text{or} \quad \frac{\Delta v}{v} = \frac{(\mathbf{u}_s - \mathbf{u}_r)}{c}\bar{\mathbf{k}},$$
(3.59)

where $\bar{\mathbf{k}} = \mathbf{k}/k$ is a unit vector parallel to $\mathbf{k}$. The relative frequency shift $\Delta\omega/\omega$ is directly given by the velocity component in the direction of the receiver, relative to the wave propagation speed in the medium.

Table 3.5: *Summary of the principle illumination types.*

Illumination type	Description
Front illumination	Illumination of an object from the front. Radiation reflected at object surface or scattered back from volumetric objects.
Rear illumination	Illumination of an object from the rear. Radiation is absorbed or scattered in forward direction.
Specular illumination	Illumination from a narrow cone of directions (in the ideal case with a beam of parallel light)
Diffuse illumination	Illumination of the object from a wide cone of directions (in the ideal case from the whole hemisphere or sphere)
Light field illumination	Illumination such that without an object the image appears equally bright. Objects diminish the radiance by reflecting, refracting, absorbing, or scattering light.
Dark field illumination	Illumination such that without objects the image is dark. Objects become visible by reflecting, refracting, or scattering light into the direction of the camera.

For electromagnetic waves, the velocity relative to a "medium" is not relevant. The theory of relativity gives the frequency

$$\nu_r = \frac{\nu_s}{\gamma\left(1 - \frac{\boldsymbol{u}}{c}\bar{\boldsymbol{k}}\right)} \quad \text{with} \quad \gamma = \frac{1}{\sqrt{1 - (u/c)^2}}. \tag{3.60}$$

Here $\boldsymbol{u}$ is the velocity difference between the receiver and source. In the limit of small difference velocity Eq. (3.60) reduces to

$$\Delta \nu \approx \boldsymbol{u}\boldsymbol{k} \quad \text{for} \quad u \ll c. \tag{3.61}$$

Thus the Doppler frequency shift for acoustic and electromagnetic waves can be treated equally in the limit of small velocities of the receiver and source relative to the propagation speed of the radiation.

3.4 Procedures

3.4.1 Introduction

The discussion of the concepts in Section 3.3 makes it evident that an enormous variety of techniques for quantitative visualization is available. In order to give useful advice, the following strategy is applied:

- The techniques are categorized, the principal possibilities are summarized, and some generally valid rules are given.
- A few characteristic examples are discussed in depth.
- References to further readings are given.

3.4.2 Types of Illumination

The setup of the *illumination* decides which optical property of the object determines the emitted radiation. Thus, it is a powerful instrument to condition the image forma-

a   b

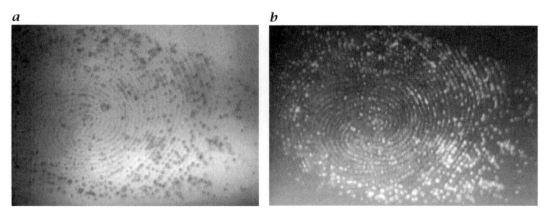

Figure 3.16: *Fingerprint on a glass plate made visible by scattering:* **a** *light field illumination: scattering reduces direct reflection, finger print appears darker;* **b** *dark field illumination: camera sees scattered light, fingerprint appears brighter.*

tion in the sense that the radiation received by the camera depends only on the object feature of interest and not on any other parameters.

The principal illumination types are summarized in Table 3.5. With *front illumination*, only objects can be illuminated that reflect or scatter light. The object radiance thus depends on the *reflection coefficient* of the object surface (Section 3.3.6c) and the object surface roughness (Section 3.3.6d). The radiance of volumetric objects depends on the *differential cross section* (Section 3.3.7b). The absorption of radiation by an object can only be measured with a light field *rear illumination* since a direct ray from the light source to the camera must be available. An important parameter of an illumination system is also the range of angles under which radiation is received by the illuminated object. With *specular illumination*, radiation is emitted only from a narrow cone of angles. The ideal case of specular illumination is given by a beam of parallel radiation. Direct sunlight is a good example of specular illumination.

With *diffuse illumination*, radiation from a wide cone of angles arrives at the object to be illuminated. An object surface element can receive radiation at most from a hemisphere. A completely overcast sky is a good example of a *hemispherical illumination*. A small object can receive radiation from the full sphere. This illumination is difficult to realize technically. A good natural example is the illumination conditions in dense fog.

In practice, the cone from which an object receives radiation is an important parameter of the illumination system. For a special setup it might be advantageous to choose a certain angular distribution of the incoming radiation.

In *light field illumination*, a direct path exists from the light source to the camera. Therefore, an object appears darker than the free background because it diminishes the radiance by absorbing, scattering, or refracting light. In contrast, with a *dark field illumination*, no direct path between the light source and the camera exists. Therefore, the background appears dark. Objects become visible only by scattering, reflecting or refracting radiatation, or emitting previously absorbed radiation (*luminsescence*, Section 3.3.7d). With dark field illumination objects can be made visible that are smaller than the resolution of the image formation system (see Section 4.3.3), provided that sufficient light is emitted by them compared to the background radiance.

As an example for dark and light field illuminations of the same object, Fig. 3.16 shows a fingerprint on a glass plate. In both cases, the fingerprint becomes visible by

Table 3.6: *Summary of illumination techniques for geometric measurements.*

Technique	Advantages and Disadvantages
Specular illumination with point light source or parallel light beam	Advantages: simple setup Disadvantages: can only be used with matte objects; deep shadows
Specular illumination with ring light source around the camera	Advantage: simple setup, no direct shadows Disadvantages: can only be used with matte objects
Diffuse illumination with extended light source	Advantage: can be used with specular reflecting objects Disadvantages: complex setup, extended light sources, requires much space
Rear illumination	Advantages: perfect for exact measurements of the shape of flat objects Disadvantages: opaque objects appear dark, no differentiation due to different surface reflectivities is possible

scattering. With dark field illumination, the fingerprint appears brighter, because the camera receives scattered light. With light field illumination, the fingerprint appears darker, because part of the light is scattered in directions not received by the camera.

Figure 3.17 shows a piece of terry towel illuminated by several techniques for illustration. The diffuse illumination (Fig. 3.17c) results in a much lower contrast than the specular illumination (Fig. 3.17a and b). A grazing specular illumination does not penetrate far into the surface (Fig. 3.17b). In contrast, diffuse rear illumination essentially measures absorption. Now, the threads appear darker but light stripes become visible in the fabric where no threads stick out (Fig. 3.17d).

3.4.3 Illumination Techniques for Geometric Measurements

For geometric measurements, the goal is to illuminate the objects in such a way that they show a uniform radiance. This ideal case is shown in the computer generated scene in Fig. 3.18a. The 3-D objects appear as flat silhouettes in constant color. Thus, they can easily be identified and segmented.

This scene is, however, far from being realistic (such images could only be acquired in the infrared where all objects are radiators). Real objects show radiance variation even if they have a uniform optical surface because of the varying angle between the surface normal and the directions of the light sources (Fig. 3.18b). This effect makes us believe that the objects have depth although they are flat in the image. While shading is thus an important clue for 3-D reconstruction (known as *shape from shading*), it is a serious difficulty for object segmentation in image processing.

Even more difficulties arise with objects that have partly specular reflecting surfaces (Section 3.3.6c). Then, parts of the objects show "highlights" that change the position at the object surface with the viewing angle. Further intensity variations are caused by shadow casting.

From the discussion above it is obvious that successful image processing and especially object segmentation require a careful setup of the illumination under consideration of the object surface properties. Table 3.6 summarizes some illumination techniques that can be used for geometric measurements.

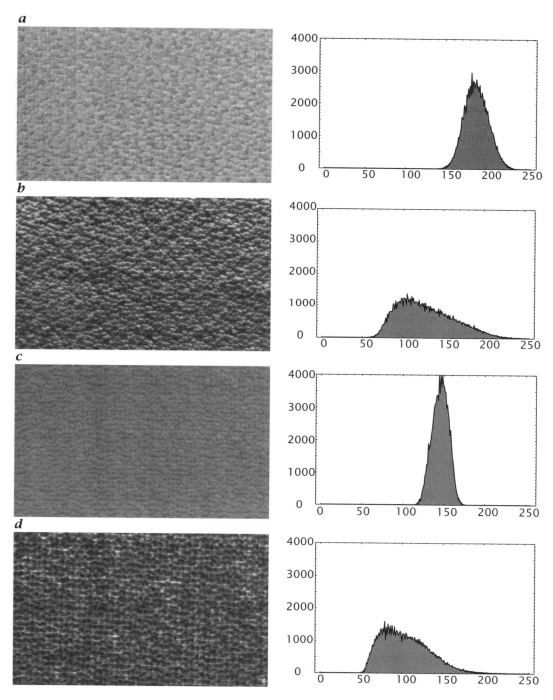

Figure 3.17: *A piece of terry towel illuminated with different setups and the corresponding histograms. a specular illumination with low incidence angle, b specular illumination with grazing incidence angle, c diffuse front illumination, d diffuse rear illumination.*

a

b

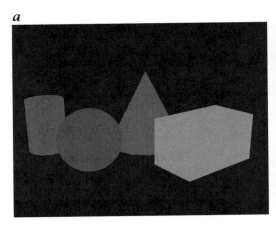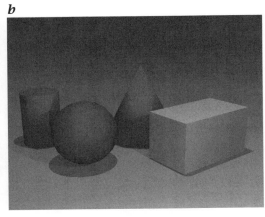

Figure 3.18: *A computer-generated (rendered) 3-D scene illustrating the complexity of the inter-actions between illumination and objects. **a** flat shading (uniformly radiating objects), the ideal world for image processing, **b** Gouraud shading (matte Lambertian surface) with multiple point light sources. Now the radiance of the objects is no longer uniform. It depends on the angle between the surface normal and the direction to the light sources. Furthermore, other objects that are between a light source and the object obstruct light of this light source to reach the object resulting in shadows. (See also Plate 12.)*

3.4.3a Specular Illumination. *Specular illumination* means that a point light source or parallel light with low divergence (e.g., direct sun light) illuminates the object. This is the simplest illumination setup and it can be used for geometric measurements only under very limited conditions. First, it does not work with specular reflecting objects. Even objects such as plant leaves that reflect only a small fraction specular show up highlights (Fig. 3.19a). Furthermore, small variances in the surface slope lead to irradiance variation that makes the object have a 3-D shape. At the lower edge of the leaf, a shadow is cast that results in a false edge detection (Fig. 3.19a).

3.4.3b Diffuse Illumination. *Diffuse illumination* with an extended light source can be used also with specular reflecting objects. This setup is, however, quite complex and still does not ensure a uniform radiance of the object. Especially edges of thick objects cause an extended *penumbra* (a partially darkened region). Its extension is proportional to the height of the edge and the angular distribution of the light source.

3.4.3c Rear Illumination. *Rear illumination* is the best illumination setup for geometric measurements of flat objects. Opaque objects appear as black objects without any structure. The technique is also suitable for many transparent and semi-transparent objects as the example of the plant leaf shows (Fig. 3.19c). The only disadvantage of the technique is that parts of opaque objects that differ, e.g., in the refraction coefficient, cannot be separated.

3.4.4 Illumination Techniques for Depth Measurements

To measure the distance of an object, basically two principles can be applied. The first is based on triangulation and the second on time of flight. For the illumination technique, this leads to the techniques of *structured light*, pulsed or modulated illumination techniques. The measurement of depth with any type of technique (including the illumination techniques discussed here) is also known as *range imaging*.

Figure 3.19: *Illustration of specular and rear illumination techniques for the measurement of the shape of a leaf. **a** specular illumination is not suitable: the leaf shows highlights and a wide distribution of the gray values of the leaf (histogram in **b**); **c** rear illumination: the leaf is put on a diffuser screen that is illuminated from the rear. The background and the object show a narrow, well separated gray value distribution (histogram in **d**).*

3.4.4a Structured Light. With the structured light technique, an object is illuminated with a striped pattern from an oblique direction. Then, a planar surface appears as an object with linear parallel stripes while a curved surface results in curved stripes. At a depth edge, the stripes are shifted. The difficulty with this technique is that the absolute depth cannot be determined, since a shift of an integer multiple of the spatial pattern period cannot be detected. This problem can be overcome in static scenes by the projection of a succession of patterns with different wavelengths resulting in a unique coding of depth.

Stripe projection techniques are widely used in industrial applications for 3-D optical gauging. Special LCD projectors are available that can project programmable patterns.

3.4.4b Pulsed Illumination. Radar altimetry as already introduced in Section 1.4.4 (Figs. 1.15 and 1.17) is the classical example of an active imaging system with a pulsed

Table 3.7: Summary of illumination techniques for depth measurements.

Technique	Description
Structured light	Illumination of the object with striped pattern
Short-time illumination	Measurement of distance by the time of flight of a short-time illumination gating the camera to receive only radiation within a narrow time interval with adjustable delay after the illumination
Modulated illumination	Measurement of the distance by the phase of the modulated illumination

illumination system. A short radar impulse is emitted from an antenna and the time interval is measured until the echo from the object is received. The difficulty of the use of *pulsed illumination* with electromagnetic waves lies in the high speed of light (Section 3.3.1). In free space, it takes only about 3 ns to travel the distance of one meter. It is much easier to apply a pulsed illumination with acoustic waves (Section 3.3.3) which need about 3 ms to travel the distance of one meter in air. Actually, many living species (e. g., bats and dolphins) send acoustic signals and sense an image of their environment by analyzing the received echoes. Another variant of pulsed imaging is to use a gated camera that takes an image at a short interval with an adjustable delay after the illumination pulse. With such an arrangement, radiation is received only from a corresponding depth range.

3.4.4c Modulated Illumination. Modulated illumination is another variant of temporally changing illumination. If we modulate the illuminating radiation with a period T, it travels a distance $\Delta X_3/(cT)$ within a period of the modulation. Therefore, the amplitude variation shows a phase shift

$$\Delta\varphi = \frac{2\pi X_3}{cT} = \frac{X_3}{c}\omega_m \tag{3.62}$$

after it has traveled the distance X_3, where ω_m is the circular frequency of the modulation. Thus, the phase shift can be used to measure depth according to Eq. (3.62).

The trouble with this technique is that the phase can be measured only in multiples of 2π. Thus, an absolute measurement of the depth is limited to a depth range of $X_3 < 2\pi c/\omega_m = cT$. The sensitivity of the modulated illumination technique is proportional to the frequency of the modulation. This is why coherent interferometric techniques which do not use a modulation but take the amplitude variation of the electromagnetic or acoustic wave itself are so sensitive. Then the modulation frequency in Eq. (3.62) can be replaced by the circular frequency of the wave, ω, and the phase shift is

$$\Delta\varphi = \frac{X_3}{c}\omega = \frac{X_3}{\lambda}, \tag{3.63}$$

where λ is the wavelength of the wave. Thus, interferometric techniques can measure depth to a fraction of the wavelength.

3.4.5   Illumination Techniques for Surface Slope Measurements

In Section 3.4.4, we discussed illumination techniques for range imaging as one technique to infer the 3-D structure of objects by measuring the distance of the objects from the camera. An alternative method measures the spatial derivative of the *depth map*. If $a(X, Y)$ denotes the surface elevation function, the spatial derivative is given as

$$s = \nabla a(X, Y) = \left[\frac{\partial a}{\partial X}, \frac{\partial a}{\partial Y} \right]^T = [s_1, s_2]^T. \tag{3.64}$$

This is a 2-D vectorial function and gives the inclination of the surface (tangent of the angle to the horizontal) in x and y direction. This function is also known as the *slope map* of a surface. Alternatively, the surface slope can also be described by the surface normal vector, a term often used in physical sciences and computer graphics to describe the orientation of surfaces. Surface normal vectors are normally expressed as unit vectors but here it is more convenient to use vectors with a unit Z component:

$$n = \left[-\frac{\partial a}{\partial X}, -\frac{\partial a}{\partial Y}, 1 \right]^T = [-s_1, -s_2, 1]^T. \tag{3.65}$$

The corresponding unit vector is

$$\bar{n} = \frac{1}{\sqrt{s_1^2 + s_2^2 + 1}} [-s_1, -s_2, 1]^T. \tag{3.66}$$

The goal of slope imaging is to illuminate the object in such a way that its radiance is only a function of its surface inclination. Note how different this goal is from the illumination techniques for geometric measurements (Section 3.4.3). For this purpose, the object radiance needs to be studied as a function of the surface slope. The 2-D space spanned by the surface slope is known as the *gradient space*. Slope imaging is a difficult task for three reasons:

1. It is an underdetermined problem. While the slope is a 2-D vector, the radiance is a scalar quantity. This problem can be overcome at the level of the illumination by multiple exposures with different light sources. This technique is known as *photometric stereo*. Alternative multispectral imaging can be used with multispectral light sources. With this technique, dynamical scenes can be studied also since a single exposure is sufficient.

2. It is in general a nonlinear problem, since the object radiance is related in a nonlinear way to the slope of its surface.

3. The object radiance is always a product of the surface reflectivity and the incident irradiance. If an object shows a variation in the surface reflectivity, it is difficult to distinguish the change in reflectivity from a change in the surface slope that results in an equal change of the radiance.

Techniques to infer the object shape from the radiance are generally known as *shape from shading* techniques.

3.4.5a   Telecentric Illumination System.   A *telecentric illumination system* is suitable to convert a radiance distribution into bundles of parallel beams. This system is the essential part of various types of illumination systems for surface slope measurements. The principle is very simple (Fig. 3.20). A light source is put into the focus of a large lens (often Fresnel lenses are used). The rays emitted from a single point at the focal plane are converted into a parallel bundle of light. The angle of the light bundle with the optical axis is determined by the position on the focal plane:

$$\tan \alpha = x/f. \tag{3.67}$$

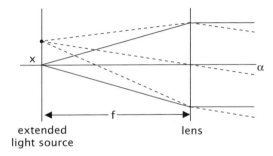

Figure 3.20: *A telecentric illumination system converts the spatial radiance distribution of a light source into bundles of parallel rays that reflect the intensity (and spectral distribution) of a single point of the light source.*

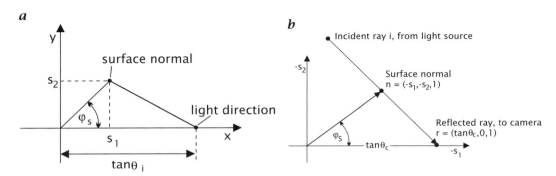

Figure 3.21: *Ray geometry in the gradient space for a shape from shading with a Lambertian surface illuminated by a distant light source under an incidence angle θ_i and an azimuthal angle ϕ_i of zero; b shape from reflection with isotropic and homogeneous daylight illumination and an oblique camera (*Stilwell photography*).*

If the radiance of the light source is isotropic within the cone gathered by the lens, a parallel bundle of light with constant intensity is emitted under each angle.

In this way, a spatial radiance distribution at the focal plane can be converted into an angular radiance distribution that is suitable for shape from shading illumination techniques. A spatial variation of the radiance distribution (also in color) can easily be achieved by putting a corresponding filter on the focal plane.

3.4.5b Shape From Shading for Lambertian Surfaces. This example stands for a typical application of the shape from shading technique with diffuse reflecting opaque objects. For the sake of simplicity, we assume that the surface of a Lambertian object is illuminated by parallel light. Then, the radiance of the surface, L, is given by:

$$L = \frac{\rho(\lambda)}{\pi} E \cos \gamma, \tag{3.68}$$

where E is the irradiance and γ the angle between the surface normal and the light direction. The geometry is illustrated in Fig. 3.21a in the gradient space. Without restriction of the generality, we can put the direction of the light source into the X direction. Then, the light direction is given by the vector $\boldsymbol{l} = [\tan\theta_i, 0, 1]^T$, and the

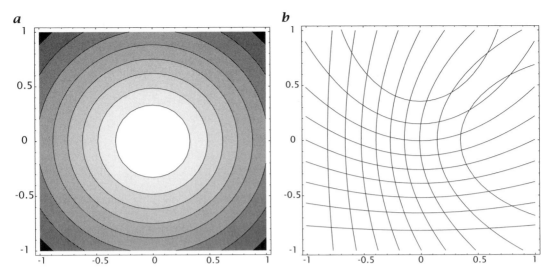

Figure 3.22: a *Contour plot of the radiance of a Lambertian surface with homogeneous reflectivity illuminated by parallel light (or a point source far above the surface) with zero incidence angle shown in the gradient space for surface slopes between -1 and 1. The radiance is normalized to the radiance for a flat surface. The distance of the contour lines is 0.05.* **b** *Superimposed contour lines for an oblique illumination with an incidence angle of 45° and an azimuthal angle of 0° and 90°. The distance of the contour lines is 0.1.*

radiance L of the surface can be expressed as

$$L = \frac{\rho(\lambda)}{\pi} E \frac{s^T l}{|s||l|} = \frac{\rho(\lambda)}{\pi} E \frac{s_1 \tan \theta_i + 1}{\sqrt{1 + \tan^2 \theta_i}\sqrt{1 + s_1^2 + s_2^2}}. \tag{3.69}$$

Contour plots of the radiance distribution in the gradient space are shown in Fig. 3.22 for light sources with 0° and 45° angle of incidence. In the case of the light source in the zenith, the contour lines of equal radiance mark lines with constant absolute slope $s = (s_1^2 + s_2^2)^{1/2}$. However, the radiance changes with slope are low, especially for low surface slopes.

An oblique illumination leads to a much higher contrast in the radiance (Fig. 3.22b). With an oblique illumination, however, the maximum surface slope in the direction opposite to the light source is limited to $\pi/2 - \theta$ when the surface normal is perpendicular to the light direction.

The curved contour line indicates that the relation between surface slope and radiance is nonlinear. This means that even if we take two different illuminations of the same surface (Fig. 3.22b), the surface slope may not be determined in a unique way. This is the case when the curved contour lines intersect each other at two points. Only a third exposure would make the solution unique. This approach has also the significant advantage in that the reflectivity of the surface can be eliminated by the use of *ratio imaging* and is illustrated in Example 3.1.

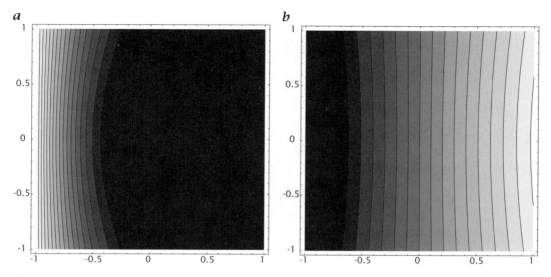

Figure 3.23: *Contour plots of the irradiance received by a camera from a specular surface in gradient space: a shape from reflection, camera incidence angle 45°, infinitely extended isotropic perpendicularly polarized light source; distance between the contour lines is 0.05. b Shape from refraction, where the radiance in a telecentric illumination source varies linearly in x_1 direction.*

Example 3.1: Unique and linear estimation of the surface vector slope of a Lambertian surface with three illuminations.

As an example system, we illuminate a Lambertian surface with the same light source from three different directions

$$l_1 = (0, 0, 1), \quad l_2 = (\tan \theta_i, 0, 1), \quad \text{and} \quad l_3 = (0, \tan \theta_i, 1). \tag{3.70}$$

Then,

$$L_2/L_1 = \frac{s_1 \tan \theta_i + 1}{\sqrt{1 + \tan^2 \theta_i}} \quad \text{and} \quad L_3/L_1 = \frac{s_2 \tan \theta_i + 1}{\sqrt{1 + \tan^2 \theta_i}}. \tag{3.71}$$

Now the equations are linear in s_1 and s_2 and they are decoupled: s_1 and s_2 depend only on L_2/L_1 and L_3/L_1, respectively. Most importantly, the normalized radiance in Eq. (3.71) does not depend on the reflectivity of the surface. The reflectivity of the surface is contained in Eq. (3.68) as a factor and is thus eliminated when the ratio of two radiance distributions of the same surface is computed.

3.4.5c   Shape From Shading for Specular Surfaces: Stilwell Photography. For *specular* reflecting *surfaces*, point light sources or parallel illumination are not adequate for shape from shading techniques, since we would receive only individual specular reflexes. A well-known example is the *sun glitter* observed on a water surface as the direct reflection of sun light.

Thus, an extended light source is required. We discuss the limitations of this approach under some simplified assumptions. Let us assume that we have an infinite light source with isotropic radiance. (A fully overcast sky is a good approximation to this ideal condition.) Then, the irradiance received from a specular surface depends only on the reflection coefficient ρ. As discussed in Section 3.3.6c, the reflection coefficient strongly depends on the angle of incidence. The geometry of an appropriate setup is illustrated in the gradient space in Fig. 3.21b. The camera is inclined in order to obtain an irradiance that increases monotonically with one slope component. Then, the angle of incidence α is given as the angle between the direction of the reflected ray (towards the camera) $\boldsymbol{r} = [\tan\theta_c, 0, 1]^T$ and the surface normal $\boldsymbol{n}$ Eq. (3.65) as

$$\alpha = \arccos(\boldsymbol{r}, \boldsymbol{n}) = \arccos\left(\frac{s_1\tan\theta_c + 1}{\sqrt{1 + s_1^2 + s_2^2}\sqrt{1 + \tan^2\theta_c}}\right). \tag{3.72}$$

The equation expresses the angle of incidence in terms of the camera inclination angle θ_c and the surface slope $\boldsymbol{s} = [s_1, s_2]^T$. Since the radiance is assumed to be homogeneous and isotropic, the surface radiance is given by

$$L = \rho(\alpha)L_s, \tag{3.73}$$

where L_s and ρ are the radiance of the isotropic and homogeneous light source and the reflection coefficient as given by Fresnel's equations Eqs. (3.35) to (3.37).

Results for unpolarized and perpendicularly polarized light with a camera inclination angle of 45° are shown in Fig. 3.23 and reveal the strong nonlinear relation between the surface slope and surface radiance, directly reflecting the dependence of the reflection coefficient on the inclination angle (Fig. 3.11). These strong nonlinearities render this technique almost useless since only a narrow range of slopes can be measured with sufficient resolution. This technique, used for some time, is known as *Stilwell photography* in oceanography to measure the fine-scale shape of the ocean surface [140], but has had — not surprisingly — only limited success.

3.4.5d Shape From Refraction for Specular Surfaces. For transparent specular surfaces, *shape from refraction* techniques can be used and are more advantageous than shape from reflection techniques because of (i) higher radiance, (ii) measurement of higher slopes, and (iii) significantly lower nonlinearities of the slope/radiance relationship. A shape from refraction technique requires a special illumination technique since — except for the small fraction of light reflected at the surface — no significant radiance variations occur. The base of the shape from refraction technique is the *telecentric illumination system* introduced in Section 3.4.5a which converts a spatial radiance distribution into an angular radiance distribution. Then, all we have to do is to compute the relation between the surface slope and the angle of the refracted beam and to use a light source with an appropriate spatial radiance distribution.

Figure 3.24 illustrates the optical geometry for the simple case when the camera is placed far above and a light source below a transparent surface of a medium with a higher index of refraction. The relation between the surface slope s and the angle γ is given by Jähne [72]

$$s = \tan\alpha = \frac{n\tan\gamma}{n - \sqrt{1 + \tan^2\gamma}} = 4\tan\gamma\left[1 + \frac{3}{2}\tan^2\gamma + O(\tan^4\gamma)\right] \tag{3.74}$$

with $n = n_2/n_1$. The inverse relation is

$$\tan\gamma = s\frac{\sqrt{n^2 + (n^2 - 1)s^2} - 1}{\sqrt{n^2 + (n^2 - 1)s^2} + s^2} = \frac{1}{4}s\left(1 + \frac{3}{32}s^2 + O(s^4)\right). \tag{3.75}$$

This technique works for slopes up to infinity (vertical surfaces). In this limiting case, the ray to the camera grazes the surface (Fig. 3.24b) and

$$\tan\gamma = \sqrt{n^2 - 1}. \tag{3.76}$$

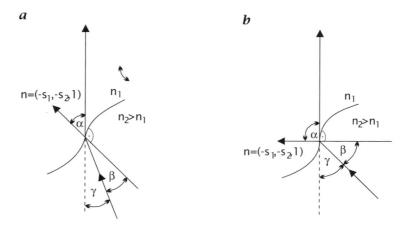

Figure 3.24: *Refraction at an inclined surface as the base for the shape from refraction technique. The camera is far above the surface. Rays that are emitted by the light source under an angle γ are refracted into the direction of the camera.* **b** *Even for a slope of infinity (vertical surface, α = 90°), rays from the light source meet the camera.*

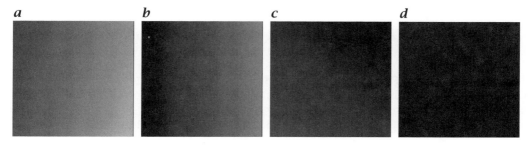

Figure 3.25: **a** *Color wedge produced by additive color mixing from the* **b** *green,* **c** *red, and* **d** *blue color wedges according to Eq. (3.78). (See also Plate 13.)*

With a telecentric illumination source, the position from which the camera receives light for a certain surface slope s can be related as

$$x/f = \frac{s}{s} \tan \gamma = s \frac{\sqrt{n^2 + (n^2 - 1)s^2} - 1}{\sqrt{n^2 + (n^2 - 1)s^2} + s^2} \tag{3.77}$$

using $\tan \gamma$ from Eq. (3.75). If the radiance of the light source is varying with the position x, we can infer the corresponding surface slope. Of course, we have again the problem that from a scalar quantity as the radiance no vector component such as the slope can be inferred. The shape from refraction technique comes, however, very close to an ideal setup. If the radiance, for instance, varies only linearly in x_1 direction, the radiance map in the gradient space is also almost linear (Fig. 3.23b). A slight influence of the cross slope results from the nonlinear terms with s^2 in Eq. (3.77) but becomes apparent only at quite high slopes.

3.4.5e Ratio Imaging for Shape from Shading Techniques. All shape from shading techniques suffer in practice from the problem that the irradiance of the surface is not homogeneous and not isotropic because of the unavoidable inhomogeneity of the illumination.

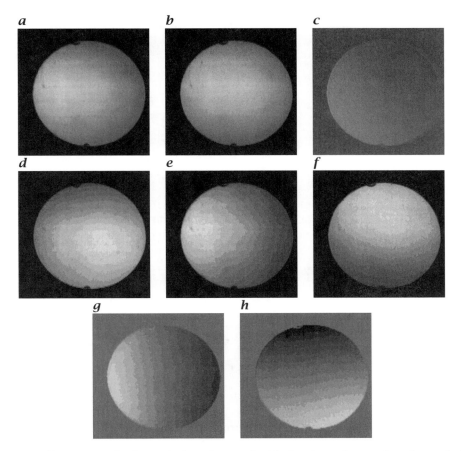

Figure 3.26: *Illustration of color ratio imaging used with the shape from refraction technique to measure the slope of water surfaces. It is demonstrated here with a calibration target, a spherical lens. **a** original color image; **b** intensity of the original image; watch the intensity fall off towards the edge of the lens (higher slopes); **c** color image normalized by **b**; **d** - **f** red, green, and blue channels of the original color image shown in discrete intensity steps; **g** and **h** x, y positions in the telecentric illumination system computed after Eq. (3.80). These quantities are according to Eq. (3.77) directly proportional to the x and y components of the surface slope. (See also Plate 14.)*

To a large extent, such effects can be compensated by *ratio imaging.* The basic idea is to take several images of the same scene with the same setup except for different illuminations. Since each of these illuminations suffers from the same additional effects listed above in a multiplicative way, they can be canceled out by dividing two images with different illuminations. We already discussed an example of this kind at the end of Section 3.4.5b.

In a static scene, sequential images can be taken by moving the illumination source. This approach is, however, not possible with dynamical scenes. In this case it is required to take all exposures with the different illumination setups at the same time. Such multiple illuminations in a single exposure can be realized by using color.

The example discussed here utilized this principle for a shape from refraction technique. Color images have three independent primary colors: red, green, and blue. With a total of three channels, we can identify the position in a telecentric illumination system (Section 3.4.5a) — and thus the inclination of the water surface — in a unique way and still have one degree of freedom left for corrections.

A unique color position coding can be achieved, for example, with the following color wedges

$$
\begin{aligned}
G(\boldsymbol{x}) &= (1/2 + x)E_0(\boldsymbol{x}) \\
R(\boldsymbol{x}) &= [1/2 - 1/2(x + y)]E_0(\boldsymbol{x}) \\
B(\boldsymbol{x}) &= [1/2 - 1/2(x - y)]E_0(\boldsymbol{x}),
\end{aligned}
\tag{3.78}
$$

where x and y are the coordinates at the focal plane of the telecentric illumination system with $|x| \leq 1/2, |y| \leq 1/2$. These wedges are shown as individual components and composite color wedges in Fig. 3.25. $E_0(\boldsymbol{x})$ is the spatially varying radiance without the color filters.

We now have three illuminations to determine two slope components. Thus, we can take one to compensate for unwanted spatial variation of E_0. This can be done by normalizing the three color channels by the sum of all channels $G + R + B$:

$$
\frac{3G}{G + R + B} = 1 + 2x \quad \text{and} \quad \frac{3(B - R)}{G + R + B} = 2y.
\tag{3.79}
$$

Then the position on the wedge from which the light originates is given as

$$
x = \frac{3}{2}\frac{G}{G + R + B} - 1/2 \quad \text{and} \quad y = \frac{3}{2}\frac{B - R}{G + R + B}.
\tag{3.80}
$$

From these position values, the x and y components of the slope can be computed according to Eq. (3.77).

How well the compensation works is shown in Fig. 3.26 with a calibration object, a convex spherical lens. This target shows a constant curvature and thus — in first order — a linear decrease of the x and y components of the slope and x and y direction, respectively. Figure 3.26d, e, and f show the significant intensity decrease towards the edge of the images which even inverts the slope of the wedges. Without normalization it would lead to considerable errors in the slope values. Division by the sum of all color channels (Fig. 3.26b) results in significantly better wedges (Fig. 3.26c, g, and h). Essentially, the normalization works well since now a position in the light source is no longer directly identified by the intensity but by a color, i.e., the ratio of intensities of differently colored illuminations.

3.4.6 Color and Multi-Spectral Imaging

Spectroscopic imaging is in principle a very powerful tool to identify objects and their properties because almost all optic material constants such as the *reflectivity* (Section 3.3.6c), the *index of refraction* (Section 3.3.6b), the *absorption coefficient* (Section 3.3.7a), the *scattering coefficient* (Section 3.3.7b), *optical activity* (Section 3.3.7c), and *luminescence* (Section 3.3.7d) depend on the wavelength of the radiation.

The trouble with spectroscopic imaging is that it adds another coordinate to imaging and the required amount of data is multiplied correspondingly. Therefore, it is important to sample the spectrum with a minimum number of samples that is sufficient to perform the required task. We introduce here several sampling strategies (Table 3.8) and discuss under this point of view also human color vision as one realization of spectral sampling in Section 3.4.7.

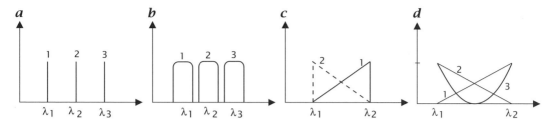

Figure 3.27: *Examples for spectral sampling: **a** line-sampling, **b** band-sampling, **c** sampling to measure the total radiative flux and the mean wavelength, **d** sampling to measure the total radiative flux, the mean wavelength, and the variance of the wavelength.*

Table 3.8: *Examples of some strategies for spectral sampling.*

Sampling Method	Description and Application
Line sampling	Channels with narrow spectral range (line); suitable for imaging of specific chemical species and/or specific processes; orthogonal base for color space.
Band sampling	Channels with wide spectral range (band) of uniform responsivity, adjacent to each other; suitable for measurements of spectral radiance with rather coarse resolution; orthogonal base for color space.
Model-guided sampling	Sampling optimized for certain components of the spectral distribution. The parameters of the spectral distribution given by the model are estimated; generally nonorthogonal base for color space.

3.4.6a   Line sampling.   With this technique, each channel picks only a narrow spectral range (Fig. 3.27a). This technique is useful if processes are to be imaged that are related to the emission or the absorption at specific spectral lines. The technique is very selective. One channel "sees" only a specific wavelength and is insensitive — at least to the degree that such a narrow bandpass filtering can be realized technically — to all other wavelengths. Thus, a very specific effect or a specific chemical species can be imaged with this technique. This technique is, of course, not appropriate to make an estimate of the total radiance from objects since it misses most wavelengths.

3.4.6b   Band sampling.   This is the appropriate technique if the total radiance in a certain wavelength range has to be imaged and still some wavelength resolution is required (Fig. 3.27b). Ideally, the individual bands have even responsivity and are adjacent to each other. Thus, band sampling gives the optimum resolution with a few channels but does not allow any distinction of the wavelengths within one band. Thus, we can measure the spectral radiance with a resolution given by the width of the spectral bands.

3.4.6c   Model-Guided Spectral Sampling.   In many cases, it is possible to make a model of the spectral radiance of a certain object. Then, a much better spectral sampling technique can be chosen that essentially does not sample certain wavelengths but rather the parameters of the model. We will illustrate this general approach with two simple examples.

Example 3.2: Measurement of total radiative flux and mean wavelength

This example illustrates a method to measure the mean wavelength of an arbitrary spectral distribution $H(\lambda)$ ("color") and the total radiative flux ("intensity") in a certain wave number range. These quantities are defined as:

$$\Phi = \frac{1}{\lambda_2 - \lambda_1} \int_{\lambda_1}^{\lambda_2} \Phi(\lambda) \, d\lambda, \quad \overline{\lambda} = \int_{\lambda_1}^{\lambda_2} \lambda \Phi(\lambda) d\lambda \Big/ \int_{\lambda_1}^{\lambda_2} \Phi(\lambda) \, d\lambda. \tag{3.81}$$

From the second equation, it becomes evident that we need a sensor that has a sensitivity that varies linearly with the wave number. We try two sensor channels with the following linear spectral responsivity as shown in Fig. 3.27c.

$$\begin{aligned} R_1(\lambda) &= \frac{\lambda - \lambda_1}{\lambda_2 - \lambda_1} R_0 = \left(\frac{1}{2} + \tilde{\lambda}\right) R_0 \\ R_2(\lambda) &= R_0 - R_1(\lambda) = \left(\frac{1}{2} - \tilde{\lambda}\right) R_0, \end{aligned} \tag{3.82}$$

where R is the responsivity of the sensor given as $R = s/\Phi$ (units A/W), s the sensor signal, usually given in units for the electronic current, and $\tilde{\lambda}$ the normalized wavelength

$$\tilde{\lambda} = \left(\lambda - \frac{\lambda_1 + \lambda_2}{2}\right) / (\lambda_2 - \lambda_1) . \tag{3.83}$$

$\tilde{\lambda}$ is zero in the middle of the interval and $\pm 1/2$ at the edges of the interval.

The sum of the responsivity of the two channels is wavelength independent, while the difference is directly proportional to the wavelength and varies from -1 for $\lambda = \lambda_1$ to 1 for $\lambda = \lambda_2$.

$$\begin{aligned} R_1'(\tilde{\lambda}) &= R_1(\tilde{\lambda}) + R_2(\tilde{\lambda}) = R_0 \\ R_2'(\tilde{\lambda}) &= R_1(\tilde{\lambda}) - R_2(\tilde{\lambda}) = 2\tilde{\lambda}R_0. \end{aligned} \tag{3.84}$$

If s_1 and s_2 are the signals received from sensor 1 and 2, respectively, we can measure the total flux Φ and the mean wavelength by

$$\Phi = \frac{s_1 + s_2}{R_0}, \quad \overline{\tilde{\lambda}} = \frac{1}{2}\frac{s_1 - s_2}{s_1 + s_2}, \quad \overline{\lambda} = \frac{\lambda_1 + \lambda_2}{2} + \frac{s_1 - s_2}{s_1 + s_2}\frac{\lambda_1 + \lambda_2}{2}. \tag{3.85}$$

This example demonstrates that it is possible to determine the total radiative flux ("intensity") and the mean wavelength ("color") with just two sensors having an adequate spectral sensitivity.

The example also illustrates that measurements are always a many-to-one mapping. The two sensors receive the same signal for all types of spectral distributions that have the same total radiative flux and mean wavelength as defined by Eq. (3.81). In contrast, with band-sampling the wavelength of a monochromatic radiative flux cannot be determined with better resolution than the bandwidth of the individual channels.

Example 3.3: Measurement of radiative flux, mean, and variance of the wavelength

The two-channel system discussed in Example 3.2 cannot measure the width of a spectral distribution ("color saturation") at all. This deficit can be overcome with a third channel that has a sensitivity which increases with the square of the distance from the mean wavelength (Fig. 3.27d).

The responsivity of the third sensor is given by

$$R_3(\lambda) = 4\tilde{\lambda}^2 R_0. \tag{3.86}$$

Consequently, the mean squared wavelength is given by

$$\overline{\tilde{\lambda}^2} = \frac{1}{4}\frac{s_3}{s_1 + s_2}. \tag{3.87}$$

The variance $\sigma_{\tilde{\lambda}}^2 = \overline{\left(\tilde{\lambda} - \overline{\tilde{\lambda}}\right)^2} = \overline{\tilde{\lambda}^2} - \overline{\tilde{\lambda}}^2$ is then given by

$$\sigma_{\tilde{\lambda}}^2 = \frac{1}{4}\left[\frac{s_3}{s_1 + s_2} - \left(\frac{s_1 - s_2}{s_1 + s_2}\right)^2\right]. \tag{3.88}$$

You can easily convince yourself that for a monochromatic distribution the variance is zero. The estimates given by Eqs. (3.85) and (3.87) are only valid as long as the spectral distribution is confined to the interval $[\lambda_1, \lambda_2]$ to which the sensors respond.

3.4.6d Measurement of Chemical Species by Imaging Spectroscopy. The absorption of various chemical species differs significantly and thus can be used to identify them and to measure their concentration (Section 1.4.6). In the optical range of the electromagnetic spectrum there are two wavelength regions (Fig. 3.3). In the ultraviolet and visible range radiation is absorbed by electronic transitions in the molecules, and in the infrared part, vibrations of molecules can be excited.

Here we will discuss how the concentration of multiple chemical species with overlapping absorption spectra can be measured. *Fluorescence* spectra (Section 3.3.7d) can be treated in a similar way.

As discussed in Section 3.3.7a absorption is governed by *Lambert Beer's law*. Let us assume that light passes a distance l through P absorbing species with the concentration c_p and the *molar absorption coefficient* $\epsilon(\lambda)$. Using Eqs. (3.46) and (3.48) the spectral intensity is then given by

$$\frac{I(l, \lambda)}{I(0, \lambda)} = \exp\left(-\sum_{p=1}^{P} c_p \epsilon_p(\lambda) l\right). \tag{3.89}$$

This nonlinear relation can be made linear by taking the natural logarithm. Using the *absorption coefficient* α from Eq. (3.44), Eq. (3.89) reduces to

$$\alpha(\lambda) = \frac{\ln I(0, \lambda) - \ln I(l, \lambda)}{l} = \sum_{p=1}^{P} c_p \epsilon_p(\lambda). \tag{3.90}$$

Provided that the molar absorption coefficients of all contributing species are known, the determination of the concentrations constitutes a linear inverse problem. If the measured absorption coefficient and the molar absorption coefficients for all species are sampled with $N > P$ samples in the same suitable wavelength range, a simple discrete linear inverse problem of the form

$$\underbrace{\boldsymbol{a}}_{N} = \underbrace{\boldsymbol{E}}_{N \times P} \underbrace{\boldsymbol{c}}_{P} \tag{3.91}$$

is obtained. The column vector $\boldsymbol{a}$ contains N samples of the measured absorption spectrum and each of the P columns of the matrix $\boldsymbol{E}$ N samples of one of the molar absorption spectra of the P species. The P unknown concentrations can be estimated by a least squares approach as outlined in Appendix B.2 by

$$\boldsymbol{c} = (\boldsymbol{E}^T\boldsymbol{E})^{-1}\boldsymbol{E}^T\boldsymbol{a}, \tag{3.92}$$

provided that the $P \times P$-matrix $\boldsymbol{E}^T\boldsymbol{E}$ can be inverted. A spectral image with a vector $\boldsymbol{a}$ at each pixel can thus be converted into a vector image with the concentrations of the P species.

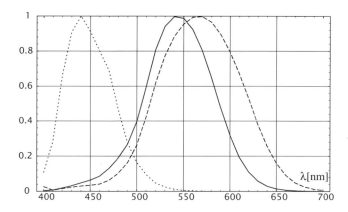

Figure 3.28: *Estimates of the relative cone sensitivities of the human eye after DeMarco et al.* [20].

3.4.7 Human Color Vision

Human color vision can be regarded in terms of the spectral sampling techniques summarized in Table 3.8 as a blend of band-sampling and model-guided sampling.

For color sensing, the human eye has three types of photopigments in the photoreceptors known as cones with different spectral sensitivities (Fig. 3.28). The sensitivities cover different bands with maximal sensitivities at 445 nm, 535 nm, and 575 nm, respectively (band sampling), but overlap each other significantly (model-based sampling). In contrast to our model examples, the three sensor channels are unequally spaced and cannot simply be linearly related. Indeed, the color sensitivity of the human eye is uneven and all the nonlinearities involved make the science of color vision rather difficult. Here, only some basic facts are given — in as much as they are useful to handle color imagery.

3.4.7a 3-D Color Space. Having three color sensors, it is obvious that color signals cover a 3-D space. Each point in this space represents one color. From the discussion on spectral sampling in Section 3.4.6, it is clear that many spectral distributions called *metameric color stimuli* or short *metameres* map onto one point in this space. Generally, we can write the signal s_i received by a sensor with a spectral responsivity $R_i(\lambda)$ as

$$s_i = \int R_i(\lambda)\Phi(\lambda)d\lambda. \tag{3.93}$$

With three primary color sensors, a triple of values is received, often called *tristimulus* and represented by the 3-D vector $\boldsymbol{s} = [s_1, s_2, s_3]^T$.

3.4.7b Primary Colors. One of the most important questions in *colorimetry* is a system of how to represent colors as linear combinations of some basic or *primary colors*. A set of three spectral distributions $\Phi_j(\lambda)$ represents a set of primary colors and results in an array of responses that can be described by the matrix P with

$$P_{ij} = \int R_i(\lambda)\Phi_j(\lambda)d\lambda. \tag{3.94}$$

Each vector $\boldsymbol{p}_j = \left[p_{1j}, p_{2j}, p_{3j}\right]^T$ represents the tristimulus of the primary colors in the 3-D color space. Then, it is obvious that any color can be represented by the

Table 3.9: *Often used primary color systems.*

Name	Description
Monochromatic primaries R_c, G_c, B_c	Adopted by C.I.E. in 1931: λ_R = 700 nm, λ_G = 546.1 nm, λ_B = 435.8 nm
NTSC Primary Receiver Standard R_N, G_N, B_N	FCC Standard, 1954, to match phosphors of RGB color monitors $$\begin{bmatrix} R_N \\ G_N \\ B_N \end{bmatrix} = \begin{bmatrix} 0.842 & 0.156 & 0.091 \\ -0.129 & 1.320 & -0.203 \\ 0.008 & -0.069 & 0.897 \end{bmatrix} \begin{bmatrix} R_C \\ G_C \\ B_C \end{bmatrix}$$
S.M.P.T.E. Primary Receiver Standard R_S, G_S, B_S	Better adapted to modern screen phosphors $$\begin{bmatrix} R_S \\ G_S \\ B_S \end{bmatrix} = \begin{bmatrix} 1.411 & -0.332 & 0.144 \\ -0.174 & 1.274 & -0.157 \\ -0.007 & -0.132 & 1.050 \end{bmatrix} \begin{bmatrix} R_C \\ G_C \\ B_C \end{bmatrix}$$
EBU Primary Receiver Standard R_e, G_e, B_e	Adopted by EBU 1974

primary colors that is a linear combination of the base vectors $\boldsymbol{p}_j$ in the following form:

$$s = R\boldsymbol{p}_1 + G\boldsymbol{p}_2 + B\boldsymbol{p}_3 \quad \text{with} \quad 0 \le R, G, B \le 1, \tag{3.95}$$

where the coefficients are denoted by R, G, and B indicating the three primary colors red, green, and blue. Note that these coefficients must be positive and smaller than one. Because of this condition, all colors can be presented as a linear combination of a set of primary colors only if the three base vectors are orthogonal to each other. This cannot be the case as soon as more than one of the color sensors responds to one primary color. Given the significant overlap in the spectral response of the three types of cones (Fig. 3.28), it is obvious that none of the color systems based on any type of real primary colors will be orthogonal. The colors that can be represented lie within the parallelepiped formed by the three base vectors of the primary colors. The more the primary colors are correlated with each other (i. e., the smaller the angle between two of them is), the smaller is the color space that can be represented by them. Mathematically, colors that cannot be represented by a set of primary colors have at least one negative coefficient in Eq. (3.95). The most often used primary color systems are summarized in Table 3.9.

3.4.7c Chromaticity. One component in the 3-D color space is intensity. If a color vector is multiplied by a scalar, only its intensity is changed but not its color. Thus, all colors could be normalized by the intensity. This operation reduces the 3-D color space to a 2-D color plane or *chromaticity diagram*:

$$r = \frac{R}{R+G+B}, \quad g = \frac{G}{R+G+B}, \quad b = \frac{B}{R+G+B} \tag{3.96}$$

with

$$r + g + b = 1. \tag{3.97}$$

It is sufficient to use only the two components r and g. The third component is then given by $b = 1 - r - g$, according to Eq. (3.97). Thus, all colors that can be represented by the three primary colors R, G, and B are confined within a triangle in

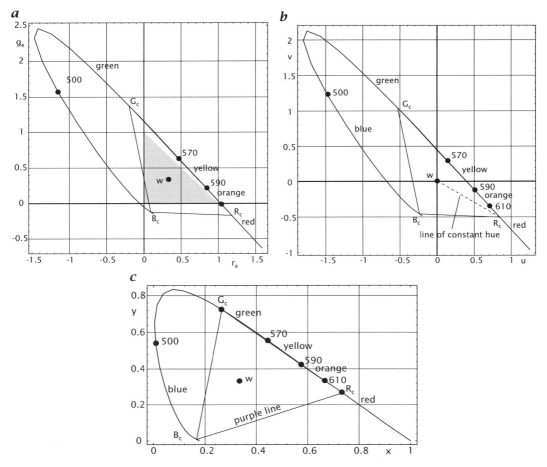

Figure 3.29: *Chromaticity diagrams:* **a** $r_e g_e$ *color space; all colors that are within the shaded triangle can be represented as linear combination of the primary receiver standard R_e, G_e, and B_e (Table 3.9).* **b** *uv color difference system centered at the white point w. The color saturation is proportional to the distance from the center and the color hue is given by the angle to the x axis.* **c** *xy color space; The u-shaped curve with the monochromatic colors (spectral curve) with wavelengths in nm as indicated and the purple line include all possible colors. Also shown are the locations of the monochromatic primaries R_c, G_c, and B_c and the white point w.*

the rg-space as shown in Fig. 3.29a. As already mentioned, some colors cannot be represented by the primary colors. The boundary of all possible colors is given by all visible monochromatic colors from deep red to blue. The line of monochromatic colors form a u-shaped curve in the rg-space (Fig. 3.29a). Note that a large part of the curve (for blue-green wavelengths) lies outside of the RGB triangle. Thus, these monochromatic colors cannot be represented by the monochromatic primaries. Since all colors that lie on a straight line between two colors can be generated as a mixture of these colors, the space of all possible colors covers the area filled by the u-shaped spectral curve and the straight mixing line between its two end points for blue and red color (*purple line*).

In order to avoid negative color coordinate values, often a new coordinate system is chosen with virtual primary colors, i. e., primary colors that cannot be realized by any physical colors. This color system is known as the *XYZ color system* and constructed in such a way that it just includes the curve of monochromatic colors with only positive coefficients (Fig. 3.29c).

3.4.7d   Hue and Saturation.   The color systems discussed so far do not directly relate to the human color sensing. From the rg or xy values, we cannot directly infer colors such as green, blue, etc. A natural type of description of colors includes besides the *luminance* (*intensity*) the type of color such as green or blue (*hue*) and the purity of the color (*saturation*). From a pure color, we can obtain any degree of saturation by mixing it with white.

Hue and saturation can be extracted from chromaticity diagrams by simple coordinate transformations. The essential point is the *white point* in the middle of the chromaticity diagram (Fig. 3.29b). If we draw a line from this point to a pure (monochromatic) color, it constitutes a mixing line for a pure color with white and is thus a line of constant hue. From the white point to the pure color, the saturation is increasing linearly. The *white point* is given in the rg chromaticity diagram by $w = (1/3, 1/3)$. A color system that has its center at the white point is called a *color difference system*. From a color difference system, we can infer a hue-saturation color system by simply using polar coordinate systems. Then, the radius coordinate is proportional to the saturation and the hue to the angle coordinate (Fig. 3.29b).

So far, color science is easy. All the real difficulties arise from the facts used to adapt the color system in an optimum way to display and print devices and for transmission by television signals or to correct for the uneven color resolution of the human visual system that is apparent in the chromaticity diagrams of simple color spaces (Fig. 3.29a and c). These facts have led to a confusing manifold of different color systems.

3.4.7e   IHS Color Coordinate System.   Here, we discuss only one further color coordinate system that is optimally suited to present vectorial image information as colors on monitors. With a gray scale image, only one parameter can be represented. In color, it is, however, possible to represent three parameters simultaneously, for instance as intensity, hue, and saturation. This representation is known as the *IHS* color coordinate system. The transformation is given by

$$\begin{bmatrix} I \\ U \\ V \end{bmatrix} = \begin{bmatrix} 1/3 & 1/3 & 1/3 \\ 2/3 & -1/3 & -1/3 \\ -1/3 & 2/3 & -1/3 \end{bmatrix} \begin{bmatrix} R \\ G \\ B \end{bmatrix}$$

$$H = \arctan\left(\frac{V}{U}\right)$$

$$S = (U^2 + V^2)^{1/2}.$$

(3.98)

This transformation essentially means that the zero point in the chromaticity diagram has been shifted to the white point. The pairs $[U, V]^T$ and $[S, H]^T$ are the Cartesian and polar coordinates in this new coordinate system, respectively.

3.4.8   Thermal Imaging

Imaging in the *infrared* and *microwave* range can be used to measure the *temperature* of objects remotely. This application of imaging is known as *thermography*. The principle is discussed in Section 3.3.6a; here, we discuss the practical aspects. Thermal imaging is complicated by the fact that objects partly reflect radiation from the surrounding. If an object has the *emissivity* ϵ, a fraction $1 - \epsilon$ originates from the environment biasing the temperature measurement. Under the simplified assumption

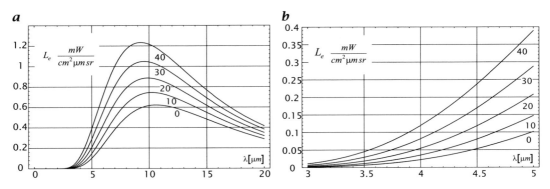

Figure 3.30: *Radiance of a blackbody at environmental temperatures as indicated in the wavelength ranges of **a** 0-20 μm and **b** 3-5 μm.*

that the environment has a constant temperature T_e, we can estimate the influence of the reflected radiation on the temperature measurement. The total radiance emitted by the object, E, is

$$E = \epsilon \rho T^4 + (1 - \epsilon)\rho T_e^4. \tag{3.99}$$

This radiance is interpreted to originate from a blackbody with the apparent temperature T':

$$\rho T'^4 = \epsilon \rho T^4 + (1 - \epsilon)\rho T_e^4. \tag{3.100}$$

Rearranging for T' yields

$$T' = T \left(1 + (1 - \epsilon)\frac{T_e^4 - T^4}{T^4}\right)^{1/4} \tag{3.101}$$

or in the limit of small temperature differences ($T_e - T \ll T$):

$$T' \approx \epsilon T + (1 - \epsilon)T_e \quad \text{or} \quad T' - T = (1 - \epsilon)(T_e - T). \tag{3.102}$$

From this simplified equation, we infer that a 1 % deviation of ϵ from unity results in a 0.01 K temperature error per 1 K difference of the object temperature from the environmental temperature. Even for an almost perfect *blackbody* as a water surface with a mean emissivity of about 0.97, this leads to considerable errors in the absolute temperature measurements. The apparent temperature of a bright sky can easily be 80 K colder than the temperature of a water surface at 300 K leading to a -0.3 · 80 K = -2.4 K bias in the measurement of the absolute temperature. This bias can, according to Eqs. (3.101) and (3.102), be corrected if the mean temperature of the environment and the emissivity of the object are known. Also relative temperature measurements are biased, although in a less significant way. Assuming a constant environmental temperature in the limit ($T_e - T$) $\ll T$, we can infer from Eq. (3.102) that

$$\partial T' \approx \epsilon \partial T \quad \text{for} \quad (T_e - T) \ll T \tag{3.103}$$

which means that the temperature differences measured are by the factor ϵ smaller than in reality.

Other corrections must be applied if radiation is significantly absorbed on the way from the object to the receiver. If the distance between the object and the camera is large, as for space borne or air borne infrared imaging of the Earth's surface, it is important to select a wavelength range with a minimum absorption. The two most

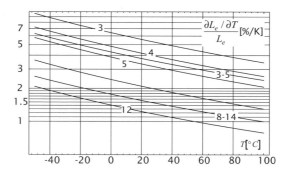

Figure 3.31: *Relative radiance — as a measure of how sensitive small temperature differences can be detected — at wavelength bands in µm and temperatures as indicated.*

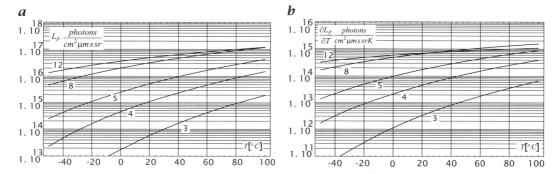

Figure 3.32: *a Photon radiance and b differential photon radiance in a logarithmic plot as a function of the temperature for wavelength bands (in µm) as indicated.*

important atmospheric windows are at 3–5 μm (with a sharp absorption peak around 4.15 μm due to CO_2) and between 8-12 μm ($\succ$ 3.15).

Figure 3.30 shows the radiance of a blackbody at environmental temperatures between 0 and 40 °C in the 0-20 μm and 3-5 μm wavelength range. Although the radiance has its maximum around 10 μm and is about 20 times higher than at 4 μm, the relative change is much larger at 4 μm than at 10 μm. This effect can be seen in more detail in Fig. 3.31 showing the relative radiance $(\partial L/\partial T)/L$ in percent as a function of the temperature in the range of -50 to 100 °C. While the radiance changes only about 1.5 %/K at 20 °C in the 8-14 μm wavelength interval, it changes about 4 %/K at 20 °C in the 3-5 μm wavelength interval. This higher relative sensitivity makes it advantageous to use the 3-5 μm wavelength range for measurements of small temperature differences although the absolute radiance is much lower.

In order to compute the number of photons incident to an infrared detector, it is useful to use the *photon radiance* (Section 3.3.5). Figure 3.32 shows the absolute and relative photon radiance in a logarithmic plot as a function of the temperature for wavelength bands as indicated in the figures.

Applications of infrared images are discussed in Section 1.4.2 (see also Fig. 1.11).

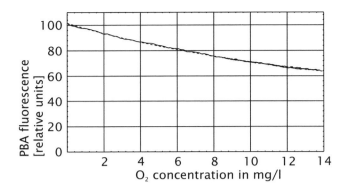

Figure 3.33: *Quenching of the fluorescence of pyrene butyric acid by dissolved oxygen: measurements and fit (dashed line, almost coincident with the data points) with the Stern-Vollmer equation.*

3.4.9 Imaging of Chemical Species and Material Properties

In order to identify a chemical species and to measure its concentration or to measure material properties such as pressure, all the surface-related and volume-related interactions of radiation (Sections 3.3.6 and 3.3.7) with matter can be considered.

One of the most useful techniques is the application of *fluorescence*. The fluorescence of a specific chemical species can be stimulated selectively by choosing the appropriate excitation wavelength. We will demonstrate the versatility of fluorescence-based techniques with several examples in this section.

3.4.9a Measurement of the Concentration of Dissolved Oxygen. Oxygen can easily be measured by fluorescence because it is able to quench the fluorescence of a dye (Section 3.3.7d). The Stern-Vollmer equation gives the dependence of the fluorescent radiance L on the concentration of the quencher, c_q:

$$\frac{L}{L_0} = \frac{1}{1 + K_q c_q},$$

(3.104)

where L_0 is the fluorescent radiance when no quencher is present and K_q is the quenching constant. A fluorescent dye that is suitable for quenching measurements must have a sufficiently long lifetime of the excited state; otherwise, the quenching effect is too weak. A fluorescent dye used for quenching by dissolved oxygen is *pyrene butyric acid (PBA)* [150] with a lifetime of the fluorescent state of 140 ns without a quencher. Other oxygen sensitive dyes include metallo-organic complexes with Ruthenium(II) as the central atom. These dyes have the advantage of excitation maxima around 450 nm and a fluorescence emission around 620 nm [57]. For an application, see Section 1.4.1.

The fluorescent radiance of PBA as a function of dissolved oxygen is shown in Fig. 3.33 [106]. Fluorescence is stimulated by a pulsed nitrogen laser at 337 nm. The data show that the change in fluorescence is rather weak but sufficiently large to make reliable oxygen measurements possible.

3.4.9b Measurement of the pH Value. It is well known that some dyes change their color with the pH value of a solution. Those dyes are used as pH indicators. Less well known is that also the fluorescence of some dyes changes with the pH value. This effect

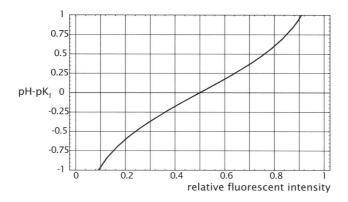

Figure 3.34: *pH measurement with a fluorescent dye. Shown is the relation between the relative radiance of the fluorescent indicator and the pH value around the pK of the indicator, pH-pK_I.*

can be explained by the fact that the dye is a weak acid or base which is fluorescent only in one of the two forms. The concentration of both forms is given by

$$\frac{[R^-][H^+]}{[HR]} = K_I \quad \text{and} \quad [HR] + [R^-] = [I], \tag{3.105}$$

where K_I is the indicator equilibrium constant and $[I]$ the total indicator concentration. Equation Eq. (3.105) can be rearranged to determine the pH value:

$$pH = pK_I' + \log \frac{[R^-]}{[I] - [R^-]}. \tag{3.106}$$

If, for example, R^- is fluorescent, and HR is not, we can determine the pH value from the measurement of the fluorescent intensity. Figure 3.34 shows that the pH value can be measured with this technique very sensitively in a range of about ± 1 around pK_I. Around the buffer point, the relation is linear and can be approximated by

$$pH \approx pK_I + \frac{4}{\ln 10} \left(\frac{R^-}{I} - \frac{1}{2} \right). \tag{3.107}$$

3.5   Advanced Reference Material

This section contains selected additional reference material that should be useful in the search to develop a new quantitative visualization technique for a specific task. The classification of radiation is detailed in Section 3.5.1 and should be useful to become familiar with the terms used in the different research areas. Technical data about radiation sources are collected in Section 3.5.2, while Section 3.5.4 contains some of the most often required optical properties for quantitative visualization.

3.5.1   Classification of Radiation

The nomenclature of electromagnetic radiation is confusing and often terms are used that are not well defined. This originates from the fact that electromagnetic radiation is used in many different scientific areas and often each community has coined its own terms. Here, the most often used terms are defined with comments and recommendations on their usage.

3.1 | **Subpartitioning of the ultraviolet part of the electromagnetic spectrum**

Name	Wavelength range	Comment
UV-A	315–400 nm	CIE standard definition
UV-B	280–315 nm	CIE standard definition
UV-C	100–280 nm	CIE standard definition
Vacuum UV (VUV)	30–180 nm	Range strongly absorbed by air; requires evacuated apparatus
Extreme ultraviolet (XUV)	1–100 nm	Partly overlaps range of soft X-rays

3.2 | **Wavelength ranges for colors; some important emission lines of elements in the gaseous state are also included**

Color	Wavelength range [nm]	Comment
Purple	360–450	Hardly visible; term also used for colors mixed from blue and red colors
	365/366, 405/407, 435.84	Mercury lines
Blue	450–500	
	479.99	Cadmium, F' line
Green	500–570	
	546.07, 577/579	Mercury lines
Yellow	570–591	
	589.59	Na D_1 line
Orange	591–610	
Red	610–830	
	643.85	Cadmium, C' line
	656.28	Hydrogen, H_α line

Subpartitioning of the infrared part of the electromagnetic spectrum 3.3

Unfortunately, no generally accepted and used nomenclature exists. Different authors even mean different wavelength ranges for terms such as near, middle, and far infrared [31, 104, 135, 139]. Avoid the fuzzy terms!

Name	Wavelength range	Comment
Very near infrared (VNIR)	0.7–1.0 μm	Wavelength range to which photographic films and standard CCD sensors respond
Near infrared (NIR)	0.7–3 μm	
Thermal infrared (TIR)	3–14 μm	Range of greatest emission at normal environmental temperatures
Middle infrared	3–100 μm	
Far infrared	100–1000 μm	
Mid-wave infrared (MWIR)	3–5 μm	Common range for thermal imaging
Long-wave infrared (LWIR)	8–14 μm	Common range for thermal imaging
IR-A	0.78–1.4 μm	CIE standard definition
IR-B	1.4–3 μm	CIE standard definition
IR-C	3 μm–1 mm	CIE standard definition
I band	~ 1.25 μm	Used by astronomers, refers to bands of optimal atmospheric transmission
H band	~ 1.65 μm	
K band	~ 2.2 μm	
L band	~ 3.6 μm	
M band	~ 4.7 μm	

Radar frequency bands 3.4

These bands were employed during the second world war to maintain secrecy but are still frequently in use.

Name	Frequency range [GHz]	Wavelength range	Comments
P band	0.230–0.390	0.77–1.3 m	
L band	0.390–1.550	0.19–0.77 m	Seasat SAR: 1.27 GHz, SIR-C SAR shuttle imaging radar: 24 cm
S band	1.550–5.2	5.8–19 cm	
C band	3.90–6.2	4.8–7.7 cm	SIR-C SAR shuttle imaging radar: 5.6 cm, ERS-1 Scatterometer and SAR: 5.3 GHz
X band	5.20–10.9	2.7–5.8 cm	X-SAR-shuttle imaging radar: 3 cm
K band	10.90–36	0.83–2.7 cm	Seasat Scatterometer: 14.6 GHz
K_u band	12.5–18	1.7–2.4 cm	
K_a band	26.5–40	0.75–1.13 cm	
Q band	36.0–46.0	6.5–8.3 mm	
V band	46.0–56.0	5.4–6.5 mm	
W band	56.0–100.0	3.0–5.4 mm	

3.5.2   Radiation Sources

3.5   **Irradiance of the sun**

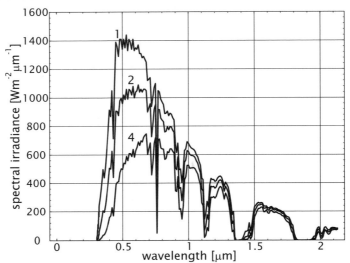

The sun with a mean radius of $6.95 \cdot 10^8$ m irradiates nearly as a blackbody at a temperature of 5900 K. From a mean distance of 1.49710^{11} m, the earth receives at the top of the atmosphere an irradiance of 1376 W/m², also known as the *solar constant*, corresponding to a blackbody at 5800 K. The spectral solar irradiance at sea levels from 0.3-2.13 μm is shown in the figure below. It depends on the length of the path the rays pass through the atmosphere, measured in multiples of the vertical path through a standard atmosphere, denoted as *ma*. The figure shows the irradiance for *ma* = 1, 2, and 4.

3.6   **Luminous efficiency K_r of radiation from some natural and artificial light sources**

Source	K_r [lm/W]
Monochromatic light, 470 nm (blue)	62
Monochromatic light, 555 nm (green)	683
Monochromatic light, 590 nm (yellow)	517
Monochromatic light, 650 nm (red)	73
Direct beam sunlight, midday	90–120
Direct beam sunlight, extraterrestrial (no absorption by atmosphere)	99.3
Overcast sky light	103–115
Tungsten wire at its melting point	53
Blackbody 2500 K ($\lambda_{max} = 1.16\mu m$)	8.71
Blackbody 3000 K ($\lambda_{max} = 0.97\mu m$)	21.97
Blackbody 6000 K ($\lambda_{max} = 0.48\mu m$)	92.9
Blackbody 10000 K ($\lambda_{max} = 0.29\mu m$)	70.6

Lighting system luminous efficiency, K_s, for some common light sources 3.7

Light Source	Lamp power	System power	Luminous flux [lm]	K_s [lm/W]
Tungsten filament				
HR 1M15	15 W	15 W	90	6
HR 1M15	60 W	60 W	730	12.2
HR 1M15	100 W	100 W	1380	13.8
HALOLINE UV-Stop halogen lamp	500 W	500 W	9500	19.0
HALOLINE IRC halogen lamp with infrared reflecting coating	400 W	400 W	9500	23.8
Fluorescent lamp DULUX-L	55 W	62 W	4800	77
Power HQI-T Halogen metal vapor discharge lamp	150 W	170 W	12500	73
Low-pressure sodium vapor lamp (590 nm)	127 W	172 W	25000	145
High-pressure sodium vapor lamp NAV-T Super	600 W	645 W	90000	140

Typical energy conversion in different light sources 3.8

	Tungsten incandescent light bulb, 100 W[1]	Cold cathode fluorescent lamp 40 W[1]	Low pressure sodium vapor lamp, Osram[2] 50X-E91 590 nm	Superbright LED, HP[3], 654 nm, $\Delta\lambda_{1/2} = 18$ nm
Energy input	100 W	40 W	127 W	37 mW
Total radiant flux	82 W	23.2 W	47.7 W	≈ 2.4 mW[6]
Radiant intensity	6.53 W/sr[4]	1.85 Ws/sr[4]	3.80 W/sr[4]	11.8 mW/sr[5]
Radiation efficiency	82 %	58 %	37.6 %	6.5 %[6]
Luminous flux	1740 lm	2830 lm	25000 lm	0.20 lm[6]
Luminous efficacy	21.21 m/W	122 lm/W	524.6 lm/W	85 lm/W
Lighting system efficacy	17.4 lm/W	71 lm/W	197 lm/W	5.5 lm/W[6]

[1] McCluney Introduction to Radiation & Photogrammetry
[2] Osram Lichtprogramm '95/96
[3] Hewlett Packard Data Sheet HLMP-810 λ Series
[4] Isotropic radiation into full sphere assumed
[5] Radiation into cone with full opening angle of 19°
[6] Computed by author

3.9 **Spectral irradiance of arc and quartz tungsten lamps**

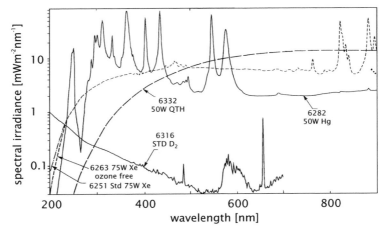

Spectral irradiance curves for Xe arc lamps, Hg arc lamps, quartz tungsten halogen lamps, and deuterium (D$_2$) lamps. Distance of 0.5 m (Courtesy Oriel Corporation).

3.10 **High-power light emitting diodes (LED)**

LEDs have become one of the most useful and versatile light sources for imaging applications. Advantages: low voltage operation, fast switching, available in wavelengths from the ultraviolet through the near infrared, high energy efficient.

Color	Wavelength [nm]	Spectral halfwidth [nm]	Radiant flux [mW]	Voltage [V]	Current [A]	Luminous flux [lm]	K_s [lm/W]
Luxeon power light source, www.lumileds.com							
White	—	—	25	3.42	0.35	25	21
Blue	455	20	150	3.42	0.35	—	—
Blue	470	25	160	3.42	0.35	10	8.4
Cyan	505	30	75	3.42	0.35	30	25
Green	530	35	50	3.42	0.35	30	25
Amber	590	14	70	2.95	0.35	36	35
Orange	617	20	190	2.95	0.35	55	53
Red	625	20	200	2.95	0.35	44	43
High power LED arrays (60 chips), TO-66, www.roithner.mcb.at							
White	—	—	60	19.0	0.24	—	—
UV	370	20	20	18.0	0.20	—	—
UV	405	20	300	18.0	0.24	—	—
Blue	450	30	40	18.0	0.24	—	—
Blue	470	30	100	18.0	0.24	—	—
Red	630	20	150	10.0	0.24	—	—
NIR	740	30	1000	9.0	0.60	—	—
NIR	850	40	1500	7.5	0.80	—	—
NIR	940	40	1200	7.1	0.80	—	—

3.5.3 Human Vision

Relative efficacy $V(\lambda)$ for photopic human vision (Fig. 3.6) $\boxed{3.11}$

The table contains the 1980 CIE values. If these figures are multiplied by 683 lm/W, radiometric quantities can be converted into photometric quantities for the corresponding wavelengths Eq. (3.23).

λ [μm]	$V(\lambda)$	λ [μm]	$V(\lambda)$	λ [μm]	$V(\lambda)$
380	0.00004	520	0.710	660	0.061
390	0.00012	530	0.862	670	0.032
400	0.0004	540	0.954	680	0.017
410	0.0012	550	0.995	690	0.0082
420	0.0040	560	0.995	700	0.0041
430	0.0116	570	0.952	710	0.0021
440	0.023	580	0.870	720	0.00105
450	0.038	590	0.757	730	0.00052
460	0.060	600	0.631	740	0.00025
470	0.091	610	0.503	750	0.00012
480	0.139	620	0.381	760	0.00006
490	0.208	630	0.265	770	0.00003
500	0.323	640	0.175	780	0.000015
510	0.503	650	0.107		

Relative efficacy $V'(\lambda)$ for scotopic human vision (Fig. 3.6) $\boxed{3.12}$

The table contains the 1980 CIE values.

λ [μm]	$V'(\lambda)$	λ [μm]	$V'(\lambda)$	λ [μm]	$V'(\lambda)$
380	0.000589	520	0.935	660	0.0003129
390	0.002209	530	0.811	670	0.0001480
400	0.00929	540	0.650	680	0.0000715
410	0.03484	550	0.481	690	0.00003533
420	0.0966	560	0.3288	700	0.00001780
430	0.1998	570	0.2076	710	0.00000914
440	0.3281	580	0.1212	720	0.00000478
450	0.455	590	0.0655	730	0.000002546
460	0.567	600	0.03315	740	0.000001379
470	0.676	610	0.01593	750	0.000000760
480	0.793	620	0.00737	760	0.000000425
490	0.904	630	0.003335	770	0.0000002413
500	0.982	640	0.001497	780	0.0000001390
510	0.997	650	0.000677		

3.5.4   Selected Optical Properties

3.13 **Reflectivity of water in the infrared**

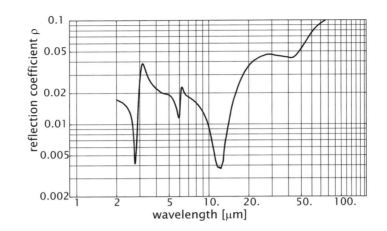

3.14 **Index of refraction for water in the infrared**

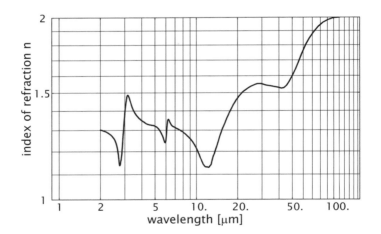

3.15 **Transmittance of the atmosphere from the ultraviolet to the radio wave range**

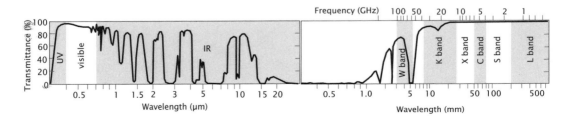

Transmissivity of water from the ultraviolet to infrared range of the electromagnetic spectrum

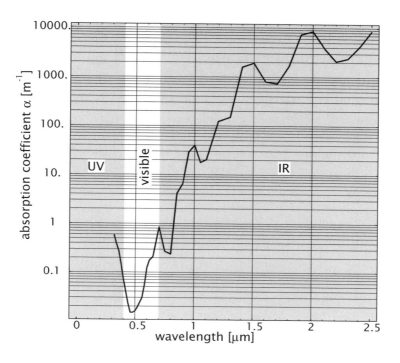

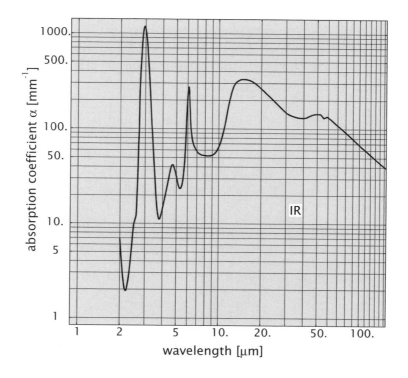

3.5.5   Further References

3.17 **Basics and general references**

W. C. Elmore and M. A. Heald, 1985. Physics of Waves. Dover, New York.

R. McCluney, 1994. Introduction to Radiometry and Photometry. Artech, Boston.

C. DeCusatis (ed.), 1997. Handbook of Applied Photometry. Springer, New York.

A. Richards, 2001. Alien Vision: Exploring the Electromagnetic Spectrum with Imaging Technology. SPIE, Bellingham, WA.

D. Malacara, 2002. Color Vision and Colorimetry. SPIE, Bellingham, WA.

3.18 **Special topics**

H. C. van de Hulst, 1981. Light Scattering by Small Particles. Dover Publications, New York.

F. J. Green, 1990. The Sigma-Aldrich Handbook of Stains, Dyes and Indicators. Aldrich Chemical Company, Milwaukee, WI, USA.

R. P. Haugland, 1996. Handbook of Fluorescent Probes and Research Chemicals. Molecular Probes, http://www.probes.com.

B. M. Krasovitskii and B. M. Bolotin, 1988. Organic Luminescent Materials. VCH, New York.

3.19 **Applications**

The following list contains references that are an excellent source to the wealth of radiation/matter interactions and their applications for imaging. Even if you are not working in one of the particular application areas to which these references refer, it might be a source of inspiration for innovative applications in your own research area.

S. A. Drury, 1993. Image Interpretation in Geology. 2nd edition, Chapman & Hall, London.

L.-L Fu and A. Cazenave (eds.), 2001. Satellite Altimetry and Earth Sciences, Academic Press, San Diego.

R. J. Gurney, J. L. Foster, and C. L. Parkinson, eds., 1993. Atlas of Satellite Observations Related to Global Change. Cambridge University Press, Cambridge.

M. Ikeda and F. W. Dobson, 1995. Oceanographic Applications of Remote Sensing. CRC, Boca Raton.

R. H. Stewart, 1985. Methods of Satellite Oceanography. University of California Press, Berkeley.

Lamps and light sources

The following World Wide Web (www) addresses of companies are a good starting point if you are searching for lamps and light sources. Please note that WWW addresses are often changing. Therefore, some of the listed addresses may be out of date.

GE Lighting (all kinds of lighting products), `http://www.gelighting.com`

Gigahertz Optik (measurement of and with light), `http://www.gigahertz-optik.de`

Heraeus Noblelight (UV and IR lamps), `http://www.heraeus-noblelight.com`

Labsphere (integrated spheres), `http://www.labsphere.com`

Lumileds (High power LEDs), `http://www.lumileds.com`

Osram (all kinds of lighting products), `http://www.osram.com`

Philips Lighting (all kinds of lighting products), `http://www.lighting.philips.com`

Roithner Lasertechnik (LEDs and lasers), `http://www.roithner.mcb.at`

4 Image Formation

4.1 Highlights

Image formation includes geometric and radiometric aspects. The geometry of imaging requires the use of several coordinate systems (Section 4.3.1): world coordinates attached to the observed scene, camera coordinates aligned to the optical axis of the optical system (Section 4.3.1a), image coordinates related to the position on the image plane (Section 4.3.1b), and pixel coordinates attached to an array of sensor elements (Section 4.3.2b).

The basic geometry of an optical system is a perspective projection. A pinhole camera (Section 4.3.2a) models the imaging geometry adequately. A perfect optical system is entirely described by the principal points, the aperture stop, and the focal length (Section 4.3.2c). For the setup of an optical system it is important to determine the distance range that can be imaged without noticeable blur (depth of field, Section 4.3.2d) and to learn the difference between normal, telecentric, and hypercentric optical systems (Section 4.3.2e).

The deviations of a real optical system from a perfect one are known as lens aberrations and include spherical aberration, coma, astigmatism, field curvature, distortion, and axial and lateral color (Section 4.3.2f).

Wave optics describes the diffraction and interference effects caused by the wave nature of electromagnetic radiation (Section 4.3.3). Essentially, the radiance at the image plane is the Fourier transform of the spatial radiance distribution of a parallel wave front at the lens aperture and the crucial parameter of the optical system is the numerical aperture.

Central to the understanding of the radiometry of imaging is the invariance property of the radiance (Section 4.3.4a). This basic property makes it easy to compute the irradiance at the image plane.

Linear system theory is a general powerful concept to describe the overall performance of optical systems (Section 4.3.5). It is sufficient to know the point spread function (PSF, Section 4.3.5a) or the optical transfer function (OTF, Section 4.3.5b). Less useful is the modulation transfer function (MTF, Section 4.3.5e).

The procedure section discusses the practical questions of the geometric and radiometric setup of optical systems (Sections 4.4.1 and 4.4.1b). Then, special imaging systems are discussed for precise optical gauging (Section 4.4.1c) and three dimensional imaging including stereo (Section 4.4.2), confocal laser scanning microscopy (Section 4.4.3), and tomography (Section 4.4.4).

The advanced reference section gives additional data on optical systems for CCD imaging (Section 4.5.1), optical design (Section 4.5.2), and further references.

Task List 2: Image Formation: I Planar Imaging

Task	Procedure
Choose imaging technique	Analyze optical properties of objects to be imaged
	Select whether planar imaging, imaging of surfaces or 3-D imaging is required
	Check whether objects can be manipulated (e. g., sliced, etc.) to be adapted for certain types of imaging techniques
Planar imaging	
Geometric setup of imaging system	Ensure mechanical stability
	Ensure proper object size in image
	Ensure that finest details of interest are resolved
	Select proper camera distance, focal length of lens, and/or perspective
	Check depth of fields
Geometric calibration	Determine magnification factor
	Determine geometrical distortion by optical system
Spatial resolution	Determine PSF and OTF of imaging system theoretically or with appropriate test target
	Check for residual aliasing
Radiometric setup of imaging system	Optimize imaging system to collect sufficient (maximum) radiant flux
	Hinder stray rays to enter optical system and to reach sensor plane
Radiometric calibration	Relative calibration: linearity
	Spectral calibration: overall spectral response
	Absolute calibration

4.2 Task

Image formation is the process where radiation emitted from objects is collected to form an image of the objects in one or the other way. Two principle types of image formation can be distinguished. With direct imaging, rays emitted from one point of an object are — except for possible aberrations — collected in one point at the image plane. This is not the case with indirect imaging. Here, a point at the image plane may collect a radiation from various points of the image objects. An image of the object can only be formed after a more or less complex reconstruction process.

Image formation includes geometric and radiometric tasks (task lists 2–3). The geometrical aspects relate the position of objects in 3-D space to the position of the object image at the image plane. The relation of object image sizes to the original sizes is considered as well as the resolution that can be obtained with a certain imaging system and possible geometrical distortions. The radiometric aspects link the radiance of the objects to the irradiance at the image plane. The question is how much of the radiation emitted by the object is collected by the imaging system and received by the sensor.

The basic difficulty of image formation results from the fact that a 3-D world is projected onto a 2-D image plane. Depending on the type of objects we are studying,

Task List 3: Image Formation II: Surfaces and 3-D Imaging

Task	Procedure
Surfaces in 3-D space (additional tasks)	
Choose depth-resolving imaging technique	Select from stereo, shape from shading, depth from focus, confocal scanning, and other techniques
Geometrical calibration	Depth calibration
Spatial resolution	Depth resolution
3-D imaging (additional tasks)	
Choose 3-D imaging technique	Select from focus series, confocal scanning, tomography, holography, and other techniques

three types of imaging can be distinguished. The simplest case is planar 2-D imaging. Actually, many important scientific and technical applications of image processing are of this type. In this case, a simple geometric relation between the object coordinates and its image exists. Each point of the object is visible and imaged onto the image plane.

The next more complex situation is given by surfaces in 3-D space. This case is typical for natural scenes which are dominated by opaque objects. The geometrical relations are significantly more complex. Part of an object may occlude other, more distant objects and parts of the surface of one object may be invisible. Thus, it is obvious that we generally obtain only incomplete information from an image of a 3-D scene. A human observer is not aware of the missing information. Unconsciously, he has learned to "fill in" the missing parts by knowledge acquired observing similar scenes previously. This makes it in the first place so hard to analyze images of such scenes with a computer. For a computer to "understand" (analyze an image of a 3-D scene in an appropriate way) it is required that the necessary knowledge is available in a form that can be handled by the machine.

The most complex situation is given by 3-D imaging. The objects are no longer opaque, so that the rays that converge at a point in the image plane do not come from a single point of the object but possibly from all points of the ray. Thus, no direct image of the object is formed. It is rather a kind of projection of the 3-D object onto an image plane. The true shape of the 3-D object can only be obtained after a reconstruction process which possibly includes projections of the same object under different directions. Thus, 3-D imaging is normally a type of indirect imaging. 3-D imaging is a challenging task since our visual system is not built to image 3-D objects but only surfaces in 3-D space. This means that new visual tools must be developed to "translate" 3-D objects imaged by 3-D techniques to a form that can be observed by man. The standard procedure is, of course, to extract surfaces from the 3-D images. But this is a very versatile procedure. An extracted surface needs not to be a real surface of the object. Furthermore, we have all the freedom to assign even unphysical optical properties to these surfaces in order to visualize the object property of interest in an optimal way.

The tasks involved in image formation depend on the type of imaging and are summarized for the three major forms in task lists 2–3. For all types of imaging, the primary task is to establish the geometric relation between the object coordinates and the co-ordinates of its image. In planar images, we have a number of choices. If we select a perspective, we have to make sure that the objects are imaged in proper size and that the finest details of interest are resolved. If we are imaging objects at different dis-

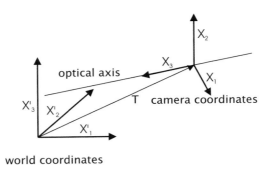

Figure 4.1: *Camera and world coordinate systems.*

tances, we have to make sure that the necessary depth range is covered by the imaging system. If the task is to determine also the depth or distance of the objects, we have to select between a wide range of surface reconstruction procedures which are listed in task list 3. In many cases, it is critical to collect sufficient radiation. Thus, it is an important task to optimize the imaging system in such a way that as much radiation as possible is collected.

In 3-D imaging, there are a number of imaging techniques (task list 3). Some of them are suitable for a wide range of conditions; others, such as focus series or confocal scanning, are only suitable for microscopic imaging. An important task in 3-D imaging is also to ensure an even and complete sampling.

This chapter is restricted to the discussion of image formation. Procedures for geometric calibration are discussed in Chapter 8, whereas procedures for radiometric calibration are discussed in Chapter 7.

4.3 Concepts

4.3.1 Coordinate Systems

4.3.1a World and Camera Coordinates. The position of objects in 3-D space can be described in many different ways (Fig. 4.1). First, we can use a coordinate system which is related to the scene observed. These coordinates are called *world coordinates* and denoted as column vectors $X' = [X_1', X_2', X_3']^T = [X', Y', Z']^T$. We use the convention that the X_1' and X_2' coordinates describe the horizontal and the X_3' coordinate the vertical positions, respectively. A second coordinate system, the *camera coordinates*

$$X = [X_1, X_2, X_3]^T = [X, Y, Z]^T, \qquad (4.1)$$

can be fixed to the (possibly moving) camera observing the scene. The X_3 axis is aligned with the *optical axis* of the camera system (Fig. 4.1).

Transition from world to camera coordinates requires a *translation* and a *rotation*. First, we shift the origin of the world coordinate system to the origin of the camera coordinate system by the translation vector T (Fig. 4.1). Then, we change the orientation of the shifted coordinate system by a sequence of rotations about suitable axes so that it coincides with the camera coordinate system. Mathematically, *translation* is described by vector subtraction and *rotation* by multiplication of the coordinate vector with a matrix:

$$X = R(X' - T). \qquad (4.2)$$

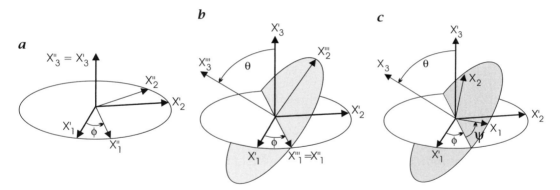

Figure 4.2: *Rotation of world coordinates X' and camera coordinates X using the three Eulerian angles (ϕ, θ, ψ) with successive rotations about the X'_3, X''_1, and X'''_3 axes.*

Rotation does not change the length or *norm* of the vectors (Section B.1). Therefore, the matrix R must be orthogonal, i.e., hold the condition

$$RR^T = I \quad \text{or} \quad \sum_{m=1}^{3} r_{km}r_{lm} = \delta_{kl} \qquad (4.3)$$

where I denotes the identity matrix. The six independent equations of the orthogonality condition leave three matrix elements out of nine independent. The relationship between the matrix elements and a set of three parameters describing an arbitrary rotation is not unique and nonlinear.

A common procedure is the three Eulerian rotation angles (ϕ, θ, ψ): The rotation from the shifted world coordinate system into the camera coordinate system is decomposed into three steps (see Fig. 4.2, [41]).

1. Rotation about X'_3 axis by ϕ, $X'' = R_\phi X'$:

$$R_\phi = \begin{bmatrix} \cos\phi & \sin\phi & 0 \\ -\sin\phi & \cos\phi & 0 \\ 0 & 0 & 1 \end{bmatrix} \qquad (4.4)$$

2. Rotation about X''_1 axis by θ, $X''' = R_\theta X''$:

$$R_\theta = \begin{bmatrix} 1 & 0 & 0 \\ 0 & \cos\theta & \sin\theta \\ 0 & -\sin\theta & \cos\theta \end{bmatrix} \qquad (4.5)$$

3. Rotation about X'''_3 axis by ψ, $X = R_\psi X'''$:

$$R_\psi = \begin{bmatrix} \cos\psi & \sin\psi & 0 \\ -\sin\psi & \cos\psi & 0 \\ 0 & 0 & 1 \end{bmatrix} \qquad (4.6)$$

A lot of confusion exists in the literature about the definition of the Eulerian angle. We use the standard mathematical approach: right-hand coordinate systems are used, rotation angles are defined positive in counterclockwise direction. Cascading the three rotations, $\mathbf{R}_\psi \mathbf{R}_\theta \mathbf{R}_\phi$, yields the matrix

$$
\begin{bmatrix}
\cos\psi\cos\phi - \cos\theta\sin\phi\sin\psi & \cos\psi\sin\phi + \cos\theta\cos\phi\sin\psi & \sin\theta\sin\psi \\
-\sin\psi\cos\phi - \cos\theta\sin\phi\cos\psi & -\sin\psi\sin\phi + \cos\theta\cos\phi\cos\psi & \sin\theta\cos\psi \\
\sin\theta\sin\phi & -\sin\theta\cos\phi & \cos\theta
\end{bmatrix}. \tag{4.7}
$$

The inverse transformation from camera coordinates to world coordinates is given by the transpose of the matrix in Eq. (4.7).

Matrix multiplication is not commutative (Section B.1). Thus, rotation is not commutative. Therefore, it is important not to interchange the order in which rotations are performed. Rotation is only commutative in the limit of infinitesimal rotations. Then, all cosine terms can be eliminated by 1 and $\sin(\varepsilon)$ reduces to ε. This limit has some practical applications since minor misalignments are common and a rotation about the X_3 axis, for instance, can be

$$
\mathbf{X} = \mathbf{R}_\varepsilon \mathbf{X}' = \begin{bmatrix} 1 & \varepsilon & 0 \\ -\varepsilon & 1 & 0 \\ 0 & 0 & 1 \end{bmatrix} \mathbf{X}' \quad \text{or} \quad \begin{array}{rcl} X_1 &=& X_1' + \varepsilon X_2' \\ X_2 &=& X_2' - \varepsilon X_1' \\ X_3 &=& X_3' \end{array} \tag{4.8}
$$

A point $(X', 0, 0)$ is rotated into the point $[X', \varepsilon X', 0]^T$. Compared with the correct position $[X'\cos\varepsilon, X'\sin\varepsilon, 0]^T$, the position error $\Delta X = [1/2\varepsilon^2 X', 1/6\varepsilon^3 X', 0]^T$. For a 512×512 image ($X' < 256$ for centered rotation) and an acceptable error of less than $1/20$ pixel, ε must be smaller than 0.02 or $1.15°$. This is still a significant rotation, since rows in an image would be displaced vertically by up to $\pm\varepsilon X' = \pm 5$ pixels.

4.3.1b   Image Coordinates and Principal Point.   Generally, an image of an optical system is formed on a plane. In some systems, the image plane could also be a curved surface. The *image coordinates* describe the position on the image plane. In order to distinguish two-dimensional image coordinates from 3-D world or camera coordinates, we denote them by lower case letters: $\mathbf{x} = [x_1, x_2]^T$ or $\mathbf{x} = [x, y]^T$. The origin of the image coordinate system is the intersection of the optical axis with the image plane at the *principal point* $\mathbf{x}_h = [x_h, y_h]^T$. An alternative choice for the origin — in a pixel-oriented view — is the upper-left corner of the image. Then, the coordinate system is left-handed and the y-axis is directed downwards. This *pixel-based coordinate system* is discussed in more detail in Section 4.3.2b.

4.3.1c   Homogeneous Coordinates.   In computer graphics and computer vision, the elegant formalism of *homogeneous coordinates* [29, 34, 101] is used to describe coordinate transformations. The significance of this formalism is related to the fact that translation, rotation, and even perspective projection (Section 4.3.2a) can be expressed by a single operation and that points and lines are dual objects.

Homogeneous or projective coordinates use four-component column vectors $\mathbf{X} = [tX_1, tX_2, tX_3, t]^T$, from which the ordinary three-dimensional coordinates are obtained by dividing the first three components of the homogeneous coordinates by the fourth. Thus, $\mathbf{X}$ and $\lambda\mathbf{X}$, $\forall \lambda \neq 0$, represent the same point. Any linear transformation can be obtained by pre-multiplying the homogeneous coordinates with a 4×4 matrix $\mathbf{M}$.

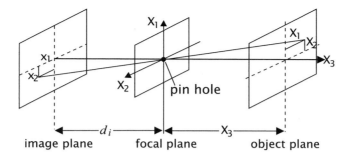

Figure 4.3: *Image formation with a pinhole camera. An image of an object is formed since only rays passing through the pinhole can meet the image plane. The origin of the camera coordinate system is located at the pinhole.*

The transformation matrices for the elementary transformations discussed in Section 4.3.1a are given by:

$$\boldsymbol{T} = \begin{bmatrix} 1 & 0 & 0 & T_1 \\ 0 & 1 & 0 & T_2 \\ 0 & 0 & 1 & T_3 \\ 0 & 0 & 0 & 1 \end{bmatrix} \qquad \text{Translation by } [T_1, T_2, T_3]^T,$$

$$\boldsymbol{R}_{x_1} = \begin{bmatrix} 1 & 0 & 0 & 0 \\ 0 & \cos\theta & -\sin\theta & 0 \\ 0 & \sin\theta & \cos\theta & 0 \\ 0 & 0 & 0 & 1 \end{bmatrix} \qquad \text{Rotation about } X_1 \text{ axis by } \theta,$$

$$\boldsymbol{R}_{x_2} = \begin{bmatrix} \cos\phi & 0 & \sin\phi & 0 \\ 0 & 1 & 0 & 0 \\ -\sin\phi & 0 & \cos\phi & 0 \\ 0 & 0 & 0 & 1 \end{bmatrix} \qquad \text{Rotation about } X_2 \text{ axis by } \phi,$$

$$\boldsymbol{R}_{x_3} = \begin{bmatrix} \cos\psi & -\sin\psi & 0 & 0 \\ \sin\psi & \cos\psi & 0 & 0 \\ 0 & 0 & 1 & 0 \\ 0 & 0 & 0 & 1 \end{bmatrix} \qquad \text{Rotation about } X_3 \text{ axis by } \psi, \qquad (4.9)$$

$$\boldsymbol{S} = \begin{bmatrix} s_1 & 0 & 0 & 0 \\ 0 & s_2 & 0 & 0 \\ 0 & 0 & s_3 & 0 \\ 0 & 0 & 0 & 1 \end{bmatrix} \qquad \text{Scaling in all coordinate directions.}$$

In these definitions, also a right-handed coordinate system is used. Positive rotations are defined such that, when looking from a positive axis towards the origin, a 90° counterclockwise rotation transforms one positive axis into the other. The last equation in Eq. (4.9) contains a matrix that scales the length in all coordinate directions by a different factor.

4.3.2 Geometrical Optics

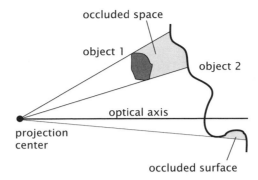

Figure 4.4: *Illustration of two types of occlusion. An object that lies in front of others occludes the whole space that is formed by the shadow of a point light source positioned at the projection center from being visible. Occlusion can also occur on steep edges of continuous surfaces.*

4.3.2a Pinhole Camera Model: Perspective Projection. The simplest possible camera is the *pinhole camera*. The imaging element of this camera is an infinitesimal small hole (Fig. 4.3). Only the light ray coming from a point of the object at $[X_1, X_2, X_3]^T$ which passes through this hole at the origin of the camera coordinate system meets the image plane at $[x_1, x_2, -d_i]^T$. Through this condition, an image of the object is formed on the image plane. The relationship between the 3-D camera and the 2-D *image coordinates* $[x_1, x_2]^T$ is given by

$$x_1 = -\frac{d_i X_1}{X_3}, \quad x_2 = -\frac{d_i X_2}{X_3}. \tag{4.10}$$

Many researchers use the dimensionless image coordinates, the *generalized coordinates*, in order to simplify the expressions. The image coordinates are divided by the image distance d_i

$$\frac{x_1}{d_i} \to \tilde{x}_1, \quad \frac{x_2}{d_i} \to \tilde{x}_2. \tag{4.11}$$

Generalized image coordinates put the image plane at a unit length distant from the focal plane and have an important meaning. They are equal to the tangent of the angle between the optical axis and the ray towards the object. Thus, they explicitly take the limitations of perspective projection into account. From these coordinates, we cannot infer absolute positions but know only the angle under which the object is projected onto the image plane.

The two world coordinates parallel to the image plane are scaled by the factor d_i/X_3. Therefore, the image coordinates $[x_1, x_2]^T$ contain only ratios of world coordinates, from which neither the distance nor the true size of an object can be inferred without additional knowledge.

A straight line in the world space is projected onto a straight line at the image plane. Perspective projection, however, does neither preserve distances between points nor the ratio of distances between points (Fig. 4.5). If A, B, and C are points on a line, and a, b, and c the corresponding projected points at the image plane, then

$$\overline{AC}/\overline{BC} \neq \overline{ac}/\overline{bc}, \tag{4.12}$$

where $\overline{AC}$ denotes the distance between the points A and C, etc.

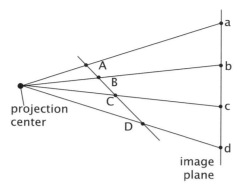

Figure 4.5: *Perspective projection does not preserve the ratio of distances* $\overline{AC} \neq \overline{BC} > \overline{ac}/\overline{bc}$ *but preserves the cross-ratio* $(\overline{AC}/\overline{BC})/(\overline{AD}/\overline{BD}) = (\overline{ac}/\overline{bc})/(\overline{ad}/\overline{bd})$.

For perspective projection, however, a more complex invariant exists, the *cross ratio*, which is the ratio of distance ratios. If A, B, C, and D are points on a line, then (Fig. 4.5):

$$\frac{\overline{AC}/\overline{BC}}{\overline{AD}/\overline{BD}} = \frac{\overline{ac}/\overline{bc}}{\overline{ad}/\overline{bd}}. \tag{4.13}$$

All object points on a ray through the pinhole are projected onto a single point in the image plane. In a scene with transparent objects, for example, in X-ray images or observing a turbulent flow in a transparent liquid made visible by small particles, the "objects" are projected onto each other. Then, we cannot infer the three-dimensional structure of the scene. We may not even be able to recognize the shape of individual objects at all.

Many scenes contain only opaque objects. In this case, the observed 3-D space is essentially reduced to a 2-D surface in 3-D space which can be described by two 2-D functions, the irradiance at the image plane, $g(x_1, x_2)$, and the *depth map* $X_3(x_1, x_2)$. A surface in space is completely projected onto the image plane provided that not more than one point of the surface lies on the same ray through the pinhole. If this condition is not met, parts of the surface remain invisible (*occlusion*, Fig. 4.4). The occluded 3-D space can be made visible if we put a point light source at the position of the pinhole. Then, the invisible parts of the scene lie in the shadow of those objects which are closer to the camera.

Direct imaging with penetrating rays such as X-rays emitted from a point source can also be modeled as a perspective projection. In this case, the object lies between the central point and the image plane (Fig. 4.6).

The projection equation corresponds to Eq. (4.10) except for the sign:

$$[X_1, X_2, X_3]^T \longmapsto [x_1, x_2]^T = \left[\frac{d_i X_1}{X_3}, \frac{d_i X_2}{X_3}\right]^T. \tag{4.14}$$

The general projection equation of perspective projection Eq. (4.14) then reduces to

$$X = [X_1, X_2, X_3]^T \longmapsto \tilde{x} = \left[\frac{X_1}{X_3}, \frac{X_2}{X_3}\right]^T. \tag{4.15}$$

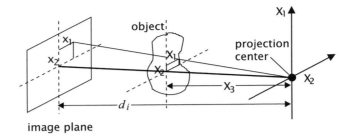

Figure 4.6: *Perspective projection with penetrating rays (shadow casting).*

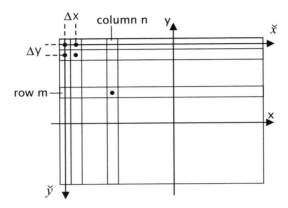

Figure 4.7: *Relation between the image coordinates, $\boldsymbol{x} = [x, y]^T$, and pixel coordinates $\check{\boldsymbol{x}} = [\check{x}, \check{y}]^T$ attached to the discrete sampling points of a sensor.*

In terms of homogeneous coordinates, the perspective projection can also be expressed by matrix multiplication:

$$\boldsymbol{P}_{X_3} = \begin{bmatrix} 1 & 0 & 0 & 0 \\ 0 & 1 & 0 & 0 \\ 0 & 0 & 1 & 0 \\ 0 & 0 & 1/d_i & 0 \end{bmatrix} \quad \text{Perspective projection in } X_3 \text{ direction.} \qquad (4.16)$$

In contrast to the matrices for rotation and translation Eq. (4.9), this matrix is not invertible as we expect from a perspective transform, since many points — all points lying on a projection beam — are projected onto one point.

4.3.2b   Pixel Coordinates and Intrinsic Camera Parameters. The irradiance at the image plane is picked up by a suitable sensor and converted into an electrical signal at discrete positions (see Chapters 5 and 6). Therefore, a last coordinate transform is required that maps the image coordinate onto a new coordinate system attached to the discrete image points, called *pixels* (Fig. 4.7). These coordinates are called *pixel coordinates*. The pixels are counted by specifying the row and column numbers in which they are located. While the columns are counted from left to right, it is common to count the rows from top to bottom, i.e., in negative direction.

We specify the position of a pixel first by the row number and then by the column number as it is standard for matrices in mathematics. It is also common to start counting with zeros (as in the programming language C). The position of the first pixel $\boldsymbol{x}_{0,0}$

is denoted by $\boldsymbol{x}_0 = [x_0, y_0]^T$, the distances of the center of two adjacent pixels in horizontal and vertical direction by Δx and Δy, respectively. Then, the position of pixel $\boldsymbol{x}_{m,n}$ in image coordinates is given by

$$\boldsymbol{x}_{m,n} = \left[\begin{array}{c} x_0 + m\Delta x \\ y_0 - n\Delta y \end{array} \right]. \tag{4.17}$$

The reverse transformation from image coordinates to pixel coordinates is given by

$$\check{\boldsymbol{x}} = \left[\begin{array}{c} m \\ n \end{array} \right] = \left[\begin{array}{c} (x - x_0)/\Delta x \\ -(y - y_0)/\Delta y \end{array} \right]. \tag{4.18}$$

With this reverse transformation, m and n may take noninteger values. The noninteger pixel coordinates are denoted by $\check{\boldsymbol{x}}$.

There are two special points. First, the intersection of the optical axis with the image plane ($\boldsymbol{x} = 0$), the *principal point* $\boldsymbol{x}_h$, is given in pixel coordinates as

$$\check{\boldsymbol{x}}_h = \left[\begin{array}{c} -x_0/\Delta x \\ y_0/\Delta y \end{array} \right]. \tag{4.19}$$

Second, the center of the pixel array $\boldsymbol{x}_c$ with $M \times N$ pixels, $\boldsymbol{x}_{M/2,N/2}$, is given in image coordinates as

$$\boldsymbol{x}_c = \left[\begin{array}{c} x_0 + (M/2)\Delta x \\ y_0 - (N/2)\Delta y \end{array} \right]. \tag{4.20}$$

Ideally, the center of the pixel array should coincide with the optical axis. In image coordinates this means $\boldsymbol{x}_c = \boldsymbol{0}$ and in pixel coordinates $\check{\boldsymbol{x}}_h = [M/2, N/2]^T$. In reality, there will be a certain misalignment between the optical axis and the center of the pixel array. Without loss of generality, however, we can always assume that the pixel rows are parallel to the X-axis of the image coordinates.

The photo-sensitive detectors of modern semiconductor sensors are almost perfectly arranged on a rectangular grid. Thus, it is almost never required to correct for geometric distortions between image and pixel coordinates. The sensor chip may, however, not be perfectly oriented perpendicularly to the optical axis introducing a slight perspective distortion. This distortion can be modeled by the angles of the sensor plane to the x axis and y axis of the image coordinate system.

More common are significant geometric distortions between world and image coordinates, either because of the projection of curved surfaces onto the image plane or because of lens aberrations (Section 4.3.2f). The four parameters x_0, y_0, Δx, and Δy are called the *intrinsic parameters* of the camera since they do not depend on the position and orientation of the camera in space.

In contrast, the 6 parameters (three for translation and three for rotation, Section 4.3.1a) describing the transformation from world coordinates to camera coordinates are called the *extrinsic parameters*. In total, we need to know at least these 10 parameters to describe the relation between the pixel and world coordinates. Further parameters are required if geometrical distortions need to be corrected.

4.3.2c Perfect Optical Systems. The pinhole camera is only a model of a camera. A real camera uses a more or less complex optical system to form an image. A perfect optical system would then still show the same geometrical relations between world and image coordinates. This statement is only true in the limit of "thin" lenses. If the extension of the optical system along the optical axis cannot be neglected (a "thick" lens), a simple modification of the pinhole camera model is required. The focal plane

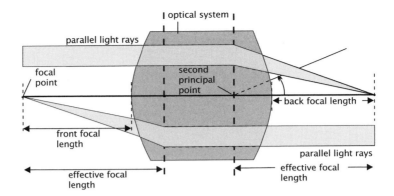

Figure 4.8: *A perfect optical system can be treated as a "black box" and modeled by its cardinal points: the first and second focal point and the first and second principal point.*

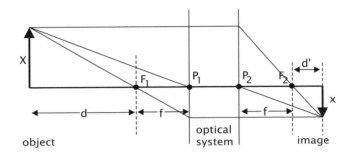

Figure 4.9: *Optical imaging with an optical system modeled by its principal points P_1 and P_2 and focal points F_1 and F_2. The system forms an image that is a distance d' behind F_2 from an object that is the (negative) distance d in front of F_1.*

has to be replaced by two *principal planes*. The two principal planes meet the optical axis at the *principal points*. A ray directed towards the first principal point appears — after passing through the system — to originate from the second principal point without angular deviation (Fig. 4.8). The distance between the two principal planes thus models the effective thickness of the optical system.

There is another significant difference between a pinhole camera and a real optical system. While a pinhole camera forms an image of an object at *any* distance, an optical system forms an image of an object only at a certain distance. Parallel rays entering an optical system from left and right meet at the second and first focal point, respectively (Fig. 4.8).

The *effective focal length* (*efl*) is the distance from the principal point to the corresponding focal point. For practical purposes, the following definitions also are useful: The *back focal length* (*bfl*) is the distance from the last surface of the optical system to the second focal point. Likewise, the *front focal length* (*ffl*) is the distance from the first surface of the optical system to the first focal point (Fig. 4.8).

The relation between the object distance and the image distance becomes very simple if they are measured from the focal points (Fig. 4.9),

$$dd' = -f^2. \tag{4.21}$$

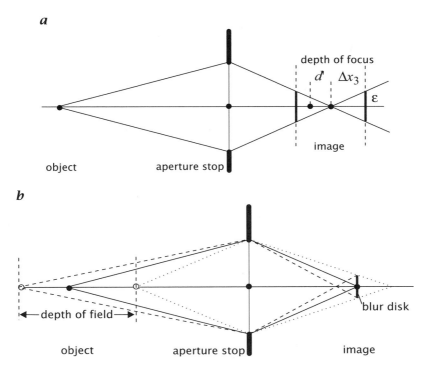

Figure 4.10: *Illustration of the concepts of **a** depth of focus and **b** depth of field with an on-axis point object. Depth of focus is the range of image distances in which the blurring of the image remains below a certain threshold. The depth of field gives the range of distances of the object in which the blurring of its image of the object at a fixed image distance remains below a given threshold.*

This is the Newtonian form of the *image equation.* The possibly better known but awkward Gaussian form is

$$\frac{1}{d' + f} = \frac{1}{f} + \frac{1}{d + f} \tag{4.22}$$

which takes the distances as to the principal points. The *magnification m* of the optical system is given by the ratio of the image size, x, to the object size, X:

$$m = \frac{x}{X} = \frac{f}{d} = -\frac{d'}{f}. \tag{4.23}$$

Note that d is negative. Thus, also the magnification is negative indicating that the image is turned upside down.

4.3.2d Depth of Focus and Depth of Field. The image equation determines the relation between object and image distances. If the image plane is slightly shifted or the object is closer to the lens system, the image is not rendered useless. It rather gets more and more blurred, the larger the deviation from the distances is, given by the image equation.

The concepts of depth of focus and depth of field are based on the fact that for a given application only a certain degree of sharpness is required. For digital image processing it is naturally given by the size of the sensor elements. It makes no sense to resolve smaller structures but to allow a certain blurring. The blurring can be described using the image of a point object as illustrated in Fig. 4.10a. At the image plane, the point object is imaged to a point. It smears to a disk with the radius ϵ with increasing distance from the image plane. Introducing the f-number of an optical system as the ratio of the focal length and diameter of lens aperture 2r:

$$n_f = \frac{f}{2r} \tag{4.24}$$

we can express the radius of the blur disk as:

$$\epsilon = \frac{1}{n_f} \frac{f}{f + d'} \Delta x_3, \tag{4.25}$$

where Δx_3 is the distance from the (focused) image plane. The range of positions of the image plane, $[d' - \Delta x_3, d' + \Delta x_3]$, for which the radius of the blur disk is lower than ϵ, is known as the *depth of focus*. Equation (4.25) can be solved for Δx_3 and yields

$$\Delta x_3 = n_f \left(1 + \frac{f}{d'} \right) \epsilon = n_f (1 - m) \epsilon, \tag{4.26}$$

where m is the negative-valued magnification as defined by Eq. (4.23). Equation (4.26) illustrates the critical role of the f-number for the depth of focus.

Of even more importance for the practical usage than the depth of focus is the *depth of field*. The depth of field is the range of object positions for which the radius of the blur disk remains below a threshold ϵ at a fixed image plane (Fig. 4.10b),

$$d \pm \Delta X_3 = \frac{-f^2}{d' \mp n_f (1 - m) \epsilon}. \tag{4.27}$$

In the limit of $|\Delta X_3| \ll d$, we obtain

$$|\Delta X_3| \approx n_f \cdot \frac{1 - m}{m^2} \epsilon. \tag{4.28}$$

If the depth of field includes infinite distance,

$$d_{min} \approx \frac{f^2}{2 n_f \epsilon}. \tag{4.29}$$

Generally, the whole concept of depth of field and depth of focus is only valid for a perfect optical system. If the optical system shows any aberrations, the depth of field can only be used for blurring significantly larger than those caused by the aberrations of the system.

4.3.2e Telecentric and Hypercentric Imaging.

In a standard optical system, a converging beam of light is entering an optical system. This setup has a significant disadvantage for optical gauging (Fig. 4.11a). If the object position is varying, the object appears larger if it is closer to the lens and smaller if it is farther away from the lens. Since the depth of the object cannot be inferred from its image either, the object must be at a precisely known depth or measurement errors are unavoidable.

A simple change of the position of the *aperture stop* from the principal point to the first focal point changes the imaging system to a telecentric lens (Fig. 4.11b). By placing the stop at this point, the *principal rays* (ray passing through the center of the aperture) are parallel to the optical axis in the object space. Therefore, slight changes in the position of the object do not change the size of the image of the object. The

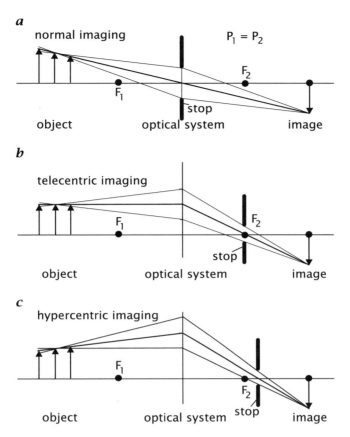

Figure 4.11: *Repositioning of the stop changes the properties of an optical system: **a** standard diverging imaging with stop at the principal point; **b** telecentric imaging with stop at the second focal point; **c** hypercentric imaging with stop before the second focal point.*

farther it is away from the focused position, the more it is blurred, of course. However, the center of the blur disk does not change the position.

Telecentric imaging has become an important principle in machine vision. Its disadvantage is, of course, that the diameter of a telecentric lens must be at least of the size of the object to be imaged. This makes telecentric imaging very expensive for large objects.

If the aperture stop is located even further to the image (between the first focal point and the image plane; Fig. 4.11c), the principal rays are a converging beam in the object space. Now, in contrast to "normal" imaging (Fig. 4.11a), an object appears larger if it is more distant! This imaging technique is called *hypercentric imaging*. It has the interesting feature that it can observe surfaces that are parallel to the optical axis. Figure 4.12 illustrates how a cylinder aligned with the optical axis with a thin wall is seen with the three imaging techniques. Standard imaging sees the cross section and the inner wall, telecentric imaging the cross section only, and hypercentric imaging the cross section and the outer wall.

The discussion of telecentric and hypercentric imaging emphasizes the importance of stops in the construction of optical systems, a fact that is often not adequately considered.

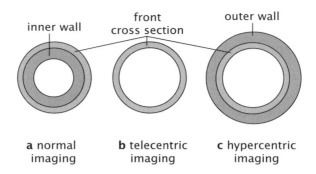

inner wall   front cross section   outer wall

a normal imaging   **b** telecentric imaging   **c** hypercentric imaging

*Figure 4.12: Imaging of a short piece of a cylindrical tube whose axis is aligned with the optical axis by **a** normal, **b** telecentric, **c** hypercentric imaging.*

Table 4.1: Aberrations as a function of lens aperture and field size x

Aberration	Aperture r	Field size x
Spherical aberration longitudinal transverse	r^2	—
Coma	r^3	—
Astigmatism	r	x
Field curvature	—	x^2
Distortion	—	x^3
Axial color	r	—
Lateral color	—	x

4.3.2f Lens Aberrations. Real optical systems are not perfect systems even in the limit of *geometric optics*, i. e., when diffraction effects inherent into the wave nature of electromagnetic radiation are neglected. A real optical system does not form a point image from a point object. The image of a point object rather results in an intensity distribution in the image. This distribution is characterized by a) the width measuring the degree of blurring and b) the deviation of the center of gravity of the distribution from its true position causing a geometrically distorted image.

All these *aberrations* have their common cause in the refraction and reflection law. First-order optics, which is the basis of the simple image equation, requires *paraxial rays* which have a low inclination α and offset to the optical axis so that

$$\tan \alpha \approx \sin \alpha \approx \alpha. \tag{4.30}$$

Only in this limit, mirrors and lenses with spherical shape focus a parallel light beam onto a single point. Additionally, so-called *chromatic aberrations* are caused by the wavelength dependency of the index of refraction (Section 3.3.1a).

For larger angles α and object height h, we can expect aberrations. Seven primary aberrations can be distinguished which are summarized in Table 4.1 and Fig. 4.13. *Spherical aberration* is related to the focusing of an axial parallel bundle of monochromatic light. Each radial zone of the lens aperture has a slightly different focal length. For a simple planoconvex lens, the focal length decreases with the distance of the ray from the optical axis. The lack of a common focus results in a certain spot size onto which a parallel light bundle is focused. Due to the larger area of the off-axis rays, the minimum blur circle is found slightly short off the focus for paraxial rays.

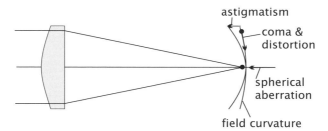

Figure 4.13: *Schematic illustration of the five primary monochromatic aberrations: spherical aberration, coma, astigmatism, field curvature, and distortion.*

Coma is an aberration that affects only off-axis light bundles. For such a light bundle, even rays piercing the lens aperture at a radial zone are not focused onto a point but — because of the lack of symmetry — onto a blur circle with a spot size increasing with the radius of the zone. Also, in contrast to spherical aberration, each radial zone focuses onto the image plane at a slightly different height. The result is a spot of comatic shape with a bright central core and a triangular shaped flare extending toward the optical axis.

While the coma describes the lateral aberration of off-axis rays, *astigmatism* characterizes the corresponding axial aberrations. Because of symmetry, tangential fans and sagittal fans are distinguished. These two types of fan beams pierce the lens aperture under different angles and thus are focused at different distances depending on the angle of the off-axis fans. Astigmatism thus not only results in a curved image plane. It is also different for tangentially and radially oriented structures. Imagine that a spoked wheel centered on the optical axis is imaged. Then, the rim of the wheel (a tangential structure) is focused at a different distance as the spokes.

If the tangential and sagittal surfaces are coincident, the lens is said to be free of astigmatism. Then, another aberration still remains, the *field curvature*. Optical systems tend to form an image on a curved surface. In a first approximation, the curved surface is a sphere. The curvature of this sphere (1/radius) is referred to as the field curvature of the optical system.

The variation of the index of refraction with the wavelength causes two additional types of aberrations. *Axial color* describes the aberration that the focal length of an optical system depends on the wavelength. For a simple lens, the focal length is shorter at shorter wavelengths. For off-axis bundles, a second chromatic aberration, *lateral color*, becomes significant. Any ray that is significantly refracted by a lens is subject to dispersion. As with a simple prism, the beam is separated into its various wavelength components, causing different wavelengths to be imaged at different heights. Chromatic aberrations can be suppressed by using lenses with different glass types. The simplest system is an achromatic doublet, or simply achromat. With this system, two wavelengths can be brought to a common focus. An apochromatic system is corrected chromatically for three wavelengths simultaneously, often extending the corrections into the near infrared.

4.3.2g Geometric Distortions. *Geometric distortions* do not affect the sharpness but the shape of an image. Because of the radial symmetry of optical systems, the radius (distance to the optical axis) and not the angle is distorted. Positive/negative distortion means that off-axis points are imaged at distances greater/shorter than nominal. Positive distortion images a square in a pincushion shape, negative distortion in a barrel shape (Fig. 4.14).

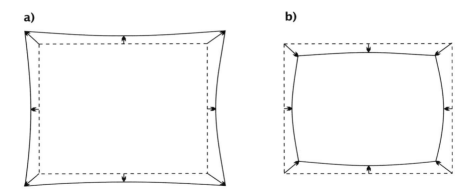

Figure 4.14: *Radial-symmetric distortion by optical imaging illustrated with the image of a square. The sides of the rectangle become curved because the distortion varies as the cube of the distance from the center.* **a** *positive or pincushion distortion;* **b** *negative or barrel distortion.*

The *radial-symmetric distortion* can be approximated by

$$x^\star = \frac{x}{1 + k_3 |x|^2}. \tag{4.31}$$

With this type of geometrical distortion, the corners of a centered square are distorted two times as much as the center of the sides (Fig. 4.14).

For higher accuracy demands (including the correction of residual distortions of lenses that are well corrected for geometrical distortions), a single quadratic correction is not sufficient. Because of Eq. (4.30) and the radial symmetry, higher-order polynomial approximations include only even powers of $|x|$:

$$x^\star \approx x \left(1 + k_3 |x|^2 + k_5 |x|^4 + k_7 |x|^6\right). \tag{4.32}$$

The radial-symmetric distortion does not include effects caused by decentering of individual lenses in an optical system or deviation of lens surfaces form the ideal spherical form. Correction of these higher-order geometrical distortions is a tricky business and must be well adapted to the optical system used [40, 47, 93]. If too many parameters are used for correction, the determination of the parameters becomes numerically instable. If a critical parameter is not included the residual error becomes unacceptably high.

For CCD camera systems, it proved to be effective to correct for the following effects in addition to radial-symmetric distortions: i) decentering distortions, ii) scale factor s, and iii) shear factor a [47]. The latter two effects are caused by a nonperfect orientation of the sensor plane perpendicular to the optical axis.

If the image coordinates are centered around the principal point, all these correction terms in addition to the radial-symmetric distortion result in a correction

$$
\begin{aligned}
x^\star - x = \Delta x \;=\; & \begin{bmatrix} s & a \\ a & 0 \end{bmatrix} x & \text{shear and scale} \\[2mm]
+ \;& \left(k_3 |x|^2 + k_5 |x|^4 + k_7 |x|^6\right) x & \text{radial-symmetric} \\[2mm]
+ \;& [p_1, p_2] \begin{bmatrix} 3x^2 + y^2 & 2xy \\ 2xy & x^2 + 3y^2 \end{bmatrix} & \text{decentering}
\end{aligned} \tag{4.33}
$$

with 7 parameters in total.

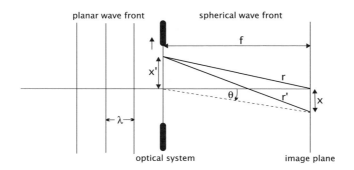

Figure 4.15: *Diffraction of a planar wave front at the aperture stop of an optical system. The optical system can be thought to convert the incoming planar wave front into spherical wave fronts in all directions converging at the image plane. The wave amplitude at the image plane is governed by the interference of all these spherical waves. Essential for the observed amplitude is the path difference at the aperture stop, δ = x' sin θ, of these superimposing spherical wave fronts.*

4.3.3 Wave Optics

4.3.3a Diffraction-limited Optics. Light is electromagnetic radiation and as such subject to wave-related phenomena. When a parallel bundle of light enters an optical system, it cannot be focused to a point even if all aberrations discussed in Section 4.3.2f have been eliminated. Diffraction at the aperture of the optical system blurs the spot at the focus to a size of at least the order of the wavelength of the light. An optical system for which the aberrations have been suppressed to such an extent that it is significantly lower than the effects of diffraction is called *diffraction-limited*. While a rigorous treatment of diffraction on the base of Maxwell's equations is mathematically quite involved (see, for example, Elmore and Heald [26, Chapters 9 and 10] or Iizuka [63, Chapter 3]), the special case of diffraction at the aperture of lenses can be treated in a simple way known as *Fraunhofer diffraction* and leads to a fundamental relation. We study the simplified case of an object at infinite distance imaged by a perfect optical system.

Then, the aperture of the optical system is pierced by a planar wave front (Fig. 4.15). The effect of a perfect lens is that it bends the planar wave front into a spherical wave front with its origin at the focal point in the optical axis. Diffraction at the finite aperture of the lens causes light also to go in other directions. In order to take this effect into account, we apply *Huygens' principle* at the aperture plane. This principle states that each point of the wave front can be taken as the origin of a new in-phase spherical wave. All these waves superimpose at the image plane to form an image of the incoming planar wave

$$r = \sqrt{x'^2 + y'^2 + f^2} \quad \text{and} \quad r' = \sqrt{(x' - x)^2 + (y' - y)^2 + f^2}. \tag{4.34}$$

Evaluating the difference between these two terms under the condition $x \ll f$ (neglecting quadratic terms in x and y) yields

$$r' - r \approx -\frac{xx' + yy'}{r}. \tag{4.35}$$

This path difference results in a phase difference of

$$\Delta\varphi = \frac{2\pi(r' - r)}{\lambda} = -\frac{2\pi(xx' + yy')}{\lambda} = -\frac{2\pi(\boldsymbol{xx'})}{\lambda}, \tag{4.36}$$

where λ is the wavelength of the wave front.

We further assume that $\psi'(x')$ is the amplitude distribution of the wave front at the aperture plane. In case of a simple aperture stop, $\psi'(x')$ is the simple box function, but we want to treat the more general case of a varying amplitude of the wave front or any type of aperture functions. Then the superimposition of all spherical waves at the image plane yields

$$\psi(x) = \int\limits_{-\infty}^{\infty} \int\limits_{-\infty}^{\infty} \psi'(x') \exp\left(-2\pi\mathrm{i}\frac{x'x}{f\lambda}\right) \mathrm{d}^2x'. \tag{4.37}$$

This equation means that the amplitude distribution at the focal plane $\psi(x)$ is simply the 2-D Fourier transform of the amplitude function $\psi'(x')$ at the aperture plane. This simple fundamental relation does not only describe diffraction-limited optical imaging but all well known diffraction phenomena such as diffraction at a slit, double slit, and grating. In the rest of this section, we apply the fundamental relation Eq. (4.37) to study the resolution of an optical system.

Example 4.1: Diffraction pattern of a lens with a circular aperture.

An optical system normally has a *circular aperture*. The amplitude distribution is then given by

$$\psi'(x') = \Pi\left(\frac{|x'|}{2r}\right), \tag{4.38}$$

where r is the radius of the aperture. The Fourier transform of Eq. (4.38) is given by the Bessel function of first order (Fig. 4.16a):

$$\psi(x) = \psi_0 \frac{I_1(2\pi xr/f\lambda)}{\pi xr/f\lambda}. \tag{4.39}$$

The irradiance distribution E on the image plane is given by the square of the amplitude (Fig. 4.16b):

$$E(x) = |\psi(x)|^2 = \psi_0^2 \left(\frac{I_1(2\pi xr/f\lambda)}{\pi xr/f\lambda}\right)^2. \tag{4.40}$$

The diffraction pattern has a central spot that contains 83.9% of the energy and encircling rings with decreasing intensity.

Example 4.2: Resolution of an optical system.

Two points imaged on the image plane can be resolved when their diffraction patterns still show two distinct peaks. The distance from the center of the disk to the first dark ring is

$$\Delta x' = 0.61 \cdot \frac{f}{r}\lambda = 1.22\lambda n_f. \tag{4.41}$$

At this distance, two points can clearly be separated (Fig. 4.17). This is the Rayleigh criterion for resolution of an optical system. The resolution of an optical system can be interpreted in terms of angular resolution and absolute resolution at the image or object plane. Taking the Rayleigh criterion Eq. (4.41), the angular resolution $\Delta\theta_0' = \Delta x/f$ is given as

$$\Delta\theta_0' = 1.22\frac{\lambda}{d}. \tag{4.42}$$

Thus, the *angular resolution* does not depend at all on the focal length but only the aperture of the optical system in relation to the wavelength of the electromagnetic radiation.

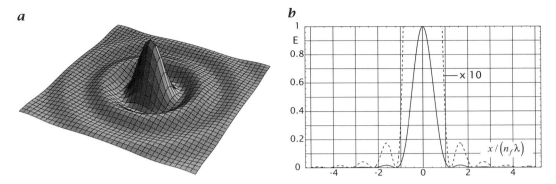

Figure 4.16: *"Airy disk", the diffraction pattern at the focal plane of a uniformly illuminated circular aperture imaged by a diffraction-limited optical system according to Eq. (4.39).* **a** *Wave amplitude ψ in a pseudo 3-D plot according to Eq. (4.39);* **b** *Radial cross section of the irradiance E distribution according to Eq. (4.40); the dashed line shows E amplified 10 times.*

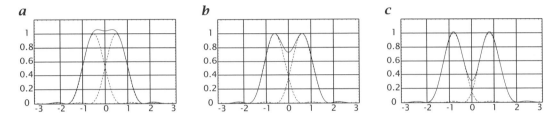

Figure 4.17: *Illustration of the resolution of the image of two points at distances $x/(n_f\lambda)$ of 1.0, 1.22, and 1.6.*

Conversely, the *absolute resolution* $\Delta x'$ Eq. (4.41) at the image plane depends only on the relation of the radius of the lens aperture to the distance f of the image of the object from the principal point. Under more general conditions (media with different indices of refraction), this ratio is the *numerical aperture* $n_a' = n \sin \theta_0'$, where n is the index of refraction and θ_0' the maximum angle under which the ray passes the lens aperture.

$$\Delta x' = 0.61\lambda/n_a'. \tag{4.43}$$

Therefore, the absolute resolution at the image plane does not at all depend again on the focal length of the system but only the numerical aperture of the image cone.

Since the light way can be reversed, the same arguments apply for the object plane. The absolute resolution of an object is given only by the numerical aperture of the object cone, i. e., the angle of the cone entering the lens aperture.

$$\Delta x = 0.61\lambda/n_a. \tag{4.44}$$

These simple relations are helpful to evaluate the maximum possible performance of optical systems. Since the maximum numerical aperture of optical systems is about one, no smaller structures than about half the wavelength can be resolved. Likewise, it is not possible to focus a parallel beam onto a spot size smaller than the same limit.

4.3.3b Gaussian Beams.

From the discussion in Section 4.3.3a, we can conclude that a beam passing through an optical system constantly changes not only its cross section but also its shape $\psi(X, Y, Z)$. At some points in the optical system, its shape is the Fourier transform of the shape at another position. Therefore, an amplitude distribution that does not change its form (except for scaling) when Fourier transformed

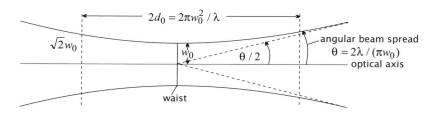

Figure 4.18: *Properties of a Gaussian beam within an amplitude distribution $\psi(x)$ = $\psi_0 \exp(-|x|^2/w^2)$.*

constitutes a special case: it keeps its form when traveling through optical systems provided that it is not truncated, i.e., the outer regions of the beam are not cut off at aperture stops. One function that does not change under the Fourier transform is the Gaussian $\psi(X, Y, Z) = \psi_0 \exp(-(X^2 + Y^2)^2/w(Z)^2)$ (Section B.3), where w is the radius of the beam.

With the advent of lasers, Gaussian beams became significant, since the shape of a laser beam is Gaussian in the fundamental TEM$_{00}$-mode. Lasers are often used for illumination purposes in imaging systems. Therefore, it is of practical importance to study their propagation. The diffractional spreading of a *Gaussian beam* is given by

$$w(Z)^2 = w_0^2 + \left(\frac{\lambda Z}{\pi w_0}\right)^2 = w_0^2 \left[1 + \left(\frac{\lambda Z}{\pi w_0^2}\right)^2\right]. \tag{4.45}$$

The Gaussian beam has a minimum diameter (called "waist") at $Z = 0$ of w_0 that is directly related to the angular spread at large Z (Fig. 4.18).

$$\theta \approx \frac{2w}{Z} = \frac{2\lambda}{\pi w_0} \qquad \text{or} \qquad w_0 = \frac{2\lambda}{\pi \theta} = \frac{2\lambda}{\pi n_a}, \tag{4.46}$$

where $n_a = \sin\theta$ is the numerical aperture of the Gaussian beam. Within a distance $d_0 = \pi w_0^2/\lambda = 4\lambda/(\pi\theta^2)$, the diameter of the Gaussian beam increases by a factor of $\sqrt{2}$ (the area increases by a factor of 2; then, intensity decreases by a factor of 2). Thus, a focused beam can be maintained only within a distance that is directly proportional to the waist diameter squared (or inversely proportional to the angular beam spread squared).

A diffraction-limited optical system translates a Gaussian beam with a waist w_1 at d in the following way:

$$\text{Position of new waist} \quad d'd\left(1 + \left(\frac{d_0}{d}\right)^2\right) = -f^2$$
$$\text{New waist size} \quad w_2^2/w_1^2 = d'/d \tag{4.47}$$

with $d_0 = \pi w_0^2/\lambda$. Equation (4.47) shows that the waist of a Gaussian beam is not imaged as objects Eq. (4.21). The new waist is formed at a distance d' which is $1 + (d_0/d)^2$ larger than the image of an object. This increase in distance is, however, only significant for small beam convergence, for which d_0 is large.

4.3.4 Radiometry of Imaging

It is not sufficient to know the geometry of imaging only. Equally important is to consider how the irradiance at the image plane is related to the radiance of the imaged objects and which parameters of an optical system influence this relationship. For a discussion of the fundamentals of radiometry, especially all terms describing the properties of radiation, we refer to Section 3.3.4.

The path of radiation from a light source to the image plane includes a chain of processes (see Fig. 3.1). The illumination path and the interaction of radiation with the object to be visualized are discussed in several sections of Chapter 3, generally in Section 3.2: interaction of radiation with matter at surfaces in Section 3.3.6 and volume-related in Section 3.3.7, various illumination techniques in Sections 3.4.3–3.4.5, and special techniques to image various object properties in Sections 3.4.8–3.4.9.

In this section, we concentrate on the observation path (compare Fig. 3.1), i. e., how the radiation emitted from the object to be imaged is collected by the imaging system.

4.3.4a Radiance Invariance. An optical system collects part of the radiation emitted by an object (Fig. 4.19). We assume that the object is a Lambertian radiator with the radiance L. The aperture of the optical system appears from the object under a certain solid angle Ω. The projected circular aperture area is $\pi r^2 \cos \theta$ at a distance $(d + f)/\cos \theta$. Therefore, a flux density $M = L\pi r^2 (\cos^2 \theta)/(d + f)^2$ enters the optical system. The radiation emitted from the projected area A is imaged onto the area $A' = A \cos^{-1} \theta$. We further assume that the optical system has a transmittance t. This leads finally to the following object radiance/image irradiance relation:

$$E' = t\pi \left(\frac{r}{f + d'} \right)^2 \cos^4 \theta \cdot L. \tag{4.48}$$

This fundamental relationship states that the image irradiance is proportional to the object radiance. The optical system is described by two simple terms: its (total) transmittance t and the ratio of the aperture radius to the distance of the image from the first principal point. For distant objects $d \gg f, d' \ll f$, Eq. (4.48) reduces to

$$E' = t\pi \frac{\cos^4 \theta}{n_f^2} \cdot L \qquad d \gg f, \tag{4.49}$$

using the f-number n_f. The irradiance falls off with $\cos^4 \theta$. For real optical systems, this term is only an approximation. If part of the entering beam is cut off by additional apertures or limited lens diameters, the fall off is even steeper at high angles θ. On the other side, a careful design of the position of the aperture can make the fall off less steep than $\cos^4 \theta$. Since also the residual reflectivity of the lens surfaces depends on the angle of incidence, the true fall off depends strongly on the design of the optical system and is best determined experimentally by a suitable calibration setup.

The astonishing fact that the image irradiance is so simply related to the object radiance has its cause in a fundamental invariance. An image has a radiance just as a real object. It can be taken as a source of radiation by further optical elements. A fundamental theorem of radiometry now states that the radiance of an image is equal to the radiance of the object times the transmittance of the optical system. The theorem can be proofed by the assumption that the radiative flux Φ through an optical system is preserved except for absorption in the system leading to a transmittance less than one. The solid angles under which the object and image see the optical system is

$$\Omega = A_0/(d + f)^2, \quad \Omega' = A_0/(d' + f)^2, \tag{4.50}$$

where A_0 is the effective area of the aperture. The flux emitted from an area A of the object is received by the area $A' = A(d + f)^2/(d' + f)^2$ on the image plane. Therefore,

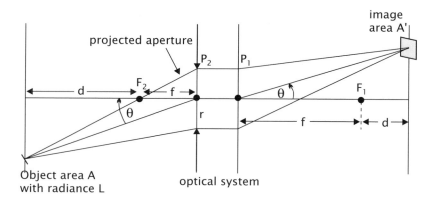

Figure 4.19: *An optical system receives a flux density that corresponds to the product of the radiance of the object and the solid angle under which the projected aperture is seen from the object. The flux emitted from the object area A is imaged onto the image area A'.*

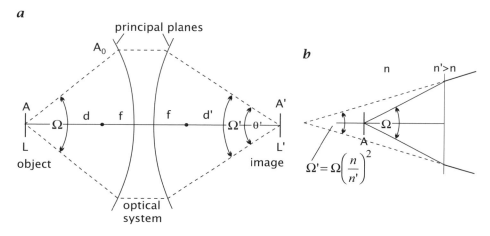

Figure 4.20: *Illustration of radiance invariance: **a** The product $A\Omega$ is the same in object and image space. Therefore, the radiance of the image is the same as the radiance of the object. **b** Change of solid angle, when a beam enters an optically denser medium.*

the radiances are

$$
\begin{aligned}
L_1 &= \frac{\phi}{\Omega A} = \frac{\Phi}{A_0 A}(d + f)^2 \\[2mm]
L' &= \frac{t\phi}{\Omega' A'} = \frac{t\Phi}{A_0 A}(d + f)^2
\end{aligned}
\tag{4.51}
$$

and the following invariance holds:

$$
L' = tL \qquad n' = n. \tag{4.52}
$$

The radiance invariance of this form is only valid if the object and image are in media of the same index of refraction. If a beam with radiance L enters into a medium with a higher index of refraction, the radiance increases since the rays are bent to the optical axis (Fig. 4.20b). In this case, the radiance divided by the squared index of refraction is invariant:

$$
L'/n'^2 = tL/n^2. \tag{4.53}
$$

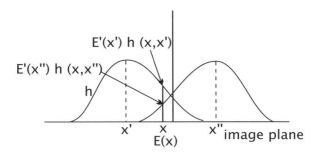

Figure 4.21: *Illustration of the linearity of the image formation process. The wave amplitude E'(x) at a point on the image plane results from the integration of the irradiance contributions by all points x'. The contribution of a single point is given by E'(x')h'(x,x').*

The fundamental importance of the radiance invariance can be compared to the principles in geometrical optics that radiation propagates in such a way that the optical path nd (real path times the index of refraction) takes an extreme value.

4.3.4b Irradiance on the Image Plane. From the radiance invariance, we can immediately infer the *irradiance* on the image plane to be

$$E' = \pi L' \sin^2 \theta' = tL\pi \sin^2 \theta'. \tag{4.54}$$

This equation does not consider the fall off with $\cos^4 \theta$ because we did not consider oblique rays. The term $\sin^2 \theta$ corresponds to $r^2/(f + d')^2$ in Eq. (4.48). The radiance invariance considerably simplifies computation of image irradiance and the propagation of radiation through complex optical systems. We only need to know the radiance of the object and the image-sided f-number (or numerical aperture) to compute the image irradiance. The image irradiance is always lower than the object irradiance.

Example 4.3: Image irradiance of a Lambertian object.

As an example we take a Lambertian object with the irradiance E. The radiance of the surface with the reflection coefficient ρ is according to Section 3.3.6d Eq. (3.43)

$$L = \frac{\rho}{\pi}E. \tag{4.55}$$

Using Eq. (4.54) the image irradiance E' then is

$$E' = \rho \cdot t \cdot \sin^2 \theta \cdot E. \tag{4.56}$$

4.3.5 Linear System Theory

4.3.5a Point Spread Function. The formation of an image from an object using an optical system is a linear process (Section 3.3.1b) and can therefore be analyzed using *linear system theory*. The image of any type of radiance distribution can be generated by superimposing it from the images of each point. Therefore, it is sufficient to know how a point source is imaged by an optical system.

We denote $E'(\boldsymbol{x'})$ as the irradiance produced by a perfect optical system with an exact point to point correspondence. Then, the image formation process can be treated by decomposing it into single points as illustrated in Fig. 4.21. The irradiance at the point $\boldsymbol{x'}$ is spread out to an irradiance distribution described by $h(\boldsymbol{x},\boldsymbol{x'})$. Therefore, it contributes to the irradiance $E'(x')h(x, x')$ at the point x. Then, the irradiance at the point x is given by integrating the contributions from all points x':

$$E(x) = \int E'(x')h(x, x')\mathrm{d}^2x'. \tag{4.57}$$

The intensity distribution function $h(x, x')$ is known as the *point spread function* of the optical system. Generally — and actually in most real optical systems — the point spread function explicitly depends on the position at the image plane. The aberrations increase with the distance from the optical axis. If, however, the point spread function does not explicitly depend on the position at the image plane, then the integral in Eq. (4.57) becomes significantly simpler since $h(x, x')$ can be replaced by $h(x - x')$:

$$E(x) = \int E'(x')h'(x - x')\mathrm{d}^2x' = (E' * h)(x). \tag{4.58}$$

Such a system is known as a *homogeneous* or *shift-invariant* system and the operation in Eq. (4.58) as a *convolution*. The performance of a *linear shift-invariant* (*LSI*) system is entirely described by the *point spread function* (PSF) $h(x)$.

For optical systems, the concept of the PSF can be extended to three dimensions. The *3-D PSF* describes not only how the image is formed at the image plane but at all distances in the image space. The 3-D PSF is significant for 3-D imaging since it describes how a point in 3-D object space is mapped onto a 3-D image space. Then, in the integral not only object points at the object plane but from the whole object space are considered to contribute to the integral in Eq. (4.58). The two-dimensional vectors are to be replaced by three-dimensional vectors.

There are two cases in which the 3-D PSF can be expressed by simple equations. In the limit of geometrical optics and for Gaussian beams (Section 4.3.3b) the PSF is

$$
\begin{aligned}
h'(x) &= \frac{M}{\pi(mn_az)^2}\Pi\left(\frac{r}{2mn_az}\right) && \text{geometrical optics} \\
h'(x) &= \frac{M}{\pi(mn_az)^2}\exp\left(\frac{-2r^2}{[\lambda/(\pi mn_a)]^2 + (mn_az)^2}\right) && \text{Gaussian beam,}
\end{aligned}
\tag{4.59}
$$

where M is the radiant flux density collected by the optical system from the object point, m the magnification, and n_a the object-sided numerical aperture. The image coordinates are written in cylinder coordinates, $x = (r, \phi, z)$ taking the rotational symmetry around the z-axis into account.

In the limit of geometrical optics, the 3-D PSF is simply a double cone. Inside the cone, the irradiance decreases with the squared distance from the image plane. Wave optics adds the important fact that at z = 0 the size of the image becomes not zero but does not decrease below a lower limit related to the wavelength and numerical aperture of the optics. Only in the case of a Gaussian beam, the transversal irradiance distribution is invariant.

The PSF can be projected back to the object space in order to relate the effects of the image formation to the true object sizes:

$$
\begin{aligned}
h(X) &= \frac{M}{\pi(n_aZ)^2}\Pi\left(\frac{R}{2n_aZ}\right) && \text{geometrical optics} \\
h(X) &= \frac{M}{\pi(n_aZ)^2}\exp\left(\frac{-2R^2}{[\lambda/(\pi n_a)]^2 + (n_aZ)^2}\right) && \text{Gaussian beam.}
\end{aligned}
\tag{4.60}
$$

The double cone has an opening angle given by the numerical aperture of the optical system.

4.3.5b   Optical Transfer Function.   The *optical transfer function (OTF)* is the Fourier transform of the point spread function and as such describes the optical system as well. The OTF says how a spatially periodic pattern is transformed by the image formation process. Convolution in the spatial domain with the PSF corresponds to multiplication with the OTF in Fourier space,

$$
\begin{aligned}
E'(X) &= h(X) & * & \quad E(X) & \text{spatial domain} \\
\hat{E}'(X) &= \hat{h}(X) & \cdot & \quad \hat{E}(X) & \text{Fourier domain.}
\end{aligned}
\tag{4.61}
$$

Since the OTF is complex-valued, both the amplitude and phase of the spatially periodic pattern can be changed. In the limit of geometrical optics, the 3-D OTF is given by

$$
\hat{h}(k) = \hat{h}(q, \bar{k}_3) = \frac{2M}{\pi |q n_a|} \left(1 - \frac{\bar{k}_3^2}{q^2 n_a^2} \right)^{1/2} \Pi \left(\frac{\bar{k}_3}{2 q n_a} \right)
\tag{4.62}
$$

with $q = (\bar{k}_1^2 + \bar{k}_2^2)^{1/2}$.

4.3.5c   Interpretation of the 3-D OTF.   A large part of the OTF is zero. Spatial structures with wave numbers in this region completely disappear and cannot be reconstructed without additional knowledge. This is particularly the case for all structures in the z direction, i. e., perpendicularly to the image plane.

3-D structures can only be seen if they also contain structures parallel to the image plane. It is, for example, possible to resolve points or lines which lie above each other. We can explain this in the spatial as well as in the Fourier domain. The PSF blurs the points and lines, but they can still be distinguished if they are not too close to each other. Points and lines correspond to a constant function and a plane, respectively, in Fourier space. Such extended objects partly intersect with the nonzero parts of the OTF and thus will not vanish entirely. Periodic structures up to an angle of α to the $k_1 k_2$ plane, which just corresponds to the opening angle of the lens, are not eliminated by the OTF. Intuitively, we can say that we are able to recognize all 3-D structures in which we actually can look into. We need at least one ray which is perpendicular to the wave number k of the structure.

From Eq. (4.62), we can conclude that the OTF is inversely proportional to the radial wave number q. Consequently, the contrast of a periodic structure is attenuated proportionally to its wave number. Since this property of the OTF is valid for all optical imaging, and thus also for the human visual system, the question arises why we can see fine structures at all.

The answer lies in a closer examination of the geometrical structure of the objects observed. Natural objects are generally opaque. Consequently, we only see their surfaces. If we image a 2-D surface onto a 2-D image plane, the PSF also reduces to a 2-D function. Mathematically, this means a multiplication of the PSF with a δ plane parallel to the observed surface. Consequently, the 2-D PSF is now given by the unsharpness disk corresponding to the distance of the surface from the lens. The convolution with the 2-D PSF preserves the intensity of all structures with wavelengths larger than the disk.

We arrive at the same conclusion in Fourier space. Multiplication of the 3-D PSF with a δ plane in the x space corresponds to a convolution of the 3-D OTF with a δ line perpendicular to the plane, i. e., an integration in the corresponding direction. If we integrate the 3-D OTF along the $\bar{k}_3$ coordinate, we actually get a constant independent

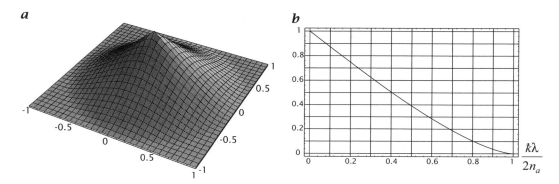

Figure 4.22: *Optical transfer function of a diffraction-limited optical system with a uniformly illuminated circular aperture. The wave number is normalized to the wave number at which the OTF vanishes $|\boldsymbol{k}|\lambda/(2n_a)$.* **a** *Pseudo 3-D plot,* **b** *radial cross section.*

of the radial wave number q:

$$\frac{2M}{\pi} \int_{-\bar{q}n_a}^{\bar{q}n_a} \frac{1}{|qn_a|} \left[1 - \frac{\bar{k}_3^2}{q^2 n_a^2} \right]^{1/2} d\bar{k}_3 = M. \tag{4.63}$$

In conclusion, the OTF for surface structures is independent of the wave number. However, for volumetric structures as they occur with transparent objects, the OTF decreases with the radial wave number. The human visual system and all optical systems that map a 3-D scene onto a 2-D image plane are not built for true 3-D imaging!

4.3.5d 2-D OTF of a Diffraction-Limited Optical System. The 2-D OTF for a diffraction-limited optical system with a circular aperture that is uniformly illuminated is given as the 2-D Fourier transform of the irradiance distribution on the image plane, the Airy disk as given by Eq. (4.40). The result is

$$\hat{h}(\boldsymbol{k}) = \left[\frac{2}{\pi} \arccos \left(|\boldsymbol{k}| \frac{\lambda}{2n_a} \right) - \frac{|\boldsymbol{k}|\lambda}{2n_a} \left(1 - \left(\frac{|\boldsymbol{k}|\lambda}{2n_a} \right)^2 \right)^{1/2} \right] \Pi \left(|\boldsymbol{k}| \frac{\lambda}{4n_a} \right) \tag{4.64}$$

where n_a is the numerical aperture of the optical system. The OTF is decreasing monotonously with the wave number (Fig. 4.22). Beyond a wave number $|\boldsymbol{k}| \geq 2n_a/\lambda$ it vanishes. This means that the smallest periodic structure that gives a nonzero irradiance variation has a wavelength of $\lambda/(2n_a)$

$$\lambda_{min} = \frac{\lambda}{2n_a} = \lambda n_f. \tag{4.65}$$

Depending on whether the numerical aperture of the optical system is referred to the object side or the image side, Eq. (4.64) refers to the real scales or the scales at the image plane, respectively. The minimum resolvable periodic structure is just a little bit smaller than the separation distance to resolve two points according to the Rayleigh criterion Eqs. (4.41) and (4.44).

4.3.5e Modulation Transfer Function. The term *modulation transfer function* (MTF) stems from the practical measurements of the resolution of optical systems. A sinusoidal pattern with an amplitude a is imaged and the amplitude in the image a' is measured. The ratio a'/a is denoted as the MTF. Thus, the MTF is just the absolute value of the OTF since the measurement of the MTF does only consider the amplitudes and no phase shifts. Negative values of the OTF (i.e., a phase shift of π) mean an inversion of the contrast. A radiance maximum is imaged onto an irradiance minimum and vice versa,

$$MTF = |OTF| = |\hat{h}(\boldsymbol{k})|. \tag{4.66}$$

In many practical cases, e.g., a well-focused diffraction limited optical system, the OTF has only positive values so that OTF and MTF are identical.

4.3.5f Coherent Versus Incoherent Imaging. So far we have considered the PSF and OTF of incoherent optical systems. For such systems, phase relations are not important and we can simply work with intensities. For coherent optical systems, however, wave amplitudes can interfere constructively and destructively. The difference between coherent and incoherent imaging can be best seen if the PSF is defined with respect to the wave amplitudes and not the intensities. Then, the irradiance E for coherent and incoherent imaging is given by

$$
\begin{aligned}
E(\boldsymbol{x}) &= |\psi * h_\psi|^2 && \text{coherent imaging} \\
E(\boldsymbol{x}) &= |\psi|^2 * |h_\psi|^2 = E' * h && \text{incoherent imaging,}
\end{aligned}
\tag{4.67}
$$

where $h = |h_\psi|^2$ and ψ is the wave amplitude function. This means that for coherent imaging the convolution with the point spread function has to be performed with the wave amplitudes and not the intensity (irradiance).

4.4 Procedures

In this section, we discuss how to set up imaging systems in an optimum way for a specific task. The first two sections deal with general practical aspects of the geometry and radiometry of imaging systems (Sections 4.4.1 and 4.4.1b). Then, a number of special imaging systems are discussed:

- telecentric imaging for optical gauging (Section 4.4.1c)
- stereo imaging (Section 4.4.2)
- confocal laser scanning microscopy (Section 4.4.3)
- tomography (Section 4.4.4)

4.4.1 Geometry of Imaging

If an object should be imaged, its size is given and the optical system must be able to include it in the imaged sector. But we can vary all other parameters of the geometry of optical systems in order to choose an optimal setup under other possible constraints. The following parameters can be varied:

- focal length
- distance of camera to object
- perspective (field of view)
- magnification
- size of image

Not all of these parameters can, of course, be chosen independently. They are interrelated by the laws of geometrical optics as discussed in Section 4.3.2.

If the experimental setup, for instance, requires a certain distance of the camera from the object, only one degree of freedom is left. We can either choose the focal length or the size of the image; all other parameters including the perspective and magnification are then fixed.

We often have no choice of the image size since the type of camera used for the application dictates the size of the images that are available. In this case, we can only choose either the focal length or the distance of the camera from the object. Then, all other parameters are fixed.

Figure 4.23 illustrates the relation between object distance and size of the imaged object for image sizes of 12.8, 8.8, 6.4, and 4.8 mm, the nominal horizontal size of 1", 2/3", 1/2", and 1/3" CCD chips (Section 5.3.6). The graphs show these relations for a standard range of focal lengths between 6 mm and 200 mm. Two different distances are shown. The solid curves show the distance between the object and image plane. This distance has a minimum for a magnification of one. The distance does not include the distance between the two principal planes of the optical system. Thus, the graphs are only valid for thin lenses. If you know, however, the distance of the two principal planes (look up the technical data sheet or ask the manufacturer of the lens), simply add this distance to the distance looked up in the graph or use the following equation

$$d_{oi} = f \left(\frac{X}{x} + \frac{x}{X} + 2 \right) + d_p, \tag{4.68}$$

where X is the object size, x the image size, and d_p the distance between the two principal planes of the optical system.

The dashed lines in Fig. 4.23 show the distance from the object to the first principal plane of the optical system given by

$$d_{op} = f \left(\frac{X}{x} + 1 \right). \tag{4.69}$$

This is approximately the distance to the object sided front surface of the lens (Fig. 4.8). More precisely, you need to know the distance of the first principal point to the front surface of the lens. Again, all good optical manufacturers list these numbers in their data sheets.

4.4.1a Depth of Field. After you have determined the object distance, you need to know the depth of field of the system. This is required to ensure that the distance range required by your setup lies within the depth of field. Otherwise, you would get partly blurred images. Figure 4.10 shows the depth of field as a function of the magnification according to Eq. (4.28) in Section 4.3.2d:

$$\Delta X_3 \approx n_f \frac{1 - m}{m^2} \epsilon. \tag{4.70}$$

This approximation is valid only as long as the depth of field is small as compared to the object distance and the blur is larger than that caused by diffraction. The dashed lines in Fig. 4.24 indicate the latter limit for wavelengths of 0.4 and 1.0 μm. In the shaded areas, the diffraction-limited resolution according to the Rayleigh criterion (Section 4.3.5d) is less than the blur circle diameter. Thus, it is not recommended to use the parameters in this area: the magnification is too high (*dead magnification*). The blur circle radii (11 μm and 6.5 μm) have been selected to be equal to the typical pixel size of 2/3" and 1/3" CCD chips (Fig. 4.24).

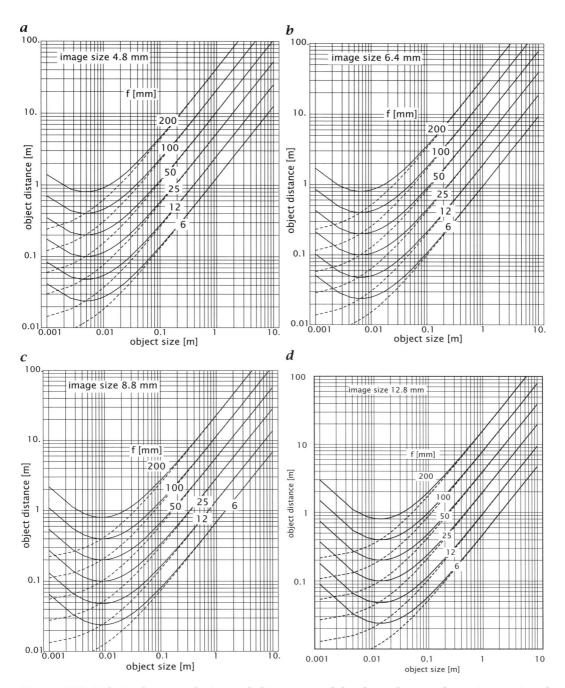

Figure 4.23: *Relation between the imaged object size and the object distance for an image size of* ***a*** *4.8 mm (nominal long side of a 1/3" CCD imager),* ***b*** *6.4 mm (nominal long side of a 1/2" CCD imager),* ***c*** *8.8 mm (nominal long side of a 2/3" CCD imager), and* ***d*** *12.8 mm (nominal long side of a 1" CCD imager). The parameter is the focal length of the lens. The solid lines refer to the distance between the object and image plane, the dashed line to the distance between the object plane and the first principal point of the optical system.*

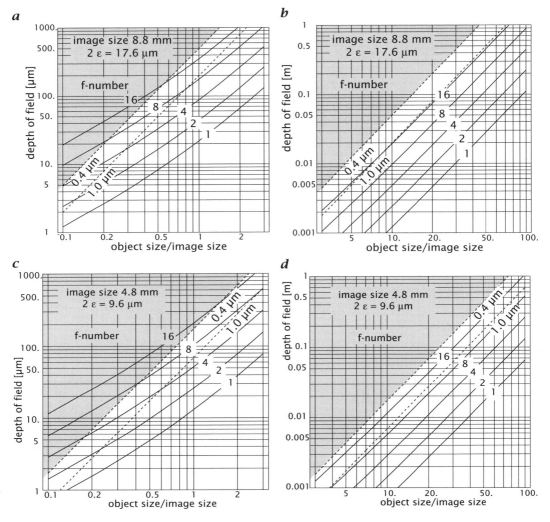

Figure 4.24: *Depth of field as a function of the object size/image size ratio (inverse magnification m^{-1}) for f-numbers from 1 to 16; **a** and **b** blur disk diameter of $17.6\,\mu m$, image size of $8.8\,mm$; **c** and **d** blur disk diameter of $9.6\,\mu m$, image size of $4.8\,mm$; **a** and **c** object/image size range from 0.1 to 3; **b** and **d** object/image size range from 3 to 100; the range of dead magnifications is shaded.*

The graphs illustrate that the depth of field — within the limits discussed above — depends only on the magnification and the f-number. If a certain object size has to be imaged onto a certain image size, the focal length has no influence.

Example 4.4:

An object of 200 mm in size has to be imaged onto a CCD chip. For a 2/3" and 1/3" CCD chip, we find that the magnification is 8.8 mm/200 mm = 1/23 and 4.8 mm/200 mm = 1/42, respectively. From Fig. 4.10 or directly from Eq. (4.70), we find that the depth of field for an f-number of 8 is ± 5 cm and ± 9 cm, respectively. As expected, the smaller image results in a higher depth of field.

In the macro range (magnification around one) and micro range (magnification larger than one), the depth of field becomes very low. It is about 100 μm for an f-number of 8

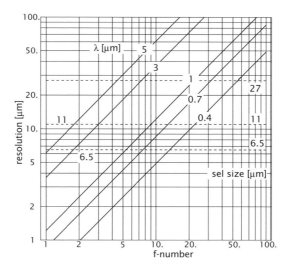

Figure 4.25: *Resolution of a diffraction-limited optical system as a function of the f-number (ranging from 1 to 100) for wavelength of 0.4, 0.7, 1.0, 3, and 5 μm. The horizontal dashed lines mark the typical sensor element (sel) sizes for a 1/3″ CCD (6.5 μm), a 2/3″ CCD (11 μm), and an InAs infrared focal plane area (27 μm). Above these lines, the diffraction-limited resolution is less than the sel size.*

at magnification one and about 2.5 μm at magnification 10 and an f-number of 2. Note that in most of the micro ranges, the simple estimates based on Eq. (4.70) are not valid (shaded areas).

4.4.1b Radiometry and Photometry of Imaging. How much or whether sufficient radiant flux at all can be collected by an optical system is a basic radiometric question. The principles of radiometry are discussed in Section 4.3.4. Here, we turn to the questions related to the practical setup. According to Eq. (4.49), the image irradiance E can be approximated by

$$E \approx t\pi \frac{L}{n_f^2},$$
(4.71)

where t is the transmittance of the optical system and L the object radiance. This equation reflects the radiance invariance (Section 4.3.4a).

The other important thing to know is that a radiation detector needs a certain *irradiation H*

$$H = \int E dt \approx E\Delta t$$
(4.72)

to respond with a sufficient signal. We take here as a good mean value $40 \, \text{nJ/cm}^2$ to obtain the maximum (saturation signal) from an imaging sensor. The minimum detectable exposure level depends on the quality of the sensor. It ranges from about $40 \, \text{pJ/cm}^2$ for standard sensors down to $0.4 \, \text{pJ/cm}^2$ for scientific-grade cooled sensors. Using the saturation irradiation H, the saturation irradiance E_s is given by

$$E_s = H_s / \Delta t$$
(4.73)

and the saturation radiance by

$$L = \frac{n_f^2}{\pi} \frac{1}{t} \frac{H_s}{\Delta t}.$$
(4.74)

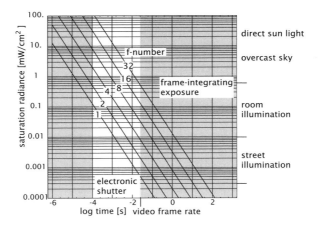

Figure 4.26: *Exposure time as a function of the radiance of the object for a sensor with a saturation exposure of $40\,nJ/cm^2$ (full sensor signal). The value represents the most sensitive visible imaging sensors well. If you know the saturation exposure of your sensor, you can correct the exposure values correspondingly. Typical values for object radiances are marked in the graph under the assumption that it is a Lambertian object with reflectivity one.*

Figure 4.27: *Telecentric lenses of the VISIONMES series offered by Carl Zeiss Jena GmbH. The f-number of these lenses is 11 and the polychromatic MTF is close to the diffraction limit. The largest lens in the figure, for example, can image objects up to 300 nm in diameter. Within a depth range of ± 75 mm, the object size varies less than 1 μm. Image courtesy of Carl Zeiss Jena GmbH, Germany.*

The required exposure time as a function of the object radiance and the f-number of the optical system is shown in Fig. 4.26. Note that this figure is only good for a rough estimate.

4.4.1c Telecentric Imaging for Optical Gauging.

Telecentric imaging has become a central tool for optical gauging, since the size of the imaged object in first order does not change with the distance from the lens (Section 4.3.2e). In a telecentric system, the principal rays enter the optical system parallel to the optical axis. Thus, the lens aperture has to be at least as large as the object diameter. Thus, telecentric imaging

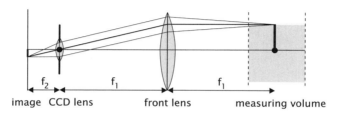

Figure 4.28: *Construction of a telecentric imaging system using a standard CCD lens and a large front lens.*

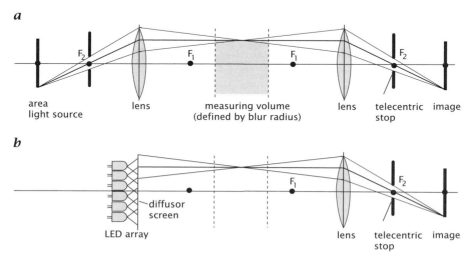

Figure 4.29: *Telecentric imaging system with a matched telecentric light field illumination system to measure gas bubbles in water by a light blocking technique; **a** with area light source; **b** with LED array light source.*

is limited to small objects and becomes increasingly expensive for large objects. Telecentric lenses are offered nowadays by all major optical companies. Figure 4.27 shows as an example a collection of telecentric lenses of the VISIONMES series offered by Carl Zeiss Jena GmbH.

It is easy to build a telecentric optical system from a standard CCD lens and a large front lens that has a sufficient performance for many applications (Fig. 4.28). The CCD lens is focused to infinity and the large second lens is put at a distance of one focal length from the object side principle plane of the CCD lens. Then the light bundle that is focused on a pixel at the image plane is parallel between the two lenses and has a diameter corresponding to the setting of the diaphragm of the CCD lens. Because the light bundle passes through the focal point at the image side, it becomes parallel to the optical axis at the object side. Because it is a parallel beam at the image side of the lens, it is focused at the object sided focal point.

Thus the focal length of the front lens determines both the length of the telecentric system and the working distance. The maximum object size to be imaged is the diameter of the front lens minus the diameter of the diaphragm of the CCD lens. Depending on the required quality, the front lens can either be a simple planoconvex lens or an achromatic lens.

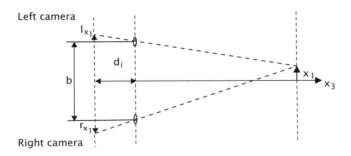

Figure 4.30: A simple stereo camera setup with parallel optical axes.

As with any imaging system, it is also important for a telecentric setup to match the illumination system carefully. As an example, Fig. 4.29 shows a do-it-yourself telecentric system built from readily available optical components to measure small particles, for instance bubbles in a light field (for an application, see Section 1.3.2). In a telecentric system with light field illumination, the illumination source should emit also parallel principal rays with a divergence (numerical aperture) matching that of the lens. If the illumination source has a smaller divergence, this divergence determines the effective f-number of the system and not that of the lens. The principles of telecentric illumination are discussed in Section 3.4.5a.

A cheap solution for a telecentric illumination system is an array of LEDs with an additional diffuser to homogenize the irradiance of the light source. The total divergence of this system is given by the angular dispersion of the LED and the diffuser.

4.4.2 Stereo Imaging

4.4.2a Stereo Setup with Parallel Camera Axes. Observation of a scene from two different points of view allows the distance of objects to be determined. A setup with two imaging sensors is called a *stereo system*. In this way, many biological visual systems perform depth perception. Figure 4.30 illustrates how the depth can be determined from a stereo camera setup. Two cameras are placed close to each other with parallel optical axes. The distance vector b between the two optical systems is called the *stereoscopic basis*.

An object will be projected onto different positions of the image plane because it is viewed under slightly different angles. The difference in the position is denoted as the *parallax* p. It is easily calculated from Fig. 4.30:

$$p = ({}^r x_1 - {}^l x_1, 0, 0) = \left(d_i \frac{X_1 + b/2}{X_3} - d_i \frac{X_1 - b/2}{X_3}, 0, 0 \right) = b \frac{d_i}{X_3}. \tag{4.75}$$

The parallax is inversely proportional to the distance X_3 of the object (zero for an object at infinity) and is directly proportional to the stereoscopic basis and the focal length of the cameras ($d_i \approx f$ for distant objects). In this simple setup, the parallax is a vector parallel to the stereoscopic basis b.

4.4.2b Stereo Setup with Verging Camera Axes. In order to gain maximum image overlap, often a stereo system with verging camera axes is used (Fig. 4.31). Its geometry is discussed here under the assumption that the size of the imaged object is much smaller than the distance of the cameras to the object. Therefore, we will compute the resolution only in the center of the image. The following computations are performed

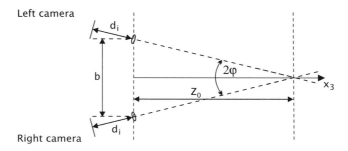

Figure 4.31: *A simple stereo camera setup with verging optical axes.*

to yield the height resolution in relation to the parameters of the stereo setup and the horizontal resolution in the images.

Let the two cameras be aligned such that the optical axes intersect each other in a point in the object space that is a distance Z_0 away from the stereo base line. Then the position difference, the parallax p, for a point at the distance $Z_0 - \Delta Z$ is given by

$$p = \frac{d_i \Delta Z}{Z_0 - \Delta Z \cos^2 \varphi} 2 \sin(\varphi) \cos(\varphi), \qquad (4.76)$$

where d_i is the distance of the image plane from the lens (the focal length) and φ the verging angle of the cameras. For $\Delta Z \ll Z_0$, the parallax p is directly proportional to ΔZ. This approximation is sufficient for the further computations of the height resolution. The relative height resolution can then be expressed as

$$\frac{\Delta Z}{Z_0} = \frac{p}{2 d_i \sin(\varphi) \cos(\varphi)}. \qquad (4.77)$$

A useful relation between the height resolution and horizontal resolution (parallel to the image plane) can be derived by projecting a horizontal line element at the object surface, ΔX, onto the oblique image plane

$$\frac{\Delta x}{d'} = -\frac{\Delta X \cos^2(\varphi)}{Z_0}. \qquad (4.78)$$

This equation relates the horizontal resolution at the object plane, ΔX, to the resolution at the image plane, Δx. Assuming that p can be determined as a fraction μ of the image resolution: $p = \mu \Delta x$, we obtain from Eqs. (4.77) and (4.78)

$$\frac{\Delta Z}{\Delta X} = \frac{p}{\Delta x} \frac{1}{2 \tan(\varphi)} = \mu \frac{Z_0}{b}. \qquad (4.79)$$

This equation shows that the ratio of the distance to the horizontal resolution is directly given by the distance to stereo basis ratio except for the factor μ. The larger this ratio is, however, the more skewed the cameras are and the less the two images overlap. Therefore, there is a practical upper limit for b/Z_0. Values between 0.3 and 0.75 are typically chosen for stereo imaging with maximum depth resolution.

4.4.2c Case Study Stereo Imaging of Ocean Surface Waves. Sometimes, you can learn more from an unsuccessful application because an analysis of the failure often leads to new techniques. In this exemplary case study for stereo imaging, we explain why the attempts of the oceanographic community to measure the wind-driven ripples

Table 4.2: *Summary of the characteristics of the stereo camera setup used in different investigations to measure the height of ocean waves.*

Reference	Height Z_0 [m]	Base b [m]	b/Z_0	Size [m × m]	z-Res. [mm]	x/y-Res. [mm]
Kohlschütter [82]	5.3	4	0.75	n/a	n/a	n/a
Schuhmacher [127]	13.54	6	0.44	150 × 175	?	?
Schuhmacher [127]	24	14.77	0.61	?	?	?
et al. [28]	1000	600	0.6	900 × x600	150	600
Dobson [21]	6	n/a	n/a	3.6 × 3.6	0.75	n/a
Holthuijsen [60]	75-450	35-220	≈0.5	54-220	n/a	n/a
Shemdin et al. [133]	9	5.0	0.4	2.5 × 2.5	3	8.5
Banner et al. [1]	5-10	2.0/3.2	0.3/0.4	2.0 × 2.0	1	?

on the ocean surface failed. At first glance, the water surface has one geometric feature which makes it an ideal target for stereo imaging: except for wave breaking, the water surface is continuous. This makes it easy to apply any kind of smoothness regularization without worrying about the difficult problems of discontinuities and occlusions. Thus it is relatively easy to compute the large-scale structures in stereo images from the water surface. The very same feature makes it, however, difficult to measure the small-scale waves: the smaller the wavelength, the smaller the wave height is. Let us make a more quantitative estimate. We assume a wave of sinusoidal shape propagating in X direction:

$$a = a_0 \sin(kX - \omega t), \tag{4.80}$$

where ω is the circular frequency of the wave, and k the wave number. Waves are limited in their slope. If the wave slope becomes too high, they become unstable. Thus it makes sense to relate the wave height to the wave slope s, which is given by spatial derivation from Eq. (4.80)

$$s = a_0 k \cos(kX - \omega t) = s_0 \cos(kX - \omega t), \tag{4.81}$$

and the wavelength $\lambda = 2\pi/k$:

$$a_0 = \frac{s_0 \lambda}{2\pi}. \tag{4.82}$$

Now we can compare the wave height with the achievable height resolution according to Eq. (4.79). To that end, we set the wavelength to be n times the horizontal resolution ΔX. Then the ratio of the height resolution ΔZ to the wave amplitude a_0 is given by

$$\frac{\Delta Z}{a_0} = \mu \frac{Z}{b} \frac{2\pi}{n s_0}. \tag{4.83}$$

This equation clearly indicates that it will be impossible to measure the height of those waves which still can be resolved horizontally. In figures, let us assume that the smallest wavelength in the stereo image is about 2π times sampled per wavelength (significant oversampling). According to Table 4.2, a typical figure for the base/distance ratio of the stereo setup used to measure ocean waves is 0.5. Even if the parallax could be measured with an accuracy of a 1/10 pixel ($\mu = 0.1$), the height resolution would be just equal to the amplitude of a wave with slope 0.2.

In conclusion, Eq. (4.83) provides the basis for an estimate of the smallest waves that can be resolved with any stereo technique if we set the error ΔZ equal to the wave amplitude a_0. Then we obtain

$$n = \frac{\lambda}{\Delta X} = \mu \frac{Z}{b} \frac{2\pi}{\Delta s_0}, \tag{4.84}$$

where Δs_0 means the required slope resolution. Let us assume that we want to resolve the wave slope with an accuracy of 0.02. This is about 10 % of a typical wave slope of 0.2. We further assume a stereo parallax accuracy of 0.1 pixels. Even with these optimistic assumptions, the smallest wavelength that could be resolved with this accuracy is about 60 pixels long according to Eq. (4.84). With a typical video resolution of 512 pixels this leaves less than a decade of wavelengths which can be measured with video images even under optimum conditions. Using a non-subpixel accurate stereo algorithm would be of no use at all, since it could just resolve the largest waves in the image with a slope accuracy of 0.02. Thus high-resolution stereo imagery is required to gain sufficient resolution.

Shemdin and Tran [132] claimed a much higher resolution. They analyzed stereo photographs which resulted in a 256×256 height map with 1.6 cm horizontal resolution. A correlation error of 1.6 mm results in a 0.3 cm error in the height estimate for a stereo base/distance relation of 0.53. Using the above estimate, it can be estimated that only waves longer than 30 cm can be resolved adequately. Actually, this estimate compares well with the smallest "wave-like" structure which can be seen in the height contour map in color plate 1 of Shemdin and Tran's [1992] paper. The smallest wave-like feature is about 1/10 of the size of the $4.1 \times 4.1 \, \text{m}^2$ map, or about 40 cm.

In sharp contrast to this estimate, Shemdin claims the resolution of waves with a wavelength of 10 cm and, if a low-pass filtering is used, even shorter waves, down to 4 cm wavelength, could be resolved. From the estimates above, it is evident that waves with 4 and 10 cm cannot be resolved. A height resolution of 0.3 cm translates (for linear waves) into a slope resolution of 0.47 and 0.19 for waves with 4 and 10 cm wavelength, respectively. Since the stereo images were taken at a low wind speed of 2 m/s, the actual slopes of these waves were certainly smaller.

In conclusion, we can state that stereo photography is a quite limited tool. The limited height resolution alone severely restricts its application. Banner et al. [1] were careful enough to limit their analysis to larger scale waves. From all the other investigations listed in Table 4.2 hardly any useful results emerged.

Besides the resolution problem, any stereo wave imaging system which uses natural illumination and thus is based on light reflection is plagued by the even more difficult *correspondence problem*. With this term we refer to the problem that a feature seen in one of the images must not necessarily appear in the other image. This leads to areas in the images where no parallax can be determined. This problem is very well known in conventional stereo photography where it occurs due to occlusions. For water surfaces, the correspondence problem is ubiquitous due to the specular reflection of light. A patch on the water surface observed from two different directions, as it is done with a stereo camera setup, receives light for each camera from a different direction and thus generally shows a different irradiance. This problem will be illustrated with several case studies.

First, we take a flat patch on the water surface. It maps different parts of the sky onto the two images. The stereo correspondence algorithm will then try to match the features seen in the sky (and not features fixed at the water surface) and consequently will not yield the distance to the water surface but the distance to the sky.

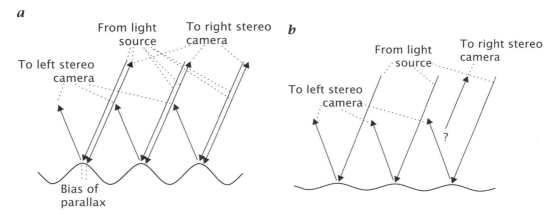

Figure 4.32: *Illustration of the stereo correspondence problem when illumination changes occur parallel to the stereo basis with a sinusoidal wave:* **a** *steep wave: bias in the parallax estimate* **b** *less steep wave: missing correspondence.*

As a second case, we assume that the sky illumination varies only in a direction perpendicular to the stereo base. Then both cameras receive the same irradiance. However, both cameras also receive irradiance changes only if the wave slope perpendicular to the stereo base is changing. Therefore, this ideal case for stereo photography has then the same disadvantages as the *Stilwell photography* (Section 3.4.5c). Waves traveling in the direction of the stereo basis remain invisible. Depending on the noise level in the images, waves traveling in a certain angular range around the direction of the stereo base therefore cannot be detected.

As a third case, we consider a change in the sky radiance in the direction of the stereo base. Figure 4.32a illustrates such a situation. Radiance from the same point in the sky will be received by the left camera reflected from the crest of the wave, while it will be received by the right camera reflected from a position of the wave where the slope is equal to the inclination of this camera. This effect biases the stereo correlation. This bias depends on both the wavelength of the wave and its steepness. The bias is small for steep waves, i.e., when the slope is much larger than the inclination of the verging stereo cameras. However, for a typical stereo setup with a stereo base / distance ratio of about 0.5, the inclination of the camera is about 0.25. This is a large value compared to typical maximum slopes of waves. The bias can reach a maximum of a quarter of the wavelength. Then the irradiance change is observed by the right camera at the position of the maximum slope of the wave while the left camera observes the same illumination discontinuity still at the crest of the wave. A quarter wavelength bias in the parallax results for $b/Z = 0.5$ in an amplitude bias of half a wavelength. If the height of larger waves is estimated by the parallax seen for shorter-scale waves between 0.5 to 10 cm, height biases between 0.25 and 5 cm may occur. This systematic error is significant as compared to the statistical height error. Shemdin and Tran [132], for instance, reported a statistical error of 0.3 cm. None of the previous stereo photography investigations did investigate this bias, nor discussed whether it could be avoided. Unfortunately, the bias occurs only for illumination inhomogeneities in the direction of the stereo base, but not for those perpendicular to it. Thus biased and unbiased feature correspondences may be obtained in the same stereo image.

When the wave slope gets less steep than the inclination of the verging stereo cameras, suddenly a new effect comes into play (Fig. 4.32b). While the left camera still

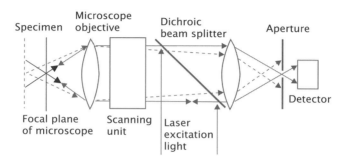

Figure 4.33: *Principle of confocal laser scanning microscopy.*

observes the irradiance change at the crest of the wave, the reflection condition cannot be met for any phase of the wave for the right camera. This effect leads to missing correspondences in the stereo images. A feature observed in one image is not present in the other. As long as the wave is steep enough, only a height bias occurs. Below the critical value which is related to the verging angle between the cameras, corresponding features no longer can be found. The sudden transition at the critical value explains an effect which has puzzled observers of stereo images of the water surface. While in certain regions of the stereo images clearly corresponding features can be found, in other regions a feature found in one image just does not appear in the other image. A strategy to disregard images with noncorresponding features from analysis is dangerous. Because missing correspondences are associated with less steep waves, such a strategy would inevitably select images with steeper and rougher waves and thus lead to a biased mean wave number spectrum.

In summary, the correspondence problem for stereo images of the ocean surface shows two different faces, a height bias and missing correspondences. While the effect of missing correspondences is an obvious hint that the corresponding stereo image cannot be analyzed, the bias effect is more subtle, but still a significant source for systematic errors.

4.4.3 Confocal Laser Scanning Microscopy

From the discussion in Sections 4.3.5b and 4.3.5c, we can conclude that it is not possible to reconstruct three-dimensional images entirely from a focus series obtained with conventional light microscopy. A wide range of wave numbers is lost completely, because of the large zero areas in the OTF. Generally, the lost structures cannot be recovered. Therefore the question arises, whether it is possible to change the image formation and thus the point spread function in such a way that the optical transfer function no longer vanishes, especially in the z direction.

One answer to this question is *confocal laser scanning microscopy*. Its basic principle is to illuminate only one point in the focal plane. This is achieved by scanning a laser beam over the image plane that is focused by the optics of the microscope onto the focal plane (Fig. 4.33). Since the same optics are used for imaging and illumination, the intensity distribution in the object space is given approximately by the point spread function of the microscope. (Slight differences occur since the light is coherent.) Only a thin slice close to the focal plane receives a strong illumination. Outside this slice, the illumination falls off with the distance squared from the focal plane. In this way contributions from defocused objects outside the focal plane are strongly suppressed and

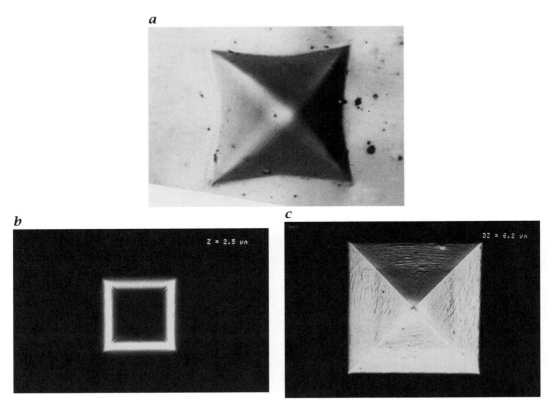

Figure 4.34: *Demonstration of confocal laser scanning microscopy (CLSM): **a** A square pyramid shaped crystal imaged with standard microscopy focused on the base of the pyramid. Towards the top of the pyramid, the edges get more blurred. **b** Similar object imaged with CLSM. Since the irradiance with CLSM is decreasing with z^{-2} from the focal plane, only a narrow height contour of the square pyramid becomes visible, here 2.5 μm above the pyramid base. **c** Image composed of a 6.5 μm depth scan of CLSM images; in contrast to **a**, the whole depth range is imaged without blur. Images courtesy of Carl Zeiss Jena GmbH, Germany.*

the distortions decrease. But can we achieve a complete distortion-free reconstruction? We will use two independent trains of thought to answer this question.

Let us first imagine a periodic structure in the z direction. In conventional microscopy, this structure is lost since all depths are illuminated equally. In confocal microscopy, however, we can still observe a periodic variation in the z direction because the irradiance decreases with the distance squared from the focal plane, provided that the wavelength in the z direction is not too small.

The same fact can be illustrated using the PSF. The PSF of confocal microscopy is given as the product of spatial intensity distribution and the PSF of the optical imaging. Since both functions fall off with z^{-2}, the PSF of the confocal microscope falls off with z^{-4}. This much sharper localization of the PSF in the z direction results in a nonzero OTF in the z direction up to the z resolution limit.

The superior 3-D imaging of confocal laser scanning microscopy is demonstrated in Fig. 4.34. An image, taken with standard microscopy, shows a crystal of the shape of a square pyramid only sharp at the base of the pyramid (Fig. 4.34a). Towards the top of the pyramid, the edges become more blurred. In contrast, in a single image taken with a confocal laser scanning microscopy, only a narrow height range is imaged at all

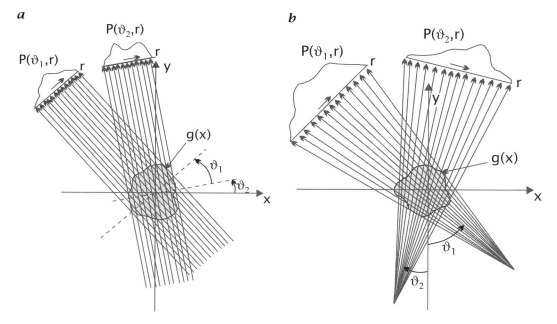

Figure 4.35: **a** *Parallel projection and* **b** *fan-beam projection in tomography.*

(Fig. 4.34b). An image composed of a 6.5 μm depth scan by adding up all images shows a sharp image for the whole depth range (Fig. 4.34c). Many fine details can be observed that are not visible in the image taken with the conventional microscope.

The laser scanning microscope has found widespread application in medical and biological sciences. Examples of confocal laser scanning microscopy can be found in Section 1.5:

1. 3-D retina imaging in Section 1.5.2: Figs. 1.23 and 1.24, Plates 6 and 7
2. 3-D imaging of cell nuclei in Section 1.5.3: Fig. 1.25, Plate 8

4.4.4   Tomography

4.4.4a   Principle. *Tomographic* imaging methods do not generate a 3-D image of an object directly, but require reconstruction of the three-dimensional shape of objects using suitable reconstruction methods. Tomographic methods can be thought of as an extension of stereoscopy. With stereoscopy, only the depth of surfaces is inferred, but not the 3-D shape of transparent objects. Intuitively, we may assume that it is necessary to view such an object from as many directions as possible for a true 3-D reconstruction.

Tomographic methods use penetrating radiation and view an object from all directions. If we use a point source (Fig. 4.35b), we observe a perspective or *fan-beam projection* on the screen behind the object just as in optical imaging (Section 4.3.2a). Such an image is taken from different projection directions by rotating the point source and the projection screen around the object. In a similar way, we can use parallel projection (Fig. 4.35a) which is easier to analyze but harder to realize. Either the whole imaging apparatus or the object to be imaged are rotated around the z-axis by at least 180°. As in other imaging methods, tomography can make use of different interactions between matter and radiation. The most widespread application is transmission tomography. The imaging mechanism is by the absorption of radiation, e. g., X-rays. Other methods include emission tomography, reflection tomography, and time-of-flight tomography (especially with ultrasound), and complex imaging methods using nuclear magnetic resonance (NMR).

The best known technique is X-ray tomography. The part of the patient's body to be examined is X-rayed from different directions. The intensity variations in the projections are related to a) the path length through the body and b) the absorption coefficient which depends on the nature of the tissue, basically the atomic weight of the elements. Emission tomography can be applied by injection of radioactive substances into the organ to be investigated.

Ultrasonic imaging is another important example. As with X-rays, we can measure the absorption for imaging. Furthermore, the speed of sound depends on the elasticity properties of the medium. The speed of sound can be investigated by measuring the time-of-flight. All this might look very simple, but ultrasonic imaging is rendered very difficult by the reflection and refraction of rays at interfaces between the layers of different speeds of sound.

Besides medical applications, tomographic methods are used in many other scientific areas. This is not surprising, since many complex phenomena can only be understood adequately if three-dimensional data can be obtained.

4.4.4b Homogeneity of Tomographic Projection. In order to reconstruct objects from projections, it is essential that they are linearly superimposed in the projections. This condition is met when the imaged property κ can be integrated along a projection beam.

$$I = \int_{\text{path}} \kappa(s)\mathrm{d}s. \tag{4.85}$$

In *emission tomography*, the emitted radiation may not be absorbed from other parts of the object. In *absorption tomography*, the imaged property is the extinction coefficient κ (Section 3.3.7b). The differential intensity loss $\mathrm{d}I$ along a path element $\mathrm{d}s$ is proportional to the extinction coefficient $\kappa(x)$, to $\mathrm{d}s$, and to the intensity $I(x)$ (see also Section 3.3.7a):

$$\mathrm{d}I = -\kappa(x)I(x)\,\mathrm{d}s. \tag{4.86}$$

Integration yields

$$\ln\frac{I}{I_0} = -\int_{\text{path}} \kappa(x)\mathrm{d}s. \tag{4.87}$$

The logarithm of the intensity is the proper quantity to be measured, since it results from a linear superposition of the absorption coefficient along the path. Generally, the intensity is not suitable, except when the total absorption is low. Then we can approximate $\ln(I/I_0)$ by $I/I_0 - 1$.

Tomographic reconstruction does not work at all if opaque objects are contained in the examined scene. In this case, we get only the shape of the opaque object in the projection, but not any information on semi-transparent objects, which lie before, in, or behind this object.

4.5   Advanced Reference Material

4.5.1   Data of Optical Systems for CCD Imaging

This section contains some additional material about commercially available optical systems for CCD imaging. They will be useful if you have to setup an optical system for imaging applications.

Standard focal lengths for CCD lenses   4.1

Focal length	Horizontal angle of view			
[mm]	1/3″	1/2″	2/3″	1″
4.8	56.4°	67.4°	92°	–
6.0	41.0°	56.1°	74.4°	–
8.0	30.4°	43.6°	58.1°	–
12.5	20.1°	28.7°	38.8°	55.2°
16.0	15.6°	22.6°	30.8°	–
25.0	10.2°	14.6°	20.0°	29.2°
50.0	5.1°	7.3°	10.1°	14.6°
75.0	3.4°	4.9°	6.7°	9.9°

Reference item 4.1 contains standard focal lengths for lenses used in CCD imaging and the horizontal angle of view for 1/3", 1/2", 2/3" and 1" imagers (Fig. **??**). Before you use a lens, make sure that it is designed for the image size. You can always use a lens for a smaller image than it is designed for but never for a larger one. The result would be a significant irradiance decrease (vignetting) and an unacceptable increase in the lens aberrations at the edges of the image.

Standard focal length of lenses that support also 1" images are 12.5, 25, 50, and 75 mm. Standard lenses are corrected only in the visible range for 400–700 nm. Standard CCD images are, however, also sensitive in the near IR up to 1000 nm. If your CCD does not include an IR cut-off filter in front of the CCD chip, significant blurring may occur if a scene is illuminated with a radiation source that includes also IR radiation. Incandescent lamps including *quartz tungsten halogen lamps* (*QTH*) radiate the dominant fraction in the near infrared (> 3.9). You can still use the standard CCD lenses for your near-IR application with some degraded performance and a slight focus shift. Then, you should use a colored filter glass cut-off filter to suppress the visible range.

Miniature CCD lenses   4.2

Because the CCD cameras are getting smaller and smaller, the optics is often the largest piece of the imager. Some companies offer miniature CCD lenses. The following table shows some lenses manufactured by Rodenstock and Neeb. The Rodenstock lenses are designed for 1/3" imagers (6 mm image diagonal) and have a diameter of only 12 mm (images courtesy of Rodenstock GmbH).

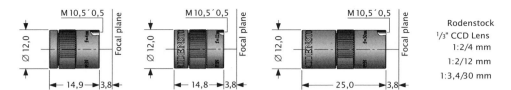

Manufacturer	Focal length [mm]	f-number	Diameter [mm]	Length
Rodenstock	4	1:2	12	14.9
(Munich,	12	1:2	12	14.8
Germany)	30	1:3.4	12	25.0
Neeb Optic GmbH	8.3	1:2.8	16	23.0
(Wetzlar,	12	1:3.5	16	14.0
Germany)	25	1:3.0	16	16.0
	36	1:4.0	16	16.0
	60	1:6.0	19	20.0
	90	1:5.6	25	43.0

4.3 Apochromatic CCD lenses

Schneider Kreuznach Optical Company offers a unique series of apochromatic lenses that are designed for the full range of wavelengths from 400–1000 nm to which CCDs are sensitive. These lenses are designed for 2/3" and 1/2" imagers (11 mm image diagonal) and also feature a low geometrical distortion, rugged stainless-steel design with fixable aperture and focus, and an outer diameter of only 31.5 mm.

Name	Focal length [mm]	f-number
CINEGON	8 mm	1:1.4
CINEGON	12 mm	1:1.4
XENOPLAN	17 mm	1:1.4
XENOPLAN	23 mm	1:1.4
XENOPLAN	35 mm	1:1.9

The MeVis-C series of apochromatic lenses offered by Linos is designed for the range of wavelengths from 400–900 nm

Name	Focal length [mm]	f-number
MeVis-C 12	12 mm	1:1.6
MeVis-C 16	16 mm	1:1.6
MeVis-C 25	25 mm	1:1.6
MeVis-C 35	35 mm	1:1.6
MeVis-C 50	50 mm	1:1.8

An interesting development to watch is hybrid diffractive/refractive lenses. As one of the first manufacturers, Melles Griot has such apochromatic doublets that offer compact lightweight lenses for CCD imagers.

Achromats as CCD lenses

4.4

Standard achromatic lenses for CCD images can be used if the field of view does not exceed ±2.5°. The following table lists the minimal focal length for the different sizes of CCD imagers.

CCD imager	1/4"	1/3"	1/2"	2/3"	1"
Image diagonal [mm]	4.0	6.0	8.0	11.0	16.0
Minimal focal length [mm]	50	70	100	125	200

Macro lenses

4.5

Standard lenses are designed only for low magnifications, typically 1/10 or lower. *Macro lenses* offer an optimal solution for magnifications around one. The technical drawings for several macro lenses including a macro zoom lens manufactured by Rodenstock follow.

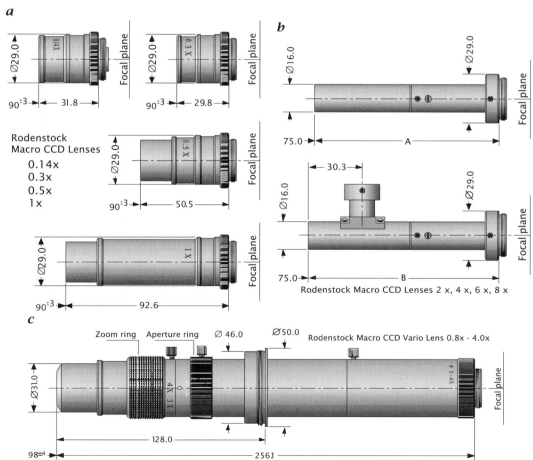

a

Rodenstock
Macro CCD Lenses
0.14x
0.3x
0.5x
1x

b

Rodenstock Macro CCD Lenses 2 x, 4 x, 6 x, 8 x

c

Rodenstock Macro CCD Vario Lens 0.8x - 4.0x

4.5.2 Optical Design

If you are not an experienced optician, it is not recommend that you design your own optical systems from scratch using optical design programs. However, one type of usage of such programs can still be advantageous. Many optical companies ($>$ 4.11) offer a lot of components that may be useful to build an imaging system. Use the optical design program to set up your system with these components and test whether they really meet the requirements. Most optical design programs contain the optical elements from the catalog of the major optical companies so that this is an easy task. Linos offers a free design program, WinLens, that is specifically developed to test their optical components in your planned configuration. `http://www.winlens.de`

4.5.3 Further References

4.6 **General references**

Handbook of Optics, 1995. Volume I: Fundamentals, Techniques, & Design, Volume
    II: Devices, Measurements, & Properties, M. Bass, E. W. Van Stryland, D. R. Williams,
    and W. L. Wolfe, editors, 2nd edition, McGraw-Hill, New York. Two-volume general
    reference work

Photonics Handbook, 49th international edition 2003. Includes four volumes:
  • The Photonics Corporate Guide to Profiles & Addresses
  • The Photonics Buyers' Guide to Products & Manufacturers
  • The Photonics Design & Application Handbook
  • The Photonics Dictionary

    The complete volume is also available on CD-ROM and on the internet
    `http://www.photonics.com`.

4.7 **Optical engineering**

W. J. Smith, 1990. Modern Optical Engineering. The Design of Optical Systems, 2nd
    edition, McGraw-Hill, New York. Excellent guide to engineering optics starting
    from the basics.

R. McCluney, 1994. Introduction to Radiometry and Photometry, Artech House,
    Boston. An excellent and easy to understand treatment of this much neglected
    subject.

4.8 **Special image formation techniques and applications**

G. Cloud, 1995. Optical Methods of Engineering Analysis, Cambridge Univ. Press,
    Cambridge, United Kingdom. The focus in this monograph is on interferometry
    methods to measure small deformations and minute displacements. Included are
    photoelasticity methods, holographic interferometry, and speckle methods.

P. Hariharan, 1984. Optical Holography, Principles, Techniques and Applications,
    Cambridge Univ. Press, Cambridge, United Kingdom. Holographic techniques.

G. F. Marshall, ed., 1991. Optical Scanning, Marcel Dekker, New York. Overview of
    optical scanning technology.

Organizations and technical societies

The following World Wide Web (www) addresses of some technical organizations and manufacturers of optical components are a good starting point if you are searching for some specific optical items. You best start your search with one of the two societies. Please note that WWW addresses are often changing. Therefore, some of the listed addresses may be out of date.

Optical Society of America (OSA), maintains the OpticsNet, `http://www.osa.org`

The International Society for Optical Engineering (SPIE), `http://www.spie.org`

European Optical Society (EOS), `http://www.europeanopticalsociety.org`

Deutsche Gesellschaft für angewandte Optik (DGaO), `http://www.dgao.de`

Magazines covering optics and photonics

LaserFocusWorld, `http://www.laserfocusworld.com`

Photonics Spectra, `http://www.photonics.com`

Vision Systems Design, `http://www.vision-systems.com`

Manufacturers and distributors of optical components

The following list contains a useful selection of manufacturers and distributors of optical components. For more complete lists see, e.g., `http://www.photonics.com` or `http://www.laserfocusworld.com`.

Edmund Scientific (optical components and instruments), `http://www.edsci.com`

Focus Software (optical design software), `http://www.focus-software.com`

Lambda Research Optics (optical components), `http://www.lambda.cc`

LightPath Technologies (Gradium lenses), `http://www.lightpath.com`

Linos (optical components), `http://www.linos.com`

Melles Griot (optical components, lasers), `http://www.mellesgriot.com`

Neeb Optik (optics), `http://www.neeb-optik.de`

New Focus (optical components), `http://www.newfocus.com`

Newport (optical components), `http://www.newport.com`

Ocean Optics (fiber optics spectrometer), `http://www.oceanoptics.com`

OptoSigma (optical components), `http://www.optosigma.com`

Oriel Instruments (optical components and instruments), `http://www.oriel.com`

piezosystem jena (piezoelectric positioning), `http://www.piezojena.com`

Polytec PI (positioning, optical components), `http://www.polytecpi.com`

Schneider Kreuznach Optics (optics and filters), `http://www.schneideroptics.com`

Sill Optics (optics), `http://www.silloptics.de`

5 Imaging Sensors

5.1 Highlights

Imaging sensors convert radiative energy into an electrical signal. The incident photons are absorbed in the sensor material and are converted into an electrical charge or current. Quantum detectors (Section 5.3.3) directly convert photons into electrons, while thermal detectors (Section 5.3.4) detect the absorbed radiative energy by secondary effects associated with the temperature change in the detector. Imaging sensors are available that cover the wide spectrum from gamma rays to the infrared.

The real difficulty of imaging detectors is to store the accumulating electrical charge in an array of detectors, to control the exposure time, and to convert the spatially distributed charge into a time-serial analog or digital data stream (Section 5.3.5). The dominating and very successful devices to perform this task are the charge coupled devices, or, for short, CCDs. However, directly addressable imaging sensors on the basis of the CMOS fabrication process become more and more promising and offer exciting new features. With this type of device, image acquisition, digitalization, and preprocessing could be integrated on a single chip. This chapter provides also a comprehensive survey of the available imaging sensors (Section 5.3.7 and $\succ$ 5.2–5.4). All the practical knowledge to understand and handle analog and digital video signals is also provided in these sections and the advanced reference material (Section 5.5).

The procedure part details as how to measure camera performance parameters (Section 5.4.1) and advises about the criteria to select the best sensor and/or camera for a wide range of imaging application (Section 5.4.2). The description of common artifacts and operation errors helps to understand the limitations of various sensor types and to avoid common handling errors and misadjustments (Section 5.4.4).

5.2 Task

An imaging sensor converts an "image", i.e., the spatially varying irradiance on the image plane into a digital data stream that can be stored and processed by a digital computer. In this chapter, we discuss the conversion of radiative energy into electrical charges and the formation of electrical signals and data streams from the spatial image. The spatial and temporal sampling of the image signal (digitization) and limitation to discrete signal levels (quantization) are treated in Chapter 6.

The first task is to get acquainted with the principles of the conversion of radiation into an electrical signal (Section 5.3.1), the terms describing the performance of imaging detectors (Section 5.3.2), and the two major classes of detectors, quantum detectors (Section 5.3.3) and the thermal detectors (Section 5.3.4, task list 4). The next step is to determine the requirements for your imaging task carefully. What irradiance levels

Task List 4: Imaging Sensors

Task	Procedure/Quantity/Device
Conversion of radiation (photons) into an electrical signal	Quantum detectors: photoemissive, photoconductive, and photovoltaic. Thermal detectors: bolometer, thermophiles, piezoelectrical
Determine requirements for imaging task	Irradiance level, irradiance resolution, dynamical range of irradiance, spectral range, spatial resolution, temporal resolution
Compare requirements with sensor performance	Minimum detectable irradiance, minimum resolvable radiance difference, dynamical range of sensor/camera system, spectral sensitivity, spatial resolution (number of sensor elements), frame rate, minimum exposure time, maximal exposure time

and dynamical range are encountered? What irradiance resolution is required? What spectral range and spatial and temporal resolution are needed? A comparison of these demands with the performance specifications of the various imaging sensor families that are available allow for an optimum selection of a sensor for a specific image task.

5.3 Concepts

5.3.1 Overview

The conversion of radiation into a digital data stream that can be read into and processed by a digital computer involves a number of steps (Fig. 5.1):

Conversion. The incident radiative energy must be converted into an electrical charge, voltage, or current. In order to form an image, an array of detectors is required. Each of the detector elements integrates the incident irradiance over a certain area.

Storage. The generated charges must be integrated and thus stored over a certain time interval, the *exposure time.* In order to control illumination, the sensor should accumulate charges only during a time interval that can be controlled by an external signal. This feature is called an *electronic shutter*.

Read-out. After the exposure, the accumulated electrical charges must be read out. This step essentially converts the spatial charge distribution (parallel data) into one (or multiple) sequential data stream that is then processed by one (or multiple) output circuit.

Signal conditioning. In a suitable electrical amplifier, the read-out charge is converted into a voltage and by appropriate signal conditioning reduced from distortions. In this stage, also the responsivity of the sensor can be controlled.

Analog-digital conversion. In a final step, the analog voltage is then digitized to convert it into a digital number for input into a computer.

Section 5.3.2 discusses the parameters that measure the performance of a photodetector. This includes many different figures. Some of them describe the features of individual detectors such as the minimum detectable irradiance, others the properties of detector arrays such as their uniformity. The two most important types of radiative detectors are *quantum detectors* in which electrical charges are generated by photons

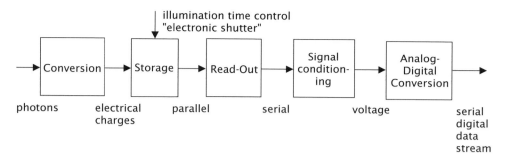

Figure 5.1: *Chain of processes that takes place in an imaging sensor that converts incident photons into a serial digital data stream.*

and *thermal detectors* that respond to the energy from the absorbed radiation (Section 5.3.4). One of the most difficult and also confusing aspects of imaging sensors is the generation of an electric signal from the sensor. While the individual sensor elements accumulate the radiation-induced charges in parallel, the signal to be generated must be (at least partially) time serial to be sent to the computer. Section 5.3.6 describes the basic read-out techniques and explains the differences between frame transfer, full frame, and interline transfer charge coupled devices (CCD), electronic shuttering, and asynchronous triggering.

5.3.2 Detector Performance

This section introduces the basic terms that describe the response of detectors to incident radiation.

5.3.2a Responsivity.   The basic quantity that describes the sensitivity of the sensor to radiation is called *responsivity R* of the detector. The responsivity is given as the ratio between the flux Φ incident to the detector area and the resulting signal *s*:

$$s = R\Phi \tag{5.1}$$

A good detector shows a constant responsivity over a wide range of radiative fluxes. A constant responsivity means that the signal is directly proportional to the incident flux. The units of responsivity are either A/W or V/W, depending on whether the signal is measured as an electrical current or voltage, respectively.

5.3.2b Quantum Efficiency.   Another basic term to describe the responsivity of a sensor is the *quantum efficiency* (*QE*). This term relates to the particulate nature of electromagnetic radiation (Section 3.3.1f) and the electric signal. The quantum of electromagnetic radiation, the photon, carries the energy $h\nu$, where h is Planck's constant and ν the frequency of the radiation. The quantum of electric charge is the elementary charge e. Then, the quantum efficiency η is the ratio of induced elementary charges e, N_e, and the number of incident photons, N_p:

$$\eta = N_e/N_p. \tag{5.2}$$

The quantum efficiency is always lower than one for photosensors where one photon can generate at most one elementary charge. Not all incident photons generate a charge unit, because some are reflected at the sensor surface, some are absorbed in a region of the sensor where no electric charges are collected, and some just transmit the sensor without being absorbed.

The elementary relation between the quanta for electromagnetic radiation and electric charge can be used to compute the responsivity of a detector. The radiative flux measured in photons is given by

$$\frac{dN_p}{dt} = \Phi_p = \frac{\Phi_e}{h\nu} \quad \text{or} \quad \Phi_e = h\nu \frac{dN_p}{dt}. \tag{5.3}$$

The resulting current I is given as the number of elementary charges per unit time times the elementary charge:

$$I = e\frac{dN_e}{dt}. \tag{5.4}$$

Then, the responsivity results with Eqs. (5.1) and (5.2) in

$$R(\lambda) = \frac{I}{\Phi_e} = \eta(\lambda)\frac{e}{h\nu} = \eta(\lambda)\frac{e}{hc}\lambda \approx 0.8066\eta(\lambda)\lambda \left[\frac{A}{W\mu m}\right]. \tag{5.5}$$

Interestingly, the responsivity increases linearly with the wavelength. This is because radiation is quantized into smaller units for larger wavelengths. Thus, a detector for infrared radiation in the 3-5 μm range is intrinsically about 10 times more sensitive than a detector for light. For practical purposes, it is important to realize that the responsivity of a detector is weakly dependent on many parameters. Among others, these are the angle of the incident light, the temperature of the detector, fatigue, and aging.

5.3.2c Signal Irradiance Relation. As illustrated in Fig. 5.1, the irradiation collected by a photosensor is finally converted into a digital number g at each sensor element of a sensor array. Thus the overall gain of the sensor element can be expressed by a single digital gain constant α that relates the number of collected charge units to the digital number g:

$$g = \alpha N_e. \tag{5.6}$$

The digital gain constant α is dimensionless and means the number of digits per unit charge. An ideal charge unit counting sensor has a digital gain constant α of one.

The incident radiative flux Φ_e is received by the sensor for an exposure time t on an area A. Therefore the received radiant energy (energy based Q_e or photon-based N_p) can be related to the irradiance incident on the sensor by

$$Q_e = \Phi_e t = AEt \quad \text{or} \quad N_p = At\frac{\lambda}{hc}E. \tag{5.7}$$

Using Eqs. (5.2) and (5.6), the relation between the incident irradiance and the digital signal g is

$$g = \alpha\frac{A}{hc}\eta(\lambda)\lambda E(\lambda)t. \tag{5.8}$$

Normally a sensor does not receive monochromatic radiation so that Eq. (5.8) must be integrated over the spectral irradiance E_λ:

$$g = \alpha\frac{At}{hc}\int_{\lambda_1}^{\lambda_2}\eta(\lambda')\lambda'E_\lambda(\lambda')d\lambda'. \tag{5.9}$$

Finally, we can define a *digital responsivity* R_d that relates the output signal g to the received spectral irradiation $H = Et$ integrated over all wavelengths:

$$R_d = \frac{g}{H} = \alpha \frac{A}{hc} \int\limits_{t_1}^{t_2}\int\limits_{\lambda_1}^{\lambda_2} \eta(\lambda')\lambda' E_\lambda(\lambda',t')\,\mathrm{d}\lambda'\,\mathrm{d}t' \left/ \int\limits_{t_1}^{t_2}\int\limits_{\lambda_1}^{\lambda_2} E_\lambda(\lambda',t')\,\mathrm{d}\lambda'\,\mathrm{d}t' \right. . \tag{5.10}$$

In this equation, an additional generalization was made by assuming a time varying irradiance that is integrated over the exposure time $t = t_2 - t_1$. If the irradiance is constant, the temporal integral can be replaced by a simple multiplication with the exposure time t as in Eq. (5.9). The digital responsivity has unit digits/$(\mathrm{J\,m^{-2}})$.

5.3.2d Dark Current. Generally, a detector generates a signal s_0 even if the incident radiative flux is zero. In a current measuring device, this signal is called the *dark current*. A sensor with a digital output will show a digital signal g_0 without illumination.

5.3.2e Noise Equivalent Exposure. The dark signal has not only a DC component but also a randomly fluctuating component with a standard deviation σ_0. Using the responsivity, σ_0 can be converted into an equivalent radiative flux which is known as the *noise-equivalent exposure* or *NEE*,

$$H_0 = \sigma_0/R. \tag{5.11}$$

The NEE is related to a certain frequency band to which the detector is responding and essentially gives the minimum exposure that can be measured with a detector in the given frequency band.

5.3.2f Saturation Equivalent Exposure. A photosensor is also limited to a maximum signal. For a digital output this is simply the largest digital number that is delivered by the analog-digital converter (ADC), which is $g_s = 2^D - 1$ for an ADC with D bits. As with the NEE, a *saturation-equivalent exposure* or *SEE* can be defined by

$$H_s = g_s/R. \tag{5.12}$$

5.3.2g Photon Noise Limited Performance. An ideal detector would not introduce any additional noise. Even then, the detected signal is not noise-free, because the generation of photons itself is a random process. Whenever the detector related noise is below the photon noise, a detector is said to have *photon noise-limited* performance. Photon generation is (as radioactive decay) a random process with a *Poisson distribution* (Section 7.3.1). The probability density function is given by

$$p_p(n) = \frac{N^n \exp(-N)}{n!}. \tag{5.13}$$

This equation gives the probability that n photons are detected in a certain time interval when in the average N are detected. The variance of the Poisson distribution, σ_n^2,

$$\sigma_n^2 = \sum_n p_p(n)(n - N)^2 = N, \tag{5.14}$$

is equal to the average value N.

For low mean values, the Poisson PDF is skewed with a longer tail towards higher values (Fig. 5.2). But even for moderate N, the Poisson distribution quickly converges

a
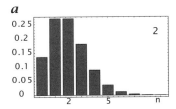

b
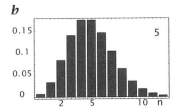

c
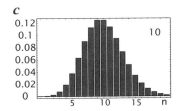

Figure 5.2: *Poisson distribution Eq. (5.13) describing the counting statistics for photons for values $\overline{n}$ as indicated. With increasing $\overline{n}$, the distribution quickly becomes symmetrical and approaches the Gaussian distribution Eq. (5.15).*

a
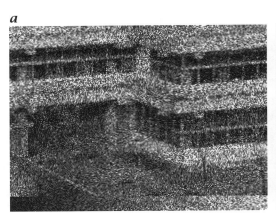

b

Figure 5.3: *Simulation of low-light images taken with a maximum of a 10 and b 100 photons.*

to the continuous *Gaussian distribution* (also known as *normal distribution*, see (Section 7.3.1c)) given by

$$p_n(x) = \exp -\frac{(x - N)^2}{2N} \tag{5.15}$$

with the equal mean value and variance.

Figure 5.3 shows some simulated low-light images that are taken with a low number of photons.

5.3.2h Noise Model for Image Sensors.   The discussion in the previous section is the basis for a simple noise model for image sensors that has gained significant practical importance. The photo signal for a single pixel is Poisson distributed as discussed in the previous section. Except for very low-level imaging conditions, where only a few electrons are collected per sensor element, a normal distribution $N(N_e, \sqrt{N_e})$ approximates the Poisson distribution accurately enough.

The electronic circuits add a number of other noise sources. For practical purposes, it is, however, only important to know that these noise sources are normal distributed and independent of the photon noise and have a variance of σ_{N_0}. According to the laws of error propagation (Section 7.3.3), the noise variances add up linearly, so that the noise variance of the total number of generated charge units $N = N_0 + N_e$ is given by

$$\sigma_N = \sigma_{N_0}^2 + \sigma_{N_e}^2 = \sigma_{N_0}^2 + N_e. \tag{5.16}$$

According to Eq. (5.6), the digital signal is given as

$$g = g_0 + \alpha N. \tag{5.17}$$

An arbitray signal g_0 is added here to account for an offset in the electronic circuits. Then the variance σ_g^2 can be described by only two terms as

$$\sigma_g^2 = \alpha^2 (\sigma_{N_0}^2 + N_e) + \sigma_{g_0}^2 = \sigma_0^2 + \alpha (g - g_0). \tag{5.18}$$

The term σ_0^2 includes the variance of all nonsignal dependent noise sources. This equation predicts a linear increase of the noise variance with the measured gray value g. In addition it can be used to measure the absolute gain factor α.

5.3.2i Dynamic Range. The *dynamic range* of an image sensor (DR) is the ratio of the maximum output signal, g_s, to the standard deviation of the dark signal, σ_0, when the sensor receives no irradiance. The dynamic range is often expressed in units of decibels (dB) as

$$DR = 20 \log \frac{g_s}{\sigma_0}. \tag{5.19}$$

A dynamic range of 60 dB thus means that the saturation signal is 1000 times larger than the standard deviation of the dark signal.

5.3.2j Signal-to-Noise Ratio. The responsivity is not a good measure for a sensor, because a sensor with a high responsivity can be very noisy. The precision of the measured irradiance rather depends on the ratio of the received signal and its standard deviation σ_g. This term is known as the *signal-to-noise ratio* (*SNR*):

$$\text{SNR} = \frac{g}{\sigma_g}. \tag{5.20}$$

The inverse SNR σ_g/g gives the relative resolution of the irradiance measurements. A value of $\sigma_g/g = 0.01$ means, e. g., that the standard deviation of the noise level corresponds to a relative change in the irradiance of 1 %.

5.3.2k Nonuniform Responsivity. For an imaging sensor that consists of an array of sensors, it is important to characterize the nonuniformity in the response of the individual sensors since it significantly influences the image quality. Both the dark signal and the responsivity can be different. The nonuniformity of the dark signal becomes evident when the sensor is not illuminated and is often referred to as the *fixed pattern noise* (*FPN*). This term is a misnomer. Although the nonuniformity of the dark current appears as noise, it is that part of the dark signal that does not fluctuate randomly but is a static pattern. When the sensor is illuminated, the variation in the responsivity of the individual sensors leads to an additional component of the nonuniformity. This variation of the image sensor under constant irradiance is called the *photoresponse nonuniformity* (*PRNU*).

If the response of the image sensors is linear, the effects of the nonuniform dark signal and responsivity can be modeled with Eq. (5.1) as

$$\boldsymbol{G} = \boldsymbol{R} \cdot \boldsymbol{H} + \boldsymbol{G}_0 \tag{5.21}$$

where all terms in the equation are matrixes.

The photoresponse nonuniformity σ_R is often simply given by the standard deviation:

$$\sigma_R^2 = \frac{1}{MN - 1} \sum_{m=0}^{M-1}\sum_{n=0}^{N-1} (r_{m,n} - \overline{R})^2 \quad \text{with} \quad \overline{R} = \frac{1}{MN} \sum_{m=0}^{M-1}\sum_{n=0}^{N-1} r_{m,n} \tag{5.22}$$

5.3.3 Quantum Detectors

The term *quantum detector* refers to a class of detectors where the absorption of the smallest energy unit for electromagnetic radiation, the *photon*, triggers the detection of radiation. This process causes the transition of an electron into a higher excited state. These variants are possible which lead to three subclasses of quantum detectors.

5.3.3a Photoemissive Detectors. By the absorption of the photon, the electron receives enough energy to be able to leave the detector and become a *free electron*. This effect is known as the (extrinsic) *photo effect*. The photo effect can be triggered only by photons below a critical wavelength (i. e., above a certain energy of the photon that is sufficient to provide enough energy to free the electron from its binding).

   Photoemission of electrons is utilized in *vacuum photo tubes* and *photo multiplier tubes* (*PMT*). PMTs are sensitive enough to count single photons, since the initial electrons generated by photon absorption are accelerated to hit another dynode with sufficient energy to cause the ejection of multiple secondary electrons. This process can be cascaded to achieve high gain factors. Because PMTs have short response times that may be less than 10^{-10} s, individual photons can be counted as short current impulses. Such a device is known as a *photon-counting device*.

   Radiation detectors based on the photo effect can be sensitive only in a quite narrow spectral range.

   The lowest wavelength is given by the minimum energy required to free an electron. For higher photon energies, the material may become more transparent, leading to a lower probability that the photon is absorbed reducing the quantum efficiency (average number of freed electrons per incident photons; see Section 5.3.2b). Quantum efficiencies peak typically at 0.3.

5.3.3b Photovoltaic and Photoconductive Detectors. Semiconductor devices that utilize the inner photo effect have largely replaced imaging detectors based on photoemission. A thorough understanding of these devices requires knowledge of condensed matter physics. Thus, the discussion of these devices is superficial and concentrating on the basic properties directly related to imaging detectors. The semiconductor devices have in common with the photoemissive detectors that threshold energy and thus a minimum frequency of the radiation is required. This is due to the fact that electrons must be excited from the valence band across a band gap to the conduction band. In the valence band, the electrons cannot move and thus cause no further effects.

   In a suitable photodetector material, the conduction band is empty. The absorption of a photon excites an electron into the conduction band. There it can move rather unrestricted. Thus, it could increase the conductance of the detector material. Detectors operating in this mode are called *photoconductive detectors*.

   In an appropriately designed detector, the generation of an electron builds up an electric tension. Under the influence of this tension, an electric current can be measured that is proportional to the rate with which the absorbed photons generate electrons. Detectors of this type are known as *photovoltaic detectors*.

5.3.4 Thermal Detectors

Thermal detectors respond to the temperature increase resulting from the absorption of incident radiation. The delivered signal is proportional to the temperature increase. Secondary effects that can be used to measure a temperature increase are

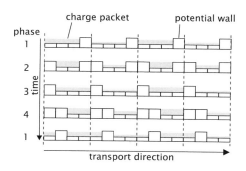

Figure 5.4: *Principle of the charge transfer mechanisms of the charge coupled device (CCD).*

Thermoelectric Effect. Two separate junctions of two dissimilar metals at different temperatures generate a voltage proportional to the temperature difference between them. One junction must be kept at a reference temperature, the other be designed to absorb the electromagnetic radiation with minimum thermal mass and good thermal insulation. Such a device is known as a *thermopile*.

Pyroelectric Effect. Pyroelectric materials have a permanent electrical polarization. Changes in the temperature cause a change in the surface charges. Pyroelectric detectors can only detect changes in the incident radiant flux and thus require a chopper. Furthermore, they lose the pyroelectric behavior above a critical temperature, the Curie temperature.

Thermoconductive Effect. Some materials feature large changes in electric conductivity with temperature. A device measuring radiation by change in conductance is called a *bolometer*. Recently, microbolometer arrays have been manufactured for uncooled infrared imagers in the 8–14 μm wavelength range that shows a noise equivalent temperature difference (NEΔT) of about 100 mK [85].

The common feature of thermal detectors is the wide spectral range to which they can be made sensible. They do not, however, reach the sensitivities of quantum detectors, since the radiation detection is based on secondary effects.

5.3.5 Imaging Detectors

The photovoltaic effects are most suitable for imaging detectors. Instead of moving the photo-induced electrons through the conduction band, they can be captured by additional potential walls in cells. In these cells, the electrons can be accumulated for a certain time.

Given this basic structure of a quantum detector, it is obvious that the most difficult problem of an imaging device is the conversion of the spatial arrangement of the accumulated charges into an electric signal. This process is referred to as *read-out*.

5.3.5a The Charge Coupled Device. The invention of the *charge coupled device* (*CCD*) in the mid seventies was the breakthrough for semiconductor imaging devices. A CCD is effectively an analog shift register.

By an appropriate sequential change of potential, the charge can move across the imaging sensor. A so-called four-phase CCD is shown in Fig. 5.4. Potential walls separate the charge packets. First, a potential wall to the left of the three-cell wide charge storage area is raised, confining the charges to two cells. Subsequently, a potential barrier to the right is lower so that the charge pockets move on to the right by one cell.

Table 5.1: *Spectral sensitivity of different CCD sensors or focal plane areas. The last column lists either the quantum efficiency or the number of electrons generated per kV for X-ray detectors.*

Sensor type	Spectral range	Typical sensitivity
Scintillator, glass-fiber coupled to Si-CCD	20-100 keV	0.5 e/keV
Fiber-coupled X-ray-sensitive fluorescence coating on Si-CCD	3-40 keV	5 e/keV
Direct detection with beryllium window on Si-CCD	3-15 keV	0.3 e/keV
Specially treated, thinned, back-illuminated Si-CCD	< 30 nm 30 eV–8 keV	
Si-CCD-X-ray detector of the MPI semiconductor lab	800 eV–10 keV	> 0.90
	100 eV–15 keV	> 0.50
Si-CCD with UV fluorescence coating and MgF_2 window	0.12–1 μm	< 0.4
Thinned, back-illuminated Si-CCD	0.25–1.0 μm	< 0.80
Standard silicon CCD or CMOS sensor	0.35–1.0 μm	0.1–0.65
GaAs focal plane array	0.9–1.7 μm	
Pt:Si focal plane array	1.4–5.0 μm	0.001–0.01
InSb focal plane array	1.0–5.5 μm	< 0.85
HgCdTe (MCT) focal plane array	2.5–5.0 μm	
"	8.0–12.0 μm	
GaAs/AlGaAs quantum well infrared photodetectors (QWIP)	3.0–5.0 μm	
"	8.0–10.0 μm	

Repeating this procedure, the charge packets can be moved all across the array. The charge transfer process can be made efficiently enough that only a few electrons are "lost" during the whole transfer.

The charge transfer mechanism is the base for one- and two-dimensional arrays of photo sensors. The details of two-dimensional transfer will be discussed in Section 5.3.6.

5.3.5b CMOS Imaging Sensors.

Despite the success of CCD imaging sensors, there are valuable alternatives [128, 129]. With the rapidly decreasing sizes for transistors on semiconductors, it is also possible to give each pixel its own preamplifier and possibly additional circuits. Such a pixel is known as an *active pixel sensor* (*APS*). Then a transfer of the accumulated photocharge is no longer required. The voltage generated by the circuits of the APS is just sensed by appropriate selection circuits. In contrast to a CCD sensor, it is possible to read out an arbitrary part of the sensor without any speed lost and with the same pixel clock.

The APS technology is very attractive because the imaging sensors can be produced in standard CMOS technology and additional analog and digital circuits can be integrated on the chip. This opens the way to integrate the complete functionality of a camera on a single chip including analog-to-digital converters (Fig. 2.10b). In addition, multiple pixels can be read out in parallel, so that it is easily possible to produce CMOS imaging sensors with high frame rates.

5.3.5c Detectable Wavelength Range.

The band gap of the semiconductor material used for the detector determines the minimum energy of the photon to excite an electron into the conduction band and thus the maximum wavelength that can be detected.

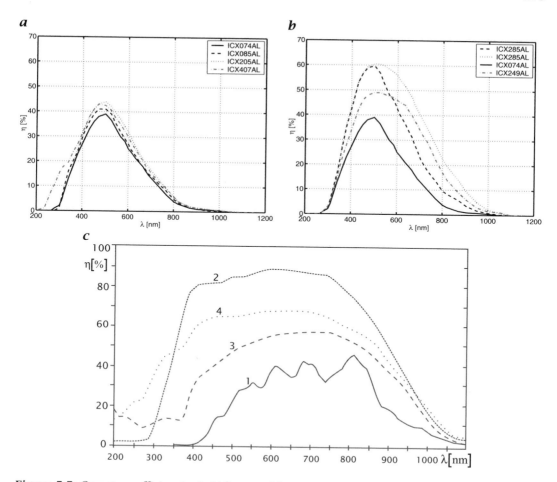

Figure 5.5: *Quantum efficiencies in % for **a** and **b** commonly used Sony CCD sensors as indicated (≻ 5.2, data courtesy PCO AG, Kelheim, Germany). **c** scientific-grade CCD sensors as indicated (data from Scientific Imaging Technologies, Inc (SITe), Beaverton, Oregon).*

For silicon, the threshold given by the band gap is at about 1.1 μm. Thus, imaging detectors on the base of silicon are sensitive also in the near infrared. Other materials are required to sense radiation at larger wavelengths (Table 5.1). The response for shorter wavelengths largely depends on the design. Standard imagers are sensitive down to about 350 nm. Then, the photons are absorbed in such a short distance that the generated electrons do not reach the accumulation regions in the imagers. With appropriate design techniques, however, imaging silicon sensors can be made sensitive for ultraviolet radiation and even X-rays (Table 5.1).

5.3.5d Quantum Efficiency. An ideal photo detector has a *quantum efficiency* of one. This means that each photon irradiating the sensor generates an electron (Section 5.3.2b). Real devices come quite close to this ideal response. Even standard commercial devices have peak quantum efficiencies of about 0.65 (Fig. 5.5a) while scientific-grade devices may reach quantum efficiencies of up to 0.95 (Fig. 5.5c). This can be achieved by illuminating a thinned sensor from the back and by using appropriate antireflection coating. In this way it is also possible to extend the sensitivity far into the ultraviolet.

5.3.5e Dark Current. Thermal energy can also excite electrons into the conduction band and thus gives rise to a *dark current* even if the sensor is otherwise perfect. The probability for thermal excitation is proportional to $\exp(-\Delta E/(k_B T))$, where ΔE is the energy difference across the band gap.

The appropriate method to limit the dark current is cooling. Because of the exponential dependency of the dark current on the absolute temperature, the dark current decreases dramatically with the temperature, one order of magnitude per 8–10 K. Imaging sensors cooled down to liquid nitrogen temperatures can be illuminated for hours without noticeable dark current.

Together with the high quantum efficiency, this type of cooled sensors is the most sensitive available and is widely used in astronomy.

5.3.5f Full-Well Capacity and Saturation Exposure. A *sensor element* () of an imaging sensor can only store a limited number of photo electrons. This limit determines the maximum possible exposure. For an ideal sensor with a quantum efficiency of one, the number of maximal electrons is equal to the number of photons that can be detected in one exposure by a sel. The ratio between the noise level and the *full-well capacity* determines also the dynamical range (Section 5.3.2i) of the sensor and the maximum irradiance resolution that can be measured due to the random nature of the photon flux (Section 5.3.2j). Note that if a sel can store even 1 000 000 electrons, the standard deviation of the irradiance measurement is not better than 0.1%, even if the only noise source of the sensor is the photon noise. CCD sensors have an electron capacity between 10 000 and 500 000 electrons; only devices sensitive in the infrared range with larger sels have a significantly higher capacity of up to 10 millions of electrons.

If we know the quantum efficiency and the electron capacity, it is easy to compute the saturation exposure H_s of a sensor element:

$$H_s = \eta^{-1} \cdot N_s h\nu, \tag{5.23}$$

where N_s is the full-well capacity of the sensor and $h\nu$ the energy of the photon. With light at a frequency of maximum sensibility to the eye ($\nu = 5.4 \cdot 10^{-14}$ Hz, $\lambda = 555$ nm), with an electron capacity of $N_e = 100\,000$ and a quantum efficiency of 80%, the saturation exposure is about $4 \cdot 10^{-14}$ J. For pixels with a size of $10\,\mu m \times 10\,\mu m$, this translates into a saturation irradiance of 10^{-6} W/cm^2 for an exposure time of 40 ms.

5.3.6 Television Video Standards

The history of television dates back to the end of the second world war. At that time the standards for television signals had been set and are still valid today. The standard set the frame rate to 30 frames/sec (US RS-170 standard) and 25 frames/sec (European CCIR standard). The image contains 525 and 625 rows, respectively, from which about 480 and 576 contain the image. The timing of the analog video signals is further detailed in Section 5.5.2.

When CCD sensors were first developed they also had to accommodate to this standard. There was no other choice because all viewing and (analog) recording equipment only worked with these standards. One of the most nasty features of the standard for digital image processing is the interlaced scanning of the images. Because of the limited bandwidth, television engineers were forced to part one frame into two fields. The first field consists of the even and the second field of the odd lines of the image.

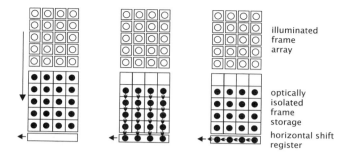

Figure 5.6: Frame integration with a frame transfer CCD sensor.

In this way the individual fields were displayed at a rate of 60 Hz reducing the image flicker without consuming more bandwidth.

Interlaced scanning caused a lot of complications for the construction of CCD sensors. It is also the source of much confusion because modern CCD sensors feature a number of different scan modes which are explained in the following section.

5.3.7 CCD Sensor Architectures

5.3.7a Frame Transfer. *Frame transfer* CCD sensors (Fig. 5.6) use the photo accumulation sites also as charge transfer registers. At the end of the exposure time, the whole frame is transferred across the whole illuminated area into an optically isolated frame storage area, where it is read out row by row with another horizontal shift register, while in the mean time the next exposure takes place in the illuminated frame array. During the transfer phase, the sensor must not be illuminated. Therefore either the illumination with flashlight or a mechanical shutter is required. The required dark period is rather short compared to the read out time for the whole area because it is approximately equal to the time required to read out just one row.

For large-area CCD sensors, the extra storage area for the frame transfer can no longer be afforded. Then, the frame is directly read out. This variant of frame-transfer sensors is called a *full-frame transfer* sensor. It has the disadvantage that the chip must not be illuminated during the whole read-out phase. This means that the exposure time and the read out time must not overlap. Most scientific-grade CCD sensors and the sensors used in consumer and professional digital cameras are full-frame transfer sensors.

5.3.7b Interline Transfer. Because frame transfer sensors are not suitable for continuous illumination and high frame rates, the most common scan mode is the more complex *interline transfer*. Such a sensor has charge storage sites between the lines of photo sensors (Fig. 5.7). At the end of an exposure period, the charges collected at the photo sensors are transferred to these storage sites. Then a two-stage transfer chain follows. First, the charge packets are shifted down the vertical interline storage registers. The charge package in the lowest cell of the vertical shift register is then transferred to a horizontal shift register at the lower end of the CCD chip, where in a second transfer an image row is shifted out to a charge-sensitive amplifier to form a time-serial analog video signal.

In order to achieve interlaced scanning, only every other row is transferred to the vertical shift register. Most interline transfer CCD chips offer several scanning modes,

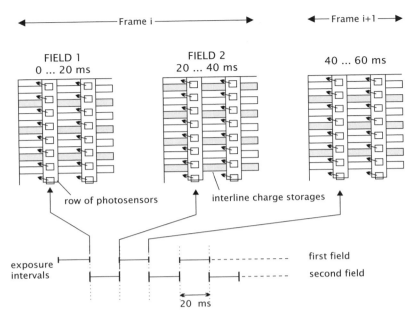

Figure 5.7: *Interlaced field integration with an interline transfer CCD sensor. The transfer to the interline charge storage areas occurs at the end of each exposure interval.*

which are still compatible with the interlaced video signal but partly offset the disadvantages.

Interlaced field integration With this scanning mode (Fig. 5.7), the charges accumulated from the even and odd row are both transferred at the end of the exposure and thus are added up.
This reduces the vertical resolution to only half the number of rows, but has the advantage of a doubled sensitivity. While in the first field rows 1 and 2, 3 and 4, and so on are added up, rows 2 and 3, 4 and 5, and so on are added up in the second field, leading to half-row vertical displacement between the two fields. It is for this displacement that the scanning mode is called interlaced.

Noninterlaced field integration The only difference of the noninterlaced field integration (Fig. 5.8) in the interlaced field integration is that in both fields the same two rows are added up. Therefore the two fields are not vertically displaced. This scanning mode is most useful for dynamic scenes and motion analysis offering twice the temporal resolution for half the vertical resolution but without a vertical offset between the frames. The exposure time in both *field integration* modes is limited to the duration of a field, i. e., 16 2/3 ms or 20 ms.

Interlaced frame integration In contrast to field integration, *frame integration* (Fig. 5.9) can only be interlaced. Now, no lines are added up as with the field integration modes. During the first field, only the odd row numbers (1, 3, 5, etc.) are transferred to the storage sites and then shifted out. During the second field, the even rows (2, 4, 6, etc.) are processed in the same way as the odd rows in field one. With frame integration, the exposure time can be extended to the full frame duration. This mode is less suitable for dynamic scenes, because the exposure between the two fields is shifted by one field time; it gives, however, full vertical resolution for static images.

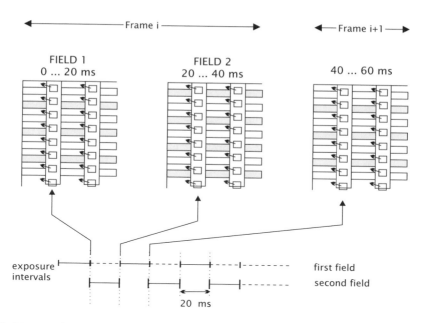

Figure 5.8: *Noninterlaced field integration with an interline transfer CCD sensor. The transfer to the interline charge storage areas occurs at the end of each exposure interval.*

5.3.7c Electronic Shutter. A very useful feature is the *electronic shutter*. Its principle is explained in Fig. 5.10. With electronic shuttering, the exposure time can be limited to shorter times than a field or frame duration. This is achieved by draining the accumulating charges at the beginning of the exposure time. Accumulation and thus the exposure starts when the draining is stopped and lasts until the end of the normal exposure time. Electronic shutter times can be as short as a few μs.

5.3.7d Micro Lens Arrays. Interline transfer sensors have the disadvantage that only a rather small fraction of the sensor element is light sensitive because the main area of the sensor is required for the interline storage areas and other circuits. Arrays of micro lenses can overcome this disadvantage. With such an area, each photo sensor is covered by a micro lens, effectively enlarging the light collecting area and thus sensitivity by a factor of 2 to 3. This enhanced sensitivity comes at the price of a smaller acceptance cone for incoming light.

An imaging sensor with micro lenses may show a lower than expected sensitivity with high-aperture lenses and show a larger fall-off towards the edge of the array for lenses with a short focal length. Still, the advantages of micro lens arrays overcome their disadvantages. They have considerably been improved recently and have boosted the effective *quantum efficiency* of interline transfer CCD sensors up to 65% (Fig. 5.5a and b).

5.3.7e Progressive Scanning. For a long time, the only major deviation from the video standard were line sensors which found wide-spread applications in industrial applications. This has meanwhile changed because analog recording of images with standard video signals is no longer required. This made it also possible to so-called *progressive scanning* cameras, which overcome the remaining disadvantages of interlaced scanning cameras because they scan the image row by row. With these cameras it is also possible to use electronic shuttering with full frames and not only fields as

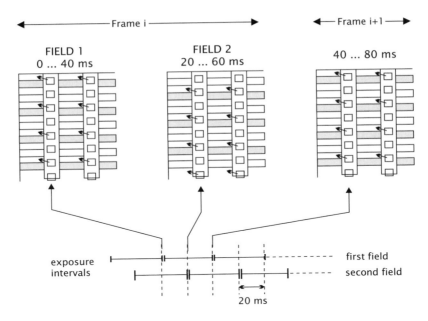

Figure 5.9: *Interlaced frame integration with an interline transfer CCD sensor. Since at the end of each field only every other row is transferred to the interline storage area, the exposure time can last for a full frame time.*

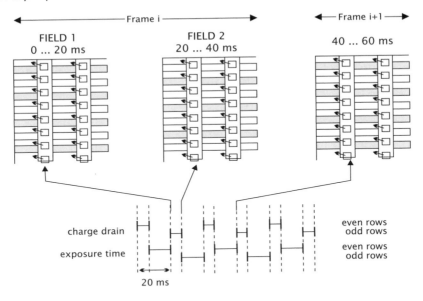

Figure 5.10: *Interlaced frame integration with electronic shutter with an interline transfer CCD sensor. The exposure begins when the charge draining stops and ends at the end of the charge accumulation period when the charges are transferred to the interline storage areas.*

with interlaced scanning cameras. The pros and cons of analog versus digital cameras is further discussed in Section 6.5.3a.

Figure 5.11: *Set up for measurement of linearity, noise, and absolute spectral responsivity.*

5.4 Procedures

5.4.1 Measuring Performance Parameters of Imaging Sensors

In this section, we discuss procedures to measure several performance parameters of imaging sensors. This includes the responsivity and linearity (Section 5.4.1a), noise and the related parameters signal-to-noise ratio (SNR) and dynamic range (DR) (Section 5.4.1b), and spatial inhomogeneities, especially the "fixed pattern noise" (FPN) and the photoresponse nonuniformity (PRNU) (Section 5.4.1c).

5.4.1a Responsivity and Linearity. Absolute measurements of the responsivity of imaging sensors require a homogeneous calibrated light source. The most suitable one is an integrated sphere as shown in Fig. 5.11. An integrated sphere consists of a hollow sphere that is coated with a highly reflective matte coating. One or several small ports take light sources to illuminate the sphere and a larger port with a radius r serves as the light output. The whole setup results in a highly homogeneous and isotropic radiance L.

An imaging sensor that is positioned at a distance x from the opening of the integrated sphere receives the irradiance

$$E = \pi L \frac{r^2}{x^2 + r^2} = \pi L \sin^2 \Theta, \tag{5.24}$$

where Θ is the half angle of light cone from the integrated sphere piercing the center of the imaging sensor. Theoretically, there is a small decrease of the irradiance towards the edge of the imaging sensor. This effect is, however, negligible for $x \geq 8r$ and when the diameter of the imaging sensor is small compared to the opening of the integrated sphere [86, 87].

The quadratic dependence of the irradiance on the distance of the sensor from the light source in Eq. (5.24) makes it easy to vary the irradiance over a wide range. In the set up shown in Fig. 5.11, a large linear optical translation stage is used to position

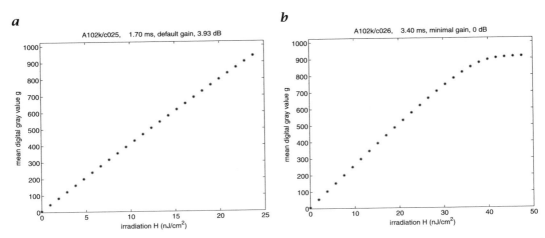

Figure 5.12: *Measurement of the responsivity by increasing the irradiance from zero in steps of 4% of the maximal irradiance. Shown are results measured with a Basler A102k digital camera (Sony ICX285AL CCD sensor and 10 bit output) at **a** default gain and a exposure time of 1.7 ms, **b** minimal gain and a exposure time of 3.4 ms; in this case the full-well capacity is reached.*

image sensors automatically at distances from 0.45–2.40 m. In this way it was possible to measure the response of imaging sensors with irradiations from zero to the maximal irradiation in steps of 4%.

The results of such measurements show that well-designed imaging sensors generally exhibit an excellent linearity (Fig. 5.12a). The slight residual nonlinearity normally poses no problem and it can easily be corrected (Section 7.4.3). The nonlinearity becomes only pronounced if the full-well capacity is reached (Fig. 5.12a). The closer the sensor is operated at the full-well capacity, the higher is the residual nonlinearity.

5.4.1b Noise and Signal-to-Noise Relation. The same set up used for linearity measurements can be used for the noise measurements. For these measurements it is required to acquire many (> 100) images at the same irradiance in order to compute the variance for each individual pixel with sufficient certainty. Note that it is not possible to compute the noise variance from multiple pixels at constant irradiance because of the *photoresponse nonuniformity* (Section 5.3.2k).

According to the noise model for image sensors discussed in Section 5.3.2h, the variance of the noise should vary linearly with the digital gray value:

$$\sigma_g^2 = \sigma_0^2 + \alpha g. \tag{5.25}$$

Once σ_0 and α are determined, it is also possible to compute the *dynamic range* (*DR*) and the *signal-to-noise ratio* (*SNR*) of the sensor according to Eqs. (5.19) and (5.20), respectively.

The example measurements shown in Fig. 5.13 verify this simple noise model both for CCD and CMOS sensors. The residual noise level σ_0 without illumination is very low for a well-designed CCD camera. This means that modern CCD cameras are dominated by photon noise and thus operate at their theoretical noise limit. CMOS sensors generally show larger values of σ_0 (Fig. 5.13b), but the noise variance at the highest irradiance is still significantly larger.

This linear increase of the noise variance with the gray values is of importance for subsequent image processing. Simple models that assume a constant noise level

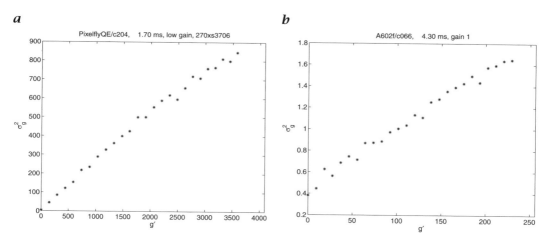

Figure 5.13: *Variance versus mean digital gray value for **a** the PCO pixelfly QE with Sony ICX285AL CCD sensor and **b** the Basler A602f with Micron MT9V403 CMOS sensor.*

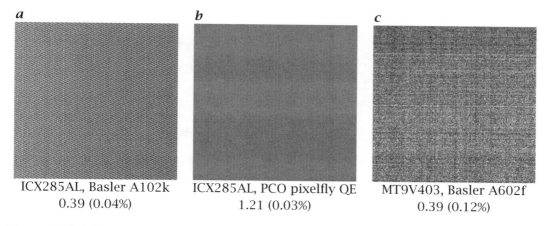

ICX285AL, Basler A102k	ICX285AL, PCO pixelfly QE	MT9V403, Basler A602f
0.39 (0.04%)	1.21 (0.03%)	0.39 (0.12%)

Figure 5.14: *Inhomogeneities in dark images (so called "fixed pattern noise") of digital imagers as indicated. The figures below give the standard deviation of the spatial variations in digits and per cent of the full range; note that the cameras in **a** to **c** have a resolution of 10, 12, and 8 bits, respectively. Contrast enhanced images with a gray value range of $\pm 3\sigma$.*

are no longer valid. But it is always possible to obtain an image signal with a gray value independent noise variance by applying a suitable nonlinear gray level transform (Section 7.4.4).

5.4.1c Spatial Inhomogeneities. As discussed in Section 5.3.2k, sensor arrays show a certain degree of nonuniform response, which has two terms. Each sensor element has a different dark signal (*fixed pattern noise, FPN*) and a different responsivity (*photoresponse nonuniformity*). The measurement of both types of inhomogeneities requires the averaging of many images (at least 100) in order to reduce the uncertainty caused by noise (Section 5.4.1b).

The measurement of the fixed pattern noise is performed most easily, because only images must be taken when the imaging sensor receives no irradiation. The fixed pattern noise for CCD sensors is remarkably low (Fig. 5.14a and b). It is only about 0.04% of the full gray value range. CMOS sensors typically show significantly higher

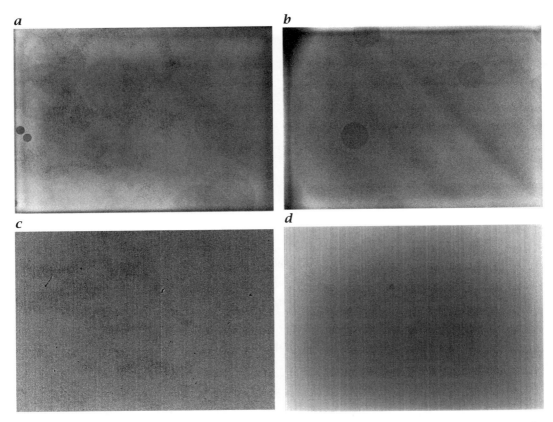

Figure 5.15: *Demonstration of the photoresponse nonuniformity (PRNU) by illuminating the imaging chips without optics using a setup as shown in Fig. 5.11:* **a** *Sony CCD area sensor ICX285AL in Basler A102k (σ = 0.29%),* **b** *Sony CCD area sensor ICX285AL in PCO pixelfly QE (σ = 0.35%),* **c** *Photonfocus MV-D1024-80 CMOS area sensor (σ = 3.33%), and* **d** *Micron CMOS area sensor MT9V403 in Basler A602f (σ = 1.64%). Contrast enhanced images with a gray value range of $\pm 3\sigma$.*

values. But they are still very low for a good CMOS camera with digital correction as shown in Fig. 5.14c.

The photoresponse nonuniformity is much more difficult to measure because a very homogeneous light source such as an integrated sphere (Fig. 5.11) is required. The imaging sensor is directly illuminated by the light source without a lens and thus receives the same irradiance across the whole sensor area.

The PRNU of CCD sensors is also very low. The standard deviation for the Sony ICX285AL is just about 0.3% (Fig. 5.15a and b). The two images also show that the PRNU is not only influenced by inhomogeneities of the sensor elements but also by small dust particles on the glass window covering the sensor chip. This effect is discussed in more detail in Section 5.4.4c. Uncorrected CMOS sensors show much larger values (Fig. 5.15c) and also much more dead pixels. The degradation of the image quality by these effects is clearly visible to the human eye. There are, however, also CMOS cameras available with a standard deviation of the PRNU that is well below the critical 2% threshold for the human eye (Fig. 5.15c, see also Section 6.3.6e) so that images taken by CMOS sensors appear to the human observer no longer degraded as compared to those taken by CCD sensors, although the PRNU is still significantly higher.

5.4.2 Sensor and Camera Selection

5.4.2a Demands from Applications.
There is no generally valid answer to the question what is the best imaging sensor. In Section 5.3.2, we have seen that there are many different quantities that describe the quality of an imaging sensor. The demands of the intended application decide which of these features are the most important ones. In the following, we discuss a number of typical demands from applications and their implication for camera selection.

Measurements at low light levels. Intuitively, one would argue that in this case, a camera with a high responsivity is required, see Eq. (5.10). What really counts, however, is the signal to noise ratio at *low irradiation* levels. As discussed in Section 5.3.2h, the noise at low irradiation levels is determined by the nonsignal dependent noise variance σ_0^2. This value depends most significantly on the read out frequency of the pixels. A higher read out frequency requires a higher bandwidth of the electronic circuits, which implies a higher noise variance. Modern CCD sensors, such as the Sony ICX285AL, show values of σ_0 that correspond to only a few electrons (about 4–12) even for high read out frequencies between 12.5 and 28 MHz. Noise levels below one electron can only be achieved with imaging sensors that amplify the generated charge unit in one or the other way, so called *intensified CCD sensors (ICCD)*.
If long exposure times can be afforded, even low irradiances result in high irradiation of the sensor. In this case the limiting factor for a good image quality is a low *dark current*. As discussed in Section 5.3.5e, this requires cooling of the sensor. The SensicamQE and PixelflyQE cameras from PCO, for example, use the same Sony ICX285AL CCD chip. The lower dark current of the SensicamQE, where the CCD chip is cooled down to $-12°C$, however, allows exposure times up to 1000 s, while the much higher dark current of the uncooled chip degrades the image quality of the Pixelfly QE already at exposure times as low as 10 s.

Measurements with high irradiance variations. In scientific and industrial applications, it is normally possible to control the illumination so that the objects to be imaged are illuminated homogeneously. This is normally not the case in outdoor scenes in application areas such as surveillance and traffic. A good sensor for such conditions requires not only a low σ_0 but also a high *full-well capacity* so that a high *dynamic range* can be obtained, see Section 5.3.2i.

Precise radiometric measurements. If small irradiance differences must be measured or irradiances with high precision, a high *signal-to-noise ratio* is the most important parameter, see Section 5.3.2j. According to the noise model discussed in Section 5.3.2h, the signal to noise ratio is higher at higher digital gray values:

$$\text{SNR}^{-1} = \frac{\sigma}{g'} = \sqrt{\frac{\alpha}{g'}}\sqrt{1 + \frac{\sigma_0^2}{\alpha g'}} \quad \text{with} \quad g' = g - g_0. \tag{5.26}$$

In addition, high accuracy irradiance measurements require a low *FPN* and *PRNU* (Section 5.3.2k).

Exact size and position measurements. In this case, it is obvious that a sensor with as many pixels as possible is useful. As the discussion in Section 6.3.5 about subpixel-accurate position measurements shows, a high SNR is also of importance for these types of measurements.

Measurements of fast processes. For fast processes, imaging sensors with high frame rates are required. This is a clear domain for CMOS sensors, because it is much

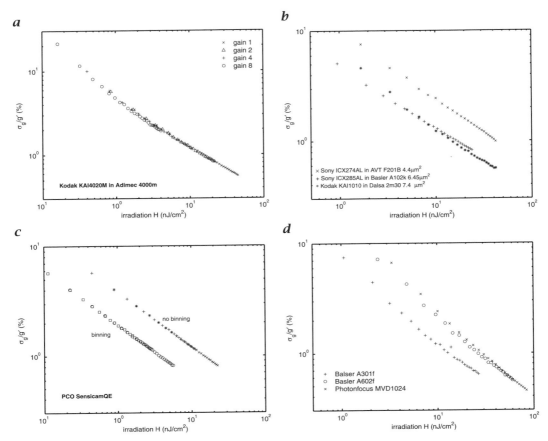

Figure 5.16: *Inverse signal-to-noise ratio σ/g' versus irradiation for different digital cameras: a Adimec A4000m at four different gains, b influence of the pixel area A, c influence of binning, and d comparison between CMOS and CCD sensors with the same pixel area.*

easier to implement parallel pixel output with these devices. In this way, the pixel clock frequency can be kept low in order to achieve a good SNR.

5.4.2b Sensitivity versus Quality. The discussion in the previous section has shown that the most important quantity of an image sensor is the relation of the signal quality expressed by the signal-to-noise ratio to the irradiation H. Using Eqs. (5.26) and (5.8), this relation can be expressed as

$$\text{SNR}^{-1} = \frac{\sigma}{g'} = \sqrt{\frac{hc/\lambda}{\eta(\lambda)AH}}\sqrt{1 + \frac{\sigma_0^2}{\alpha g'}}. \tag{5.27}$$

At high irradiation ($\alpha g' \gg \sigma_0^2$) this equation reduces to

$$\text{SNR}^{-1} = \frac{\sigma}{g'} = \sqrt{\frac{hc/\lambda}{\eta(\lambda)AH}} \tag{5.28}$$

and at low irradiation to

$$\text{SNR}^{-1} = \frac{\sigma}{g'} = \frac{\sigma_0}{g'}. \tag{5.29}$$

These equations give surprisingly simple and general answers for the most important quality parameters of imaging sensors. They clearly show that the SNR increases at high irradiance levels with the square root of the quantum efficiency and the area of the sensor element. It is especially important to note that the SNR does not depend on the amplification factor α. Increasing the amplification does only increase the signal but not its quality!

Unfortunately manufacturers mostly fail to specify the parameters to establish σ / g as a function of the irradiation. Therefore an extensive test measurements were conducted in cooperation with a number of camera manufacturers using the procedures described in Section 5.4.1.

Figure 5.16 shows some results. First, Fig. 5.16a proves that the gain does not influence the relation between the SNR and irradiation. At higher gains lower irradiations are reached, but at the price of a worse SNR. The values at the highest possible irradiation give the best SNR that is possible with a given camera. Figure 5.16b demonstrates that sensors with large pixel areas generally result in a higher SNR.

The sensitivity of an imaging sensor can significantly be increasing by a technique called *binning* (Fig. 5.16c). Here the charges of pixels of a small neighborhood, e. g., a 2×2 neighborhood, are collected together. This reduces the spatial resolution in horizontal and vertical direction by a factor of two, but increases the sensitivity by a factor of four. Note that binning does generally not increase the SNR.

Finally, Fig. 5.16d compares CCD and CMOS sensors with the same pixel size. The two CMOS sensors show an about two times lower sensitivity, because these two sensors do not use microlens arrays to increase the light collecting area. However, the SNR is significantly better. This indicates that these two CMOS sensors have a higher full-well capacity than the corresponding CCD sensor with the same pixel area.

5.4.3 Spectral Sensitivity

As already discussed in Section 5.3.5d (Fig. 5.5), modern CCD sensors do not show much differences in the spectral sensitivity in the visible range (Fig. 5.17). There are, however, significant differences in the ultraviolet and infrared spectral range.

The sensitivity in the ultraviolet (*UV*) range is generally low but there are special sensors available with enhanced UV sensitivity (Fig. 5.5). Note that the UV sensitivity is also influenced by the material used for the glass window protecting the CCD sensor. Standard CCD sensors do not use a quartz glass window and thus typically cut off the sensitivity for wavelengths below 360 nm.

Concerning the sensitivity in the infrared (*IR*) range, there is a general tendency that the IR sensitivity is reduced with the size of the pixels. Again, there are special sensors available with enhanced IR sensitivity (Fig. 5.5). It is important to note that many camera vendors integrate an infrared cut-off filter to limit the sensitivity of the CCD sensor to the visible range.

As shown in Fig. 5.17b, the spectral sensitivity of CMOS sensors differs somewhat from that of CCD sensors. The peak sensitivity is shifted from about 550 nm to about 650 nm and the sensitivity is higher in the infrared. Thus one cannot generally say that current CMOS sensors are less sensitive than CCD sensors. A direct comparison of the Sony ICX074AL CCD sensor with the Micron MT9V403 CMOS sensors shows that the CMOS sensor is about two times less sensitive at 550 nm but more sensitive at wavelengths larger than 750 nm.

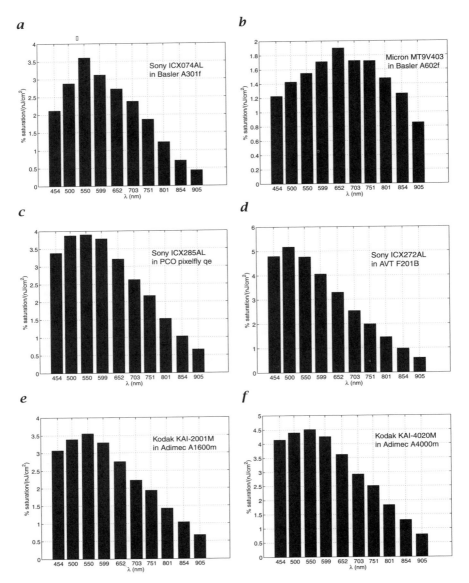

Figure 5.17: *Spectral sensitivity for wavelengths between 450 and 905 nm for imaging sensors as indicated.*

5.4.4 Artifacts and Operation Errors

Given the complexity of modern imagers, the quality of the acquired images can seriously suffer by misadjustments or by simply operating them incorrectly. It can also happen that a vendor delivers a sensor that is not well adjusted. In this section, we therefore show some common misadjustments and discuss a number of artifacts limiting the performance of imaging sensors.

5.4.4a Offsets and Nonlinearities by Misadjustments. A serious misadjustment is a strong nonlinearity shown in Fig. 5.18a. The trouble with such nonlinearity is that is can hardly be detected by just observing images of the camera. In order to discover

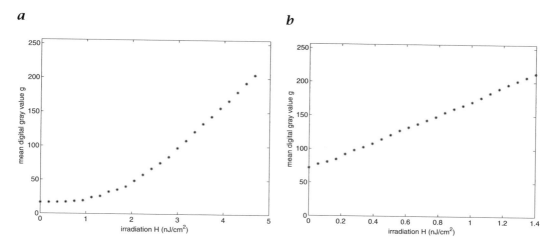

Figure 5.18: *Bad digital camera signals by misadjustments: **a** Nonlinear response at low irradiance, **b** Too high values for the dark image.*

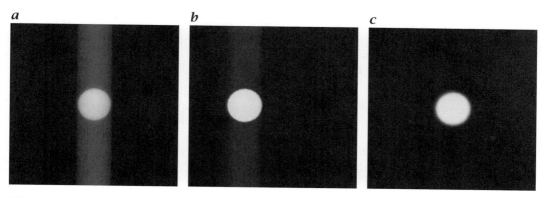

Figure 5.19: *Smear effects observed by constant illumination and a short exposure time of 20 μs with **a** PCO pixelfly scientific (Sony ICX285AL sensor, diameter of spot about 80 pixel); **b** PCO pixelfly qe (Sony ICX285AL sensor, diameter of spot about 80 pixel); **c** Basler A602f CMOS camera (Micron MT9V403 sensor, diameter of spot about 54 pixel).*

such a misadjustment, a sophisticated linearity measurement is required as described in Section 5.4.1. It is much easier to detect and avoid a too high offset value as shown in Fig. 5.18b.

Especially disturbing is an underflow. Then the digital gray values are zero below a critical irradiation and it appears that the dark image shows no noise.

Figure 5.18b shows also wiggles in the linearity that indicate that the camera electronics is not well adjusted.

5.4.4b Blooming and Smear.

When short exposure times in the μs range are used with constant illumination, an artifact known as *smear* can be observed. This effect is caused by the tiny residual light sensitivity of the interline storage areas. At exposure times equal to the read out time, this causes no visible effects. When the exposure time is only a small fraction of the read out time (e. g., 20 μs versus 100 ms), additional charges are accumulated when a charge packet is transferred through an illuminated area. Therefore vertical streaks are observed when a small area is illuminated (Fig. 5.19a

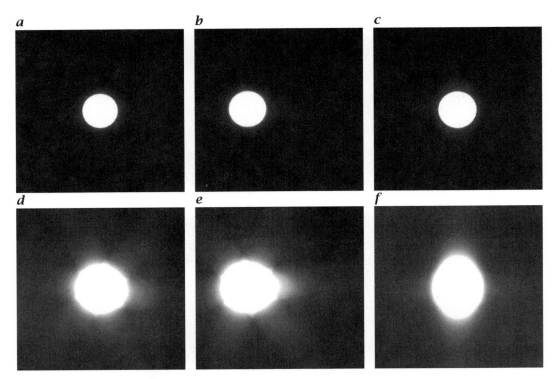

Figure 5.20: *Blooming effects observed by 20-fold (upper row) and 400-fold (lower row) overexposure with **a** PCO pixelfly scientific (Sony ICX0285AL sensor, diameter of spot about 80 pixel); **b** PCO pixelfly qe (Sony ICX0285AL sensor, diameter of spot about 80 pixel); **c** Basler A602f CMOS camera (Micron MT9V403 sensor, diameter of spot about 54 pixel). In all exposures a Schneider Kreuznach Xenon 0.95/25 lens was used.*

and b). In contrast to CCD sensors, CMOS sensors do not show any smear effects (Fig. 5.19c).

Another artifact, known as *blooming*, occurs in overexposed areas. Then a part of the generated charges leaks to neighboring pixels. Modern CCD imaging sensors include antiblooming circuits so that this effect occurs only with high overexposure (Fig. 5.20d and e). Some CMOS sensors also show blooming (Fig. 5.20f).

5.4.4c Dirt on the CCD Cover Glass. A trivial but ubiquitous flaw is dirt on the cover glass of the CCD sensor. Dirt can significantly contribute to the inhomogeneity of the sensor responsivity (*PRNU*, see Section 5.3.2k). The images in Fig. 5.21 have been taken by removing the lens from the camera and illuminating the chip with an integrated sphere. In this way, the dirt is directly projected onto the chip surface. The visibility of dirt in images strongly depends on the f-number. At lower f-numbers it is less visible because the projection onto the chip is very blurred due to the more convergent light beams.

Cleaning of the window of a CCD imager is not easy. Thus the first rule is to keep dirt from settling on the window by removing the cover or lens of the camera only when required and then only in a clean environment.

5.4.4d Motion Artifacts by Interlaced Exposure. An important practical limitation for the analysis of image sequences with standard CCD sensors is the interlaced ex-

a   b

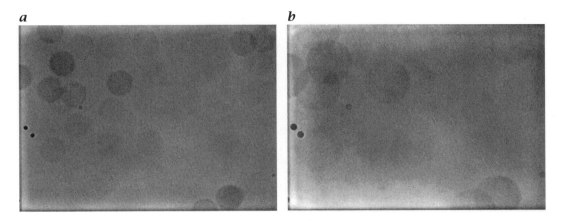

Figure 5.21: *Small dirt particles on the cover glass of a CCD sensor and the IR cutoff filter made visible by illuminating the sensor with diverging light corresponding to an aperture of **a** 10.6 and **b** 6.1. The contrast ranges are ±2% and ±1%, respectively.*

a   b

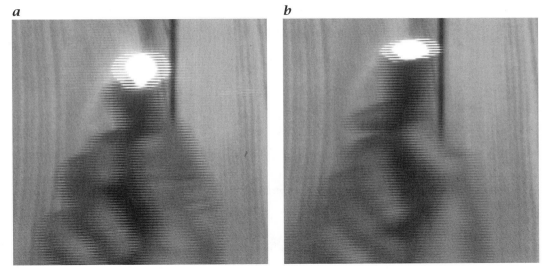

Figure 5.22:
*Separate exposure of the two fields of an image with interlaced CCD sensors: **a** interlaced field integration (20 ms exposure time for each field with 20 ms offset); **b** interlaced frame integration (40 ms exposure time for each field with 20 ms offset).*

posure of the two fields of an image (Section 5.3.6). This means that moving objects appear in a frame twice (Fig. 5.22). The exposure of the second field is delayed by one field time (20 ms with the European and 16 2/3 ms with the US video norms). The two example images also show nicely that with interlaced field integration (Fig. 5.7) the two fields are exposed one after the other, while with interlaced frame integration the exposure overlaps by one field time (Fig. 5.9). There is no way to avoid this problem with standard CCD sensors. For image sequence analysis, the two fields must be regarded as individual images with half the vertical resolution. Full resolution can only be gained with progressive scanning sensors.

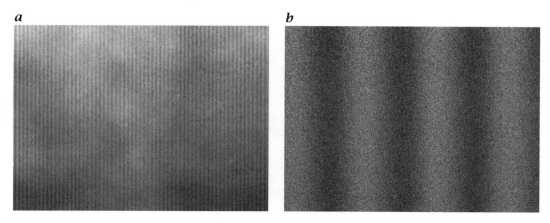

Figure 5.23: *Mean and standard deviation of the blue channel of a 3-CCD color camera computed from a series of 100 images.*

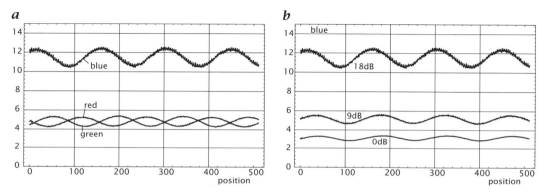

Figure 5.24: *Row profiles of the standard deviation of signals of a 3-CCD color camera:* **a** *red, green, and blue channel with 18 dB amplification;* **b** *blue channel with 0 dB, 9 dB, and 18 dB amplification.*

5.4.4e Electronic Interferences. Various types of electronic interferences can degrade the quality of image signals. Figures 5.23 and 5.24 shown significant periodic interference patterns in the mean and the standard deviation of the dark signals of an analog 3-CCD color camera, possibly caused by an electromagnetic interference in the electronic circuits of the camera. Figure 5.24a demonstrates that the noise level of the blue channel is more than two times higher than that of the red and green channels. Figure 5.24b shows the increase of the noise level with the video gain for values between 0 dB and 18 dB.

Other forms of disturbances in dark images are shown in Fig. 5.25.

5.4.4f Color Carrier Signal in Gray Scale Images. In a composite color video signal (Section 5.5.3) the color information is carried in a modulated carrier signal at the high frequency end (4 MHz) of the video signal. Some gray scale frame grabbers filter the frequency around the carrier frequency; others do not. If not filtered, the pattern caused by the color signal in gray scale frame grabbers renders these images almost useless (Fig. 5.26). Thus it cannot be recommended to digitize the composite signal of a color video camera with a gray scale frame grabber.

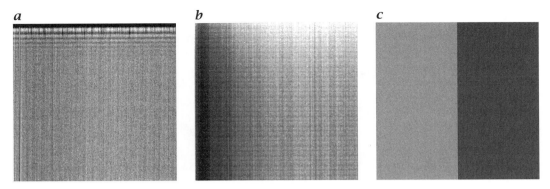

Figure 5.25: *Various examples of disturbances in dark images:* **a** *and* **b** *electronic interferences,* **c** *Offset in a dual-tap camera with two parallel outputs.*

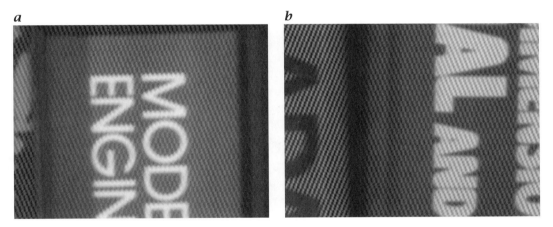

Figure 5.26: *PAL color signal visible in a composite video signal grabbed with a gray scale frame grabber without color carrier filter.*

5.5 Advanced Reference Material

5.5.1 Basic Properties of Imaging Sensors

5.1 **Nominal sizes of imaging sensors**

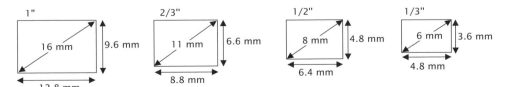

Nominal sizes of the imaged area for CCD cameras for the four standards of 1", 2/3", 1/2", and 1/3" sensors. Imaged sectors of real sensors may slightly deviate from these sizes (≻ 5.2–5.4).

5.2   Basic data of commonly used Sony interline CCD sensors

Camera manufacturers often do not specify the CCD sensor that is contained in their products. With the help of the following three tables, especially the pixel size, it is mostly possible to identify the used sensor.

Chip	Format	Area	Pixel size $(H) \times (V)$ [µm]	Pixel area [µm²]	Number of effective pixels	Pixel clock [MHz]
RS170 (American) video norm, interlaced, 30 frames/s						
ICX422AL	2/3"	8.91×6.67	11.6×13.5	156.6	768×494	
ICX428ALL	1/2"	6.45×4.84	8.4×9.8	82.3	768×494	
ICX254AL	1/3"	4.90×3.69	9.6×7.5	72.0	510×492	
ICX258AL	1/3"	4.88×3.66	6.35×7.4	47.0	768×494	
ICX278AL	1/4"	3.65×2.74	4.75×5.55	26.4	768×494	
CCIR (European) video norm, interlaced, 25 frames/s						
ICX249AL	2/3"	8.72×6.52	11.6×11.2	129.9	752×582	
ICX249AL	1/2"	6.47×4.83	8.6×8.3	71.4	752×582	
ICX259AL	1/3"	4.89×3.64	6.50×6.25	40.6	752×582	
ICX279AL	1/4"	3.65×2.71	4.85×4.65	22.6	752×582	
Progressive scanning, VGA resolution						
ICX074AL	1/2"	6.52×4.89	9.9×9.9	98.0	659×494	
ICX415AL	1/2"	6.40×4.83	8.3×8.3	68.9	782×582	29.5
ICX424AL	1/3"	4.88×3.66	7.4×7.4	54.8	659×494	24.5
Progressive scanning, 1.3 Mpixel						
ICX085AL	2/3"	8.71×6.90	6.7×6.7	44.9	1300×1030	
ICX285AL	2/3"	8.98×6.71	6.45×6.45	44.9	1392×1040	
ICX205AL	1/2"	6.47×4.84	4.65×4.65	21.6	1392×1040	
Progressive scanning, 2 Mpixel						
ICX274AL	1/1.8"	7.16×5.44	4.40×4.40	19.4	1628×1236	

5.3   Basic data of some CMOS imaging sensors

Chip	Format	Area	Pixel size $(H) \times (V)$ [µm]	Pixel area [µm²]	Number of effective pixels	Pixel clock [MHz]
Micron MT9V403	1/2"	6.49×4.86	9.9×9.9	98.0	656×491	66
Photonfocus D1024	1"	10.85×10.85	10.6×10.6	112.4	1024×1024	2×40
Fillfactory Lupa1300		17.92×14.34	14.0×14.0	196.0	1280×1024	16×40
Micron MV40		16.46×12.10	7.0×7.0	49.0	2352×1728	16×40

Basic data of some Kodak interline CCD imaging sensors | 5.4 |

Chip	Format	Area	Pixel size $(H) \times (V)$ [μm]	Pixel area [μm^2]	Number of effective pixels	Pixel clock [MHz]
Progressive scanning, VGA resolution						
KAI-0330D	1/2"	5.83×4.36	9.0×9.0	81.0	648×484	
Progressive scanning, 1.0 Mpixel						
KAI-1003M		13.1×13.1	12.8×12.8	163.8	1024×1024	
KAI-1010M		9.07×9.16	9.0×9.0	81.0	1008×1018	
KAI-1020M		7.4×7.4	7.4×7.4	54.8	1000×1000	2×40
Progressive scanning, multi-Mpixel						
KAI-2001M		11.84×8.88	7.4×7.4	54.8	1600×1200	2×40
KAI-2093M		14.21×7.99	7.4×7.4	54.8	1920×1080	2×40
KAI-4020M		15.16×15.16	7.4×7.4	54.8	2048×2048	2×40
KAI-10000M		43.20×24.05	9.0×9.0	81.0	4008×2672	2×28

5.5.2 Standard Video Signals; Timing and Signal Forms

The standard analog video signal contains the image information in a time-serial form. Besides the image information itself, the signal contains synchronization information indicating the start of a new image and of a new row of the image.

Timing of analog video signals | 5.5 |

Timing of the RS-170 (US norm), CCIR (European norm), and the VGA (video graphics adapter) analog video signals.

	RS-170	CCIR	VGA
Horizontal scan time	63.5556 μs	64.0 μs	63.556 μs
Horizontal scan frequency f_H	15.7343 kHz	15.625 kHz	15.7343 kHz
Vertical scan time	16.6833 ms	20.0 ms	16.6833 ms
Vertical scan frequency f_V	59.9401 Hz	50.0 Hz	59.9401 Hz
f_H / f_V	262.5	312.5	
Number of rows	525	625	525
Visible rows	21–263[2]	22[1]–310	484
	20[1]–262	336–623[2]	
	total: 485	total: 577	
Scan type	interlaced	interlaced	noninterlaced progressive
Number of fields	2	2	1

[1] only second half of row is visible; [2] only first half of row is visible

5.6   **Video timing diagram for the RS170 norm.**

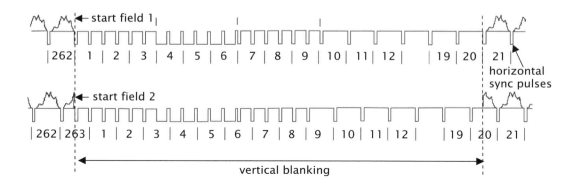

5.7   **Video timing diagram for the CCIR norm.**

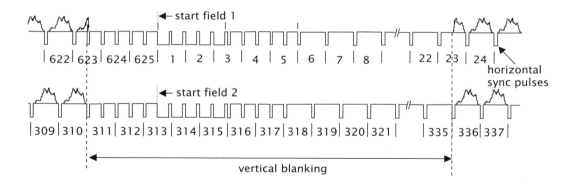

5.5.3 Color Video Signals

Color bar

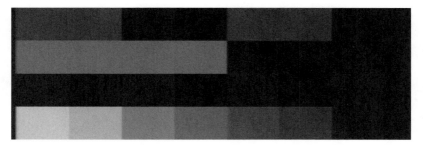

The color bar is used as a standard test pattern for analog video signals: The four horizontal stripes in the figure contain from top to bottom the following components of the signal: red channel only; green channel only; blue channel only; red, green, and blue channel together.

In this figure, the color bar signal is shown as a PAL composite video signal. It was digitized with a gray scale frame grabber that does not suppress the color carrier frequency and, thus, shows the colors as high frequency patterns. Notice that the white stripe is free of the high frequency patterns since it contains no color (the U and V components are zero). Compare also Fig. 5.26 and Plate 15.

The following diagrams show the representation of the color bar in different video signal standards:

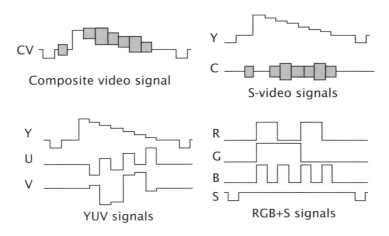

Figure 5.27: *a Red, **b** green, and **c** blue channels of a color image (**d**) taken with a 3-CCD RGB color camera. (See also Plate 16.)*

The rest of the reference section on color video signals shows example images in various color systems. If you study and compare these images carefully, you will get a good impression on what type of information is contained in which signal and how these signals are interrelated.

- All sensors take color images with three sensors that are sensitive to red, green, and blue colors as is the human eye (Section 3.4.7). Such an image is referred to as an *RGB image*. Figure 5.27 shows the three channels, red, green, and blue, separately and combined to an RGB color image.

- Normally, video signals are not stored as RGB images, although this would be the best way for scientific applications. To reduce the bandwidth of television signals, the color images are adapted to the human visual system which has a significantly lower spatial resolution for colors than for luminance. Therefore, the signal is transformed into a *color difference system* (Figure 5.28). The Y channel contains the luminance signal, i. e., the sum of the red, green, and blue signals, while the U and V signals contain color differences.

- Color can also be presented in a more natural way, i. e., how we sense colors (Section 3.4.7d). Figure 5.29 shows the hue and the color saturation signals.

Figure 5.28: *Demonstration of the YUV color coordinate system with the same image as shown in Fig. 5.27:* **a** *Y channel,* **b** *U channel,* **c** *V channel.*

Figure 5.29: **a** *Hue and* **b** *saturation signals of the image shown in Fig. 5.27.*

5.5.4 Cameras and Connectors

As a final part of the reference section in this chapter, some digital solid state cameras and their standard connectors are shown for the readers who are not yet familiar with these devices.

| 5.9 | **Firewire (IEEE1394)**

a Back of a digital camera with Firewire connector and a trigger connector (Basler A602f) compared to a standard analog video camera (Sony XC75). **b** Two more examples of Firewire cameras (Teli and Allied Vision)

a  *b*

| 5.10 | **Camera link cameras**

Examples of camera link cameras; note that only the connector for digital data transmission and camera control is standardized but not the power connector. **a** Pulnix TM1020-CL and Photonfocus MV-D1024-80-CL-8 **b** Balser A102k and Dalsa 2M30-SA CL.

a  *b*

5.5.5 Further References

General References 5.11

If you want to get up to date on recent developments on imaging sensors and related topics, one or more of the following monthly published magazines with free subscription to qualified readers are an excellent source.

Vision Systems Design, PennWell, Nashua, NH, USA,
    `http://www.vision-systems.com`

Laser Focus World, Pennwell Publishing Company, Nashua, NH, USA,
    `http://www.lfw.com`

Photonics Spectra, Laurin Publishing Company, Pittsfield, MA, USA,
    `http://www.photonics.com`

Manufacturers of imaging sensors and cameras 5.12

Here is a collection of manufacturers of imaging sensors including the infrared.

Adimec Electronic Imaging, Eindhoven, The Netherlands (industrial CCD cameras),
    `http://www.adimec.com`

Allied Vision Technologies, Stadtroda, Germany (industrial CCD cameras),
    `http://www.sticksel.de`

AIM, Heilbronn, Germany (Infrared cameras), `http://www.aim-ir.de`

Andor Scientific, Belfast, Northern Ireland (scientific-grade CCD cameras),
    `http://www.andor-tech.com`

Basler Vision Technologies, Ahrensburg, Germany (industrial CCD cameras),
    `http://www.baslerweb.com`

CEDIP Infrared Systems, France (infrared cameras),
    `http://www.cedip-infrared.com`

Indigo Systems, Goleta, CA (infrared cameras),
    `http://www.indigosystems.com`

Jai (industrial CCD cameras), `http://www.jai.dk`

KAPPA opto-electronics, Gleichen, Germany, `http://www.kappa.de`

Kodak (image sensors), `http://www.kodak.com/global/en/digital/ccd/`

Micron (CMOS image sensors), `http://www.micron.com/products/imaging/`

Photonfocus, Lachen, Switzerland (CMOS image sensors),
    `http://www.photonfocus.com`

Pixera digital cameras, `http://www.pixera.com`

PixelVision (scientific-grade CCD cameras), `http://www.pv-inc.com`

Pulnix America, Inc., Industrial Products Division, `http://www.pulnix.com`

Roper Scientific, (scientific-grade CCD cameras),
    `http://www.roperscientific.com`

Sony Electronics (image sensors),
    `http://www.sony.net/Products/SC-HP/index.html`

SITe (scientific-grade CCD sensors), `http://www.site-inc.com`

Thermosensorik, Erlangen, Germany (infrared cameras),
    `http://www.thermosensorik.com`

5.13   References to solid-state imaging

G. C. Holst, 1998. CCD Arrays, Cameras, and Displays, 2nd edition, SPIE, Bellingham, WA.

S. B. Howell, 2000. Handbook of CCD Astronomy, Cambridge University Press, Cambridge, UK. *Monograph on high-end CCD technology used in astronomy.*

J. R. Janesick, 2001. Scientific Charge-Coupled Devices, SPIE, Bellingham, WA. *Detailed account on CCD technology.*

5.14   References to special topics

A. F. Inglis, 1993. Video engineering, McGraw-Hill, New York. *Introduction to the basics of modern video technology including HDTV and digital video.*

G. Gaussorgues, 1994. Infrared Thermography, Chapman & Hall, London. *A well written introduction to infrared thermography with a lot of useful reference material.*

J. L. Miller, 1994. Principles of Infrared Technology. A Practical Guide to the State of the Art, Van Nostrand Reinhold, New York. *Technical reference to infrared detectors and systems.*

G. C. Holst, 2000. Common Sense Approach to Thermal Imaging, SPIE, Bellingham, WA.

P. W. Kruse, 2001. Uncooled Thermal Imaging, Arrays, Systems, and Applications, SPIE, Bellingham, WA. *Tutorial text on principles, state, and trends of uncooled thermal imaging.*

6 Digitalization and Quantization

6.1 Highlights

Images need to be digitized before they can be processed by a digital computer. In order to form a discrete image, the irradiance at the image plane is sampled in regular spatial (and for image sequences also in temporal) intervals (Section 6.3.1). Thus, each point of the digital image, a pixel or voxel, represents the irradiance of a rectangular cell.

The sampling procedure leads not only to a limitation in the spatial and temporal resolution of images but also can introduce significant distortions by aliasing finer to coarser spatial structures (Section 6.3.1d).

The sampling theorem (Section 6.3.2) states that each periodic structure contained in an image must be sampled at least twice per wavelength. Then, no distortions are introduced by the sampling, no information contained in the image is lost, and the original continuous image can completely be reconstructed from the sampled digital image.

Quantization (Section 6.3.6) of the irradiance into discrete levels to form digital numbers results in a similar resolution reduction in the gray scale.

The following procedures are required to measure the overall quality of the digital image as a result of the image formation and digitalization process:

1. Measurement of the point spread function (PSF) and transfer function (TF) (Section 6.4.1).
2. Measurement of the quantization error (Section 6.4.2).

The reference part (Section 6.5) gives a survey of the classes of frame grabbers available to digitize images and to transfer them to the computer for further processing. The discussion includes the advantages and disadvantages of cameras with digital output (Section 6.5.3) and the real-time acquisition and image processing (Section 6.5.4).

6.2 Task

Digitalization and quantization are the processes that convert a continuous image irradiance as it is projected by an optical system onto the image plane into a matrix of digital numbers that can be stored and processed by a digital computer. This process is not unique to image processing. It occurs with any type of signals and is generally known as *sampling*. For scalar signals, e. g., a temperature record, a time series is formed. This time series is sampled at discrete temporal intervals. For images, a similar spatial sampling strategy has to be applied. Finally, image sequences require spatial and temporal sampling.

Generation of a *digital image* from a continuous image requires first of all knowing how to number the elements (pixels) of the digital image and becoming familiar with the peculiarities of digital images such as neighborhood relation, and the definition of discrete grids, distances, and directions in digital images. This is the topic of Section 6.3.1. In Section 6.3.1d, we make ourselves familiar with the sampling process and the basic problem that we not only lose information but can also introduce false structures into the sampled digital images that are not present in the original image.

Thus, the central task of this chapter is to understand under what conditions we can expect that the sampled image does not contain false structures and what information content of the image is lost in the sampling process. The definitive answer to this question is given by the *sampling theorem* in Section 6.3.2. In Section 6.3.3, sampling is extended to image sequences.

Sampling limits the spatial resolution significantly and even high-resolution images with $2k \times 2k$ or $4k \times 4k$ pixels can be regarded as low-resolution in comparison with many precision measuring techniques available in physical sciences. Therefore, it is an important issue to discuss whether the measurement of position, distance, and size is limited to the digital grid of the sensor or whether we can do better and gain *subpixel accuracy* (Section 6.3.5).

The quantization issue is discussed in Section 6.3.6. Quantization limits the resolution with which the radiant flux can be measured and inevitably results in additional errors.

6.3 Concepts

6.3.1 Digital Images

Digitization means sampling the gray values at a discrete set of points. Sampling may already be happening in the sensor that converts the collected photons into an electrical signal. In a conventional tube camera, the image is already sampled in lines, as an electron beam scans the imaging tube line by line. A CCD camera already has a matrix of discrete sensors (Section 5.3.6). Each sensor is a sampling point on a 2-D grid. The standard video signal is, however, an analog signal (Section 5.5.2). Consequently, we lose the horizontal sampling again, as the signal from a line of sensors is converted back into an analog signal.

Mathematically, digitization is described as the mapping from a continuous function in $\mathbb{R}^2$ onto an array with a finite number of elements:

$$g(x_1, x_2) \xrightarrow{D} G_{m,n} \qquad x_1, x_2 \in \mathbb{R} \quad \text{and} \quad m, n \in \mathbb{Z}. \tag{6.1}$$

The array of discrete points can take different geometries. Solid state physicists, mineralogists, and chemists are familiar with problems of this kind. Crystals show periodic 3-D patterns of the arrangements of their atoms, ions, or molecules which can be classified due to their symmetries. In 3-D this problem is quite complex. In 2-D we have fewer choices. For a two-dimensional grid of the image matrix, a rectangular basis cell is almost exclusively chosen.

6.3.1a Pixel or Pel. A point on the 2-D grid is called a *pixel* or *pel*. Both words are abbreviations of the word picture element. A pixel represents the gray value at the corresponding grid position. The position of the pixel is given in the common notation for matrices. The first index, m, denotes the position of the row, the second, n, the position of the column (Fig. 6.1a). If the image is represented by an $M \times N$ matrix, the

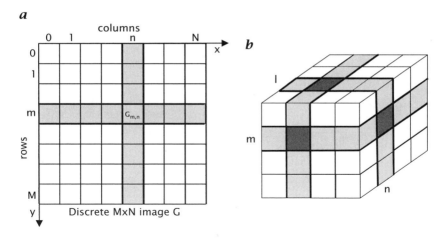

Figure 6.1: *Representation of digital images by arrays of discrete points with rectangular cells:* **a** *2-D image,* **b** *3-D image.*

index n runs from 0 to $N-1$, and the index m from 0 to $M-1$. M gives the number of rows, N the number of columns. Note also that the vertical axis runs from top to bottom.

A point on a 3-D grid is called a *voxel*, an abbreviation of volume element. On a rectangular grid, each voxel represents the mean gray value of a cuboid. The position of a voxel is given by three indices. The first, l, denotes the depth, m the row, and n the column (Fig. 6.1b). The coordinate transformations between image coordinates and pixel coordinates are discussed in Section 4.3.2b.

6.3.1b Neighborhood Relations. On a *rectangular grid* in two dimensions, there are two possibilities to define neighboring pixels (Fig. 6.2a and b). We can either regard pixels as neighbors when they have a joint edge or when they have at least one joint corner. Thus four and eight neighbors exist, respectively, and we speak of a *4-neighborhood* or an *8-neighborhood*.

Both definitions are needed for a proper definition of objects as connected regions. A region is called connected or an object when we can reach any pixel in the region by walking to neighboring pixels. The black object shown in Fig. 6.2c is one object in the 8-neighborhood, but constitutes two objects in the 4-neighborhood. The white background, however, shows the same property. Thus the inconsistency arises that we may have either two connected regions in the 8-neigborhood crossing each other or two separated regions in the 4-neighborhood. This difficulty can be overcome if we declare the objects as 4-neighboring and the background as 8-neighboring, or vice versa.

These complications are a special feature of the *rectangular grid*. They do not occur with a *hexagonal grid* (Fig. 6.2d), where we can only define a *6-neighborhood*, since pixels which have a joint corner, but no joint edge, do not exist. Neighboring pixels have always one joint edge and two joint corners. Despite this advantage, hexagonal grids are hardly used in image processing, as the imaging sensors generate pixels on a rectangular grid.

In three dimensions, the neighborhood relations are more complex. Now, there are three ways to define a neighbor: voxels with joint faces, joint edges, and joint corners. These definitions lead to a 6-, 18-, and 26-neighborhood, respectively (Fig. 6.3). Again,

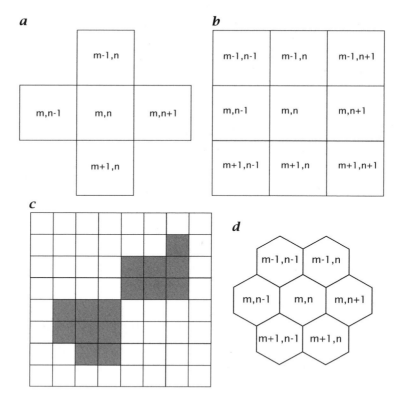

Figure 6.2: *Neighborhoods are defined by adjacent pixels. On a rectangular grid, neighbors could have a common edge (**a** , 4-neighborhood) or common corners (**b** , 8-neighborhood). **c** The black region is one object (connected region) in an 8-neighborhood but two objects in a 4-neighborhood. **d** On a hexagonal grid, only one type of neighborhood exists, the 6-neighborhood.*

we have to define two different neighborhoods for objects and the background in order to achieve a consistent definition of connected regions. The objects and background must be declared as a 6-neighborhood and a 26-neighborhood, respectively, or vice versa.

6.3.1c Discrete Geometry.   The discrete nature of digital images makes it necessary to redefine elementary geometrical properties such as distance, slope of a line, translation, rotation, and scaling of the coordinates. In order to discuss the discrete geometry properly, we introduce the *grid vector* which represents the position of the pixel and which has already been used to define the pixel coordinates in Section 4.3.2b. The grid vector is defined in 2-D as

$$\boldsymbol{x}_{m,n} = \left[\begin{array}{c} n\Delta x \\ m\Delta y \end{array} \right],$$ (6.2)

in 3-D as

$$\boldsymbol{X}_{l,m,n} = \left[\begin{array}{c} n\Delta x \\ m\Delta y \\ l\Delta z \end{array} \right],$$ (6.3)

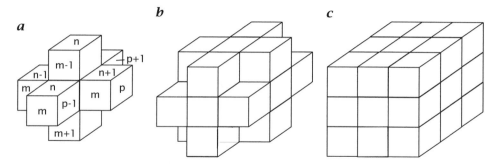

Figure 6.3: *Neighborhoods on a 3-D rectangular grid. **a** 6-neighborhood: voxels with joint faces. **b** 18-neighborhood: voxels with joint edges. **c** 26-neighborhood: voxels with joint corners.*

and for an image sequence as

$$X_{k,m,n} = \begin{bmatrix} n\Delta x \\ m\Delta y \\ k\Delta t \end{bmatrix}. \tag{6.4}$$

Translation on a discrete grid is only defined in multiples of the pixel or voxel distances

$$x'_{m,n} = x_{m,n} + x_{m',n'} \qquad \text{discrete translation,} \tag{6.5}$$

i. e., by addition of a grid vector.

Likewise, scaling is possible only for integer multiples of the scaling factor

$$x'_{m,n} = x_{pm,qn} \qquad p,q \in \mathbb{Z} \qquad \text{discrete scaling} \tag{6.6}$$

by taking only every q^{th} pixel on every p^{th} line. Since this discrete scaling operation subsamples the grid, it remains to be seen whether the scaled version of the image is still a valid representation.

Rotations on a discrete grid are not possible except for some trivial angles. The condition is that all points of the rotated grid must coincide with points on the original grid. On a rectangular grid, only rotations by multiples of 180° are possible, on a square grid by multiples of 90°, and on a hexagonal grid by multiples of 60°.

Generally, the correct representation even of simple geometric objects such as lines and circles is not trivial. Lines are well defined only for angles with values of multiples of 45°, whereas for all other directions they appear as jagged, staircase-like sequences of pixels (Fig. 6.4).

These difficulties do not only lead to "ugly" representations of boundaries of objects and contour lines but also cause errors in the determination of the position of objects, edges, the slope of lines, and the shape of objects in general.

Problems related to the discrete nature of digital images are common to both image processing and *computer graphics*. In computer graphics, the emphasis is on a better appearance of images, e. g., techniques to avoid jagged lines. In contrast, image processing focuses on the accurate representation and analysis of the position and form of objects.

6.3.1d Moiré-Effect and Aliasing. Digitization of a continuous image constitutes an enormous loss of information, since we reduce the information about the gray values from an infinite to a finite number of points. Therefore, the crucial question which

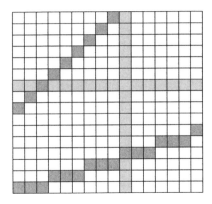

Figure 6.4: *A digital line is only well defined in the directions of axes and diagonals. In all other directions, a line appears as a staircase-like jagged pixel sequence.*

a   b   c

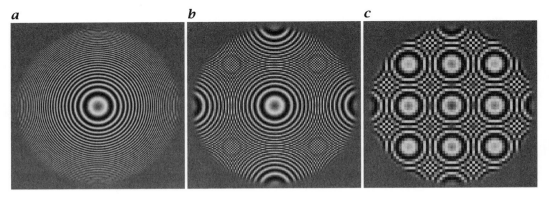

Figure 6.5: *Illustration of the Moiré-effect: The original image with the ring test pattern in **a** is subsampled by taking every second pixel and line (**b**) and every fourth pixel and line (**c**).*

arises now is if we can ensure that the sampled points are a valid representation of the continuous image. This means two things. First, no significant information should be lost. Second, no distortions should be introduced. We also want to know to which extent we can reconstruct a continuous image from the sampled points. We will approach these questions by first studying the distortions which result from inadequate sampling.

Intuitively, it is clear that sampling leads to a reduction in resolution, i. e., structures about the scale of the sampling distance and finer will be lost. It might come, however, as a surprise to know that considerable distortions occur if we sample an image, which contains fine structures. Figure 6.5 shows a simple example, the sampling of a striped pattern of different directions and orientations. Digitization is simulated by taking only every second or fourth pixel and row. After sampling, fine structures appear as coarser structures with different periodicity and direction. From Figure 6.5c it is obvious that structures with different periodicity and direction are mapped to the same pattern since the ring patterns appear in periodic replications. This kind of image distortion is called the *Moiré-effect*.

The same phenomenon is known for one-dimensional signals such as time series as *aliasing*. Figure 6.6 shows a signal with a sinusoidal oscillation. It is sampled with a sampling distance which is slightly smaller than its wavelength. As a result we can

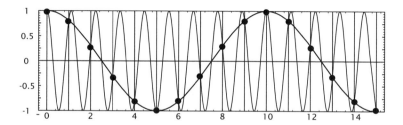

Figure 6.6: *Demonstration of the aliasing effect: an oscillatory signal (thin line) is sampled with a sampling distance Δx equal to 9/10 of the wavelength (thick dots). The result (thick line) is an aliased wavelength which is 10 times the sampling distance.*

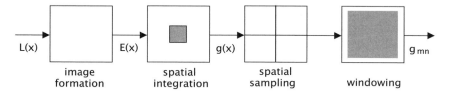

Figure 6.7: *The chain of operations from object radiance L to a finite discrete image matrix G.*

observe a much larger wavelength. Whenever we digitize analog data, these problems occur. It is a general phenomenon of signal processing. In this respect, image processing is only a special case in the more general field of signal theory.

Since the aliasing effect has been demonstrated with periodic signals, the key to understand and thus to avoid it is by an analysis of the digitization process in Fourier space. In the following, we will perform this analysis step by step. As a result, we can formulate the conditions under which the sampled points are a correct and complete representation of the continuous image in the so-called *sampling theorem*.

6.3.2 The Sampling Theorem

Our starting point is an infinite, continuous image $g(\boldsymbol{x})$, which we want to map onto a finite matrix $g_{m,n}$. In this procedure we will include the image formation process, which we discussed in Section 4.3.5. We can then distinguish four separate steps (Fig. 6.7):

1. image formation (Section 6.3.2a),
2. spatial and temporal integration by the photo sensor elements (Section 6.3.2b),
3. sampling (Section 6.3.2c), and
4. limitation to a finite image matrix (Section 6.3.2d).

6.3.2a Image Formation. Digitization cannot be treated without the *image formation* process. The optical system influences the image signal so that we should include the effect in this process. In Section 4.3.5 we discussed in detail that an optical system can be regarded as a linear shift invariant system in good approximation. Therefore the optical system can be described by a point spread function and the irradiance on the image plane $E(\boldsymbol{x})$ is given by convolving the object radiance projected onto the image plane, $L(\boldsymbol{x})$, with the point spread function, $h_o(\boldsymbol{x})$:

$$E(\boldsymbol{x}) = t\pi \sin^2(\Theta') \int\limits_{-\infty}^{\infty} \int\limits_{-\infty}^{\infty} L(\boldsymbol{x} - \boldsymbol{x}') h_o(\boldsymbol{x}') \mathrm{d}^2 x'. \tag{6.7}$$

In this equation, Eq. (4.54) was used to relate the irradiance on the image plane to the radiance of an object observed by an optical system with an image-sided numerical aperture $\sin(\Theta')$.

6.3.2b Spatial and Temporal Integration by Sensor Element.

The photosensors further modify the irradiance $E(\boldsymbol{x})$ by integrating it both in the spatial and temporal domain. We focus here only on the spatial integration. In order to separate this effect from the sampling, we just think of a single photosensor that is moved around over the whole image plane. As the simplest example, we take an individual sensor of an ideal sensor array, which consists of a matrix of directly neighboring photodiodes, without any light insensitive strips in between. We further assume that the photodiodes are uniformly and equally sensitive. Then, the irradiance distribution $E(\boldsymbol{x})$ at the image plane will be integrated over the area of the individual photodiode. This corresponds to the operation

$$g(\boldsymbol{x}) = \alpha\eta\lambda\frac{t}{hc}\int_{x-1/2\Delta x}^{x+1/2\Delta x}\int_{y-1/2\Delta y}^{y+1/2\Delta y} E(\boldsymbol{x})\,\mathrm{d}x\,\mathrm{d}y. \tag{6.8}$$

This equation also includes the conversion of the irradiance on the image plane into a digital signal by a sensor with a quantum efficiency η, the area $A = \Delta x \Delta y$, and the exposure time t according to Section 5.3.2d, Eq. (5.9).

This operation in Eq. (6.8) constitutes a *convolution* with a two-dimensional box function. Because convolution is an associative operation, we can combine the averaging process of the CCD sensor with the *PSF* of the optical system (Section 4.3.5b) in a *single* convolution process. Therefore a single convolution operation describes the whole image formation from the object radiance to the digital sensor including the effects of the optical system and the sensor:

$$g(\boldsymbol{x}) = t\pi\sin^2(\Theta')\alpha\eta\lambda\frac{t}{hc}L(\boldsymbol{x}) * h_o(\boldsymbol{x}) * h_s(\boldsymbol{x}), \tag{6.9}$$

or in Fourier space

$$\hat{g}(\boldsymbol{k}) = t\pi\sin^2(\Theta')\alpha\eta\lambda\frac{t}{hc}\hat{L}(\boldsymbol{k})\hat{h}_o(\boldsymbol{k})\hat{h}_s(\boldsymbol{k}), \tag{6.10}$$

where $h_s(\boldsymbol{x})$ and $\hat{h}(\boldsymbol{k})$ are the *point spread function* and *transfer function* of the spatial integration by the sensor elements.

Generally, the whole image formation process results in a blurring of the image; fine details are lost. In Fourier space this leads to an attenuation of high wave numbers. The resulting gray scale image g is called *bandlimited*.

6.3.2c Sampling.

Now we perform the *sampling* as a next step. Sampling means that all information is lost except at the grid points. Mathematically, this constitutes a multiplication with a function that is zero everywhere except for the grid points. This operation can be performed by multiplying the image function $g(\boldsymbol{x})$ with the sum of δ functions located at the grid points $\boldsymbol{x}_{m,n}$, Eq. (6.2). This function is called the two-dimensional δ comb, or "nail-board function". Then sampling can be expressed as

$$g_s(\boldsymbol{x}) = g(\boldsymbol{x})\sum_{m,n}\delta(\boldsymbol{x} - \boldsymbol{x}_{m,n}) \circ\!\!-\!\!\bullet \hat{g}_s(\boldsymbol{k}) = \sum_{u,v}\hat{g}(\boldsymbol{k} - \boldsymbol{k}_{u,v}), \tag{6.11}$$

where

$$\boldsymbol{k}_{u,v} = \left[\begin{array}{c} u k_1 \\ v k_2 \end{array} \right] = \left[\begin{array}{c} u/\Delta x_1 \\ v/\Delta x_2 \end{array} \right] = \left[\begin{array}{c} u \square k_1 \\ v \square k_2 \end{array} \right] \tag{6.12}$$

are the points of the so-called *reciprocal grid*, which plays a significant role in solid state physics and crystallography. The symbols $\square k_1$ and $\square k_2$ denote the grid constant of the reciprocal grid in horizontal and vertical direction. According to the convolution theorem, multiplication of the image with the 2-D δ comb corresponds to a convolution of the Fourier transform of the image, the image spectrum, with another 2-D δ comb, whose grid constants are reciprocal to the grid constants in x space (see Eqs. (6.2) and (6.12)). A dense sampling in x space yields a wide mesh in the k space, and vice versa. Consequently, sampling results in a reproduction of the image spectrum at each point of the grid.

Now we can formulate the condition where we get no distortion of the signal by sampling. The nonzero image spectrum may cover such a large area that parts of it overlap with the periodically repeated copies. Then we cannot distinguish in overlapping areas whether the spectral amplitudes come from the original spectrum at the center or from one of the periodically repeated copies. The spectral amplitudes of different wave numbers are then mixed at one wave number and consequently lead to distorted patterns.

A safe condition to avoid overlapping is as follows: the spectrum must be restricted to the area which extends around the central grid point up to the lines parting the area between the central grid point and all other grid points. In solid state physics this zone is called the first *Brillouin zone*. On a rectangular grid, this results in the simple condition that the maximum wave number for the nonzero image spectrum must be restricted to less than half of the grid constants of the reciprocal grid. This is the condition formulated in the *sampling theorem*:

If the spectrum $\hat{g}(\boldsymbol{k})$ of a continuous function $g(\boldsymbol{x})$ is bandlimited, i. e.,

$$\hat{g}(\boldsymbol{k}) = 0 \ \forall \ |k_p| \geq \square k_p/2, \tag{6.13}$$

then it can be reconstructed exactly from samples with a distance

$$\Delta x_p = 1/\square k_p. \tag{6.14}$$

In simple words, we will obtain a periodic structure correctly only if we take at least two samples per wavelength. The maximum wave number which can be sampled without errors is called the *Nyquist* or *limiting* wave number. In the following, we will often use wave numbers which are scaled to the limiting wave number. We denote this dimensionless, scaled wave number with a tilde:

$$\tilde{k}_p = 2 k_p \Delta x_p = \frac{2 k_p}{\square k_p}. \tag{6.15}$$

All components of the scaled wave number $\tilde{k}_p$ fall into the $]-1, 1[$ interval.

Example 6.1: Explanation of the Moiré-Effect

The periodic replication of the spectrum caused by the sampling process gives a direct explanation of the Moiré- and aliasing effects. We start with a sinusoidal structure which does not meet the sampling condition. Because of the periodic replication of the sampled spectrum, there is exactly one peak, at $\boldsymbol{k}'$, which lies in the central cell. This peak does not only have another wavelength than the original pattern, but in general also another direction, as observed in Fig. 6.5.

The observed wave number k' differs from the true wave number k by a grid translation vector $k_{u,v}$ on the reciprocal grid. u and v must be chosen to meet the condition

$$|k_1 + u \,\square k_1| \quad < \quad \square k_1/2$$
$$|k_2 + v \,\square k_2| \quad < \quad \square k_2/2. \tag{6.16}$$

In the one-dimensional example of Fig. 6.6, the wave number of the original periodic signal is 9/10 wavelengths per sampling interval. According to the condition in Eq. (6.16), we yield an alias wave number

$$k' = k - \square k = 9/10 \,\square k - \square k = -1/10 \square k. \tag{6.17}$$

This is just the wave number we observed in Fig. 6.6.

Example 6.2: Superresolution

The sampling theorem, as formulated in Section 6.3.2, is actually too strict a requirement. A sufficient and necessary condition is that the periodic replications of the nonzero parts of the image spectra must not overlap. If the spectrum is sparse in the Fourier domain, higher wave numbers than restricted by the sampling theorem can be contained in the signal without the occurrence of overlap. If also the original wave number is known, a complete reconstruction of the original signal is possible. This works even though not the original but only the aliased wave number is observed. In higher dimensional signals, sparse spectra are more likely. Thus superresolution is an interesting possibility in image processing.

6.3.2d   Limitation to a Finite Window.   So far, the sampled image is still infinite in size. In practice, we can only work with finite image matrices. Thus the last step in the sampling process is the limitation of the image to a finite window size. The simplest case is the multiplication of the sampled image with a box function. More generally, we can take any *window function* $w(x)$ which is zero for sufficient large x values:

$$g_l(x) = g_s(x) \cdot w(x) \circ\!\!-\!\!\bullet \hat{g}_l(k) = \hat{g}_s(k) * \hat{w}(k). \tag{6.18}$$

In Fourier space, the spectrum of the sampled image will be convolved with the Fourier transform of the window function.

Example 6.3: Box window function

Let us consider the example of the box window function in detail. If the window in the spatial domain includes $M \times N$ sampling points, its size is $M \Delta x_1 \times N \Delta x_2 = \square x_1 \times \square x_2$. The Fourier transform of the 2-D box function is the 2-D sinc function (Appendix B.3.3):

$$\Pi(x_1/\square x_1)\Pi(x_2/\square x_2) \circ\!\!-\!\!\bullet \frac{\sin(\pi k_1/(\square k_1/M))}{\pi k_1/(\square k_1/M)} \cdot \frac{\sin(\pi k_2/(\square k_2/N))}{\pi k_2/(\square k_2/N)}. \tag{6.19}$$

The main peak of the sinc function has a half-width of $\square k_1/M \times \square k_2/N$. This means that a narrow peak in the Fourier domain will be widened by the convolution with the sinc function to a width that is about M and N times smaller than the horizontal and vertical grid constant of the reciprocal grid, respectively. Thus the resolution in the Fourier domain is about $\square k_1/M$ and $\square k_2/N$ in the horizontal and vertical direction, respectively.

In summary, sampling leads to a limitation of the wave number, while the limitation of the image size determines the wave number resolution. The scales in the spatial and

Fourier domain are reciprocal to each other. The resolution in the x space determines the size in the k space, and vice versa:

$$\Delta k_p \square x_p = 1 \quad \text{and} \quad \square k_p \Delta x_p = 1. \tag{6.20}$$

Example 6.4: Sampling the continuous Fourier transform

The fundamental reciprocal relation in Eq. (6.20) can also be illustrated by converting the continuous Fourier transform into the discrete Fourier transform by sampling. We start with the continuous one-dimensional Fourier transform given by

$$\hat{g}(k) = \int_{-\infty}^{\infty} g(x)\exp(-2\pi ikx)\,dx. \tag{6.21}$$

We approximate the integral by a sum and sample the x at N points with a distance Δx: $x = n\Delta x$. Likewise, we take N samples in the wave number domain with a distance Δk: $k = v\Delta k$. But what is the correct sampling distance in the Fourier domain? We cannot choose it arbitrarily because the sampling in the spatial domain fixes the grid constant of the reciprocal grid to $\square k = 1/\Delta x$. As we take also N samples in the Fourier domain, we obtain

$$\Delta k = \square k/N = 1/(N\Delta x). \tag{6.22}$$

Sampling then yields

$$\hat{g}(v\Delta k) = \sum_{n=0}^{N-1} g(n\Delta x)\exp(-2\pi inv\Delta x\Delta k) = \sum_{n=0}^{N-1} g(n\Delta x)\exp\left(\frac{-2\pi inv}{N}\right). \tag{6.23}$$

The sampling intervals Δx and Δk cancel each other and the index v (*wave number*) of the samples in the Fourier domain gets a clear meaning. It says how many periods of the complex-valued periodic function $\exp(-2\pi inv/N)$ fit into the sampled spatial interval $\square x$.

6.3.3 Sampling Theorem in xt Space

As an intuitive introduction, we reconsider a periodic gray value structure with the wavelength λ_0 (and wave number k_0), moving with the velocity u_0. Physically speaking, this is a planar wave

$$g(x,t) = g_0\exp[-2\pi i(k_0 x - k_0 u_0 t)].$$

We take images at temporal intervals Δt. Because of the motion, the phase of the planar wave changes from image to image. The inter-image phase shift is then given by

$$\Delta\phi = 2\pi k_0 u_0 \Delta t. \tag{6.24}$$

Displacements and velocities cannot be determined unambiguously from phase shifts, since they are additively ambiguous by integer multiples of the wavelength. In order to avoid this ambiguity, the magnitude of the phase shift must be smaller than π:

$$|\Delta\phi| < \pi.$$

Together with Eq. (6.24), this yields a condition for the temporal sampling of the image sequence

$$\Delta t < \frac{1}{2k_0 u_0} = \frac{1}{2v_0} = \frac{T_0}{2}. \tag{6.25}$$

This condition says that we must sample each temporal pattern at least two times per period T_0 to avoid an aliased signal.

Example 6.5: Avoiding temporal aliasing by tracking

Using standard video equipment, the temporal sampling rate of 30 and 25 frames/s for the RS-170 (US) or CCIR (European) television norms, respectively, seriously limits the analysis of fast events. The temporal sampling rate can be doubled if each field of the interlaced video image is exposed individually as is the case with standard CCD cameras (Section 5.3.6). In many cases, it is easy to avoid temporal aliasing without the need to use expensive high speed image acquisition systems. As an example, we discuss water surface waves. The smallest waves (called capillary waves because they are governed by surface tension) have wavelengths down to 2 mm and frequencies exceeding 200 Hz. At first glance, it appears that temporal aliasing cannot be avoided with standard video equipment, and that it is not possible to determine the speed of these waves.

However, because the small waves propagate mainly in one direction, namely the wind direction, there is a simple solution. It is just necessary to move the imaging system with the mean speed of propagation, u_0. Thus we observe wave motion from a moving coordinate system with $x' = x - u_0 t$. The frequencies are shifted down to $\omega' = \omega - k^T u_0$ or $v' = v - u_0/\lambda$. Let us assume that waves with a wave number of $12 \, \text{cm}^{-1}$ (wavelength $\lambda = 5 \, \text{mm}$) move with $30 \pm 10 \, \text{cm/s}$. Then the frequency of these waves is between 40 and 80 Hz. In a coordinate system moving with a speed of 30 cm/s, the frequencies are shifted down to $\pm 20 \, \text{Hz}$ and thus are lower than the Nyquist frequency of 25 or 30 Hz.

Observing from a *tracking* system is an example of an *active vision* system. It has further advantages. The object of interest remains much longer in the observed image area than in an image sequence with a fixed image sector and can be studied in detail. Let us illustrate this fact again with the example of the wave image sequences. In a fixed image sector of $24 \times 32 \, \text{cm}^2$ a wave train remains visible only for about one second, while it can be tracked for several seconds with a moving observation system.

6.3.4 Reconstruction from Sampling

Because of the central importance of the correct sampling, we now derive the sampling theorem for image sequences in a more formal way. The basic task is to prove that the sampled values are a complete representation of the continuous space-time image. Therefore we start with an expression to reconstruct the original continuous function from the sampled points. Reconstruction is performed by a suitable *interpolation* of the sampled points. Generally, the interpolated points $g_r(x)$ are calculated from the sampled values $g_s(x_{k,m,n})$ weighted with suitable factors depending on the distance from the interpolated point x:

$$g_r(x) = \sum_{k,m,n} g_s(x_{k,m,n}) h(x - x_{k,m,n}), \qquad (6.26)$$

where $x_{k,m,n}$ are vectors pointing to the sampling grid in the xt space:

$$x_{k,m,n} = \begin{bmatrix} n\Delta x_1 \\ m\Delta x_2 \\ k\Delta t \end{bmatrix}, \qquad (6.27)$$

while $x = (x_1, x_2, t)$ is a 3-D vector with one time and two space coordinates. Thus $\Delta x_1, \Delta x_2$, and Δt are the sampling intervals for the image sequence. Using the integral properties of the δ function, we can substitute the sampled points on the right side by the continuous values:

$$g_r(\boldsymbol{x}) = \sum_{k,m,n} \int_{-\infty}^{\infty} g(\boldsymbol{x}')h(\boldsymbol{x}-\boldsymbol{x}')\delta(\boldsymbol{x}_{k,m,n}-\boldsymbol{x}')\,\mathrm{d}^3x'$$

$$= \int_{-\infty}^{\infty} h(\boldsymbol{x}-\boldsymbol{x}')\left[\sum_{k,m,n}\delta(\boldsymbol{x}_{k,m,n}-\boldsymbol{x}')g(\boldsymbol{x}')\right]\mathrm{d}^3x'.$$

The last integral means a convolution of the weighting function h with the image function g sampled at the grid point in the spatial domain. In Fourier space, the convolution becomes a multiplication of the corresponding Fourier transformed functions:

$$\hat{g}_r(\boldsymbol{k}) = \hat{h}(\boldsymbol{k})\sum_{s,u,v}\hat{g}(\boldsymbol{k} - {}^p\boldsymbol{k}_{s,u,v}), \tag{6.28}$$

where $\boldsymbol{k}_{s,u,v}$ are points on the reciprocal grid in the Fourier space:

$$\boldsymbol{k}_{s,u,v} = \left[\begin{array}{c} v/\Delta x_1 \\ u/\Delta x_2 \\ s/\Delta t \end{array}\right] = \left[\begin{array}{c} v\,\square\,k_1 \\ u\,\square\,k_2 \\ s\,\square\,v \end{array}\right]. \tag{6.29}$$

Similarly, as $\boldsymbol{x}$, $\boldsymbol{k} = (k_1, k_2, v)$ contains the frequency v and two components of the wave number vector.

The interpolated image function is only equal to the original image function if the weighting function is a box function with the width of the elementary cell of the reciprocal grid in Fourier space. Then only one term unequal to zero remains at the right side of Eq. (6.28):

$$\hat{g}_r(\boldsymbol{k}) = \Pi(k_1\Delta x_1, k_2\Delta x_2, v\Delta t)\hat{g}(\boldsymbol{k}), \tag{6.30}$$

provided the sampling condition is met:

$$\hat{g}(\boldsymbol{k}, v) = 0 \qquad \forall\,|k_p| \ge \Delta k_p/2 \quad \text{and} \quad v \ge \Delta v/2. \tag{6.31}$$

In the last two equations $\boldsymbol{k}$ and $\boldsymbol{x}$ are 2-D vectors. By Eq. (6.30) we know that the transfer function of the correct interpolation filter is a 3-D box-function. Consequently, the correct *interpolation* function is a sinc function of the form:

$$h(\boldsymbol{x}) = \frac{\sin\pi x_1/\Delta x_1}{\pi x_1/\Delta x_1}\,\frac{\sin\pi x_2/\Delta x_2}{\pi x_2/\Delta x_2}\,\frac{\sin\pi t/\Delta t}{\pi t/\Delta t}. \tag{6.32}$$

Unfortunately, this filter kernel is infinitely large and the coefficients drop only inversely proportional to the distance from the center point. Thus a correct interpolation requires a large image area; mathematically, it must be infinitely large. This condition can be weakened if we "overfill" the sampling theorem, i. e., ensure that $\hat{g}(\boldsymbol{k})$ is already zero before we reach the Nyquist wave number. According to Eq. (6.28), we can then choose $\hat{h}(\boldsymbol{k})$ arbitrarily in the region where $\hat{g}(\boldsymbol{k})$ vanishes. We can use this freedom to construct an interpolation function which decreases more quickly in the $\boldsymbol{x}$ space, i. e., it has a minimum-length interpolation mask. We can also start from a given interpolation formula. Then the deviation of its Fourier transform from a box function tells us to what extent structures will be distorted as a function of the wave number.

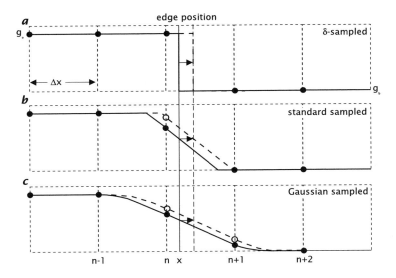

Figure 6.8: *Illustration of the determination of the position of a step edge:* **a** *δ sampling without any smoothing;* **b** *standard sampling;* **c** *Gaussian sampling.*

6.3.5 Sampling and Subpixel Accurate Gauging

In this section, it will be illustrated that proper sampling and accurate position and *subpixel-accurate* edge detection are directly related. We consider a step edge in a one-dimensional image. If the step edge is sampled without any prior smoothing (δ sampling, Fig. 6.8a), the position of the edge cannot be known more accurately than to the width of the pixel. If, however, the edge is smoothed prior to sampling, its position can be determined much more accurately. Figure 6.8b and c shows two examples: *standard sampling* and *Gaussian sampling*.

The situation is especially simple for standard sampling. It is possible to derive the position of the edge directly from the one pixel with an intermediate gray value at x_p.

$$x = x_p + \frac{g_p - \overline{g}}{\Delta g} \tag{6.33}$$

where $\Delta g = g_2 - g_1$ and $\overline{g} = (g_1 + g_2)/2$ are the gray value difference at the edge (counted positive for an edge ascending in positive x direction) and the mean gray value, respectively. From Eq. (6.33), the standard deviation for the position determination is derived as

$$\sigma_x = \frac{\sqrt{2}\sigma_g}{\Delta g}\Delta x \tag{6.34}$$

and is thus directly related to the ratio of the standard deviation of the gray value noise to the gray value change at the edge.

For Gaussian sampling, more points result in intermediate gray values (Fig. 6.8c). Thus one might suspect that the estimate of the subpixel accurate edge position is even more robust in this case. If the image is further blurred, however, the position estimate becomes less accurate because the gray value differences become smaller and eventually become blurred in the noise level. Optimum subpixel accuracy seems to be achieved at a certain blurring or scale. In general, we can conclude: when the sampling theorem is met, we can reconstruct the original sample, and only the systematic errors

a　　*b*

c　　*d*

Figure 6.9: *Illustration of quantization. An image is digitized with different quantization levels:*
a 2, b 4, c 8, d 16. Too few quantization levels produce false edges and make features with low
contrast partly or totally disappear.

in the sampling procedure and the statistical uncertainty in the sampled signal values
hinder us to do this with arbitrary accuracy.

6.3.6 Quantization

6.3.6a Uniform Quantization. After digitalization (Section 6.3.2), the pixels still
show continuous gray values. For use with a computer we must map them onto a
limited number Q of discrete gray values:

$$[0, \infty[\xrightarrow{Q} \{g_0, g_1, \ldots, g_{Q-1}\} = G.$$

This process is called *quantization*. The number of quantization levels in image processing should meet two criteria.

First, no gray value steps should be recognized by our visual system. Figure 6.9
shows images quantized with 2 to 16 levels. It can be clearly seen that a low number
of gray values leads to false edges and makes it very difficult to recognize objects
which show no uniform gray values. In printed images, 16 levels of gray values seem
to be sufficient, but on a monitor we would still be able to see the gray value steps.
Generally, image data are quantized with 256 levels. Then each pixel occupies 8 bit or

one byte. This bit size is well adapted to the architecture of standard computers which can address memory bytewise. Furthermore, the resolution is good enough that we have the illusion of a continuous change in the gray values, since the relative intensity resolution of our visual system is only about 2 %.

The other criterion for the number of quantization levels is related to the imaging task. For a simple application in machine vision, where the objects show a uniform brightness which is different from the background, or for particle tracking in flow visualization (Section 1.6.1), two quantization levels, i. e., a *binary image*, might be sufficient. Other applications might require the resolution of faint changes in the intensity. Then, an 8-bit gray scale resolution would be certainly too coarse.

6.3.6b Quantization Error. Quantization always introduces errors, since the true value g is replaced by one of the quantization levels g_q. If the quantization levels are equally spaced with a distance Δg and all gray values are equally probable, the variance introduced by the quantization is given by

$$\sigma_q^2 = \frac{1}{\Delta g} \int_{g_q-\Delta g/2}^{g_q+\Delta g/2} (g - g_q)^2 \mathrm{d}g = \frac{1}{12}(\Delta g)^2. \tag{6.35}$$

This equation shows how we select a quantization level. We choose the level g_q for which the distance from the gray value g, $|g - g_q|$, is smaller than from the neighboring quantization levels q_{k-1} and q_{k+1}. The standard deviation σ_q is about 0.3 times the quantization resolution distance Δg.

Quantization with unevenly spaced quantization levels may be appropriate if the gray values cluster in a narrow gray scale range and are much less frequent in other ranges. Unevenly spaced quantization levels are hard to realize in any image processing system. An easier way to yield unevenly spaced levels is to use equally spaced quantization but to transform the intensity signal before quantization with a nonlinear amplifier, e. g., a logarithmic amplifier. In the case of a logarithmic amplifier we would obtain levels whose widths increase proportionally with the gray value.

6.3.6c Accuracy and Precision of Gray Value Measurements. With respect to the quantization, the question arises, with which accuracy could we measure a gray value? At first glance, the answer to this question seems to be trivial and given by Eq. (6.35): the maximum error is half a quantization level and the mean error is about 0.3 quantization levels.

But what if we measure the value repeatedly? This could happen if we take many images of the same object or if we have an object with uniform radiance and want to measure the mean radiance of the object by averaging over many pixels.

From the laws of statistical error propagation, it is well known that the error of the mean value decreases with the number of measurements according to

$$\sigma_{\mathrm{mean}} \approx \frac{1}{\sqrt{N}}\sigma \tag{6.36}$$

where σ is the standard deviation of the individual measurement and N the number of measurements taken (Example 7.4). This equation tells us that if we take 100 measurements, the error of the mean should be just about 1/10 of the error of the individual measurement.

Does this law apply to our case? Yes and no — it depends and the answer appears to be a paradox.

a

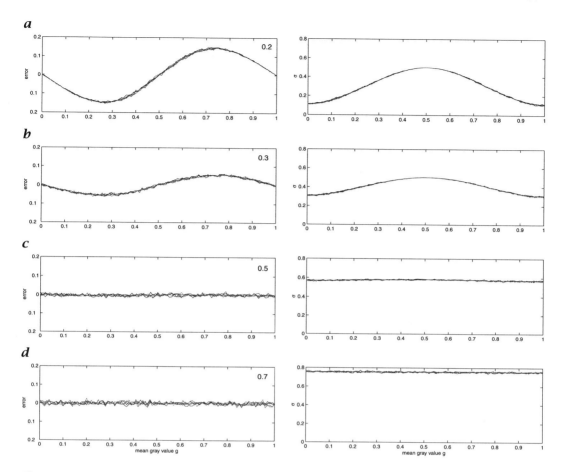

Figure 6.10: *Simulation of the influence of quantization on the estimation of mean gray values from 10000 noisy gray values with standard deviations of 0.2, 0.3, 0.5, and 0.7 for true mean gray values between 0 and 1: left column: deviation of the estimated gray value from the true value; right column: estimated standard deviation.*

If we measure with a perfect system, i.e., without any noise, we would always get the same quantized value and, therefore, the result could not be more accurate than the individual measurement. If, however, the measurement is noisy, we would obtain different values for each measurement. The probability for the different values reflects the mean and variance of the noisy signal and because we can measure the distribution, we can estimate both the mean and the variance. In images we can easily obtain many measurements by spatial averaging. Therefore, the potential is there for much more accurate mean values than given by the quantization level.

A simulation gives some more insight into these relations. Ten thousand normally distributed random samples with standard deviations of 0.2, 0.3, 0.5, and 0.7 and with mean values between 0 and 1 were taken. From these samples the mean and standard deviation were computed. The results are shown in Fig. 6.10. The left column shows the deviation of the estimated mean gray value from the true mean value. If the true value is an integer (0 or 1) or 0.5, no bias occurs in the estimation of the mean gray values because of symmetry reasons. For the values 0 and 1 the mean value lies in the middle of the quantization interval, while for 0.5 it is just at the border between two

a

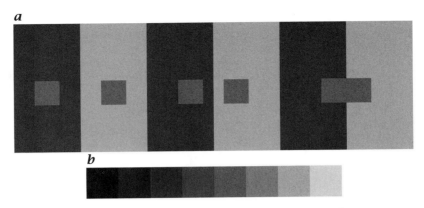

b

*Figure 6.11: Perception of gray values: **a** small rectangular areas of constant gray value are placed in different arrangements in a darker and brighter background; **b** a linear stepwise increase in brightness.*

quantization intervals so that probability is equal that the values fall either into the lower or higher quantization interval. In between, the bias is up to about 0.15 for a standard deviation of 0.2. The standard deviation is overestimated at mean gray values close to 0.5 because it is equally likely that a value of 0 and 1 is measured. At a mean value of 0 and 1, the standard deviation is underestimated, because with a standard deviation of 0.2 the likelihood that the gray value is either larger than 0.5 or lower than 0.5 is quite low and thus in most cases a 0 is measured.

The bias in the estimate of the mean gray value quickly diminishes with increasing standard deviation. The bias is already well below 0.01 for a standard deviation of 0.5 (Fig. 6.10c). Thus it appears to be possible to measure mean values with an accuracy that is well below a hundredth of the width of the quantization interval.

With a real system the accuracy is, however, limited also by systematic errors. The most significant source of error is the unevenness of the quantization levels. The quantization levels in real analog to digital converter are not equally distant but show systematic deviations, which may be up to half of the width of the quantization interval. Thus, a careful investigation of the analog to digital converter is required to estimate the limits of the accuracy of the gray value measurements.

6.3.6d Unsigned and Signed Representation. Normally we think of radiometric terms as nonnegative quantities. Thus, the quantized values are unsigned numbers ranging from 0 to 255 in 8-bit images. As soon as we perform operations with images, e.g., if we subtract two images, negative values may appear which cannot be represented. Thus we are confronted with the problem of two different representations for image data, as unsigned and signed 8-bit numbers. Correspondingly, we must have two versions of algorithms, one for unsigned and one for signed gray values.

A simple solution to this problem is to handle gray values principally as signed numbers. For 8-bit data, this can be done by subtracting 128. Then the mean gray value intensity of 128 becomes the gray value zero. Gray values lower than this mean value are negative. For display, we must convert the gray values to unsigned values by adding 128 again.

6.3.6e Human Perception of Luminance Levels. We observe and evaluate the images we are processing with our visual system. Based on this, we draw conclusions about the quality and properties of images and image processing operators. With respect to gray values, it is, therefore, important to know how the human visual system perceives them and what luminance differences can be distinguished. Figure 6.11a demonstrates that the small rectangular area with a medium luminance appears brighter in the dark background than in the light background, though its absolute luminance is the same. This deception only disappears when the two areas merge. The stair case-like increase in the luminance in Fig. 6.11b shows a similar effect. The brightness of one step appears to increase towards the next darker step.

Because of the low luminance resolution of printed images, we cannot perform experiments to study the luminance resolution of our visual sense. It shows a logarithmic rather than a linear response. This means that we can distinguish relative but not absolute luminance differences. In a wide range of luminance values, we can resolve relative differences of about 2%.

These characteristics of the human visual system are quite different from those of a machine vision system. Typically only 256 gray values are resolved. Thus a digitized image has much lower dynamics than the human visual system. This is the reason why the quality of a digitized image, especially of a scene with high luminance contrast, appears inferior to us compared to what we see directly. Although the *relative* resolution is far better than 2% in the bright parts of the image, it is poor in the dark parts of the images. At a gray value of 10, the relative luminance resolution is only 10%.

In order to cope with this problem, video cameras generally convert the measured irradiation E not linearly, but with some nonlinear functions, as the following two examples demonstrate.

Example 6.6: Gamma correction

An often applied nonlinear transform is the *gamma transform* that converts the gray value g according to an exponential law:

$$g' = a_1 g^\gamma. \tag{6.37}$$

The exponent γ is denoted the *gamma value*. Typically, γ has a value of 0.4. With this exponential conversion, the logarithmic characteristic of the human visual system may be approximated. Here the contrast range is significantly enhanced. If we presume a minimum relative luminance resolution of 10%, we get useable contrast ranges of 25 and 316 with $\gamma = 1$ and $\gamma = 0.4$, respectively. For many scientific applications, however, it is essential that a linear relation is maintained between the radiance of the observed object and the gray value in the digital image ($\gamma = 1$). Many CCD cameras provide a control to switch or adjust the gamma value. For a modern CCD sensor with low dark noise, a gamma value of 0.5 results in an image with a gray value independent noise level (Section 7.4.4).

Example 6.7: Logarithmic transform

Another useful nonlinear gray value transform is a logarithmic transform:

$$g' = a_0 + a_1 \ln g. \tag{6.38}$$

The parameters a_0 and a_1 control the *dynamic range* (Section 5.3.2i) that can be mapped to the output gray values. The logarithmic transform has the significant advantage that object contrast becomes illumination independent. In Section 4.3.4 we found that the irradiance on the image plane, E', is related in the following way to the irradiance and reflectivity of an object with Lambertian reflectivity, see Eq. (4.56):

$$E' = \rho t E \sin^2 \Theta. \tag{6.39}$$

If the object irradiance is low, the contrast at the image plane caused by differences in the object reflectivity is also low. However, after a logarithmic transform

$$g' \propto \ln E' = \ln \rho + \ln(t E \sin^2 \Theta) \tag{6.40}$$

changes in the reflectivity cause changes in the gray values that do not depend on the illumination. In this way, a logarithmic transform is very useful for further image processing (e. g., edge detection, see Chapter 12) in scenes with largely varying irradiation. Such scenes are often encountered outdoors (traffic, surveillance, etc.).

6.4 Procedures

A digital image is formed as the result of a chain of processes including the setup of the illumination, the image formation with an optical system, the conversion of the irradiance into an electric signal, and the digitalization. We have discussed these processes in this and the last two chapters. All the processes can contribute to errors and distortions in the final image. Therefore, it is an important task to measure the overall performance of the acquisition system. For a good approximation it can be modeled as a linear system and is thus described by the point spread function and the transfer function. In this section, it is shown how the transfer function of the entire acquisition system from the image formation to the digitization can be measured.

6.4.1 The Transfer Function of an Image Acquisition System

The measurement of the system transfer function is often required since it is difficult to estimate it theoretically. Too many parts of the system may influence it:

1. The optical transfer function of the lens (Sections 4.3.5b and 4.3.5d).
2. Possible misalignment, defocusing, motion blur, or vibration may further decrease the optical transfer function.
3. Analog or digital processing of the video signal in the camera.
4. Attenuation of the high frequencies in the video signal, especially if analog signals are transmitted over long cables.
5. Transfer function of a recording medium such as a video tape recorder if the image is first recorded before it is digitized for further processing.
6. Filtering and conditioning of the analog video signal before it is digitized.

   All parts that influence the time-serial video signal cause an anisotropy in the transfer function because mainly the horizontal resolution is affected.

   Of great practical importance for possible distortions in the digitized images — and thus errors in the further processing of the image — is to know to which extent aliasing, i. e., Moiré patterns (Section 6.3.1d), is still present in the system. A system is only free of Moiré-effects when the system transfer function vanishes for wave numbers larger than the Nyquist wave number (Section 6.3.2c). This condition is not met automatically. In fact, even an ideal CCD camera that averages the irradiance evenly over the area of the sensor cell (*standard sampling*) still shows significant aliasing.

6.4.1a Test Pattern for OTF Measurements.
An ideal test target to measure the transfer function of the entire formation process of the digital image is the *ring test pattern* described in Section 10.4.7a and shown in Fig. 6.12. This pattern maps the *Fourier domain* into the *spatial domain*. The local wave number of the concentric rings is proportional to the distance from the center of the pattern. Therefore, it directly shows the system transfer function in phase and amplitude. This ring test pattern can easily be computed using Eq. (10.77) and then be printed on a high-resolution laser printer. For higher accuracy demands, you can generate a Postscript file from the ring test pattern and ask a service bureau to produce a film output.

Figure 6.12: *Test pattern with concentric rings for the measurement of the system transfer function of an image acquisition system. The local wave number goes with the distance from the center and thus maps the Fourier domain into the spatial domain.*

Since in most cases only the amplitude response is of interest, the following simple processing can be applied to determine the amplitude of the sinusoidal patterns and, thus, the *modulation transfer function* (*MTF*, Section 4.3.5e):

Subtraction of the mean gray value. This can be performed by computing a *Laplacian pyramid* (Section 14.3.5), and merging all levels of the pyramid except for the highest level which contains the coarsest image structures and the mean gray value.

Squaring and averaging. If the smoothing is over a distance larger than the wavelength of the patterns, it results in the squared amplitude of the periodic patterns.

Square root. This final step computes the square root and thus yields the amplitude.

6.4.1b Moiré Pattern. One disturbing artifact in images are Moiré patterns (Section 6.3.1d). These patterns lead to significant errors for subsequent image processing since too fine scales appear as coarser scales. It is also not possible to remove the Moiré patterns by a smoothing filter since such an operation cannot distinguish Moiré patterns from true patterns at the same scale and, thus, would remove both of them. To investigate Moiré patterns, the ring test pattern (Fig. 6.12) is imaged with two different CCD cameras. One of the cameras, an early model (Siemens K210, manufactured around 1986), shows significant Moiré patterns in the horizontal direction. A modern CCD camera (2/3" Hitachi KP-M1, manufactured 1995) does not show significant Moiré patterns. The CCD sensor in this camera features a *lenticular array* that puts a microlens on top of each sensor element focusing the incident light onto its light sensitive area and, thus, achieving a larger degree of spatial averaging (see Section 6.3.2).

6.4.1c Modulation Transfer Function and Depth of Field. A good way to investigate the system transfer function of an image acquisition system is to take images of the test ring pattern at different degrees of defocusing. In this way, we not only gain the transfer function but also the depth of field at which the optical system can be used without noticeable blurring. Figure 6.14 shows images taken of the test ring pattern focusing the optical system at different distances. The test ring pattern had a distance of 57 cm

a   b

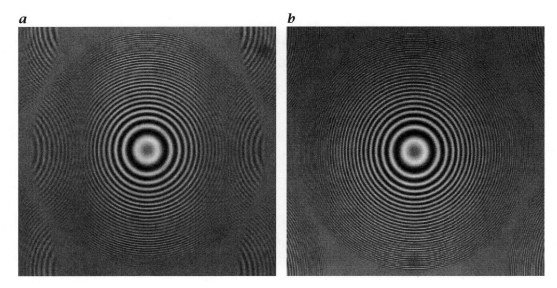

*Figure 6.13: Measurements of the system transfer function of a CCD camera and the image digitization with a frame grabber using the ring pattern in Fig. 6.12 as a test pattern. The images were taken with a 16 mm lens and **a** an early CCD camera (Siemens K210, manufactured around 1986, Sony 021-L CCD sensor) showing significant Moiré patterns, **b** a modern CCD camera (2/3" Hitachi KP-M1, manufactured 1995, Sony ICX082AL/ICX083AL CCD sensor).*

from the front lens. The lens was focused at the test pattern and at a distance of 6, 10, and 20 cm further away. The images show the quick degradation with increasing defocusing.

A more quantitative analysis is given in Fig. 6.15. The figure shows a profile of the amplitude of the sinusoidal patterns, as evaluated with the simple procedure described at the beginning of this section, at a row through the center of the image. Disregard the dip at the center of the profiles. Here the simple analysis gives no correct determination of the amplitude because of the small wave number of the periodic patterns. Outside of the central dip, the curves decrease monotonically, indicating a monotonous decrease of the MTF with increasing wave number. The higher the defocusing is, the more rapidly the decrease in the MTF is. Note, however, that even at the best focus, the MTF decreases by about a factor of two. Furthermore, the decrease in the MTF is nearly linear, starting at quite low wave numbers. This is exactly what is theoretically predicted for a diffraction-limited optical system (Section 4.3.5d).

In conclusion, great care should be taken when measuring amplitudes at scales which are still far away from the minimum possible scale given by the sampling theorem. Without a careful measurement of the MTF and without correcting for it, the measured amplitudes might easily be too small.

6.4.2 Quality Control of Quantization

Histograms (Sections 7.3.1f and 7.4) allow a first examination of the quality of image digitization if an image is taken of a gradual change in radiance, i. e., a distribution where all gray values occur with about the same probability. We might then expect to also obtain a smooth histogram. However, it could be that the histogram shows large variations from gray value to gray value up to several 10 %. What could be the reason for this surprising effect? First, we check which statistical variations are likely to occur.

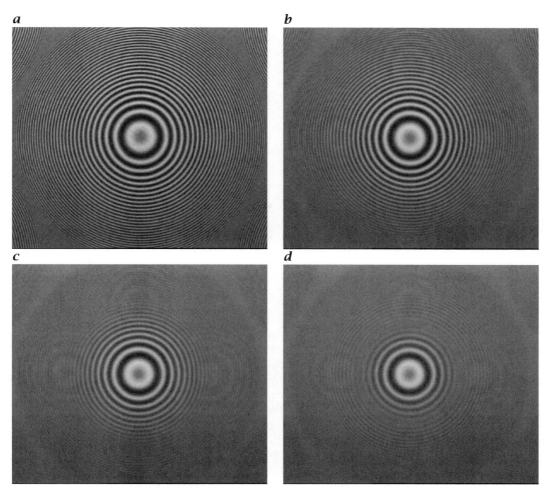

Figure 6.14: *Sequence of ring test patterns taken with increasing distance from the focal plane to demonstrate the decreasing OTF as a function of the distance from the focal plane and to measure the depth of field: **a** sharpest possible image, 0 cm, **b** 6 cm, **c** 10 cm, and **d** 20 cm distance from the focal plane.*

A 512×512 image has 1/4 million pixels, so that, on average, 1000 pixels show the same gray values. Consequently, the relative standard deviation σ_r of the statistical fluctuations should be $\sqrt{1000}$ or about 3 % and we should see at most $3\sigma_r \approx 10\%$ fluctuation.

There is, however, another reason for fluctuations in the histogram. It is caused by the varying widths of the quantization levels. Imagine that the decision levels of the video analog-digital converter are accurate to 1/8 least significant bit. Then the widths of a quantization level might vary from 3/4 to 5/4 least significant bit. Consequently, the probability distribution may vary by ±25 %. From these considerations it is obvious that we should detect also defects in the analog digital converter, such as missing quantization levels, easily with the histogram.

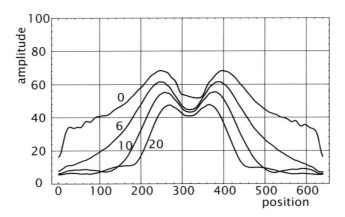

Figure 6.15: *Radial cross section through the images of the test ring pattern in Fig. 6.14 showing the amplitude of the sinusoidal patterns as determined with the simple procedure described at the beginning of this section. The numbers indicate the distance from the focal plane in cm.*

6.5   Advanced Reference Material

This section gives an overview of the hardware required to acquire digital images. Because the hardware has changed considerably in the last decade, Section 6.5.1 first discusses the evolution of the image acquisition hardware in the last twenty years.

The only hardware components that are specific for image acquisition are the circuits to digitize the video signals. There are two principal possibilities. First, the camera produces an analog video signal that is transmitted to a frame grabber in the PC, where the analog signal is converted into digital data. Second, the camera includes electronic circuits to digitize the image data and sends the digital data via a proprietary or standardized digital connection to the PC.

Because both possibilities are commonly used nowadays, we give the necessary background knowledge about analog and digital video input in Sections 6.5.2 and 6.5.3. We also discuss the pros and cons of both approaches.

6.5.1   Evolution of Image Acquisition Hardware

6.5.1a   Hardwired Frame Grabbers.   This type of image processing cards emerged first on the micro computer market at the end of the 1970s and initiated the widespread use of digital image processing in scientific applications. The functions of these boards were hardwired and controlled via a set of registers in the input/output memory space of the host. Display of video images required a separate RGB monitor. The frame grabber included a proprietary frame memory to store two images. The transfer of image data to and from the PC main memory was slow, so that no real-time storage of image sequences was possible. A very successful example of this type of frame grabbers was the PCVISION*plus* board from Imaging Technology (Fig. 6.16).

6.5.1b   Modular Image Processing Systems with a Pipelined Video Bus.   A video bus system allows the integration of special purpose processing elements in a modular way to adapt the hardware to the processing needs. Certain types of image processing operations can then be performed much faster, most of them in real time. Examples for high-end modular image processing systems with a pipelined video bus include the Modular Vision Computers (MVC) from Imaging Technology, the IMAGE series from

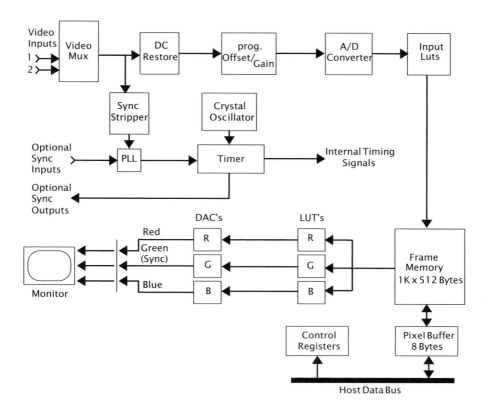

Figure 6.16: *Block diagram of a classical frame grabber board that made history in image processing, the PCVISIONplus frame grabber from Imaging Technology.*

Matrox, and the MaxPCI from Datacube. Such boards are still used in some industrial applications where high-speed real-time image processing is required for specific tasks.

6.5.1c Frame Grabber with Programmable Processor. This type of image processing boards includes its own graphics, digital signal or general purpose processor to enhance image processing capabilities. The rapid increase in processing speed of these devices made this solution more and more attractive over dedicated hardware. Standard tools such as C-compilers and development tool kits are available to program the boards in a much more flexible and portable way than it would ever be possible with a hard-wired system. One of the first examples was the AT-Vista board from Truevision. Modern systems use fast digital signal processors.

6.5.1d Frame Grabbers with Direct Image Transfer to PC RAM. A significant break through in the performance of image acquisition hardware came with the introduction of fast bus systems for the PC such as the PCI bus. This paved the way for a new generation of frame grabbers. With a peak rate of 132 MB/s the PCI bus system is fast enough to transfer digitized image data in real-time to the DRAM of the PC. Since the bottleneck of slow transfer of image data is removed, processing of image data on the PC CPU is now much more attractive.

The frame grabber hardware also got much less expensive. It now needs to include only video input circuits and a DMA controller, and no extra frame buffer, because

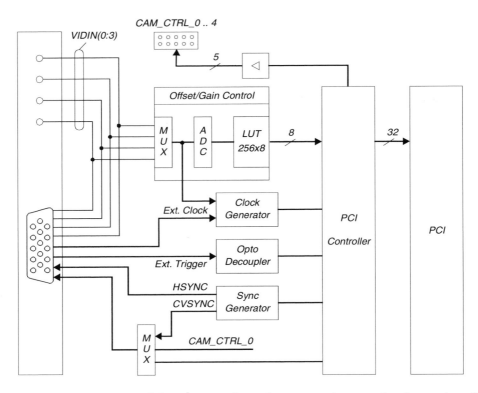

Figure 6.17: *Block diagram of the PC_EYE1 from Eltec, an early example of a modern frame grabber without frame buffer and with direct DMA transfer of image data to the PC DRAM.*

the image data is transferred directly to the PC memory where it can be displayed and processed. An early example of this class of frame grabbers is shown in Fig. 6.17.

6.5.2 Analog Video Input

The *video input* component is very similar in all frame grabbers (Figs. 6.16 and 6.17). First, the analog video signal is processed and the video source is synchronized with the frame buffer in a suitable way. Then the signal is digitized and stored after pre-processing with the input look-up table in the frame buffer or sent via direct memory access to the PC memory.

6.5.2a Analog Video Signal Processing. All the systems, which only process black-and-white video signals, include a video multiplexer. This allows the frame buffer to be connected to more than one video source. Via software, it is possible to switch between the different sources. Before the analog video signal is digitized, it passes through a video amplifier with a programmable gain and offset. In this way, the incoming video signal can best be adapted to the input range of the analog-digital converter.

Digitization of color images from RGB video sources, which provide a separate red, green, and blue signal, requires three analog-digital converters. In order to capture composite color video signals in which the luminance and color information is composed in one video signal, a decoder is required which splits the composite signal into the RGB video signals. (For an explanation of the different types of color video signals,

Table 6.1: *Typical resolutions of digitized video images.*

	number of rows	number of columns	Pixel clock [MHz]	aspect ratio	comment
US video norm RS-170 30 frames/s	512	480	10.0699	1.10	
	640	480	12.28	1.01	square pixel (almost)
	740	480	14.3181	0.87	
European video norm CCIR 25 frames/s	512	512	10.00	1.48	
	768	572	15.00	1.00	square pixel

see Section 5.5.3.) These devices can also decode Y-C signals according to the *S-VHS* video standard.

Some color frame grabbers also feature real time color space conversion. Color video signals can be converted to and from various color coordinate systems. Of most importance for image processing is the HSI (hue, saturation, intensity) color space. In composite, Y-C, and YUV video signals, the resolution for the color signal is considerably lower than for the luminance signal. Consequently, only color cameras with RGB video output can be recommended for most scientific applications. Color acquisition modules can also be used to capture video signals from up to three synchronized black-and-white video cameras. In this way, *stereo images* or other types of *multichannel images* can be acquired and processed.

6.5.2b Synchronization. Besides the image contents, a video signal contains *synchronization* signals which mark the beginning of an image (vertical synchronization) and the image rows (horizontal synchronization); see Section 5.5.2. The synchronization signals are extracted in the video input by a synchronization stripper and used to synchronize the frame buffer (image display) with the video input via a *phase-locked loop* (PLL).

Some PLL circuits do not work properly with unstable video sources such as video recorders. If digitization of recorded images is required, it should be tested whether the images are digitized without distortions. Many frame grabbers do not support digitizing from videotapes. Digitization of still images is especially critical. It cannot be done without a time-base corrector, which is built into some high-end video recorders. If a particular image from videotape must be digitized, it is necessary that a time code is recorded together with the images either on the audio track or with the video signal itself. This time code is read by the computer and used to trigger the digitization of the image.

Another difficulty with analog video signals is the difference between the European and American video standards. Most modern frame grabbers can process video signals according to both the American RS-170 and the European CCIR norms with 30 and 25 frames/s, respectively. The details of the video timing for both standards can be found in Section 5.5.2.

6.5.2c Digitization. Generally, analog video signals are *digitized* with a resolution of 8 bits, i.e., 256 gray value levels at a rate of 10 to 20 million pixels per second. The resulting digitized image contains 512×512 and 480×512 pixels in the European and American formats, respectively. For a long time, this has been a standard. Modern frame grabbers, however, are much more flexible. Thus the number of rows and columns and the pixel clock can be programmed in a wide range. Some typical resolutions are shown in Table 6.1. With modern frame grabbers, it is no problem to digitize nonstandard video signals, e. g., from electron microscopes, ultrasonic and thermal sensors, line scan cameras, and from high-resolution CCD-cameras.

The digitized image can be processed via an input *look-up table* (*LUT*). Most boards incorporate several such tables, which can be selected by software. LUTs provide a real-time implementation of *homogeneous point operations* (Section 7.3.4) before the pixels are stored in the frame buffer. The LUTs can be accessed from the host either by hardware registers or by mapping the LUT memory into the address space of the PC. Writing to an LUT for 8-bit images, which contains just 256 entrances, is fast enough so that it can be performed interactively.

6.5.2d   Pros and Cons of Analog Video Input.   This section summarizes the pros and cons of analog video input, while digital video input is discussed in Section 6.5.3a. The decision, which system is the best, depends on many aspects of an application, so that there is no general answer. The long-term trend, however, is clear. Digital cameras will gradually replace analog cameras, in the same way as digital photography is currently replacing photography with films.

Nonsquare Pixels.   Generally, the pixels are not squares but rectangles (see Table 6.1). This fact is very important for videometry and filter operations. Isotropic filters become nonisotropic. In recognition of the importance of square pixels, most modern frame grabbers can operate in a square pixel mode.

Sel ≠ Pel.   The digitization rate generally does not coincide with the rate at which the collected charges are read out from CCD sensors. Consequently, sensor elements (*sels*) and pixels (*pels*) are not identical. This may lead to vertically oriented disturbance patterns in the digitized images. Many modern frame grabbers (e. g., the PC_EYE1, Fig. 6.17) include a *pixel clock* input which allows the sel and pel clock rates to be synchronized to avoid such disturbances. Pixel-synchronous digitization is also important for sub-pixel accurate position, displacement, and size measurements (Sections 1.3 and 6.3.5).

Interlaced Scanning.   Standard video cameras operate in the *interlaced* mode (Section 5.3.6). This means that a frame is composed of two *fields* which either contain the even or odd lines. The two fields are scanned one after the other. Consequently, the two fields are illuminated with a time shift of half a frame time. A single frame containing moving objects flickers because it is composed of the two fields illuminated at different times (Section 5.4.4d). For image sequence analysis this means that we must work with fields. In this way, the temporal resolution is doubled at the cost of half the vertical image resolution.

6.5.3   Digital Video Input

For a long time, imaging sensors were locked to the video standard of 30 (25) frames/s and a limited spatial resolution of about 480 (580) lines/frame. Such a standard was required in order to be able to view images on monitors and to record them to video tape, because there were no other possibilities. For digital image processing, however, severe disadvantages are linked to this standard since it has been developed for television broadcasting.

One of the most disturbing features is the interlacing of a frame into even and odd fields (Section 6.5.2d). Another severe limitation is, of course, that applications had to be adapted to the temporal and spatial resolution of video imagery. The only major deviation from the video standard were *line sensors*, which found widespread usage in industrial applications.

With *digital video input*, there are no longer such narrow restrictions to frame rates and image sizes, as discussed in Section 5.3.6. Of course, the rate at which pixels can be read out is still limited. Therefore, a tradeoff exists between spatial and temporal resolution. A similar tradeoff exists between the dynamics in brightness and pixel read-out frequency. Cameras are also available with resolutions of up to 16 bits. These cameras find applications in spectroscopy, photometry, and radiometry. They can be used only with much lower pixel read-out frequencies. This limit can, however, be overcome with multiple digital outputs.

6.5.3a Pros and Cons of Digital Video Input.
The benefits of digital cameras are significant.

Higher signal quality. A much higher signal quality can be achieved since it is no longer required to transmit analog video signals from the camera to the frame buffer.

Sel = pel. A pixel in the digital image really corresponds to a sensor element in the camera. Thus, the image preserves the exact geometry of the sensor allowing for precise geometrical measurements. This emerging field is called *videometry* in analogy to photometry.

Free choice of spatial, temporal, and irradiance resolution. A sensor can be chosen that fits the requirements of the application concerning spatial, temporal, and irradiance resolution. In contrast to CCD sensors, CMOS sensors (Section 5.3.5b) can take full advantage of this feature.

Simpler computer interface. Digital cameras require a much simpler interface to computers. It is no longer required to use a complex frame buffer and to get it running with your camera. A much simpler device is sufficient that takes the digital data from the camera.

The advantages of digital cameras are significant. Therefore the question arises why digital cameras have not already replaced analog cameras. The reason is simple and has to do with the failure of the industry to develop a standard for transmission of digital video signals from cameras to computers and to control cameras from the computer for a long time. Without such a standard, digital cameras show two significant disadvantages.

No direct image. A digital camera always requires a computer to display and view an image. It is just not possible to take the video signal and display the image for adjustments and focusing. You first have to install the driver and get it running and to get familiar with the proprietary way to control the camera parameters such as frame rate, exposure time, binning, etc. Only if you have mastered all these steps, you will be able to see an image.

Nonstandard connectors and expensive cables. Each manufacturer used to have its own connectors and pin out of the signal. Except for the signal level (TTL single-ended or differential RS-422 signals), there was no standard. Cables for digital cameras require many wires, are expensive, and a common source for failures.

6.5.3b Standards for Digital Video Input. It is obvious that the general acceptance of digital cameras is directly related to well defined and easy to handle standards for digital cameras. The standardization must include two levels. First, on the physical layer the communication protocol and the cable that connect the camera with the PC must be defined. Secondly, the software layer specifies the communication between the computer and the camera and the image data formats. Currently, two standards are available, *Camera Link* and *FireWire*, and some more are emerging, including USB (2.0) and Gigabit Ethernet.

The camera link standard (http://www.pulnix.com) uses the serial channel link transmission standard with low voltage differential signaling (*LVDS*) and standard MDR-26 pin connectors. The standard comes in three configurations (base, medium, and full with up to 24, 48, and 64 bits in parallel). The base configuration requires one connector, the medium and full configuration two connectors. The maximum image data transmission rate depends on the clock rate used. With a 40 MHz pixel clock the maximum data rates for 8-bit images are 80 MB/s, 160 MB/s, and 320 MB/s for base, medium, and full configuration, respectively. Two extra LVDS cable pairs provide asynchronous and bidirectional communication between the frame grabber and the camera.

The FireWire standard (http://www.1394ta.org) is much slower with a maximum isosynchronous data rate of 40 MB/s for image data. But it has the advantage that it is a standard serial bus system connecting up to 63 peripherals and thus also multiple cameras, while Camera Link is a point-to-point connection. Furthermore, the communication protocol is standardized, so that it should be easy to control all camera features with the same software. Unfortunately, manufacturers are very ingenious in making it difficult to use cameras of the competition with their software. Therefore camera manufacturers have still not succeeded in establishing easy to use standards for digital cameras, although the use of standard cables is already a significant progress.

6.5.4 Real-Time Image Processing

Section 2.3 and Table 2.2 showed computers have become powerful enough that many image processing tasks can be done in real time. This requires, however, that image acquisition and processing runs in parallel. Synchronizing these tasks is not easy. Progress in software engineering makes it possible, however, that real-time programming tasks can be performed without writing a single line of code.

This is made possible by image processing development systems that include a script interpreter and compiler. An example of such a software is heurisko.

The software includes the acStart command. This command starts a real-time acquisition and returns immediately, so that acquisition runs in the background until it is stopped with acStop. The acWait command can then be used to synchronize the processing with the DMA transfer. For further details see http://www.heurisko.de.

Example 6.8: Real-time processing with background DMA image transfer

> The following piece of C-like script code of the image processing software heurisko demonstrates that it is possible to perform even real-time image processing in a flexible way with a high-level software interface. All the named objects used in the example are dynamically allocated, when the software executes a workspace.

```
# Frame grabber definition and initialization
# Buffer description: default size
struct buffer {
    string name, long dim,
    long size[3], long offs[3],
    string format
```

```
};
# Set buffer for sequence of two images of default size
buffer = "buf".{3}.{-1,-1, 2}.{-1,-1}." %3i";
# Get size of image buffer
long n[3]; long nx = n[0]; long ny = n[1];
n = GetCoordinates(fg.buf);

# Open acquisition device and allocate resources
device fg, type "bcam", buffers buffer;

# Object definitions
ubyte b0 = fg.buf[0];           # Alias to first buffer
ubyte b1 = fg.buf[1];           # Alias to second buffer
ubyte row1 = b0[128];           # row 256 in buffer b0
ubyte row2 = b1[128];           # row 256 in buffer b1
ubyte xt[100][nx];              # Space-time image
ushort s[ny][nx];               # Buffer for image addition


# Real-time image addition of n byte images into a short buffer
operator grabadd(n);
    short i;
    s = Clr();             # Zero accumulator buffer
    acStart(fg.buf);       # Start continuous acquisition
    while (n);
        # Wait until next acquisition is finished
        acWait(fg.buf);
        if (i == 0);       # Toggle between two buffers
            s = Sum(b0); # Add image from first buffer
            i=1;
        else;
            s = Sum(b1); # Add image from second buffer
            i=0;
        endif;
    endwhile;
    acStop(fb);                 # Stop background acquisition
endoperator;

# Generation of space-time image from row 128 of image
operator grabxt();
    buf = acStart(fg.buf);  # Start continuous acquisition
    # Scan through xt image row by row
    scan(xt|0);
        # Wait until next acquisition is finished
        acWait(fg.buf);
        if (i==0);              # Toggle between two buffers
            xt = row1;          # Copy row 128 into xt image
            i=1;
        else;
            xt = row2;          # Copy row 128 into xt image
            i=0;
        endif;
    endscan;
    acStop(fg.buf);             # Stop background acquisition
endoperator;
```

6.5.5 Further References

6.1   **Graphics file formats**

C. W. Brown and B. J. Shepherd, 1995. Graphics File Formats, Reference and Guide. Manning Publications Co., Greenwich, CT.

J. D. Murray and W. vanRyper, 1996. Encyclopedia of Graphics File Formats. 2nd edition, O'Reilly & Associates, Inc., Sebastopol, CA.

6.2   **List of some manufacturers of image acquisition hardware**

Active Imaging, Inc., USA, `http://www.ActiveImaging.com`

BitFlow Inc., Woburn, MA, USA, `http://www.bitflow.com`

Alacron, Nashua, NH, USA, `http://www.alacron.com`

Coreco Imaging Inc., St.-Laurent, Quebec, Canada, `http://www.imaging.com`

Data Cube, USA, `http://www.datacube.com`

Data Translation, USA, `http://www.datatranslation.com`

ELTEC Electronic GmbH, Mainz, Germany, `http://www.eltec.de`

Epix Inc., Ottawa, ON, Canada, `http://www.epixinc.com/vs`

Leutron Vision AG, Glattbrugg, Switzerland, `http://www.leutron.com`

Matrix Vision GmbH, Oppenweiler, Germany,
   `http://www.matrix-vision.de`

Matrox Electronic Systems Ltd., Dorval, Quebec, Canada,
   `http://www.matrox.com/imaging`

National Instruments, Austin, TX, `http://www.ni.com/`

Silicon Software, Mannheim, Germany
   `http://www.siliconsoftware.de/`

Part II

Handling and Enhancing Images

7 Pixels

7.1 Highlights

A pixel is the elementary unit in a digital image. It holds a digital number — in multichannel images a set of numbers — that represent the irradiance in a cell at the image plane corresponding to the sampling grid.

A pixel can be regarded as a random variable as any other measured quantity. Therefore, it can be described by first-order statistics (Section 7.3.1) and a probability density function characterizes the distribution of the values. From this distribution, mean values and measures for the scatter of the values, the variance, or standard deviation can be computed.

The normal distribution (Section 7.3.1c) and its discrete counterpart, the binomial distribution (Section 7.3.1d), are the most important distributions for image processing. Estimates of the probability density distribution in images are given by histograms.

Advanced concepts in this chapter include point operations with multichannel images (Section 7.3.6). Homogeneous point operations with two 8-bit channels or images (dyadic point operations, Section 7.3.6c) can still be performed with fast look-up table operations.

On the level of an individual pixel, a number of useful and required low-level image processing tasks are discussed in the procedure section:

- Interactive gray value evaluation (Section 7.4.1). These techniques are essential to improve image acquisition and to get a first impression of the image quality. It includes the evaluation of the inhomogeneity of the illumination (Section 7.4.1a), the detection of underflow or overflow during the analog to digital conversion (Section 7.4.1b), and various interactive techniques for the inspection of images (Section 7.4.1c).
- Correction of inhomogeneous illumination (Section 7.4.2)
- Radiometric calibration (Section 7.4.3)
- Noise reduction by image averaging (Section 7.4.6)
- Windowing (Section 7.4.7)

These tasks can be performed with two classes of point operations. Homogeneous point operations are the same for all pixels and can be performed efficiently in software and hardware using look-up tables (LUT). Inhomogeneous point operations differ from pixel to pixel and require more computations.

Task List 5: Pixel

Task	Procedures
Evaluate image acquisition conditions: uniform illumination, full usage of digitalization range	Perform homogeneous point operations interactively using techniques such as pseudo color display; compute histograms
Analyze the statistical processes and sources for noise in the acquired images	Measure the noise level under various conditions (zero and full illumination); measure the statistical properties of the noise
Inspect contents of acquired images	Interactively perform homogeneous point operations, compute histograms
Correct inhomogeneous illumination	Perform a correction procedure using inhomogeneous point operations
Perform relative (and, if required, absolute) radiometric calibration	Perform a calibration procedure using an appropriate experimental setup and inhomogeneous point operations
Reduce noise level in images	Perform image averaging
Mask image (required by global transformations such as the Fourier Transform)	Multiply image with window function (inhomogeneous point operation)
Convert multichannel images into other representations	Perform linear or nonlinear multichannel point operations

7.2 Task

Pixels are the elementary units in digital image processing. A number of image processing tasks can be performed by simply handling these pixels as individual objects or measuring points. As long as the statistical properties of a pixel do not depend on its neighbor pixels, we can apply the classical concepts of *first-order statistics* which are used to handle measurements of single scalar quantities, e. g., temperature, pressure, and size. The random nature of the signal measured at individual pixels can be related to one or more of the following processes:

1. The imaging sensor introduces various types of noise into the measured irradiance. For most — but not all — noise sources, the individual pixels are not correlated to each other and can thus be treated individually.

2. In low-light level application, the quantization of radiative energy introduces randomness (Section 3.3.1f). Then, a sensor does not measure a continuous signal but rather counts individual events, the absorption of photons in the detector material. The rate at which the photons are counted is a statistical process that can be described by a *Poisson distribution* (Section 5.3.2g).

3. The process or object observed may exhibit a statistical nature. Examples are wind waves (Section 1.4.3), turbulent flows (Sections 1.6.1 and 1.6.2), and the positions and size distributions of small particles (Sections 1.3.1–1.3.3).

At the level of individual pixels, a number of simple but important image processing tasks are required to optimize the acquired images before they can be used for further processing. Most of them are directly related to the optimization, control, and calibration of image formation and digitalization, as summarized in task list 5.

7.3 Concepts

7.3.1 Random Variables and Probability Density Functions

7.3.1a Continuous and Discrete Random Variables. We consider an experimental setup in which we are measuring a certain quantity. In this process, we also include the noise introduced by the sensor. The measured quantity is the radiative flux or — converted to digital numbers — the gray value of a pixel. Because of the statistical nature of the process, each measurement will give a different value. Therefore, the observed process cannot be characterized by a single value but rather by a *probability density function* $p(g)$ or *PDF* indicating how often we observe the gray value g. A *random variable*, or short *RV*, denotes a measurable quantity which is governed by a random process — such as the gray value g of a pixel in image processing.

In the following sections, we discuss both continuous and discrete random variables and probability functions. We need discrete probabilities as only discrete values can be handled by a digital computer. Discrete values result from *quantization* (Section 6.3.6). All formulas in this section contain continuous formulation on the left side and their discrete counterparts on the right side. In the continuous case, a gray value g is measured with the probability $p(g)$. In the discrete case, we can only measure a finite number, Q, of gray values g_q ($q = 0, 1, \ldots, Q-1$) with the probability p_q. Normally, the gray value of a pixel is stored in one byte so that we can measure $Q = 256$ different gray values. Since the total probability to observe any gray value is 1, the PDF meets the requirement

$$\int_{-\infty}^{\infty} p(g)\, \mathrm{d}g = 1, \quad \sum_{q=0}^{Q-1} p_q = 1. \tag{7.1}$$

The integral of the PDF is known as the *distribution function*

$$P(g) = \int_{-\infty}^{g'} p(g')\mathrm{d}g'. \tag{7.2}$$

The distribution function increases monotonically from 0 to 1 because the PDF is a nonnegative function.

7.3.1b Mean, Variance, and Moments. The *expected* or *mean gray value* μ is defined as

$$\mu = \overline{g} = \int_{-\infty}^{\infty} p(g)g\, \mathrm{d}g, \quad \mu = \sum_{q=0}^{Q-1} p_q g_q. \tag{7.3}$$

The computation of the expectation value is denoted by a bar over the corresponding term. The *variance* is a measure to which extent the measured values deviate from the mean value

$$\sigma_g^2 = \operatorname{var} g = \overline{(g - \overline{g})^2} = \int_{-\infty}^{\infty} p(g)(g - \overline{g})^2\, \mathrm{d}g, \quad \sigma^2 = \sum_{q=0}^{Q-1} p_q (g_q - \overline{g})^2. \tag{7.4}$$

The probability function can be characterized in more detail by quantities similar to the variance, the *central moments*:

$$\mu_n = \overline{(g - \overline{g})^n} = \int_{-\infty}^{\infty} p(g)(g - \overline{g})^n\, \mathrm{d}g, \quad \mu_n = \sum_{q=0}^{Q-1} p_q (g_q - \overline{g})^n. \tag{7.5}$$

The first central moment is — by definition — zero. The second moment corresponds to the variance. The third moment, the *skewness*, is a measure for the asymmetry of the probability function around the mean value. If a distribution function is symmetrical with respect to the mean value, i. e.,

$$p(-(g - \overline{g})) = p(g - \overline{g}), \tag{7.6}$$

the third and all higher-order odd central moments vanish.

7.3.1c Normal Distribution. The probability function depends on the nature of the underlying process. Many processes with continuous random variables can be adequately described by the *normal* or *Gaussian probability density function*

$$p(g) = \frac{1}{\sqrt{2\pi}\sigma} \exp\left(-\frac{(g - \overline{g})^2}{2\sigma^2}\right). \tag{7.7}$$

The normal distribution is completely described by the two elementary statistical parameters, mean and variance. Many physical random processes are governed by the normal distribution because they are a linear superposition of many (n) individual processes. The *central limit theorem* of statistics states that in the limit $n \to \infty$ the distribution tends to a normal distribution provided certain conditions are met by the individual processes. As an example of the superposition of many processes, the measurement of the slope distribution at the ocean surface is discussed in Example 7.1.

Example 7.1: Distribution of the slope of the ocean surface.

The ocean surface is undulated by surface waves, which incline the water surface. The elementary processes are sinusoidal waves. Such a single wave shows a slope distribution very different from that of a normal distribution (Fig. 7.1a). The maximum probability occurs with the maximum slopes of the wave. Let us assume that waves with different wavelengths and direction superpose each other without any disturbance and that the slope of the individual wave trains is small. The slopes can then be added up. The resulting probability distribution is given by convolution of the individual distributions since, at each probable slope s_1 of the first wave, the second can have all slopes s_2 according to its own probability distribution, so that the PDF of the sum $s = s_1 + s_2$ is given by

$$p(s) = \int_{-\infty}^{\infty} p_{s_1}(s_1) p_{s_2}(s - s_1) ds_1. \tag{7.8}$$

The variable $s - s_1$ for p_{s_2} ensures that the sum of the two slopes s_1 and s_2 is s.

The superposition of two waves with equal slope results in a distribution with the maximum at slope zero (Fig. 7.1b). Even for quite a small number of superpositions, we can expect a normal distribution (Fig. 7.1c).

Deviations from the normal distribution occur when the elementary processes do not superpose each other randomly and without interaction. A simple example are phase-coupled waves (Fig. 7.1d). The distribution becomes asymmetric. The maximum is shifted to small negative slopes; high positive slope values are much more likely than high negative slopes and than those expected from a normal distribution. In consequence, deviations from the normal distribution generally provide some clues about the strength and the kind of nonlinear interactions.

7.3.1d Binomial Distribution. For discrete values, the equivalent to the Gaussian distribution is the *binomial distribution*

$$B(Q, p): \quad p_q = \frac{Q!}{q!(Q - q)!} \beta^q (1 - \beta)^{Q - q}, \quad \text{with } 0 < \beta < 1. \tag{7.9}$$

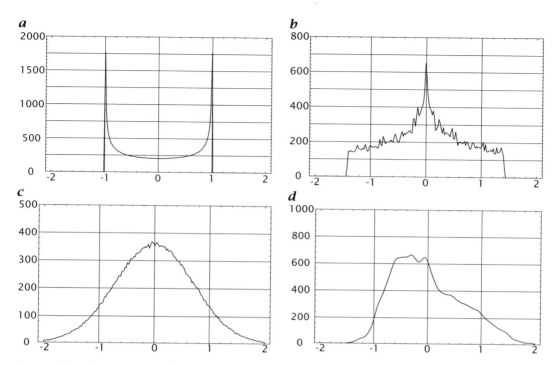

Figure 7.1: *Illustration of the superposition of the probability functions with the slope distribution on the undulated ocean surface: **a** slope distribution of a single sinusoidal wave; **b** slope distribution of the superposition of two statistically independent sinusoidal waves; **c** Gaussian distribution for the linear superposition of many sinusoidal waves; **d** skewed distribution for a nonlinear wave with phase coupled harmonics (all results from Monte Carlo simulations).*

Again Q denotes the number of quantization levels or possible outcomes. The parameter β determines the mean and the variance

$$\mu = Q\beta \tag{7.10}$$

$$\sigma^2 = Q\beta(1 - \beta). \tag{7.11}$$

For large Q, the binomial distribution quickly converges to the Gaussian distribution (see Section 11.4.2a).

7.3.1e Poisson Distribution.   As already discussed in Section 5.3.2g, another PDF is of importance for image acquisition. An imaging sensor element that is illuminated with a certain irradiance receives within a time interval Δt, the *exposure time*, on average N electrons by absorption of photons. Because of the random nature of the stream of photons, a different number of photons $n \geq 0$ arrive during each exposure with a probability density function that is governed by a *Poisson process* $P(N)$:

$$P(N): \quad p_n = \exp(-N)\frac{N^n}{n!}, \quad n \geq 0 \tag{7.12}$$

with the mean and variance

$$\mu = N \quad \text{and} \quad \sigma^2 = N. \tag{7.13}$$

The Poisson process has the following important properties:

1. The standard deviation σ is equal to the square root of the number of events. There-fore the noise level is signal-dependent.

2. Nonoverlapping exposures are statistically independent events [110, Section. 3.4]. This means that we can take images captured with the same sensor at different times as independent RVs.

3. The Poisson process is additive: the sum of two independent Poisson-distributed RVs with the means μ_1 and μ_2 is also Poisson distributed with the mean and variance $\mu_1 + \mu_2$.

7.3.1f Histograms. Generally, the probability density function is not known a priori. Rather, it is estimated from measurements. If the observed process is *homogeneous*, that is, it does not depend on the position of the pixel in the image, there is a simple way to estimate the probability distribution with the so-called *histogram*.

A histogram of an image is a list that contains as many elements as quantization levels. In each element, the number of pixels is stored that show the corresponding gray value. Histograms can be calculated straightforwardly. First, we set the whole list to zero. Then, we scan all pixels of the image, take the gray value as the index to the list, and increment the corresponding element of the list by one. The actual scanning algorithm depends on how the image is stored.

7.3.2 Functions of Random Variables

Any image processing operation changes the signal g at the individual pixels. In the simplest case, g at each pixel is transformed into g' by a function p: $g' = f(g)$. Because g is a random variable, g' will also be a RV and we need to know its PDF in order to know the statistical properties of the image after processing it.

The PDF $p_{g'}$ of g' has the same form as the PDF p_g of g if f is a linear function $g' = f_0 + f_1 g$:

$$p_{g'}(g') = \frac{p_g(g)}{|f_1|} = \frac{p_g((g' - f_0)/f_1)}{|f_1|}, \tag{7.14}$$

where the inverse linear relation $g = f^{-1}(g') : g = (g' - f_0)/f_1$ is used to express g as a function of g'.

In the general case of a nonlinear function $p(g)$, the slope f_1 in Eq. (7.14) will be replaced by the derivative $p'(g_p)$ of $p(g_p)$. Further complications arise if the inverse function has more than one branch. A simple and important example is the function $g' = g^2$ with the two inverse functions $g_{1,2} = \pm\sqrt{g'}$. In such a case, the PDF of g' needs to be added from all branches of the inverse function:

$$p_{g'}(g') = \sum_{p=1}^{P} \frac{p_g(g_p)}{|f'(g_p)|}, \tag{7.15}$$

where g_p are the P real roots of $g' = p(g)$.

Because a strictly monotonic function f has a unique inverse function $f^{-1}(g')$, Eq. (7.15) reduces in this case to

$$p_{g'}(g') = \frac{p_g(f^{-1}(g'))}{|f'(f^{-1}(g'))|}. \tag{7.16}$$

The following two examples further illustrate the use and handling of functions of random variables.

Example 7.2: Conversion to a specific PDF.

In image and signal processing, often the following problem is encountered. We have a signal g with a certain PDF and want to transform g by a suitable transform into g' in such a way that g' has a specific probability distribution. This is the inverse problem to what we have discussed so far and it has a surprisingly simple solution when we use the distribution functions P as introduced in Section 7.3.1a. The transform

$$g' = P_{g'}^{-1}(P_g(g)) \tag{7.17}$$

converts the $p_g(g)$-distributed random variable g into the $p_{g'}(g')$-distributed random variable g'. The solution is especially simple for a transformation to a *uniform distribution* because then P^{-1} is a constant function and $g' \propto P_g(g)$).

Example 7.3: Mean and variance of the function of a RV.

Intuitively, you may assume that the mean of g' can be computed from the mean of $g : \overline{g'} = f(\overline{g})$. This is, however, only possible if f is a linear function.

By definition according to Eq. (7.3), the mean of g' is

$$\overline{g'} = \mu_{g'} = \int_{-\infty}^{\infty} g' f_{g'}(g') \mathrm{d}g'. \tag{7.18}$$

We can, however, also express the mean directly in terms of the function $f(g)$ and the PDF $p_g(g)$:

$$\overline{g'} = \mu_{g'} = \int_{-\infty}^{\infty} p(g) f_g(g) \mathrm{d}g. \tag{7.19}$$

If $f(g)$ is approximated by a polynomial

$$f(g) = f(\mu_g) + f'(\mu_g)(g - \mu_g) + f''(\mu_g)(g - \mu_g)^2/2 + \ldots \tag{7.20}$$

then

$$\mu_{g'} \approx f(\mu_g) + f''(\mu_g)\sigma_g^2/2. \tag{7.21}$$

From this equation we see that $\mu_{g'} = f(\mu_g)$ is only a good approximation if both the curvature of $f(g)$ and the variance of g are small.

The first-order estimate of the variance of g' is given by

$$\sigma_{g'}^2 \approx \left| f'(\mu_g) \right|^2 \sigma_g^2. \tag{7.22}$$

This expression is only exact for linear functions p.

The following simple relations for means and variances follow directly from the discussion above (a is a constant):

$$\overline{ag} = a\overline{g}, \quad \mathrm{var}(ag) = a^2 \,\mathrm{var}\,g, \quad \mathrm{var}\,g = \overline{g^2} - \overline{g}^2. \tag{7.23}$$

7.3.3 Multiple Random Variables and Error Propagation

In image processing, we have many pixels and thus many random variables and not just one. Many image processing operations compute new values from values at many pixels. Thus, it is important to study the statistics of multiple RVs in order to learn how the statistical properties of processed images depend on the statistical properties of the original image data.

7.3.3a   Joint Probability Density Functions. First, we need to consider how the random properties of multiple RVs can be described. Generally, the random properties of two RVs, g_1 and g_2, cannot be described by their individual PDFs, $p(g_1)$ and $p(g_2)$. It is rather necessary to define a *joint probability density function* $p(g_1, g_2)$.

Only if the two random variables are *independent*, i.e., if the probability that g_1 takes a certain value does not depend on the value of g_2, can we compute the joint PDF from the individual PDFs, known as *marginal PDFs*:

$$p(g_1, g_2) = p_{g_1}(g_1) p_{g_2}(g_2) \quad \Leftrightarrow \quad g_1, g_2 \text{ independent.} \tag{7.24}$$

For R random variables g_k, the random vector $\boldsymbol{g}$, the joint probability density function is $p(g_1, g_2, \dots, g_R) = p(\boldsymbol{g})$. The P RVs are called independent if the joint PDF can be written as a product of the marginal PDFs

$$p(\boldsymbol{g}) = \prod_{r=1}^{R} p_{g_r}(g_r) \quad \Leftrightarrow \quad g_r \text{ independent.} \tag{7.25}$$

7.3.3b   Covariance and Correlation. The covariance measures to which extent the fluctuations of two RVs, g_r and g_s, are related to each other. In extension of the definition of the *variance* in Eq. (7.4), the *covariance* is defined as

$$\Sigma_{rs} = \overline{((g_r - \mu_r)(g_s - \mu_s))} = \overline{g_r g_s} - \overline{g_r} \cdot \overline{g_s}. \tag{7.26}$$

For R random variables, the covariances form an $R \times R$ symmetric matrix, the *covariance matrix* $\Sigma = \text{cov}\,\boldsymbol{g}$. The diagonal of this matrix contains the variances of the R RVs.

The *correlation coefficient* relates the covariance to the corresponding variances:

$$c_{rs} = \frac{\Sigma_{rs}}{\sigma_r \sigma_s} \quad \text{with} \quad |c_{rs}| \leq 1. \tag{7.27}$$

Two RVs g_p and g_q are called *uncorrelated* if the covariance C_{rs} is zero. Then according to Eqs. (7.26) and (7.27) the following relations are true for uncorrelated RVs:

$$C_{rs} = 0 \Leftrightarrow c_{rs} = 0 \Leftrightarrow \overline{g_r g_s} = \overline{g_r} \cdot \overline{g_s} \Leftrightarrow g_r, g_s \text{ uncorrelated.} \tag{7.28}$$

From the last of these conditions and Eq. (7.24), it is evident that independent RVs are uncorrelated.

At first glance it appears that only the statistical properties of independent RVs are easy to handle. Then we only need to consider the marginal PDFs of the individual variables together with their mean and variance. Generally, the interrelation of random variations of the variables as expressed by the covariance matrix $\boldsymbol{C}$ must be considered. Because the covariance matrix is symmetric, however, we can always find a coordinate system, i.e., a linear combination of the RVs, in which the covariance matrix is diagonal and thus the RVs are uncorrelated.

7.3.3c   Functions of Multiple Random Variables. In extension to the discussion of functions of a single RV in Section 7.3.2, we can express the mean of a function of multiple random variables $g' = p(g_1, g_2, \dots, g_R)$ directly from the joint PDF:

$$\overline{g'} = \int_{-\infty}^{\infty} p(g_1, g_2, \dots, g_R) f(g_1, g_2, \dots, g_R) \mathrm{d}g_1 \mathrm{d}g_2 \dots \mathrm{d}g_R. \tag{7.29}$$

From this general relation it follows that the mean of any linear function

$$g' = \sum_{r=1}^{R} a_r g_r \qquad (7.30)$$

is given as the linear combination of the means of the RVs g_r:

$$\overline{\left(\sum_{r=1}^{R} a_r g_r\right)} = \sum_{r=1}^{R} a_r \overline{g_r}. \qquad (7.31)$$

Note that this is a very general result. We did not assume that the RVs are independent, and this is not dependent on the type of the PDF. As a special case Eq. (7.31) includes the simple relations

$$\overline{g_1 + g_2} = \overline{g_1} + \overline{g_2}, \quad \overline{g_1 + a} = \overline{g_1} + a. \qquad (7.32)$$

The variance of functions of multiple RVs cannot be computed that easy even in the linear case. Let g be a vector of R RVs, g' a vector of S RVs that is a linear combination of the R RVs g, M a $S \times R$ matrix of coefficients, and a a column vector with S coefficients. Then

$$g' = Mg + a \quad \text{with} \quad \overline{g'} = M\overline{g} + a \qquad (7.33)$$

in extension to Eq. (7.31). If $R = S$, Eq. (7.33) can be interpreted as a coordinate transformation in a R-dimensional vector space. Therefore it is not surprising that the symmetric covariance matrix transforms as a second-order tensor [110]:

$$\text{cov}(g') = M \, \text{cov}(g) M^T. \qquad (7.34)$$

To illustrate the application of this important general relation, we apply it to several examples.

Example 7.4: Variance of the mean of uncorrelated RVs.

First, we discuss the computation of the variance of the mean $\overline{g}$ of R RVs with the same mean and variance σ^2. We assume that the RVs are uncorrelated. Then the matrix M and the covariance matrix cov g are

$$M = \frac{1}{R}[1, 1, 1, \ldots, 1] \quad \text{and} \quad \text{cov}(g) = \begin{bmatrix} \sigma^2 & 0 & \ldots & 0 \\ 0 & \sigma^2 & \ldots & 0 \\ \vdots & \vdots & \ddots & \vdots \\ 0 & 0 & \ldots & \sigma^2 \end{bmatrix} = \sigma^2 I.$$

Using these expressions in Eq. (7.34) yields

$$\sigma_{\overline{g}}^2 = \frac{1}{R}\sigma^2. \qquad (7.35)$$

Thus the variance $\sigma_{\overline{g}}^2$ is proportional to R^{-1} and the *standard deviation* $\sigma_{\overline{g}}$ decreases only with $R^{-1/2}$. This means that we must take four times as many measurements in order to double the precision of the measurement of the mean. This is not the case for correlated RVs. If the RVs are fully correlated ($r_{rs} = 1$, $\Sigma_{rs} = \sigma^2$), according to Eq. (7.34), the variance of the mean is equal to the variance of the individual RVs. In this case it is not possible to reduce the variance by averaging.

Example 7.5: Variance of the sum of uncorrelated RVs.

In a slight variation, we take R uncorrelated RVs with unequal variances σ_r^2 and compute the variance of the sum of the RVs. From Eq. (7.32), we know already that the mean of the sum is equal to the sum of the means (even for correlated RVs). Similar as for the previous example, it can be shown that for uncorrelated RVs the variance of the sum is also the sum of the individual variances:

$$\text{var} \sum_{r=1}^{R} g_r = \sum_{r=1}^{R} \text{var}\, g_r. \tag{7.36}$$

Example 7.6: Variance of linear combination of uncorrelated RVs.

As a third example we take S RVs g_s' that are a linear combination of R uncorrelated RVs g_r with equal variance σ^2:

$$g_s' = \mathbf{a}_r^T \mathbf{g}. \tag{7.37}$$

Then the vectors $\mathbf{a}_q^T$ form the rows of the $S \times R$ matrix $\mathbf{M}$ in Eq. (7.33) and the covariance matrix of $\mathbf{g}'$ results according to Eq. (7.34) in

$$\text{cov}(\mathbf{g}') = \sigma^2 \mathbf{M}\mathbf{M}^T = \sigma^2 \begin{bmatrix} \mathbf{a}_1\mathbf{a}_1 & \mathbf{a}_1\mathbf{a}_2 & \dots & \mathbf{a}_1\mathbf{a}_S \\ \mathbf{a}_1\mathbf{a}_2 & \mathbf{a}_2\mathbf{a}_2 & \dots & \mathbf{a}_2\mathbf{a}_S \\ \vdots & \vdots & \ddots & \vdots \\ \mathbf{a}_1\mathbf{a}_S & \mathbf{a}_2\mathbf{a}_S & \dots & \mathbf{a}_S\mathbf{a}_S \end{bmatrix}. \tag{7.38}$$

From this equation, we can learn two things. First, the variance of the RV g_s' is given by $\mathbf{a}_s\mathbf{a}_s$, i.e., the sum of the squares of the coefficients

$$\sigma^2(g_s') = \sigma^2 \mathbf{a}_s\mathbf{a}_s. \tag{7.39}$$

Second, although the RVs g_r are uncorrelated, two RVs g_s' and $g_{s'}'$ are only uncorrelated if the scalar product of the coefficient vectors, $\mathbf{a}_s\mathbf{a}_{s'}$, is zero, i.e., the coefficient vectors are orthogonal. Thus, only orthogonal transform matrixes $\mathbf{M}$ in Eq. (7.33) leave uncorrelated RVs uncorrelated.

Example 7.7: Variance of nonlinear functions of RVs

The above analysis of the variance for functions of multiple RVs can be extended to nonlinear functions provided that the function is sufficiently linear around the mean value. A Taylor expansion of the nonlinear function $p_s(\mathbf{g})$ around the mean value yields

$$g_s' = p_s(\mathbf{g}) \approx p_s(\boldsymbol{\mu}) + \sum_{r=1}^{R} \frac{\partial p_s}{\partial g_r}(g_r - \mu_r). \tag{7.40}$$

We compare this equation with Eq. (7.33) and find that the $S \times R$ matrix $\mathbf{M}$ has to be replaced by the matrix $\mathbf{J}$

$$\mathbf{J} = \begin{bmatrix} \dfrac{\partial p_1}{\partial g_1} & \dfrac{\partial p_1}{\partial g_2} & \dots & \dfrac{\partial p_1}{\partial g_R} \\ \dfrac{\partial p_2}{\partial g_1} & \dfrac{\partial p_2}{\partial g_2} & \dots & \dfrac{\partial p_2}{\partial g_R} \\ \vdots & \vdots & \ddots & \vdots \\ \dfrac{\partial p_S}{\partial g_1} & \dfrac{\partial p_S}{\partial g_2} & \dots & \dfrac{\partial p_S}{\partial g_R} \end{bmatrix}, \tag{7.41}$$

known as the *Jacobian matrix* of the transform $\mathbf{g}' = \mathbf{p}(\mathbf{g})$. Thus the covariance of $\mathbf{g}'$ is given by

$$\text{cov}(\mathbf{g}') \approx \mathbf{J}\,\text{cov}(\mathbf{g})\mathbf{J}^T. \tag{7.42}$$

7.3.4   Homogenous Point Operations

Point operations are a class of very simple image processing operations. The gray values at individual pixels are modified depending on the gray value and the position of the pixel. Generally, such a kind of operation is expressed by

$$g'_{mn} = P_{mn}(g_{mn}). \tag{7.43}$$

The indices at the function P denote the explicit dependence on the position of the pixel. If the point operation is independent of the position of the pixel, we call it a *homogeneous point operation* and write

$$g'_{mn} = P(g_{mn}). \tag{7.44}$$

It is important to note that the result of the point operation does not depend at all on the gray value of neighboring pixels. A point operation maps the set of gray values onto itself. Generally, point operations are not invertible, as two different gray values may be mapped onto one. Thus, a point operation generally results in a loss of information which cannot be recovered. Only a point operation with a one-to-one mapping of the gray values is invertible.

Example 7.8: Invertible and noninvertible point operations

The point operation

$$P(q) = \begin{cases} 0 & q < t \\ 255 & q \geq t \end{cases}, \tag{7.45}$$

for example, performs a simple threshold evaluation. All gray values below the threshold are set to zero (black), all above and equal to the threshold to 255 (white). Consequently, this point operation cannot be inverted. An example for an invertible point operation is the image negation computing an image with an inverted gray scale:

$$P(q) = Q - 1 - q. \tag{7.46}$$

The inverse operation of a negation is another negation. Another example for an invertible point operation is the conversion between signed and unsigned representation of gray values discussed in Section 6.3.6d.

7.3.4a   Look-Up Tables.   The direct computation of homogeneous point operations according to Eq. (7.44) may be very costly as demonstrated in Example 7.9.

Example 7.9: Logarithmic look-up table

A 512×512 image should be presented in an 8-bit logarithmic gray scale covering 5 decades from 1 to 100 000. This requires the following point operation:

$$P(q) = 51 \log q. \tag{7.47}$$

A straightforward implementation would require the following operations per pixel:

- integer to double conversion,
- computation of logarithm,
- multiplication with 51.0, and
- double to 8-bit integer conversion.

All these operations must be computed 262 144 times for a 512×512 image.

The key point for a more efficient implementation lies in the observation that the definition range of any point operation consists of only very few gray values, typically 256. Thus, we would have to calculate the very same values many times, in the mean

1000 times for a 512×512 image! We can avoid this by precalculating $P(g_q)$ for *all* 256 possible gray values and store the computed values in a 256-element table. Then, the computation of the point operation is reduced to a replacement of the gray value by the element in the table with an index corresponding to the gray value. Such a table is called a *look-up table* or *LUT*. As a result, homogeneous point operations are equivalent to *look-up table operation*s.

A cautionary note is necessary for all kinds of LUT operations: *any LUT operation makes the image look better, but does not actually improve it.* This is why we should use them thoughtfully. A careful preparation of images using an LUT operation is very important for printouts which have a lower contrast range than images on monitors. However, for further processing of images, especially if we are interested in a *quantitative* analysis of gray values, they are not of much help. On the contrary, they may introduce additional errors, because of the rounding errors introduced by nonlinear LUT functions. They may lead to missing gray values in the output or mapping of two consecutive gray values onto one.

In most image processing systems, look-up tables are implemented in hardware (Section 6.5). Generally, a look-up table, the *input LUT*, is located between the *analog-digital converter* and the frame buffer (Fig. 6.16). Another, the *output LUT*, is located between the frame buffer and the digital-analog converter for output of the image in the form of an analog video signal, e. g., to a monitor. The input LUT allows a point operation to be performed *before* the image is stored in the frame buffer. With the output LUT, a point operation can be performed and observed on the monitor. In this way, we can interactively perform point operations *without* modifying the stored image (Section 7.4.1).

The use of input LUTs is limited. Input LUTs would only be valuable if the digitization precision were higher than the storage precision. Imagine that we digitize with 12 bits, pass the data through a 12-bit input LUT, and store them with 8 bits. Then, rounding errors would be reduced by a factor of 16. In addition, we could compress a larger dynamic range with a nonlinear LUT onto 8 bits.

In contrast to the input LUT, the output LUT is a tool much more widely used, since it does not change the stored image. With LUT operations, we can also convert a gray-value image into a *pseudo-color image*. Again, this technique is common even with the simplest image processing boards, since not much additional hardware is needed. Three digital analog converters are used for the primary colors red, green, and blue. Each channel has its own LUT. In this way, we can map each individual gray value q to any color by assigning a color triple to the corresponding LUT addresses $r(q)$, $g(q)$, and $b(q)$. Formally, we now have a *vector point operation*

$$\boldsymbol{P}(q) = \begin{bmatrix} r(q) \\ g(q) \\ b(q) \end{bmatrix}. \tag{7.48}$$

As long as all three point functions $r(q)$, $g(q)$, and $b(q)$ are identical, a gray value image will be displayed. If two of them vanish, the image will appear in the remaining color. For more details on color vision, see Section 3.4.7.

7.3.5 Inhomogeneous Point Operations

Although often used, homogeneous point operations are only a subclass of point operators. In the more general case, the point operation depends also on the position of the image. This general class of operations is called *inhomogeneous point operations*.

Inhomogeneous point operations are mostly related to calibration procedures. Simple examples are the subtraction of a background image (Example 7.10) and a two-point calibration (Example 7.11).

Example 7.10: Subtraction of a background image

The subtraction of a background image without objects or illumination is a simple example of an inhomogeneous point operation. It can be written as:

$$g'_{mn} = P_{mn}(g_{mn}) = g_{mn} - b_{mn}, \tag{7.49}$$

where B_{mn} is the background image.

Example 7.11: Two-point calibration

Often it is required to translate the gray scale in an image into the object property it represents, e. g., a temperature, concentration, reflection, etc. If the relation between the feature and the gray value is linear, a two-point calibration can be performed. We assume that images are taken under two different calibration conditions, g'_{mn} and g''_{mn} with features f' and f'', where the object features are well known. Then, any image of the same class can be converted into a feature image by the following inhomogeneous point operation:

$$f_{mn} = f' + \frac{g_{mn} - g'_{mn}}{g''_{mn} - g'_{mn}}(f'' - f'). \tag{7.50}$$

Computation of an *inhomogeneous point operation* is much more time consuming. We cannot use look-up tables since the point operation depends on the pixel position and we are forced to calculate the function for each pixel.

7.3.6   Point Operations with Multichannel Images

Point operations can be generalized to multichannel point operations in a straightforward way. The operation still depends only on the values of a single pixel. The only difference is that it depends on a vectorial input instead of a scalar input. Likewise, the output image can be a multichannel image. For homogeneous point operations that do not depend on the position of the pixel in the image, we can write

$$\boldsymbol{G}' = \boldsymbol{P}(\boldsymbol{G}) \tag{7.51}$$

with

$$
\begin{aligned}
\boldsymbol{G}' &= [\boldsymbol{G}'_0 \; \boldsymbol{G}'_1 \; \ldots \; \boldsymbol{G}'_{L-1}] & \text{L-channel output image,} \\
\boldsymbol{G} &= [\boldsymbol{G}_0 \; \boldsymbol{G}_1 \; \ldots \; \boldsymbol{G}_{K-1}] & \text{K-channel input image,}
\end{aligned}
\tag{7.52}
$$

where $\boldsymbol{G}'_l$ and $\boldsymbol{G}_k$ are the components l and k of the multichannel images $\boldsymbol{G}'$ and $\boldsymbol{G}$ with L and K components, respectively.

7.3.6a   Linear Multicomponent Point Operations.   An important subclass of multi-component point operators is linear operations. This means that each component of the output image $\boldsymbol{G}'$ in Eq. (7.51) is a linear combination of the components of an input image $\boldsymbol{G}$:

$$\boldsymbol{G}'_l = \sum_{k=0}^{K-1} P_{lk}\boldsymbol{G}_k \tag{7.53}$$

,where P_{lk} are constant coefficients. Therefore, a general linear multicomponent point operation is given by a matrix (or tensor) of coefficients P_{lk}. Then, we can write Eq. (7.53) in matrix notation as

$$\boldsymbol{G}' = \boldsymbol{P}\boldsymbol{G}, \tag{7.54}$$

where P is the matrix of coefficients.

If the components of the multichannel images are not interrelated to each other, all point operations except those on the diagonal become zero. For K-channel input and output images, just K different point operations remain, one for each channel. The matrix of point operations can finally collapse to a scalar point operation when to each channel of a multicomponent image the same point operation is applied.

For a K-channel output and input image, linear point operations can be interpreted as coordinate transforms. The rows in P_{lk} Eq. (7.53) contain the base vectors of the new coordinate system after the transform. If the matrix has a rank R lower than K, the tensorial point operation projects the K-dimensional space to an R-dimensional subspace or hyperplane. In conclusion, linear multichannel point operations are quite easy to handle as they can be described in a straightforward way with the concepts of linear algebra (matrix algebra). For square matrices, for instance, we can easily give the condition when an inverse operation exists and compute it.

7.3.6b Nonlinear Multicomponent Point Operations. For nonlinear multicomponent point operations, the linear coefficients in Eqs. (7.53) and (7.54) have to be replaced by nonlinear functions:

$$G'_l = P_l(G_0, G_1, \ldots, G_{K-1}). \tag{7.55}$$

Nonlinear multicomponent point operations cannot be handled in a general way as it is the case with linear operations. Thus, they must be considered individually. The complexity can be reduced if it is possible to separate a multichannel point operation into its linear and nonlinear parts.

7.3.6c Dyadic Point Operations. Operations in which only two images are involved are termed *dyadic point operations*. In this section, we discuss how dyadic homogeneous point operations can be implemented as LUT operations and consider some examples. Generally, any dyadic image operation can be expressed as

$$g'_{mn} = P(g_{mn}, h_{mn}) \tag{7.56}$$

and performed as an LUT operation. Let the gray values of each image in P take Q different values. In total, we have to calculate Q^2 combinations and, thus, have Q^2 elements in the LUT table L. For 8-bit images, 64k values need to be calculated; that is still a quarter less than with a direct computation for each pixel in a 512×512 image. We can store all the results of the dyadic operation in a large LUT with $Q^2 = 64k$ entries in the following manner:

$$L(2^8 p + q) = P(p, q), \quad 0 \le p, q < Q. \tag{7.57}$$

High and low bytes of the LUT address are given by the gray values in the images G and H, respectively.

Some image processing systems contain a 16-bit LUT as a modular processing element. Computation of a dyadic point operation either with a hardware or software LUT is often significantly faster than a direct implementation especially if the operation is complex. It is also easier to control exceptions such as division by zero or underflow and overflow. Example 7.12 shows how a dyadic point operation can be used to perform two point operations simultaneously.

Example 7.12: Phase and amplitude computation

The phase and magnitude of a complex-valued image, such as the DFT of an image, can be computed simultaneously with one LUT operation if we also restrict the output values to 8 bits:

$$L(2^8 r_p + i_q) = 2^8 \sqrt{r_p^2 + i_q^2} + \frac{128}{\pi} \arctan\left(\frac{i_p}{r_q}\right), \quad 0 \le r_p, i_q < Q. \tag{7.58}$$

The magnitude is returned in the high byte and the phase, scaled to ± 128, in the low byte.

7.4 Procedures

This section deals with the practical aspects of point operations with pixels as summarized in task list 5 at the beginning of this chapter. Almost all procedures are related to the image acquisition process.

7.4.1 Gray Value Evaluation and Interactive Manipulation

The first task of point operations is to aid in gaining the optimal adjustment of the image acquisition conditions. Two tasks are of importance:

- Homogeneous illumination resulting in a constant object radiance and, thus, gray level of the digital images. The better this condition is met, the easier subsequent image processing, especially segmentation, will be.
- Optimal usage of the limited number of quantization levels without under- or overflow. Given the limited resolution of image data — normally only 8 bits or 256 gray levels — it is important to use this limited dynamic range in an optimum way.

7.4.1a Evaluation of Homogeneous Illuminance. With the naked eye, it is not possible to estimate the homogeneity of an illuminated area as demonstrated in Fig. 7.2a and b. Using an objective tool, such as a histogram, reveals the gray scale distribution but not its spatial variation (Fig. 7.2c and d). Therefore, it is not of much help to optimize the illumination interactively. This requires techniques that mark in one or the other way gray scales such that absolute gray levels become perceivable for the human eye. If the radiance distribution is continuous, it is sufficient to use equidensities. This technique uses a staircase type of homogeneous point operation, causing false edges in the images (Fig. 7.2e and f). This point operation can be achieved very easily. The resolution is limited artificially by zeroing the least significant bits with a logical and operation:

$$q' = P(q) = q \wedge \overline{(2^p - 1)}, \tag{7.59}$$

where $\wedge$ and overlining denote the logical (bitwise) *and* and *negation*, respectively. This point operation limits the resolution to $Q - p$ bits and, thus, 2^{Q-p} quantization levels. Note that the "fuzzyness" of the false edges gives also a direct visual impression of the noise level in the image.

Another way to mark absolute radiance is the so-called *pseudo-color* display. With this technique, a gray level q is mapped onto an RGB triple for display. Since color is much better recognized by the eye, it helps marking absolute gray levels. Equidensities and pseudo-color mapping are both suitable techniques to optimize illumination interactively in an objective way. Figure 7.4 shows a number of commonly used pseudo-color mappings.

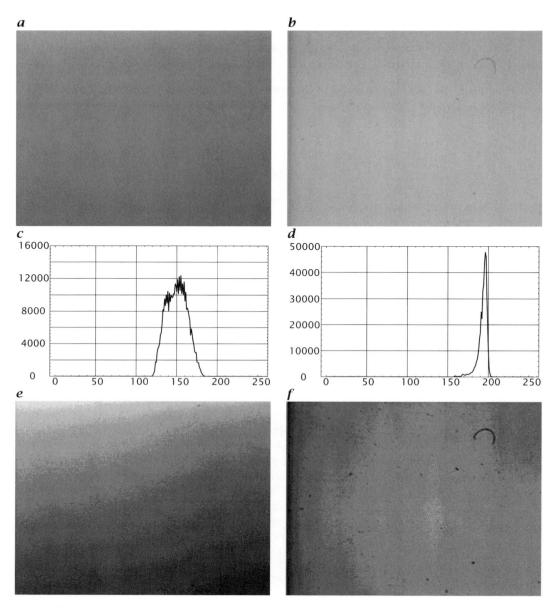

Figure 7.2: *The area in **a** shows an intensity that is slowly decreasing from the top to the bottom which is not recognized by the eye, while **b** shows a more homogeneously illuminated area. Histograms (**c** and **d**) reveal the gray scale distribution but not its spatial variation. **e** and **f**: The artificial edges generated by a stair-case LUT with a step height of 8 help to achieve a visual impression of the spatial distribution of the absolute radiance. The images are contrast enhanced.*

7.4.1b Detection of Underflow and Overflow. A dangerous error of image acquisition is under- and overflows in the gray values because it is hard to be detected directly. It may be that it becomes apparent by a surprisingly low gray level variance. But in an image with a low noise level, the low variance goes unnoticed. Over- and underflow are detected easily in histograms by strong peaks at the minimum and/or maximum gray values (Fig. 7.3). Again, pseudo-color mapping is very useful. The few lowest and

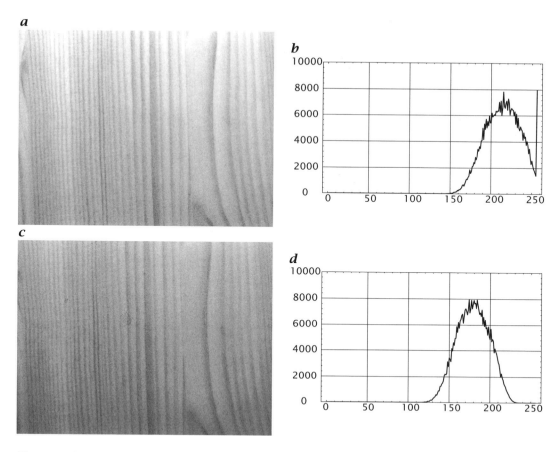

Figure 7.3: *Histograms are a sensitive indicator to detect underflow or overflow in digitized images:* **a** *Overexposed image and* **b** *its histogram.* **c** *Correctly illuminated image and* **d** *its histogram.*

highest gray values could be displayed with colors and thus become immediately visible when the dangerous lower and upper gray-scale thresholds are reached (Fig. 7.4d). Gray scale under or overflow is a common error which often goes unnoticed and causes a serious bias in further processing, for instance for mean gray values of objects, the center of gravity of an object, etc.

7.4.1c Interactive Gray Scale Manipulation. Homogeneous point operators or LUT operators are a very useful tool to interactively manipulate the gray scale in such a way that the information of interest can be observed in an optimum way. Here we demonstrate three examples.

Contrast Enhancement. As a first example of LUT operations, we will consider contrast enhancement. Because of poor illumination conditions, it often happens that images are underexposed. Then, the image is too dark and of low contrast (Fig. 7.5a). The histogram (Fig. 7.5b) shows that the image contains only a low range of gray values at low gray values. We can improve the appearance of the image considerably if we apply a point operation which maps a small gray scale range to the full contrast range (for example, $q' = 4q$ for $q < 64$, and 255 for $q \geq 64$). Values above and below the selected range must then be set to 0 and 255, respectively (Fig. 7.5c). It is important

Figure 7.4: *Illustration of pseudo-color display of gray scale images: **a** Original gray scale image, pseudo-color display, **b** rainbow colors, **c** glow colors (from red via yellow to white), **d** under/overflow marking: blue underflow, green low, yellow high, red overflow. (See also Plate 17.)*

to recognize that we only improve the appearance of the image with this operation but not the image quality itself. The gray value resolution is still the same, as the histogram shows (Fig. 7.5d).

The right way to improve the image quality is to optimize the lighting conditions. If this is not possible, we can increase the gain of the analog video amplifier. All modern image processing boards include an amplifier whose gain and offset can be set by software (see Section 6.5.2). By increasing the gain, we can improve the brightness and resolution of the image but only at the expense of an increased noise level.

Contrast Stretching. It is often required to analyze faint irradiance differences which are beyond the resolution of the human visual system or the display equipment used. This is especially the case if images are printed. Therefore, it is a useful operation to stretch a small gray scale range to the maximum possible gray scale range. This operation is demonstrated in Fig. 7.6a and b. The wedge at the bottom of the images, ranging from 0 to 255, directly shows which gray value range is stretched.

Range Compression. It is a common problem that digital images appear to have a low dynamical range. In comparison to the human visual system, a digital image has a considerably smaller dynamical range. If a minimum resolution of 10% is demanded, the maximum contrast ratio in an 8-bit image is $255/10 \approx 25$.

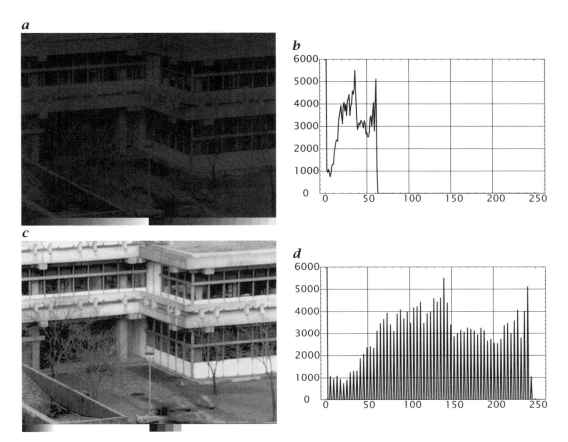

Figure 7.5: *Demonstration of interactive LUT operations: **a** Underexposed image and **b** its histogram. **c** Interactively contrast enhanced image and **d** its histogram.*

A possible cure to increase the dynamical range is a *gamma transformation* as discussed in Section 6.3.6e. This is a nonlinear homogeneous point operation of the form

$$q' = 255/(255^\gamma)q^\gamma. \tag{7.60}$$

A gamma transformation of Fig. 7.6a with $\gamma = 0.5$ and 0.25 is shown in Fig. 7.6c and d. This transformation allows a larger dynamic range to be recognized at the cost of resolution in the bright parts of the image. The dark parts become brighter and show more details. This contrast transformation is better adapted to the logarithmic characteristics of the human visual system which can detect relative intensity differences over a wide range of intensities (Section 6.3.6e).

7.4.2 Correction of Inhomogeneous Illumination

Every real-world application has to contend with *uneven illumination* of the observed scene. Even if we spend a lot of effort optimizing the lighting system, it is still very hard to obtain a perfectly even illumination. A more difficult problem is small dust particles in the optical path especially on the glass window close to the CCD sensor. These particles are not sharply imaged but absorb some light and, thus, cause a drop in the illumination level in a small area. These effects are not easily visible in a scene

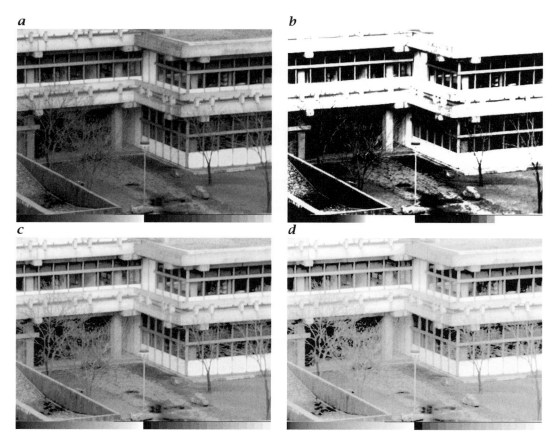

Figure 7.6: *Demonstration of interactive LUT operations: **b** Contrast stretching of the image shown in **a**. The stretched range can be read from the transformation of the gray scale wedge at the bottom of the image. **c** and **d**: Gamma transformed version of the image shown in **a** with* $\gamma = 0.5$ *and 0.25, respectively.*

with high contrast and many details, but become very apparent in a scene with a uniform background (Fig. 7.2a and b). CCD sensors also show an uneven sensitivity of the individual photo receptors. These distortions severely limit the quality of the images. Additional noise is introduced, it is more difficult to separate an object from the background, and additional systematic errors have to be considered concerning the accuracy of gray values.

Nevertheless, it is possible to correct these effects if we can take a reference image. We might either be able to take a picture without the objects, or, if they are distributed randomly, we can calculate a mean image from the many different images. The reference image R_{mn} can be used to correct the uneven illumination and sensitivity of our sensor. We just divide the image by the background image:

$$G'_{mn} = c \cdot G_{mn}/R_{mn}. \tag{7.61}$$

Since the gray values of the divided image again have to be represented by integers, multiplication with an appropriate constant is necessary. Figure 7.7 demonstrates that an effective suppression of an uneven illumination is possible using this simple method.

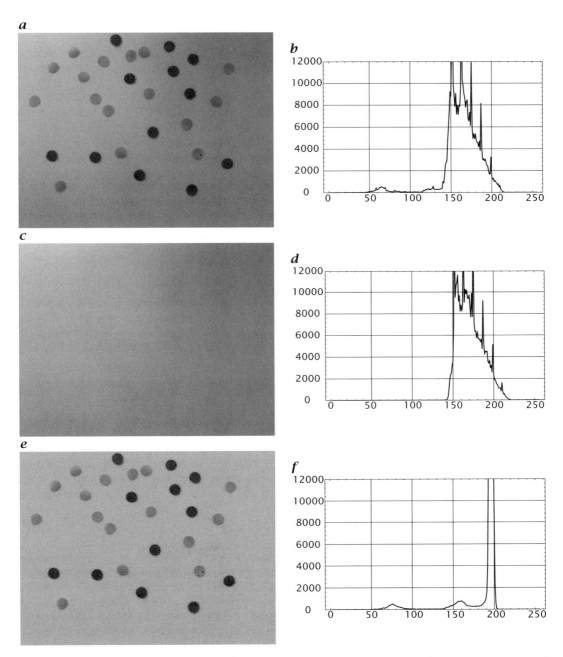

Figure 7.7: *Correction of uneven illumination with an inhomogeneous point operation: **a** original image and **b** its histogram; **c** background image and **d** its histogram; **e** division of the image by the background image and **f** its histogram.*

a   **b**   **c**

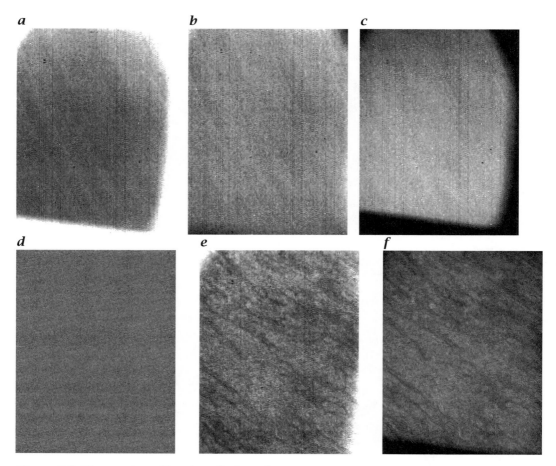

d   **e**   **f**

Figure 7.8: *Three-point calibration of infrared temperature images:* **a** - **c** *show images of calibration targets made out of aluminum blocks at temperatures of 13.06, 17.62, and 22.28° centigrade. The images are stretched in contrast to a narrow range of the 12-bit digital output range of the infrared camera:* **a***: 1715–1740,* **b***: 1925–1950,* **c***: 2200–2230, and show some residual inhomogeneities (especially vertical stripes).* **d** *Calibrated image using the three images* **a** - **c** *with quadratic interpolation.* **e** *Original and* **f** *calibrated image of the temperature microscale fluctuations at the ocean surface (area about 0.8 × 1.0 m²).*

7.4.3 Radiometric Calibration

Many image measuring tasks require an absolute radiometric calibration of the measured irradiance at the image plane. Once such a calibration is obtained, we can infer the radiance of the objects from the irradiance in the image. One obvious example is thermography. Here, the radiance itself is not of so much interest as the temperature of the emitted object which is directly related to the radiance via Planck's equations. Sections 3.3.6a and 3.4.8 detail these relations.

Here, we will show a practical calibration procedure for ambient temperatures. Because of the nonlinear relation between radiance and temperature, a simple two-point calibration with linear interpolation is not sufficient. Haußecker [54] showed that a quadratic relation is accurate enough for a small temperature range from 0 to 40° centigrade. Therefore, three calibration temperatures are required.

The calibration delivers three images of objects with constant temperature. From these three calibration images G_1, G_2, and G_3 with temperatures T_1, T_2, and T_3, the temperature image T of an image G can be computed by quadratic interpolation as

$$T = \frac{\Delta G_2 \cdot \Delta G_3}{\Delta G_{21} \cdot \Delta G_{31}} T_1 - \frac{\Delta G_1 \cdot \Delta G_3}{\Delta G_{21} \cdot \Delta G_{32}} T_2 + \frac{\Delta G_1 \cdot \Delta G_2}{\Delta G_{31} \cdot \Delta G_{32}} T_3, \tag{7.62}$$

with

$$\Delta G_k = G - G_k \quad \text{and} \quad \Delta G_{kl} = G_k - G_l. \tag{7.63}$$

The symbol $\cdot$ indicates pointwise multiplication of the images in order to distinguish it from matrix multiplication. Figure 7.8a, b, and c shows three calibration images. The infrared camera looks at the calibration targets via a mirror which limits the field of view at the edges of the images. This is the reason for the sharp temperature changes seen at the image borders in Fig. 7.8a and c. The calibration procedure removes the residual inhomogeneities, especially the vertical stripes that can be observed in the original images.

7.4.4 Noise Variance Equalization

From the discussion of a simple linear noise model of an image sensor in Section 5.3.2h, we know that the variance of the noise generally depends on the image intensity according to

$$\sigma_g^2(g) = \sigma_0^2 + \alpha g. \tag{7.64}$$

The statistical analysis of images and image operations in the previous sections of this chapter assumed that the noise variance is independent of the gray value. Therefore it can be advantageous to apply a nonlinear gray value transform $h(g)$ in such a way that the noise variance becomes constant.

In first order, the variance of $h(g)$ is

$$\sigma_h^2 \approx \left(\frac{dh}{dg}\right)^2 \sigma_g^2(g) \tag{7.65}$$

according to Eq. (7.42). If we set σ_h^2 to be constant, we obtain [35]

$$dh = \frac{\sigma_h}{\sqrt{\sigma^2(g)}} dg.$$

Integration yields

$$h(g) = \sigma_h \int_0^g \frac{dg'}{\sqrt{\sigma^2(g')}} + C \tag{7.66}$$

with two free parameters σ_h and C. Using Eq. (7.64), the integral in Eq. (7.66) yields

$$h(g) = \frac{2\sigma_h}{\sqrt{\alpha}} \sqrt{\sigma_0^2 + \alpha g} + C. \tag{7.67}$$

We can use the two free parameters σ_h and C to map the transformed signal h to the same interval $[0, g_m]$ as the original signal. Then the transform becomes

$$h(g) = g_m \frac{\sqrt{\sigma_0^2 + \alpha g} - \sigma_0}{\sqrt{\sigma_0^2 + \alpha g_m} - \sigma_0} \quad \text{with} \quad \sigma_h = g_m \frac{\alpha/2}{\sqrt{\sigma_0^2 + \alpha g_m} - \sigma_0}. \tag{7.68}$$

a  *b*

c   *d*

:lements (section 4.3.2b).   **:lements (section 4.3.2b).**

f an optical system is a perspectiv   **f an optical system is a perspectiv**

a) models the imaging geometry a   **a) models the imaging geometry a**

y described by the position of the   **y described by the position of the**

focal length (section 4.3.2c). For th   **focal length (section 4.3.2c). For th**

determine the distance range that   **determine the distance range that**

of field, section 4.3.2d) and to lear   **of field, section 4.3.2d) and to lear**

d hypercentric optical systems (se   **d hypercentric optical systems (se**

aal ontical system from a nerfect   **aal ontical system from a nerfect**

Figure 7.9: *Demonstration of histogram equalization:* **a** *and* **c** *original images;* **b** *and* **d** *histogram equalized images.*

The nonlinear transform becomes particularly simple for an ideal imaging sensor with $\sigma_0 = 0$. Then a square root transform must be applied to obtain an intensity independent noise variance:

$$h(g) = \sqrt{g_m g} \quad \text{with} \quad \sigma_h = \sqrt{\alpha g_m}/2. \tag{7.69}$$

7.4.5 Histogram Equalization

Instead of simply trying to optimize the appearance of an image with the interactive tools described in Section 7.4.1 it is sometimes useful to have an automatic tool. One such tool is *histogram equalization*. It applies a nonlinear gray value transform so that the *histogram* becomes flat. Then the image is spread out over all possible gray values in an optimal way.

According to Example 7.2 the procedure is straightforward. The nonlinear transform that transfers a given PDF p to a flat PDF is given by the distribution function:

$$g' \propto P_g(g). \tag{7.70}$$

Thus we just compute the histogram h as an approximation to the PDF and integrate this function by recursively adding up the histogram with Q bins:

$$h_i = h_{i-1} \quad 0 < i < Q. \tag{7.71}$$

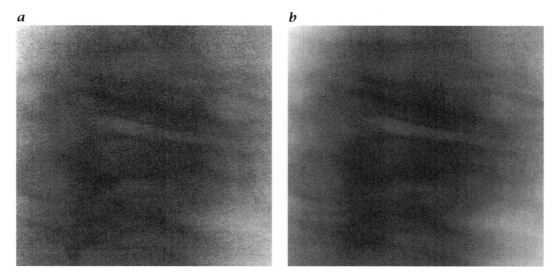

Figure 7.10: *Noise reduction by image averaging:* **a** *single thermal image of small temperature fluctuations on the water surface in a wind-wave facility cooled by evaporation at a wind speed of 1.8 m/s;* **b** *same, averaged over 16 images; temperature range corresponding to full gray value range: 1.1 K.*

Then an approximation of the discrete distribution function H is obtained. If we want the output range of the transformed image in the range from $[g_1, g_2]$ an additional scaling must be performed and the nonlinear transform becomes

$$g' = g_1 + \left(H_g - H_0\right) \frac{g_2 - g_1}{H_{Q-1} - H_0}. \tag{7.72}$$

Here H_g is the value of the distribution function with an index equal to the discrete gray value g in the interval $[0, Q - 1]$. Generally, this transform delivers the wanted result as demonstrated in Fig. 7.9a and b. It is, however, not suitable for all classes of images, especially when it makes no sense to spread out the gray values equally. This is the case, for example, with images that are basically limited to two small gray value ranges (Fig. 7.9c and d).

7.4.6 Noise Reduction by Image Averaging

An application of first-order statistics is shown in the handling of noisy images. There are a number of imaging sensors available which show a considerable noise level. Prominent examples include *thermal imaging* (Section 3.4.8) and all applications with slow-scan CCD imagers or image amplifiers where only a limited number of photons are collected.

Figure 7.10a shows the temperature of the water surface of a wind-wave facility cooled by evaporation. The small temperature fluctuations can be detected but the noise level is also substantial. Taking the mean over several images significantly reduces the noise level.

An estimate of the error of the mean taken from N samples is given by

$$\sigma_{\bar{g}}^2 \approx \frac{1}{(N-1)} \sigma_g^2 = \frac{1}{N(N-1)} \sum_{n=1}^{N} (g - \bar{g})^2. \tag{7.73}$$

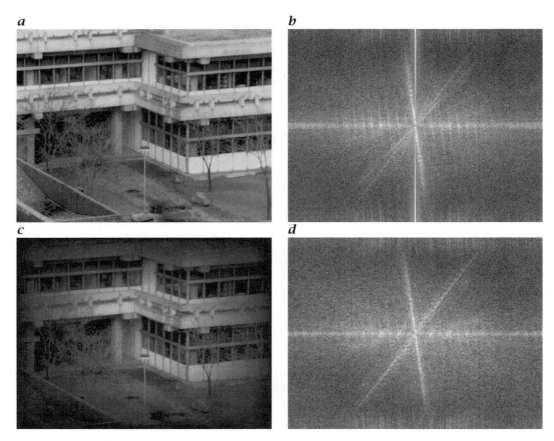

a   *b*

c   *d*

Figure 7.11: *Effect of windowing on the discrete Fourier transform:* **a** *Original image;* **b** *DFT of* **a** *without using a window function;* **c** *image multiplied with a cosine window;* **d** *DFT of* **c** *using a cosine window.*

If we take the average of N images, the noise level is reduced by $1/\sqrt{N}$ compared to a single image. Taking the mean over 16 images thus reduces the noise level by a factor of four (Fig. 7.10b).

7.4.7 Windowing

Before we can calculate the *discrete Fourier transform* (*DFT*) of an image, the image must be multiplied with a *window function*. If we omit this step, the spectrum will be distorted by the convolution of the image spectrum with the Fourier transform of the box function, the sinc function (see Appendix B.3), which causes spectral peaks to become star-like patterns along the coordinate axes in Fourier space (Fig. 7.11b). We can also explain these distortions with the periodic repetition of finite area images, an effect that was discussed in conjunction with the sampling theorem in Section 6.3.2. The periodic repetition in the spatial domain leads to discontinuities in horizontal and vertical directions which cause correspondingly high spectral densities along the x and y axes in the Fourier domain. In order to avoid these disturbances, we must multiply the image with a window function which approaches zero towards the edges of the image. An optimum window function should preserve a high spectral resolution and show minimum distortions in the spectrum, that is, its DFT should fall off as fast as possible. These are two contradictory requirements. A good spectral resolution requires a broad window function. Such a window, however, falls off steeply at the edges causing a slow fall-off of the sideslopes of its spectrum.

A carefully chosen window is very crucial for a spectral analysis of time series [97, 108]. However, in digital image processing it is not that critical because of the much lower dynamic range of the gray values. A simple cosine window

$$w_{mn} = \sin\left(\frac{2\pi m}{M}\right) \sin\left(\frac{2\pi n}{N}\right), \quad 0 \le m < M, \, 0 \le n < N \qquad (7.74)$$

performs this task well (Fig. 7.11c and d).

A direct implementation of the windowing operation is very time consuming because we would have to calculate the cosine function MN times. It is much more efficient to perform the calculation of the window function once, store it in the frame buffer, and use it for the calculation of many DFTs. The computational efficiency can be further improved by recognizing that the window function Eq. (7.74) is separable, i.e., a product of two functions $W_{mn} = {}^c w_m \cdot {}^r w_n$. Then, we need to calculate only the M plus N values for the column and row function ${}^c w_m$ and ${}^r w_n$, respectively. As a result, there is no need to store the whole window image. It is sufficient to store only the row and column functions at the expense of an additional multiplication per pixel when using the window operation.

7.5 Advanced Reference Material

Statistics and random processes
[7.1]

J. A. Rice, 1995. Mathematical Statistics and Data Analysis. Duxbury Press, Belmont, CA. *Excellent introduction to random signals and data analysis*

P. R. Bevington, 2002. Data Reduction and Error Analysis, 3rd ed. McGraw-Hill. *Another excellent introduction to data analysis*

A. Papoulis, 1991. Probability, Random Variables, and Stochastic Processes, 2nd ed. McGraw-Hill, New York. *Detailed account of the theory of probability and random variables*

A. Rosenfeld and A. C. Kak, 1982. Digital Picture Processing, 2nd ed. Academic Press, 1982. *Includes introduction to stochastic processes with respect to image processing*

Radiometric calibration of sensors and cameras
[7.2]

G. C. Holst, 1998. CCD Arrays, Cameras, and Displays. SPIE, Bellingham, WA.

G. C. Holst, 2000. Common Sense Approach to Thermal Imaging. SPIE, Bellingham, WA.

L. M. Biberman, ed., 2001. Electro Optical Imaging: System Performance and Modeling. SPIE, Bellingham, WA.

8 Geometry

8.1 Highlights

Geometric operations are required to transform the pixel coordinates back to world coordinates for geometric measurements. With the advent of semi-conductor image sensors such as the CCD sensor (Section 5.3.5), geometrically stable sensors have become available. This has opened the possibilities for accurate geometric measurements in electronic images. Sometimes this new field is named *videometry* with reference to *photogrammetry*. Accurate geometric operations on digital images depend on several factors.

First, precise models are required about the relations between the coordinates of the objects in 3-D space and the pixel coordinates. The chain of geometric transformations that leads from world coordinates to pixel coordinates has been discussed in detail as part of the image formation process in Sections 4.3.1 and 4.3.2.

Second, calibration techniques are required that allow inference of the parameters of the geometric transformations. This is performed by measuring the position of points in images whose world coordinates or other reference coordinates are known. In Sections 8.3.1b and 8.3.1c it is discussed how the parameters of affine and perspective transformations can be computed from such point correspondencies.

Third, the pixels of digital images are on a discrete grid. After a nontrivial geometric transformation, they will no longer lie on this grid and the question arises how to interpolate appropriate values for the new grid points from the surrounding grid points.

What sounds like a rather simple operation is, in reality, one of the most difficult problems of geometric operations on digital images. We do not only want that the transformed images look "nice", i. e., that they do not show jagged lines or other forms of aliasing, but that also no distortions are introduced into the signal that are not directly visible. The most significant problem for further processing that wants to maintain position accuracy well into subpixel distances (say 0.01 pixel distances) is the phase shift introduced by imperfect interpolation algorithms. Therefore, interpolation techniques are discussed in detail in Section 8.3.2.

Finally, we want to be able to perform geometric transformations fast. Fast algorithms for image scaling, translation, rotation, affine and perspective transformation, and general warping are discussed in Section 8.4.

Task List 6: Geometry

Task	Procedures
Transform image to different resolution (size) for display and inspection (zooming)	Scale image
Transform image to a different coordinate system	Perform one (or a combination) of the following transformations: translation (shift), rotation, affine transformation, perspective transformation
Geometric calibration including correction of geometric distortion	(1) Determine parameters of distortion using a calibration procedure; (2) Warp image
Image registration	(1) Determine corresponding points or features; (2) Transform images to a common coordinate system

8.2 Task

Generally, the need for geometric operations on images arises from the fact that we want to relate the pixel coordinates to the world coordinates. This is required to be able to measure the position, size, distance, and other geometric parameters of the imaged objects. In Sections 4.3.1 and 4.3.2, the chain of transformations from world coordinates via camera coordinates and image coordinates to pixel coordinates was discussed. Section 4.3.2f introduced the simplest *geometric distortion* — radial distortion — as one of the basic seven third-order *lens aberrations*. These sections set the stage for which kind of geometric operations are required.

Task list 6 summarizes the tasks related to geometric operations in image processing. Scaling of images is an important tool for inspection of the contents of images. We want to be able to display an image with different magnification, i. e., resolution. Thus, we require procedures which magnify and minify images without distortions.

In order to transform images back from pixel coordinates to world coordinates, various transformations are required. In the simplest case, this includes only scaling, translation (shifting), and rotation of images. More general are affine (Section 8.3.1b) and perspective (Section 8.3.1c) transformations. For precise geometric measurements, it is also required to correct for the residual geometric distortion introduced by even well-corrected optical systems. Modern imaging solid-state sensors are geometrically very precise and stable. Therefore, the potential of a position accuracy of better than 1/100 pixel distance is there. To maintain this accuracy, all geometric transformations applied to digital images must preserve this high position accuracy. This demand goes far beyond the fact that no distortions are visible in the transformed images.

Geometric calibration includes two tasks. First, the parameters of the geometric transformations and distortions must be determined by applying an appropriate geometric calibration procedure. Then, as a second step, the geometric transformation is applied to transform the measured pixel coordinates of the objects into another coordinate system. Often, a scene is observed with more than one camera, for instance a stereo camera setup or multiple cameras imaging in different wavelength ranges. Then, it is required to transform the images into one common coordinate system. This procedure is called *image registration*.

Geometric transformation required for the above tasks should be fast and accurate. Fast algorithms for image scaling (Section 8.4.1), translation (Section 8.4.2), rotation (Section 8.4.3), affine and perspective transforms (Section 8.4.4) are discussed.

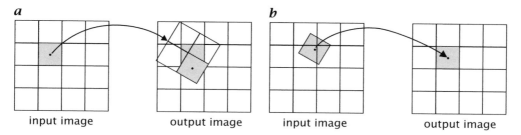

Figure 8.1: *Illustration of **a** forward mapping and **b** inverse mapping for spatial transformation of images. With forward mapping, the value at an output pixel must be accumulated from all input pixels that overlap it. With inverse mapping, the value of the output pixel must be interpolated from a neighborhood in the input image.*

The problem of accuracy is common to all of the above transformations and related to the interpolation procedures. On a discrete grid, any geometric transformation results generally in points that do not any longer lie on the original grid. Therefore, suitable algorithms are required to interpolate the values at transformed points from the neighboring pixels. The high demands for position accuracy make it necessary to discuss image interpolation in detail in Section 8.3.2. Interpolation techniques play also a major role in the formation of multigrid data structures such as pyramids (Chapter 14).

8.3 Concepts

8.3.1 Geometric Transformations

8.3.1a Forward and Inverse Mapping. A geometric transform defines the relationship between the points in two images. This relation can be expressed in two ways. Either the coordinates of the output image, x', can be specified as a function of the coordinates of the input image, x, or vice versa:

$$\begin{aligned} x' &= M(x) &&\text{forward mapping} \\ x &= M^{-1}(x') &&\text{inverse mapping} \end{aligned}$$

(8.1)

where M specifies the mapping function and M^{-1} its inverse. The two expressions in Eq. (8.1) give rise to two principal ways of spatial transformations: *forward mapping* and *inverse mapping*.

With *forward mapping*, a pixel of the input image is mapped onto the output image (Fig. 8.1a). Generally, the pixel of the input image lies in between the pixels of the output image. With forward mapping, it is not appropriate to write the value of the input pixel just to the nearest pixel in the output image (point-to-point mapping). Then, it may happen that a value is never written to some of the pixels of the output image (holes) while others receive a value more than once (overlap). Thus, an appropriate technique must be found to distribute the value of the input pixel to several output pixels. The easiest procedure is to regard pixels as squares and to take the fraction of the area of the input pixel that covers the output pixel as the weighting factor. Each output pixel accumulates the corresponding fractions of the input pixels which — if the mapping is continuous — add up to cover the whole output pixel.

With *inverse mapping*, the coordinates of the output pixels are mapped back onto the input image (Fig. 8.1b). It is obvious that this scheme avoids any holes and overlaps

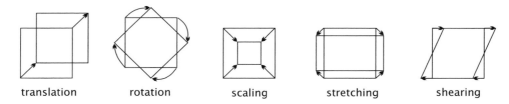

| translation | rotation | scaling | stretching | shearing |

Figure 8.2: *Elementary forms of transformations for a planar surface element, translation, rotation, scaling, stretching, and shearing.*

in the output image since all pixels are scanned sequentially. The interpolation problem occurs now in the input image. The coordinate of the output image does in general not hit a pixel in the input image but lies in between the pixels. Thus, its correct value must be interpolated from the surrounding pixels in the input image.

8.3.1b Affine Transform. The *affine transform* is a linear coordinate transformation that includes the elementary transformations *translation, rotation, scaling, stretching,* and *shearing* and can be expressed by vector addition and matrix multiplication:

$$\left[\begin{array}{c} X' \\ Y' \end{array}\right] = \left[\begin{array}{cc} a_{11} & a_{12} \\ a_{21} & a_{22} \end{array}\right]\left[\begin{array}{c} X \\ Y \end{array}\right] + \left[\begin{array}{c} t_x \\ t_y \end{array}\right]. \tag{8.2}$$

With *homogeneous coordinates* (Section 4.3.1c), the affine transform is written with a single matrix multiplication:

$$\left[\begin{array}{c} X' \\ Y' \\ 1 \end{array}\right] = \left[\begin{array}{ccc} a_{11} & a_{12} & t_x \\ a_{21} & a_{22} & t_y \\ 0 & 0 & 1 \end{array}\right]\left[\begin{array}{c} X \\ Y \\ 1 \end{array}\right]. \tag{8.3}$$

An affine transform has six degrees of freedom: two for translation, one for rotation, one for scaling, one for stretching, and one for shearing. These elementary transformations are depicted in Fig. 8.2. The affine transform maps a triangle into an arbitrary triangle and a rectangle into a parallelogram. Therefore, it is sometimes also referred to as *three-point mapping*. Thus, more general distortions (mapping of a rectangle into an arbitrary quadrilateral) cannot be handled by an affine transformation.

The coefficients of an affine transform can be inferred from a corresponding set of three noncolinear points (mapping of a triangle into a triangle) resulting in the following linear equation system:

$$\left[\begin{array}{ccc} X_1' & X_2' & X_3' \\ Y_1' & Y_2' & Y_3' \\ 1 & 1 & 1 \end{array}\right] = \left[\begin{array}{ccc} a_{11} & a_{12} & t_x \\ a_{21} & a_{22} & t_y \\ 0 & 0 & 1 \end{array}\right] \cdot \left[\begin{array}{ccc} X_1 & X_2 & X_3 \\ Y_1 & Y_2 & Y_3 \\ 1 & 1 & 1 \end{array}\right] \tag{8.4}$$

or

$$\boldsymbol{X'} = \boldsymbol{A}\boldsymbol{X} \tag{8.5}$$

from which A can be computed as

$$\boldsymbol{A} = \boldsymbol{X'}\boldsymbol{X}^{-1}. \tag{8.6}$$

The inverse of the matrix X exists when the three points X_1, X_2, X_3 are linear independent. This means geometrically that they must not lie on one line.

With more than three corresponding points, the parameters of the affine transform can be solved by the following equation system in a least square sense:

$$A = \begin{bmatrix} \sum x_i' x_i & \sum x_i' y_i & \sum x_i' \\ \sum y_i' x_i & \sum y_i' y_i & \sum y_i' \\ \sum x_i & \sum y_i & N \end{bmatrix} \begin{bmatrix} \sum x_i^2 & \sum x_i y_i & \sum x_i \\ \sum x_i y_i & \sum y_i^2 & \sum y_i \\ \sum x_i & \sum y_i & N \end{bmatrix}. \tag{8.7}$$

For details of solving overdetermined equation systems, see Appendix B.2. The inverse of an affine transform is itself affine. The transformation matrix of the inverse transform is given by the inverse of A.

8.3.1c Perspective Transform. The *perspective transform* of projection is the base of optical imaging as discussed in Section 4.3.2a. The affine transform corresponds to parallel projection and can only be used as a model for optical imaging in the limit of a small *field of view*. The general perspective transform is most conveniently written with homogeneous coordinates as

$$\begin{bmatrix} w'x' \\ w'y' \\ w' \end{bmatrix} = \begin{bmatrix} a_{11} & a_{12} & a_{13} \\ a_{21} & a_{22} & a_{23} \\ a_{31} & a_{32} & 1 \end{bmatrix} \begin{bmatrix} wx \\ wy \\ w \end{bmatrix} \quad \text{or} \quad X' = PX. \tag{8.8}$$

The two additional coefficients, a_{31} and a_{32}, in comparison to the affine transform Eq. (8.3) describe the perspective projection (compare Eq. (4.16) in Section 4.3.2a). Written in standard coordinates, perspective transform reads as

$$x' = \frac{a_{11}x + a_{12}y + a_{13}}{a_{31}x + a_{32}y + 1}$$

$$y' = \frac{a_{21}x + a_{22}y + a_{23}}{a_{31}x + a_{32}y + 1}. \tag{8.9}$$

In contrast to the affine transform, the perspective transform is nonlinear. It can be reduced, however, to a linear transform by using homogeneous coordinates. A perspective transform maps lines into lines but only lines parallel to the projection plane remain parallel. Therefore, the perspective transform is sometimes also referred to as *four-point mapping*. Given four or more corresponding points, the coefficients of the perspective transform can be determined. To that end, we rewrite Eq. (8.9)

$$x' = a_{11}x + a_{12}y + a_{13} - a_{31}xx' - a_{32}yx'$$

$$y' = a_{21}x + a_{22}y + a_{23} - a_{31}xy' - a_{32}yy'. \tag{8.10}$$

For N points, this leads to a linear equation system of the form

$$\begin{bmatrix} x_1' \\ y_1' \\ x_2' \\ y_2' \\ \vdots \\ x_N' \\ y_N' \end{bmatrix} = \begin{bmatrix} x_1 & y_1 & 1 & 0 & 0 & 0 & -x_1 x_1' & -y_1 x_1' \\ 0 & 0 & 0 & x_1 & y_1 & 1 & -x_1 y_1' & -y_1 y_1' \\ x_2 & y_2 & 1 & 0 & 0 & 0 & -x_2 x_2' & -y_2 x_2' \\ 0 & 0 & 0 & x_2 & y_2 & 1 & -x_2 y_2' & -y_2 y_2' \\ & & & & \vdots & & & \\ x_N & x_N & 1 & 0 & 0 & 0 & -x_N x_N' & -y_N x_N' \\ 0 & 0 & 0 & x_N & y_N & 1 & -x_N y_N' & -y_N y_N' \end{bmatrix} \begin{bmatrix} a_{11} \\ a_{12} \\ a_{13} \\ a_{21} \\ a_{22} \\ a_{23} \\ a_{13} \\ a_{23} \end{bmatrix} \tag{8.11}$$

or

$$d = M \cdot a. \tag{8.12}$$

It can be solved as a least squares problem (Appendix B.2) by

$$\mathbf{a} = (\mathbf{M}^T\mathbf{M})^{-1}\mathbf{M}^T\mathbf{d} \tag{8.13}$$

with

$$\mathbf{M}^T\mathbf{d} = \begin{bmatrix} \sum x_i x_i' \\ \sum y_i x_i' \\ \sum x_i' \\ \sum x_i y_i' \\ \sum y_i y_i' \\ \sum y_i' \\ -\sum x_i(x_i'^2 + y_i'^2) \\ -\sum y_i(x_i'^2 + y_i'^2) \end{bmatrix} \tag{8.14}$$

and the symmetric 8×8 matrix $\mathbf{M}^T\mathbf{M}$

$$\begin{bmatrix} \sum x_i^2 & \sum x_i y_i & \sum x_i & 0 & 0 & 0 & -\sum x_i^2 y_i x_i' & -\sum x_i y_i x_i' \\ \sum x_i y_i & \sum y_i^2 & \sum y_i & 0 & 0 & 0 & -\sum x_i y_i x_i' & -\sum y_i^2 x_i' \\ \sum x_i & \sum y_i & N & 0 & 0 & 0 & -\sum x_i x_i' & -\sum y_i x_i' \\ 0 & 0 & 0 & \sum x_i^2 & \sum x_i y_i & \sum x_i & -\sum x_i^2 y_i' & -\sum x_i y_i y_i' \\ 0 & 0 & 0 & \sum x_i y_i & \sum y_i^2 & \sum y_i & -\sum x_i y_i y_i' & -\sum y_i^2 y_i' \\ 0 & 0 & 0 & \sum x_i & \sum y_i & N & -\sum x_i y_i' & -\sum y_i y_i' \\ -\sum x_i^2 x_i' & -\sum x_i y_i x_i' & -\sum x_i y_i' & -\sum x_i^2 x_i' & -\sum x_i y_i y_i' & -\sum x_i y_i' & \sum x_i^2(x_i^2 + y_i^2) & \sum x_i y_i(x_i^2 + y_i^2) \\ -\sum x_i y_i x_i' & -\sum y_i^2 x_i' & -\sum y_i x_i' & -\sum x_i y_i y_i' & -\sum y_i^2 y_i' & -\sum y_i y_i' & \sum x_i y_i(x_i^2 + y_i^2) & \sum y_i^2(x_i^2 + y_i^2). \end{bmatrix}$$

8.3.2 Interpolation

The second important aspect of discrete geometric operations is *interpolation*. Interpolation is required since the transformed grid points of the input image do in general no longer coincide with the grid points of the output image and vice versa.

The base of interpolation is the sampling theorem (Section 6.3.2). If the image formation process meets the conditions of the sampling theorem, the digital image is a *complete* representation of the continuous image. Thus, we can reconstruct the continuous image and perform a new sampling to the new grid points. It is obvious that this procedure only works as long as the new grid has narrower or equally narrow spacing of the grid. If it is wider, aliasing will occur. In this case it is, therefore, required to prefilter the image removing fine details before it is resampled.

Although these procedures sound simple and straightforward, they are not. The problem is related to the fact that the reconstruction of the continuous image from the sampled image — which is in principle possible as stated above — in practice is quite involved and can be performed only approximately. Thus, we need to consider how to optimize the interpolation given certain constraints. In this section, we will first illuminate why ideal interpolation is not possible and then discuss various practical approaches in Sections 8.3.2a-8.3.2d.

In Section 6.3.3, we stated that reconstruction of a continuous function from sampled points can be considered as a convolution operation

$$g_i(\mathbf{x}) = \sum_{m,n} g(\mathbf{x}_{m,n})h(\mathbf{x} - \mathbf{x}_{m,n}), \tag{8.15}$$

where h is the continuous interpolation mask

$$h(\boldsymbol{x}) = \frac{\sin \pi x_1/\Delta x_1}{\pi x_1/\Delta x_1} \frac{\sin \pi x_2/\Delta x_2}{\pi x_2/\Delta x_2} \tag{8.16}$$

with the *transfer function*

$$\hat{h}(\boldsymbol{k}) = \Pi(k_1\Delta x_1/2\pi, k_2\Delta x_2/2\pi). \tag{8.17}$$

The interpolation mask in Eq. (8.15) provides ideal interpolation but it has an infinite extension and decreases only slowly within an envelope of $|x_p|^{-1}$ from the center point at the origin in each coordinate direction. Thus, the interpolation function cannot simply be chopped off at a certain distance without introducing significant interpolation errors. The ideal interpolation function Eq. (8.15) is separable. Therefore, it can as easily be formulated for higher dimensional images. We can expect that appropriate approximations to the interpolation problem will also be separable. Consequently, we need only to consider the 1-D interpolation problem.

An important special case is the interpolation to intermediate grid points half way between the existing grid points. This interpolation scheme doubles the resolution and image size in all directions in which it is applied. In this case, the continuous interpolation kernel reduces to a discrete convolution mask. Since the interpolation kernel Eq. (8.16) is separable, we first can interpolate the intermediate points in the row in horizontal direction before we interpolate the intermediate rows by vertical interpolation. In three dimensions, a third 1-D interpolation is added in the z or t direction. The interpolation kernels are the same for all directions. We need the continuous kernel $h(x)$ at only half integer values for $x/\Delta x$

$$\cdots -5/2 \ -3/2 \ -1/2 \ \ 1/2 \ \ 3/2 \ \ 5/2 \ \cdots \tag{8.18}$$

and obtain the infinite kernel with coefficients of alternating sign

$$h = \left[\cdots \ \frac{(-1)^{m-1}2}{(2m-1)\pi} \ \cdots \ -\frac{2}{3\pi} \ \frac{2}{\pi} \ \frac{2}{\pi} \ -\frac{2}{3\pi} \ \cdots \ \frac{(-1)^{m-1}2}{(2m-1)\pi} \ \cdots \right]. \tag{8.19}$$

8.3.2a Interpolation in Fourier space. Interpolation reduces to a simple operation in the Fourier domain. The transfer function of an ideal interpolation kernel is a rectangular function which is zero outside the wave numbers that can be represented, i. e., wave numbers larger than the Nyquist wave number. This basic fact suggests the following interpolation procedure in the Fourier space:

1. Enlarge the matrix of the Fourier transformed image. If an $M \times M$ matrix is increased to an $M' \times M'$ matrix, the image in the spatial domain is also increased to an $M' \times M'$ image. However, because of the reciprocity of the Fourier transform, the image size is not changed but the spacing of the pixels is reduced:

$$M\Delta k \to M'\Delta k \quad \circ\!\!-\!\!\bullet \quad \Delta x = \frac{2\pi}{M\Delta k} \to \Delta x' = \frac{2\pi}{M'\Delta k} \tag{8.20}$$

2. Fill the padded space with zeroes and perform an inverse Fourier transform.

The Fourier transform of the new $M' \times M'$ image spectrum results in a perfectly interpolated image. Unfortunately, this simple procedure has three serious drawbacks.

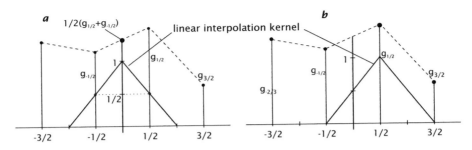

Figure 8.3: *Illustration of linear interpolation:* **a** *at* $x = 0$, *the mean of* $g_{1/2}$ *and* $g_{-1/2}$ *is taken;* **b** *at* $x = 1/2$, $g_{1/2}$ *is replicated; the interpolated curve (dashed line) goes through the discrete grid points but is not continuous in the first derivative at the grid points.*

1. The Fourier transform of a finite $M' \times M'$ image implies a cyclic repetition of the image both in the spatial and Fourier domain (Appendix B.4). Thus, the convolution performed by the Fourier transform is also cyclic. When the mask reaches the right or left edge of the image, it continues to convolve with the image at the other side. This may lead to significant distortions of the interpolation at the edges of the image when the gray values at opposing edges are different.

2. The Fourier transform can efficiently be computed only for a limited number of values for M'. Most well known are the fast radix-2 algorithms that can be applied to images of the size $M' = 2^{N'}$. Therefore, the Fourier transform interpolation is limited to scaling factors of powers of two.

3. Since the Fourier transform is a global transform, it can be applied only to scaling. In principle, it could also be applied to rotation and affine transforms. But then, the interpolation problem is only shifted from the spatial to the wave number domain.

8.3.2b Polynomial Interpolation. *Polynomial interpolation* is the classical approach to interpolation. $N + 1$ grid points are taken to fit a polynomial of degree N through the grid points. The values of the polynomial give the interpolated values.

The simplest case is linear interpolation ($N = 1$). We locate the two grid points at $-\Delta x/2$ and $\Delta x/2$. This yields the interpolation

$$g(x) = \frac{g_{1/2} + g_{-1/2}}{2} + (g_{1/2} - g_{-1/2})x \quad \text{for} \quad |x| \le 1/2. \tag{8.21}$$

By comparison with Eq. (8.15), we can conclude that the interpolation mask for linear interpolation is

$$h_1(x) = \begin{cases} 1 - |x| & |x| \le 1 \\ 0 & \text{otherwise} \end{cases}. \tag{8.22}$$

To interpolate the intermediate grid point $x = 0$, the simple discrete convolution mask $1/2\,[1\ 1]$ has to be applied. The mask Eq. (8.22) is to be applied only in the interval $[-1/2, 1/2]$. Its interpolatory nature is illustrated in Fig. 8.3.

The transfer function of the interpolation mask for linear interpolation $h_1(x)$ Eq. (8.22) is the squared sinc function (Fig. 8.4b)

$$\hat{h}_1(\tilde{k}) = \frac{\sin^2 \pi \tilde{k}/2}{(\pi \tilde{k}/2)^2}. \tag{8.23}$$

A comparison with the ideal transfer function, $\Pi(\tilde{k}/2)$, shows that two distortions are introduced by linear interpolation:

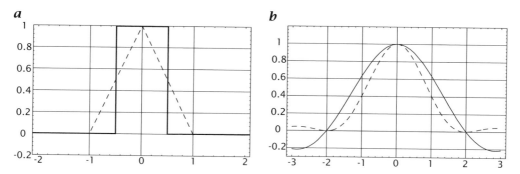

Figure 8.4: *Continuous interpolation from discrete samples: nearest neighbor and linear interpolation (dashed lines):* **a** *interpolation kernels (point spread functions),* **b** *transfer functions of the interpolation kernels.*

1. While low wave numbers (and especially the mean value, $\tilde{k} = 0$) are interpolated correctly, high wave numbers are reduced in amplitude resulting in some degree of smoothing. At $\tilde{k} = 1$, $\hat{h}_1(1) = (2/\pi)^2 \approx 0.4$.

2. Since $\hat{h}_1(\tilde{k})$ is not zero at wave numbers $\tilde{k} > 1$, some spurious high wave numbers are introduced. If the continuously interpolated image is resampled, this yields moderate aliasing. The first side lobe has an amplitude of $(2/3\pi)^2 \approx 0.045$.

The discrete interpolation kernel $H_1 = 1/2\,[1\ 1]$ already includes the resampling at a grid with distances $\Delta x/2$. Therefore, the transfer function for the interpolation of the intermediate grid points is (Fig. 8.6)

$$\hat{h}_1(\tilde{k}) = \cos \pi \tilde{k}/2. \tag{8.24}$$

The interpolated grid points do not experience any phase shift as the transfer function is real. The amplitude attenuation at higher wave numbers, however, is significant. Structures with the highest wave number ($\tilde{k} = 1$) completely disappear at the intermediate grid points. You can easily verify this by applying the interpolation kernel to the signal $[1, -1, 1, -1, \ldots]$, which is sampled twice per wavelength. Phase shifts do occur at other fractional shifts, as illustrated in Example 8.1.

Example 8.1: Linear interpolation with fractional integer shifts

In this example, we investigate fractional shifts $\Delta x \in [-1/2, 1/2]$ with linear interpolation. This operation is of practical importance since it is required if images must be shifted by fractions of a pixel distance. We locate the two grid points for symmetric purposes at $\Delta x = -1/2$ and $1/2$.

The interpolation mask is $[-1/2 - \epsilon, 1/2 + \epsilon]$. The mask contains a symmetric part $[1/2, 1/2]$ and an antisymmetric part $[-\epsilon, \epsilon]$. Therefore, the transfer function is complex

$$\hat{h}_1(\epsilon, k) = \cos \pi \tilde{k}/2 + 2i\epsilon \sin \pi \tilde{k}/2. \tag{8.25}$$

In order to estimate the error in the phase shift, it is useful to subtract the linear phase shift caused by the displacement ϵ: $\Delta \varphi = \epsilon \pi \tilde{k}$. This is done by multiplying the transfer function by $\exp(i\epsilon\pi\tilde{k})$:

$$\hat{h}_1(\epsilon, k) = (\cos \pi \tilde{k}/2 + 2i\epsilon \sin \pi \tilde{k}/2) \exp(-i\epsilon\pi\tilde{k}). \tag{8.26}$$

For $\epsilon = 0$ and $\epsilon = 1/2$, the transfer function is real: $\hat{h}_1(0, \tilde{k}) = \cos(\pi\tilde{k}/2)$, $h_1(1/2, \tilde{k}) = 1$; but at all other fractional shifts, a nonzero phase shift remains, as illustrated in

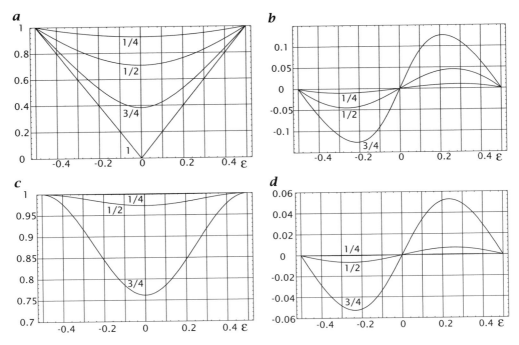

Figure 8.5: *Amplitude attenuation (left column) and phase error (right column) for wave numbers $\tilde{k} = 1/4, 1/2, 3/4$ (and 1), as indicated, displayed as a function of the fractional position from $-1/2$ to $1/2$ for linear interpolation (**a** and **b**) and cubic B-spline interpolation (**c** and **d**). The phase error is given as position shift for corresponding periodic structures, i. e., $\Delta\varphi \cdot \lambda/2\pi = \Delta\varphi/(\pi\tilde{k})$.*

Fig. 8.5. In this figure, the phase shifts are expressed as position shifts of the corresponding periodic structure which is given by

$$\Delta = \Delta\varphi \, \lambda/2\pi = \Delta\varphi/(\pi\tilde{k}). \tag{8.27}$$

Note that linear interpolation does not only cause an amplitude damping but also a significant signal shift of up to 0.13 pixel at $\tilde{k} = 3/4$ (Fig. 8.5b). These errors are too high to be used for further image processing with subpixel accurate algorithms.

The extension to higher-degree polynomials is straightforward. A polynomial of degree N requires $N + 1$ grid points. For symmetry reasons, it is obvious that only polynomials of uneven degree are suitable since they have the same number of grid points $(N + 1)/2$ on both sides of the interpolated value.

We discuss the following practically important case to interpolate the intermediate grid point at 0, when the original grid points are at $\Delta x = -\frac{N+1}{2}, \dots -1/2, 1/2, 3/2, \dots \frac{N+1}{2}$. For symmetry reasons, the coefficient at $\Delta x = n/2$ and $\Delta x = -n/2$ must be equal. Therefore, the transfer function is given by

$$\hat{h}(\tilde{k}) = \sum_{u=1}^{r} 2h_u \cos \frac{(2u-1)}{2}\pi\tilde{k}. \tag{8.28}$$

For a given filter length r, we now have to choose a set of coefficients, so that the sum in Eq. (8.28) approximates the ideal transfer function, the box function $\Pi(\tilde{k}/2)$, as close as possible. One approach is to expand the cosine function in $u\pi\tilde{k}$ and then choose the coefficients h_u so that as many terms as possible with powers of $\tilde{k}$ vanish

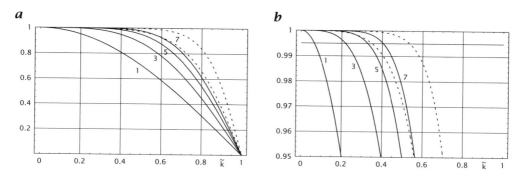

Figure 8.6: *Transfer function of discrete polynomial interpolation filters to interpolate the value between two grid points. The degree of the polynomial (1 = linear, 3 = cubic, etc.) is marked in the graph. The dashed lines mark the transfer function for cubic and quintic B-spline interpolations (Section 8.3.2c). a Shows the full range; b a 5 % margin below the ideal response $\hat{h}(\tilde{k}) = 1$.*

except for the constant term in the expansion. Example 8.2 illustrates this approach for cubic interpolation.

Example 8.2: Discrete cubic interpolation for intermediate grid points

Cubic interpolation uses four sampling points; thus $r = 2$. A second order Taylor expansion of Eq. (8.28) in $\tilde{k}$ yields

$$\hat{h}(\tilde{k}) = 2h_1 - \frac{h_1}{4}(\pi\tilde{k})^2$$
$$+ \ 2h_2 - \frac{9}{4}h_2(\pi\tilde{k})^2$$

or

$$\hat{h}(\tilde{k}) = 2(h_1 + h_2) - \frac{1}{4}(h_1 + 9h_2)(\pi\tilde{k})^2.$$

Since the factor of the $\tilde{k}^2$ term should vanish and the constant factor be equal to one, we have two equations with the two unknowns h_1 and h_2. The solution is $h_1 = 9/8$ and $h_2 = -1/8$. This yields the filter mask

$$h = \frac{1}{16}[-1 \ 9 \ 9 \ -1].$$

The general approach is as follows. The cosine functions are expanded in Taylor series. We can then collect all the terms with equal powers in $\tilde{k}$. Our aim is to have a filter with a transfer function which is constant as long as possible. Thus all coefficients of the Taylor expansion, except for the constant term, should vanish and we obtain the linear equation system

$$\begin{bmatrix} 1 & 1 & 1 & \cdots & 1 \\ 1 & 9 & 25 & \cdots & (2r-1)^2 \\ 1 & 81 & 625 & \cdots & (2r-1)^4 \\ \vdots & \vdots & \vdots & \ddots & \vdots \\ 1^{2r} & 3^{2r} & 5^{2r} & \cdots & (2r-1)^{2r} \end{bmatrix} \begin{bmatrix} h_1 \\ h_2 \\ h_3 \\ \vdots \\ h_r \end{bmatrix} = \begin{bmatrix} 1 \\ 0 \\ 0 \\ \vdots \\ 0 \end{bmatrix}. \quad (8.29)$$

The solution of this linear equation system leads to the family of interpolation masks summarized in Table 8.1. With increasing r the transfer function approaches the box function better. However, convergence is slow. For an accurate interpolation, we must either take a large interpolation mask or limit the wave numbers to smaller $\tilde{k}$ by smoothing the image before we apply the interpolation.

Table 8.1: *Interpolation masks for discrete polynomial interpolation of the points between two samples. The last column contains the Taylor expansion in $\tilde{k}$ for the transfer function. For comparison, the last row contains the mask and transfer function for cubic B-spline interpolation (Section 8.3.2c).*

Order	Mask	Transfer function
1	1/2 [1 1]	$1 - 1/8(\pi\tilde{k})^2 + O(\tilde{k}^4)$
3	1/16[-1 9 9 -1]	$1 - 3/128(\pi\tilde{k})^4 + O(\tilde{k}^6)$
5	1/256[3 -25 150 150 -25 3]	$1 - 5/1024(\pi\tilde{k})^6 + O(\tilde{k}^8)$
7	1/2048[-5 49 -245 1225 1225 -245 49 -5]	$1 - 35/32768(\pi\tilde{k})^8 + O(\tilde{k}^{10})$
Cubic B-spline	1/48[1 23 23 1] and $\alpha = 2 - \sqrt{3}$	$1 - 1/384(\pi\tilde{k})^4 + O(\tilde{k}^6)$
Quintic B-spline	1/3840 [1 237 1682 1682 237 1] and $\alpha_1 = 0.430575, \alpha_2 = 0.043096$	$1 - 1/15360(\pi\tilde{k})^6 + O(\tilde{k}^8)$

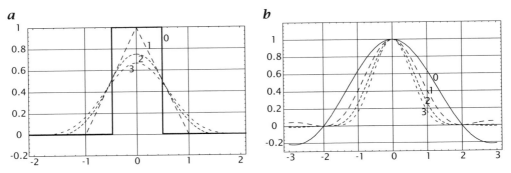

Figure 8.7: *Interpolation with B-splines of order 0 (nearest neighbor), 1 (linear interpolation), 2 (quadratic B-spline), and 3 (cubic B-spline): **a** interpolation kernels generated by cascaded convolution with the box kernel; **b** corresponding transfer functions.*

8.3.2c Spline-Based Interpolation. Polynomial interpolation has the significant disadvantage that the interpolated curve is not continuous in all its derivatives even for higher degrees of polynomial interpolation (Fig. 8.3). This is due to the fact that for each interval between grid points another polynomial is taken. Thus, only the interpolated function is continuous at the grid points but not its derivatives.

Splines avoid this disadvantage by additional constraints for the continuity of derivatives at the grid points. From the wide classes of splines, we will discuss here only one class, the *B-splines*. Since B-splines are separable, it is sufficient to discuss the properties of 1-D B-splines. From the background of image processing, the easiest access to B-splines is their convolution property. The kernel of an pth-order B-spline curve is generated by convolving the box function n times with itself (Fig. 8.7a):

$$\beta_n(x) = \underbrace{\Pi(x) * \ldots * \Pi(x)}_{p+1-times} \qquad (8.30)$$

Consequently, the transfer function of the pth-order B-spline is

$$\hat{\beta}_p(\hat{k}) = \left(\frac{\sin \pi\tilde{k}/2}{(\pi\tilde{k}/2)} \right)^{n+1}. \qquad (8.31)$$

Figure 8.7b shows that direct B-spline interpolation makes no suitable interpolation kernel. The side lobes for $|\tilde{k}| > 1$ are low and the transfer function decreases too early indicating that B-splines perform too much averaging. Moreover, the B-spline curve no longer is interpolating for $n > 1$, since also at the grid points an average with neighboring pixels is taken. Thus, the interpolation constraint $\beta_p(0) = 1$ and $\beta_p(n) = 0$ for $n \neq 0$ is not satisfied.

B-splines can still be very useful for interpolation if first the discrete grid points are transformed in such a way that a B-spline interpolation gives the original values at the grid points. This transform is known as the *B-spline transform* and constructed from the following condition:

$$g_p(x) = \sum_n c_n \beta_p(x - x_n) \qquad \text{with} \qquad g_p(x_n) = g(x_n). \tag{8.32}$$

The computation of the coefficients c_n requires the solution of a linear equation system with as many unknowns as the length of the image rows. An important progress was, therefore, the discovery that the B-spline transform can be performed effectively by recursive filtering. Unser et al. [144] showed that for a cubic B-spline this operation can be performed by the following simple *recursive filter* (Section 10.3.8)

$$g'_m = g_m - (2 - \sqrt{3})(g'_{m\pm1} - g_m) \tag{8.33}$$

applied first in forward (g'_{m-1}) and then in backward direction (g'_{m+1}). The whole operation takes only two multiplications and four additions per pixel. The transfer function of the B-spline function is then (Section 10.3.8)

$$\tilde{\beta}_T(\tilde{k}) = \frac{1}{2/3 + 1/3\cos\pi\tilde{k}}. \tag{8.34}$$

After the B-spline transformation, the B-spline interpolation is applied or vice versa. For a continuous interpolation, this yields the effective transfer function using Eqs. (8.31) and (8.34)

$$\tilde{\beta}_I(\tilde{k}) = \frac{\sin^4(\pi\tilde{k}/2)/(\pi\tilde{k}/2)^4}{(2/3 + 1/3\cos\pi\tilde{k})}. \tag{8.35}$$

Essentially, the B-spline transformation performs an amplification of high wave numbers (at $\tilde{k} = 1$ by a factor 3) which compensates the smoothing of the B-spline interpolation to a large extent. We investigate this compensation at the grid points and at the intermediate points. From the equation of the cubic B-spline interpolating kernel Eq. (8.30) (see also Fig. 8.7a) the interpolation coefficients for the grid points and intermediate grid points are:

$$1/6\,[1\ 4\ 1] \qquad \text{and} \qquad 1/48\,[1\ 23\ 23\ 1] \tag{8.36}$$

respectively. Therefore, the transfer functions are

$$2/3 + 1/3\cos\pi\tilde{k} \qquad \text{and} \qquad 23/24\cos(\pi\tilde{k}/2) + 1/24\cos(3\pi\tilde{k}/2). \tag{8.37}$$

At the grid points, the transfer function exactly compensates the application of the B-spline transformation Eq. (8.34). Thus the B-spline curve goes through the values at the grid points. At the intermediate points the effective transfer function for cubic B-spline interpolation is then

$$\hat{\beta}_I(\tilde{k}) = \frac{23/24\cos(\pi\tilde{k}/2) + 1/24\cos(3\pi\tilde{k}/2)}{2/3 + 1/3\cos\pi\tilde{k}}. \tag{8.38}$$

The amplitude attenuation and the phase shifts expressed as a position shift in pixel distances for the cubic B-spline interpolations for fractional shifts ε in the interval [-0.5, 0.5] are shown in Fig. 8.5c and d. The shift and amplitude damping is zero at the grid points ($\varepsilon = -0.5$ and 0.5). While the amplitude damping is maximal for the intermediate point, the position shift is also zero at the intermediate point because of symmetry reasons. Also, at the wave number $\tilde{k} = 3/4$ the phase shift is about two times smaller than for linear interpolation (Fig. 8.5b). It is still significant with a maximum of about 0.05. This value is too high for algorithms that ought to be accurate in the 1/100 pixel range. If no more sophisticated interpolation technique can be applied, it means that the maximum wave number should be lower than 0.5. Then, the maximum shift is lower than 0.01 and the amplitude damping less than 3%.

For higher accuracy demands, higher-order B-splines can be used. The transfer function of a quintic B-spline interpolation kernel is given by [144, 145, 146]

$$\tilde{\beta}_I^5(\tilde{k}) = \frac{\sin^6(\pi\tilde{k}/2)/(\pi\tilde{k}/2)^6}{33/60 + 26/60\cos\pi\tilde{k} + 1/60\cos 2\pi\tilde{k}}. \tag{8.39}$$

The quintic B-spline transform now includes either two simple recursive filters with one coefficient

$$\begin{aligned} g'_m &= g_m - 0.430575(g'_{m\pm1} - g_m) \\ g''_m &= g'_m - 0.043096(g''_{m\pm1} - g'_m) \end{aligned} \tag{8.40}$$

that are applied after each other in forward and backward direction or that can be combined into a single filter with two coefficients

$$g''_m = g_m - 0.473671(g''_{m\pm1} - g_m) + 0.018556(g''_{m\pm2} - g_m). \tag{8.41}$$

The interpolation coefficients for grid points and intermediate grid points are

$$1/120\,[1\ 26\ 66\ 26\ 1] \quad \text{and} \quad 1/3840\,[1\ 237\ 1682\ 1682\ 237\ 1]. \tag{8.42}$$

Two final notes regarding the application of B-spline based interpolation. First, convolution is commutative. Therefore we can first apply the B-spline transform with the recursive filter and then perform the interpolation with the FIR filter, or vice versa. Second, phase shifts do only apply for arbitrary fractional shifts. For intermediate pixels, no position shift occurs at all. In this special case — which often occurs in image processing, for instance for pyramid computations (Chapter 14) — optimization of interpolation filters is quite easy because only the amplitude damping must be minimized over the wave number range of interest, as will be discussed in the next section.

8.3.2d Least Squares Optimal Interpolation. Filter design for interpolation can — as any filter design problem — be treated in a mathematically more rigorous way as an optimization problem. The general idea is to vary the filter coefficients in such a way that the derivation from the ideal transfer function reaches a minimum. For nonrecursive filters, the transfer function is linear in the coefficients h_u:

$$\hat{h}(\tilde{k}) = \sum_{u=1}^{r} h_u \hat{f}_u(\tilde{k}). \tag{8.43}$$

The ideal transfer function is $\hat{h}_I(\tilde{k})$. Then, the optimization procedure should minimize the integral

$$\int_0^1 w(\tilde{k}) \left| \left(\sum_{u=1}^{r} h_u \hat{f}_u(\tilde{k}) \right) - \hat{h}_I(\tilde{k}) \right|^n d\tilde{k} \to \text{minimum}. \tag{8.44}$$

Table 8.2: *Optimal nonrecursive interpolation masks for discrete interpolation of the points between two samples; after Eq. (8.47) (upper half) and Eq. (8.48) (lower half). Only one half of the coefficients of the symmetrical masks are shown, starting in the center. The last column lists the mean weighted standard deviation in % from the ideal interpolation mask.*

Order	Coefficients h_u	σ [%]
2	0.583885, -0.11383	2.17
3	0.613945, -0.138014, 0.0386661	0.73
4	0.6164, -0.163849, 0.055314, -0.0163008	0.30
5	0.624405, -0.173552, 0.073189, -0.0264593, 0.00785855	0.14
6	0.625695, -0.183723, 0.0826285, -0.0382784, 0.0140532, -0.00416625	0.076
2	0.589071, -0.089071	3.08
3	0.606173, -0.135766, 0.0295927	0.92
4	0.617404, -0.158185, 0.053219, -0.0124378	0.36
5	0.622276, -0.17277, 0.0695685, -0.0251005, 0.00602605	0.16
6	0.626067, -0.18166, 0.081616, -0.0360213, 0.0132193, -0.00322136	0.085

In this expression, a weighting function $w(\tilde{k})$ has been introduced which allows control of the optimization for a certain wave number range. In equation Eq. (8.44) an arbitrary n-norm is included. Mostly the 2-norm is taken which minimizes the sum of squares.

The minimization problem results in a linear equation system for the r coefficients of the filter which can readily be solved:

$$\boldsymbol{Mh} = \boldsymbol{d} \tag{8.45}$$

with

$$\boldsymbol{d} = \begin{bmatrix} \overline{h_I \hat{f}_1} \\ \overline{h_I \hat{f}_2} \\ \vdots \\ \overline{h_I \hat{f}_r} \end{bmatrix} \quad \text{and} \quad \boldsymbol{M} = \begin{bmatrix} \overline{\hat{f}_1^2} & \overline{\hat{f}_1 \hat{f}_2} & \cdots & \overline{\hat{f}_1 \hat{f}_r} \\ \overline{\hat{f}_1 \hat{f}_2} & \overline{\hat{f}_2^2} & \cdots & \overline{\hat{f}_2 \hat{f}_r} \\ \vdots & & \ddots & \vdots \\ \overline{\hat{f}_1 \hat{f}_r} & \overline{\hat{f}_2 \hat{f}_r} & \cdots & \overline{\hat{f}_r^2} \end{bmatrix}$$

where the abbreviation

$$\overline{\hat{e}(\tilde{k})} = \int_0^1 w(\tilde{k}) \hat{e}(\tilde{k}) \mathrm{d}\tilde{k} \tag{8.46}$$

has been used, where $\hat{e}(\tilde{k})$ stands for any of the terms in the equations. We demonstrate the least squares optimization technique in Example 8.3 for optimal interpolation with nonrecursive filters.

Example 8.3: Optimum interpolation for intermediate grid points

In this example, we compare two approaches to optimum interpolation filters. We use either

$$\hat{h}(\tilde{k}) = \sum_{u=1}^{r} h_u \cos\left(\frac{2u-1}{2}\pi\tilde{k}\right) \tag{8.47}$$

or

$$\hat{h}(\tilde{k}) = \cos\left(\frac{1}{2}\pi\tilde{k}\right) + \sum_{u=2}^{r} h_u \left[\cos\left(\frac{2u-3}{2}\pi\tilde{k}\right) - \cos\left(\frac{1}{2}\pi\tilde{k}\right)\right]. \tag{8.48}$$

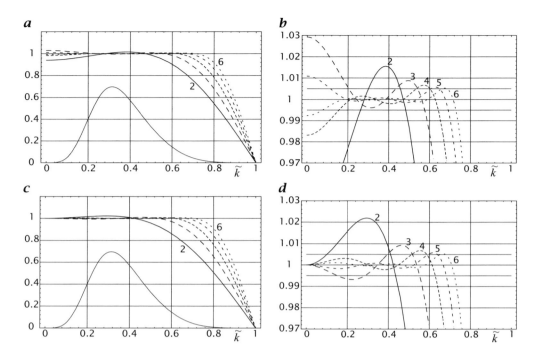

Figure 8.8: *Transfer function of interpolation kernels optimized with the weighted least squares technique. The weighting function peaking at $\tilde{k} \approx 0.3$ is shown as a thin curve in **a** and **c**. Two families of solutions are shown according to **a** Eq. (8.47) and **b** Eq. (8.48) with R = 2–6. The plots in **b** and **d** show a narrow sector of the plots in **a** and **c**, respectively. The corresponding mask coefficients are listed in Table 8.2.*

The second formulation Eq. (8.48) ensures that $\hat{h}(0) = 1$, i.e., mean gray values are preserved by the interpolation. This is done by forcing the first coefficient to be one minus the sum of all others. The first formulation Eq. (8.47) does not apply this constraint. For all optimizations a weighting function peaking at $\tilde{k} \approx 0.3$ has been taken as shown in Fig. 8.8a and c. Figure 8.8 compares the optimal transfer functions with both approaches for R = 2–6; the corresponding interpolation mask coefficients are shown in Table 8.2.

The solutions are significantly better than those obtained by the Taylor series method and the cubic B-spline method (Fig. 8.6). At a low number of filter coefficients, the additional degree of freedom for Eq. (8.48) leads to significantly better solutions for the wave number range where the weighting function is maximal. For a larger number of filter coefficients the advantage quickly gets insignificant. Note that all solutions resemble the alternating signs of the ideal interpolation mask Eq. (8.19).

The best interpolation masks can be obtained by using recursive filters. Here the result of a technique is shown that combines the nonrecursive interpolation masks with a recursive correction term based on the $\mathcal{U}$ filter introduced in Section 10.4.4 (see Fig. 10.18, Table 10.2):

$$g'_m = g_m + \alpha(g'_{m\pm1} - g_m). \tag{8.49}$$

Figure 8.9 and Table 8.3 show the results from the following formulation:

$$\hat{h}(\tilde{k}) = \frac{\cos\left(1/2\,\pi\tilde{k}\right) + \sum_{u=2}^{r} h_u \left[\cos\left((2u-3)/2\,\pi\tilde{k}\right) - \cos\left(1/2\,\pi\tilde{k}\right)\right]}{1 - \alpha + \alpha\cos\left(\pi\tilde{k}\right)}. \tag{8.50}$$

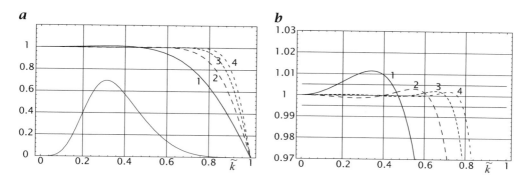

Figure 8.9: *Transfer functions of interpolation kernels with recursive correction according to Eq. (8.50) for R = 1–4, optimized with a nonlinear weighted least squares technique. The weighting function peaking at $\tilde{k} \approx 0.3$ is shown as a thin curve in **a**. The plots in **b** show a narrow sector of the plots in **a**. The corresponding mask coefficients are listed in Table 8.3.*

Table 8.3: *Optimal recursive interpolation masks according to Eq. (8.50) for discrete interpolation of the points between two samples. The last column lists the mean weighted standard deviation in % from the ideal interpolation mask.*

Order	α	Coefficients h_u	σ [%]
1	-0.210704	0.5	1.43
2	-0.425389	0.456028, 0.043972	0.18
3	-0.544516	0.442106, 0.061381, 0.00348714	0.044
4	-0.621455	0.435643, 0.069612, 0.0059977, 0.000742795	0.015

With recursive filters, the least squares optimization becomes nonlinear because Eq. (8.50) is no longer linear in the recursive filter coefficient α. Then, iterative techniques are required to solve the optimization problem.

8.4 Procedures

With the extensive discussion on interpolation we are well equipped to devise fast algorithms for the different geometric transforms. Basically there are two common tricks to speed up the computation of geometric transforms.

First, many computations are required to compute the interpolation coefficients for fractional shifts. For each shift, different interpolation coefficients are required. Thus we must devise the transforms in such a way that we need only constant shifts for a certain pass of the transform. If this is not possible, it might still be valid to precompute the interpolation coefficients for different fractional shifts and to reuse them later.

Second, we learned in Section 8.3.2 that interpolation is a separable procedure. Taking advantage of this basic fact considerably reduces the number of operations. In most cases it is possible to part the transforms in a series of 1-D transforms that operate only in one coordinate direction.

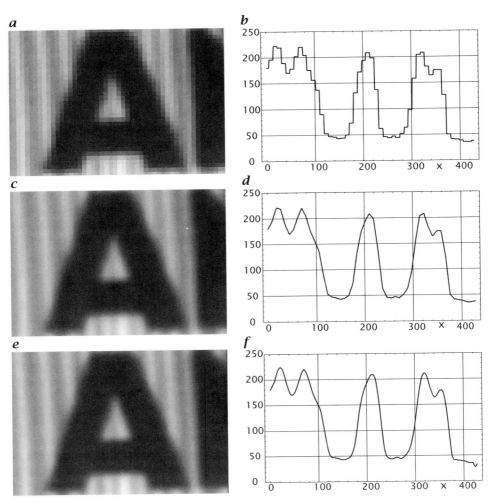

Figure 8.10: *A* 35×50 *sector of the image shown in Fig. 5.28a, scaled to a size of* 300×428 *using* ***a*** *pixel replication,* ***c*** *linear interpolation, and* ***e*** *cubic interpolation. The diagrams to the right of each image show a row profile through the center of the corresponding images.*

8.4.1 Scaling

Scaling is a common geometric transform that is mostly used for interactive image inspection. Another important application is multigrid image data structures such as pyramids (Sections 14.3.4 and 14.3.5).

From the algorithmic point of view, magnification and diminishing must be distinguished. In the case of magnification, we can directly apply an appropriate interpolation algorithm. In the case of diminishing, however, it is required to presmooth to such an extent that no *aliasing* occurs (Section 6.3.1d). According to the sampling theorem (Section 6.3.2), all fine structures must be removed that are sampled less than two times at the resolution of the scaled image.

The algorithms for scaling are rather simple because scaling in the different directions can be performed one after the other. In the following, we discuss scaling of a 2-D image using inverse mapping. In the first step, we scale the N points of a row to N' points in the scaled image. The coordinates of the scaled image mapped back to the original image size are given by

$$x'_{n'} = \frac{N-1}{N'-1} n' \quad \text{with} \quad 0 \le n' < N'. \tag{8.51}$$

Note that the first and last points of the original image at 0 and $N'-1$ are mapped onto the first and last point of the scaled image: $x'_0 = 0$, $x'_{N'-1} = N-1$. All points in between do in general not meet a grid point on the original grid and thus must be interpolated. The horizontal scaling can significantly be sped up since the position of the points and thus the interpolation coefficients are the same for *all* rows. Thus we can precompute all interpolation coefficients in a list and then use them to interpolate all rows. We illustrate the approach with linear interpolation. In this case, we only need to use two neighboring points for the interpolation. Thus, it is sufficient to store the index of the point, given by $n = \text{floor}(x_{n'})$ (where the function floor computes the largest integer lower than or equal to $x_{n'}$), and the fractional shift $f_{n'} = x_{n'} - \text{floor}(x_{n'})$. Then we can interpolate the gray value at the n'th point in the scaled row from

$$g'_{n'} = g_n + f_{n'}(g_{n+1} - g_n). \tag{8.52}$$

Thus 1-D linear interpolation is reduced to 3 arithmetic operations per pixel. The computation of n and $f_{n'}$ must be performed only once for all rows.

In a very similar way, we can proceed for the interpolation in the other directions. The only difference is that we can now apply the same interpolation factor to all pixels of a row. If we denote rows n and rows n' in the original and scaled image with $\boldsymbol{g}_n$ and $\boldsymbol{g}'_{n'}$, respectively, we can rewrite Eq. (8.52) as the following vector equation

$$\boldsymbol{g}'_{n'} = \boldsymbol{g}_n + f_{n'}(\boldsymbol{g}_{n+1} - \boldsymbol{g}_n). \tag{8.53}$$

For higher-order interpolation nothing changes with the general computing scheme. Instead of Eqs. (8.52) and (8.53), we just have to take more complex equations to interpolate the gray value in the scaled image from more than two pixels of the original image. This also implies that more interpolation coefficients need to be precomputed before we apply the horizontal scaling. It is also straightforward to extend this approach to higher-dimensional images.

Figure 8.10 compares scaling with zero-order (pixel replication), linear, and cubic interpolation. A small sector is blown up by a scale factor of about 8.6. Simple pixel replication leads to a pixelation of the image resulting in jagged edges. Linear interpolation gives smooth edges. But some visual disturbance remains. The piecewise linear interpolation between the pixels results in discontinuities in the slope as can be best seen in the profile (Fig. 8.10d) through the center row of Fig. 8.10c. The best results are obtained with cubic interpolation.

As already mentioned in Section 8.3.2a, scaling causes aliasing effect, if the size of an image is shrinking and no appropriate smoothing is applied beforehand. Figure 8.11b shows image Figure 8.11a scaled down from a size of 256×256 to a size of 100×100 using linear interpolation. An elegant way to perform smoothing and scaling in a single step is to perform the scaling in the Fourier domain as described in Section 8.3.2a. Then, however, other artifacts become visible (Fig. 8.11c). Processing in the discrete Fourier domain assumes that the images are repeated periodically at the edges. Thus there can be step jumps in the gray values at the edges, if the gray values at the upper and lower or left and right edge are significantly different. This causes periodic disturbances parallel to the edges penetrating into the image.

a

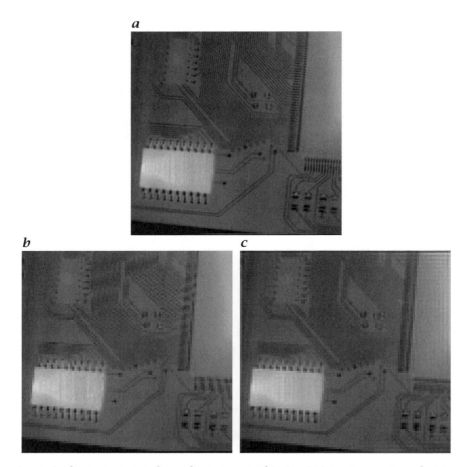

b   *c*

Figure 8.11: *Scaling an image down from a size of* 256×256 *in **a** to a size of* 100×100 *by* **b** *linear interpolation without presmoothing (observe aliased structures),* **c** *interpolation in the Fourier domain.*

8.4.2 Translation

Translation is the simplest of all geometric transformations and needs not much consideration. The shift is the same for all pixels. Thus, we need only one set of interpolation coefficients that is applied to each pixel. As for scaling, we perform first the horizontal shift and then the vertical shift. Note that interpolation algorithms applied in this way have the same computational structure as a convolution operator. Therefore, we can use the same efficient storage scheme for the horizontally shifted or scaled rows as we will discuss in Section 10.4.5 for neighborhood operators.

8.4.3 Rotation

With rotation it is less obvious how it can be decomposed into a sequence of one-dimensional geometrical transforms. Indeed there are several possibilities. Interestingly, these algorithms have not been invented in the context of image processing but for computer graphics.

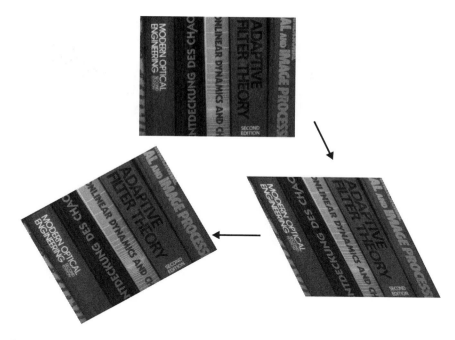

Figure 8.12: *Fast image rotation by combined shearing and scaling in horizontal and vertical directions.*

Catmull and Smith [16] suggested the following 2-pass shear-scale transform

$$R = \begin{bmatrix} \cos\theta & \sin\theta \\ -\sin\theta & \cos\theta \end{bmatrix} = \begin{bmatrix} 1 & 0 \\ -\tan\theta & 1/\cos\theta \end{bmatrix} \begin{bmatrix} \cos\theta & \sin\theta \\ 0 & 1 \end{bmatrix}. \tag{8.54}$$

The first matrix performs a horizontal shear/scale operation, the second a vertical shear/scale operation. Written in components, the first transform is

$$\begin{aligned} x' &= x\cos\theta + y\sin\theta \\ y' &= y \end{aligned} \tag{8.55}$$

The image is shifted horizontally by the offset $y\sin\theta$ and is diminished by the factor of $x\cos\theta$. Thus Eq. (8.55) constitutes a combined shear-scale transform in horizontal direction only (Fig. 8.12).

Likewise, the second transform

$$\begin{aligned} x'' &= x' \\ y'' &= y'/\cos\theta - x'\tan\theta \end{aligned} \tag{8.56}$$

performs a combined shear-scale transform in vertical direction only. This time, the image is enlarged in vertical direction by the factor of $1/\cos\theta$.

This usage of scaling operations for image rotation has two disadvantages. First, the scale operations require extra computations since the image must not only be shifted but also be scaled. Second, although the size of an image does not change during a rotation, the first scaling reduces the size of the image in horizontal direction and thus could cause aliasing. If a smoothing is applied to avoid aliasing, resolution is lost. Fortunately, the problem is not too bad since any rotation algorithm must only be applied for angles $|\theta| < 45°$. For larger rotation angles, first a corresponding transposition

Figure 8.13: *Fast image rotation by decomposition into three one-dimensional shear transforms only: first in horizontal, then in vertical, and again in horizontal direction.*

and/or mirror operation can be applied to obtain a rotation by a multiple of 90°. Then the residual rotation angle is always smaller than 45°.

With a 3-pass transform [109, 142] (Fig. 8.13) rotation can be decomposed into three 1-D shear transforms avoiding scaling (Fig. 8.13):

$$
\begin{aligned}
\boldsymbol{R} &= \begin{bmatrix} \cos\theta & \sin\theta \\ -\sin\theta & \cos\theta \end{bmatrix} \\
&= \begin{bmatrix} 1 & \tan(\theta/2) \\ 0 & 1 \end{bmatrix} \begin{bmatrix} 1 & 0 \\ -\sin\theta & 1 \end{bmatrix} \begin{bmatrix} 1 & \tan(\theta/2) \\ 0 & 1 \end{bmatrix}.
\end{aligned}
\tag{8.57}
$$

8.4.4 Affine and Perspective Transforms

A 2-D affine transform adds three more degrees of freedom to a simple transform that includes only rotation and translation. These are two degrees of freedom for scaling in x and y direction and one degree of freedom for shearing. Together with the degree of freedom for rotation, we have (without the translation) four degrees of freedom which can generally be described by a 2×2 matrix as discussed in Section 8.3.1b:

$$
\boldsymbol{A} = \begin{bmatrix} a_{11} & a_{12} \\ a_{21} & a_{22} \end{bmatrix}.
\tag{8.58}
$$

One way to perform an affine transform efficiently is to decompose it into a rotation, shear, and scaling transform:

$$
\boldsymbol{A} = \begin{bmatrix} a_{11} & a_{12} \\ a_{21} & a_{22} \end{bmatrix} = \begin{bmatrix} s_x & 0 \\ 0 & s_y \end{bmatrix} \begin{bmatrix} 1 & s \\ 0 & 1 \end{bmatrix} \begin{bmatrix} \cos\theta & \sin\theta \\ -\sin\theta & \cos\theta \end{bmatrix}.
\tag{8.59}
$$

The parameters s_x, s_y, s, and θ can be computed from the matrix coefficients by

$$
\begin{aligned}
s_x &= \sqrt{a_{11}^2 + a_{21}^2} \\
s_y &= \frac{\det(A)}{s_x} \\
s &= \frac{a_{11}a_{12} + a_{21}a_{22}}{\det(A)} \\
\tan\theta &= -\frac{a_{21}}{a_{11}},
\end{aligned}
\tag{8.60}
$$

where $\det(A)$ is the determinant of the matrix A. The shear transform and the rotation can be computed simultaneously by the 3-pass shear transform discussed in Section 8.4.3 with the following modification to Eq. (8.57):

$$
\begin{aligned}
R &= \begin{bmatrix} 1 & s \\ 0 & 1 \end{bmatrix} \begin{bmatrix} \cos\theta & \sin\theta \\ -\sin\theta & \cos\theta \end{bmatrix} \\
&= \begin{bmatrix} 1 & s + \tan(\theta/2) \\ 0 & 1 \end{bmatrix} \begin{bmatrix} 1 & 0 \\ -\sin\theta & 1 \end{bmatrix} \begin{bmatrix} 1 & \tan(\theta/2) \\ 0 & 1 \end{bmatrix}.
\end{aligned}
\tag{8.61}
$$

Thus the affine transform can be computed with a 3-pass shear transform Eq. (8.61) followed by a scaling transform as discussed in Section 8.4.1. Fast algorithms for perspective and more general transforms are treated in detail in Wolberg [155].

8.5 Advanced Reference Material

G. Wolberg, 1990. Digital Image Warping, IEEE Computer Society Press, Los Alamitos. *This book has become a standard reference for geometric image operations from the perspective of computer vision. But it is an excellent and useful survey also for the image processing community.*

O. Faugeras, 1993. Three-Dimensional Computer Vision, A Geometric Viewpoint, The MIT Press, Cambridge. *This book looks at geometric operations from the viewpoint of computer vision. It handles projective geometry, and modeling and calibration of cameras to apply it for the classical computer vision tasks of stereo vision, motion determination, and 3-D reconstruction.*

P. Thévenaz, T. Blu, and M. Unser, 2000. Interpolation revisited, IEEE Trans. Medical Imaging, *19*, 739–758. *Recent review article on interpolation techniques.*

9 Restoration and Reconstruction

9.1 Highlights

No image formation system is perfect because of inherent physical limitation. Since scientific applications always push the limits, there is a need to correct for the distortions caused by the imaging system. Humans also make errors. Images blurred by a misadjustment in the focus, smeared by the motion of objects, the camera, or a mechanically unstable mounting, or degraded by faulty or misused optical systems are more common than we may think. A famous recent example was the flaw in the optics of the Hubble space telescope where an error in the test procedures for the mirror grinding resulted in a significant residual aberration of the telescope. The correction of known and unknown image degradation is called restoration.

There are also many indirect imaging techniques which do not deliver a directly usable image. The generation of an image requires reconstruction techniques. In most cases, many indirect images are required to form a direct image of the object.

The most important technique of this kind is tomography. Computer tomography is a well-known 3-D imaging technique used for diagnosis in medical sciences. It is less well known that tomographic techniques are applied meanwhile throughout all sciences. Just two examples are mentioned here. Acoustic tomography is used in oceanography to reveal 3-D flow patterns. X-ray tomography has become an important tool in nondestructive 3-D material inspection.

Restoration and reconstruction are treated together in this chapter. They have in common that the relation between the "perfect" image and the degraded or indirect image is in general well modeled by a linear shift-invariant system (Section 4.3.5). Thus, it can be described by a point spread function (PSF) in the space domain and a transfer function (TF) in the wave number domain. The PSF is a blur disk for defocused images (Section 9.3.2), a streak function for velocity blurring (Section 9.3.3), and a delta line for tomographic imaging (Section 9.3.6).

The basic concept to restore images is inverse filtering (Section 9.4.2); however, only nonlinear and model-based restoration techniques really yield promising results (Section 9.3.5). The most critical parameter that determines the success of any restoration is the noise level in the image.

Furthermore, the reconstruction of surfaces (depth map) from focus series is discussed (Section 9.4.1). Confocal laser scanning microscopy (Section 4.4.3) has become the leading research tool for 3-D imaging in cell biology. This technique overcomes the limits implied in 3-D reconstruction techniques from focus series in conventional microscopy.

In contrast to restoration, the techniques for reconstruction from tomographic projections are well established. We discuss here the fastest available technique, the filtered back projection technique (Section 9.4.3). The beauty of this technique is that it directly builds on the basic relations between images and their projection, the Radon transform and the Fourier slice theorem.

9.2 Task

During image formation and acquisition, many types of distortions limit the quality of images. The question arises if, and if yes, to which extent the effects of the degradation can be reverted. It is obvious, of course, that information that is no longer present at all in the degraded image cannot be retrieved. To make this point clear, let us assume the extreme case that only the mean gray value of an image is retained. Then, it will not be possible by any means to reconstruct its content.

However, images contain a lot of redundant information. Thus, we can hope that a distortion only partially removes the information of interest even if we can no longer "see" it directly.

Inverting the effects of known or unknown distortions in images is known as *restoration*. In this chapter, we will not handle noise suppression since it requires a solid knowledge about spatial structures in images which will be treated in Chapters 10–14. Here, we only discuss degradation that can be modeled by *linear system theory*. In Sections 4.3.5 and 6.4.1, we discussed that generally any optical system including digitization can be regarded as a *linear shift-invariant system* and, thus, described to a good approximation by a *point spread function* and a *transfer function*.

The first task is to determine and describe the image degradation as accurately as possible. This can be done by analyzing the image formation system either theoretically or experimentally by using some suitable test images. If this is not possible, the degraded image remains the only source of information. After the analysis, a suitable restoration procedure can be selected.

Closely related to image restoration is image *reconstruction* from indirect imaging techniques. These techniques, for example, tomography, deliver no direct image but collect the image in such a way that each point of the observed spatial information contains information from many points in the image.

This process can be described by a point spread function just as a degradation. The difference between reconstruction and restoration is that with reconstruction we generally have many different indirect images, each convolved by a different point spread function. They are selected in such a way that an optimum reconstruction is possible.

9.3 Concepts

9.3.1 Types of Image Distortions

Given the enormous variety of image formation (Chapter 4), there are many reasons for image degradation. Table 9.1 gives a summary. Directly related to imperfections of the optical system, lens aberrations (Section 4.3.2f) limit the sharpness of images. Even with a perfect optical system, the sharpness is limited by the diffraction of electromagnetic waves at the aperture stop of the lens (diffraction-limited optics, Section 4.3.3a).

While these types of degradation are inherent properties of a given optical system, blurring by *defocusing* is a common misadjustment that limits the sharpness in images.

Table 9.1: *Summary of different types of image distortions and degradation.*

Reason	Descriptive comments
Defocusing	More or less isotropic blurring of the image by a misadjustment of the optical system.
Lens aberration	Inherent limit in sharpness of a given optical system; often increases significantly towards the edge of images.
Internal reflections and scattering	Cause unwanted additional irradiance on the image plane, especially if bright objects are images.
Velocity smearing	Directed blurring by motion of the camera or object during the exposure time.
Vibration	Periodic or random motion of the camera system during exposure.
Jitter	Electronic instability of the video signal causing random or periodic fluctuation of the beginning of image rows and, thus, corresponding position errors.
Echo	Reflection of video signals in damaged or improperly designed lines causing echo images.
Fixed pattern	Caused by electromagnetic interference from other sources or inhomogeneity of the responsivity of the sensor elements.
Moving pattern	Caused by electromagnetic interference from other sources.
Smear and blooming	Artifacts of CCD sensors resulting from short exposure times or overflow by bright objects.
Transmission errors	Individual bits that are flipped during digital transmission of images cause false gray values at randomly distributed positions in the image.

Another reason for blurring in images is unwanted motions and vibrations of the camera system during the exposure time. Systems with telelenses (narrow field of view) are very sensitive to this kind of image degradation. Likewise, fast moving objects (more than a pixel per exposure time at the image plane) cause the same type of blurring. In contrast to defocusing, motion related distortions normally cause a blurring only in one direction.

Mechanical instabilities are not the only reason for image degradation. They can also have electric reasons. A well-known example is *row jittering* caused by a misadjusted or faulty *phase-locked loop* in the synchronization circuits of the frame buffer (Section 6.5.2b). The start of the rows shows random fluctuations causing corresponding position errors. Bad transmission lines or video lines can cause echoes in images by reflections of the video signals. Electromagnetic interference by other sources may cause fixed or moving patterns in video images. Fixed patterns in images can also be caused by variations in the responsivity or by a dark current of the sensor elements (Section 5.3.2). Finally, with digital transmission of images over noisy lines, individual bits may flip and cause erroneous values at random positions. This type of error was one of the most serious problems at the beginning of space exploration when distant unmanned space crafts sent image data back to earth in faint signals. Meanwhile, communication technology has matured to a point that this type of image distortion is no longer a major concern.

9.3.2 Defocusing and Lens Aberrations

Defocusing and *lens aberrations* are discussed together in this section since they are directly related to the optical system. Indeed, the concept of the *3-D point spread function (PSF)* and the *3-D optical transfer function (OTF)* as discussed in Section 4.3.5 unifies these two effects into one common concept.

Lens aberrations are generally more difficult to handle. Most aberrations increase strongly with the distance from the optical axis (Section 4.3.2f) and are, thus, not shift invariant. However, the aberrations change only slowly and continuously with the position in the image. As long as the resulting blurring is limited to an area in which we can consider the aberration to be constant, we can still treat them with the theory of linear shift invariant systems. The only difference is that the PSF and OTF vary slowly with position.

The effect of blurring or an aberration is given by the point spread function $h(x)$ which says how a point of the object space is spread out in the image. Thus, the relation between the ideal image $g(x)$ and the measured image $g'(x)$ is

$$g'(x) = (h * g)(x). \tag{9.1}$$

In Fourier space, the object function $\hat{g}(k)$ is multiplied with the OTF $\hat{h}(k)$:

$$\hat{g}'(k) = \hat{h}(k)\hat{g}(k). \tag{9.2}$$

While the PSF $h(k)$ and the OTF $\hat{h}(k)$ may have different forms, two limiting cases are of special importance and, thus, deserve more attention.

Gaussian function. In practice, many different effects contribute to the system PSF of an optical system. This means that the system PSF is formed by the convolution of many individual functions. The *central limit theorem* of statistics states that multiple convolutions tend to shape functions to become more and more equal to the bell-shaped Gaussian function [110].

Since the Fourier transform of a Gaussian function is also a Gaussian function, this is valid for both the spatial and Fourier domains. The relation between the two Gaussian functions in one dimension is:

$$\frac{1}{\sigma\sqrt{2\pi}} \exp\left(-\frac{x^2}{2\sigma^2}\right) \circ\!\!-\!\!\bullet \exp\left(-\frac{k^2}{2/\sigma^2}\right). \tag{9.3}$$

In higher dimensions, the 1-D equations can be multiplied with each other since the Gaussian function and the Fourier transform are separable. In 2-D, we have

$$\frac{1}{2\pi\sigma_x\sigma_y} \exp\left(-\frac{x^2}{2\sigma_x^2}\right) \exp\left(-\frac{y^2}{2\sigma_y^2}\right) \circ\!\!-\!\!\bullet \exp\left(-\frac{k_x^2}{2/\sigma_x^2}\right) \exp\left(-\frac{k_y^2}{2/\sigma_y^2}\right). \tag{9.4}$$

If the PSF of a degradation is unknown or only approximately known, it is therefore the best idea to approximate it with a Gaussian function.

Blur disk. If the defocusing effect is the dominant blurring effect, the PSF becomes a constant function inside the aperture stop and is zero outside. Since most aperture stops can be approximated by a circle, the function is a disk. The Fourier transform of a disk with radius r is a Bessel function (Appendix B.3)

$$\frac{1}{\pi r^2} \Pi\left(\frac{|x|}{2r}\right) \circ\!\!-\!\!\bullet \frac{J_1(2\pi|k|r)}{\pi|k|r}. \tag{9.5}$$

The 2-D OTF of a diffraction-limited system is discussed in Section 4.3.5d. Examples of defocused images are shown in Fig. 9.1.

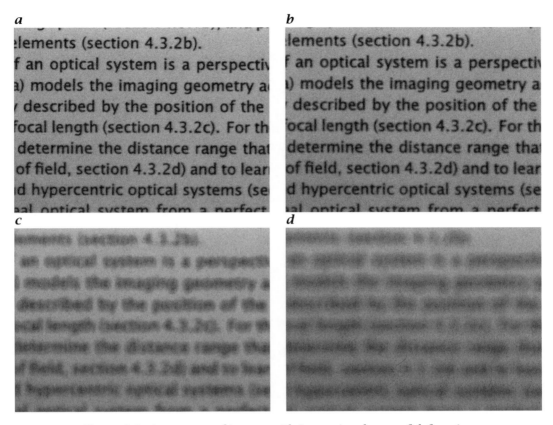

Figure 9.1: A sequence of images with increasing degree of defocusing.

9.3.3 Velocity Smearing

While blurring by defocusing and lens aberrations tends to be isotropic, all blurring effects by motion are only a one-dimensional blurring as demonstrated in Fig. 9.2. In the simplest case, motion is constant during the exposure. Then, the PSF of motion blur is a one-dimensional box function. Without loss of generality, we first assume that the direction of motion is along the x axis. Then (Appendix B.3),

$$h(x) = \frac{1}{u\Delta t}\Pi\left(\frac{x}{2u\Delta t}\right) \circ\!\!-\!\!\bullet \hat{h}(k) = \frac{\sin(\pi k u \Delta t)}{\pi k u \Delta t} = \mathrm{sinc}(ku\Delta t), \qquad (9.6)$$

where u is the magnitude of the velocity, Δt the exposure time, and $\Delta x = u\Delta t$ the blur length. If the velocity $\boldsymbol{u}$ is oriented in another direction, Eq. (9.6) can be generalized to

$$h(\boldsymbol{x}) = \frac{1}{|\boldsymbol{u}|\Delta t}\Pi\left(\frac{x\bar{\boldsymbol{u}}}{2|\boldsymbol{u}|\Delta t}\right)\delta(\boldsymbol{u}x) \circ\!\!-\!\!\bullet \hat{h}(\boldsymbol{k}) = \mathrm{sinc}(\boldsymbol{k}u\Delta t), \qquad (9.7)$$

where $\bar{\boldsymbol{u}} = \boldsymbol{u}/|\boldsymbol{u}|$ is a unit vector in the direction of the motion blur.

9.3.4 Inverse Filtering

Common to defocusing, motion blur, and 3-D imaging by techniques as focus series or confocal laser scanning microscopy (Section 4.4.3) is that the object function $g(\boldsymbol{x})$

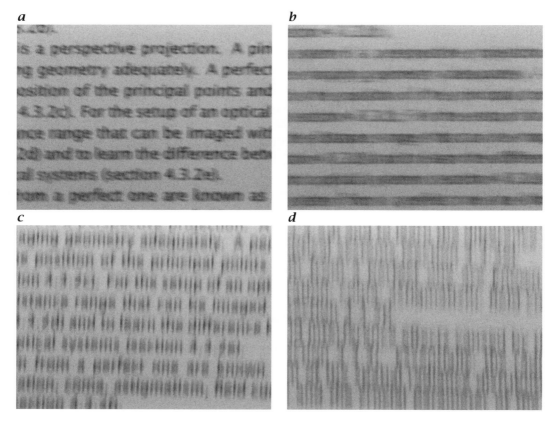

Figure 9.2: *Examples of images blurred by motion:* **a** *small and* **b** *large velocity blurring in horizontal direction;* **c** *small and* **d** *large velocity blurring in vertical direction.*

is convolved by a point spread function. Therefore, the principal procedure to reconstruct or restore the object function is the same. Essentially, it is a *deconvolution* or an *inverse filtering* since the effect of the convolution process by the PSF is to be inverted. Given the simple relations Eqs. (9.1) and (9.2), inverse filtering is in principle an easy procedure. The effect of the convolution operator $\mathcal{H}$ is reversed by the applications of the inverse operator $\mathcal{H}^{-1}$. This means in the Fourier space:

$$\hat{G}_R = \frac{\hat{G}'}{\hat{H}'} = \hat{H}^{-1} \cdot \hat{G}' \tag{9.8}$$

and in the spatial domain

$$G_R = \mathcal{F}^{-1} \hat{H}^{-1} \cdot (\mathcal{F} G'). \tag{9.9}$$

The Fourier transformed image, $\mathcal{F} G'$, is multiplied by the inverse of the OTF, $\hat{H}^{-1}$, and then transformed back to the spatial domain.

The inverse filtering can also be performed in the spatial domain by convolution with a mask that is given by the inverse Fourier transform of the inverse OTF:

$$G_R = (\mathcal{F}^{-1} \hat{H}^{-1}) * G'. \tag{9.10}$$

At first glance, inverse filtering appears simple and straightforward. In most cases, however, it is useless or even impossible to apply it as given by Eqs. (9.9) and (9.10).

The reason for the failure is related to the fact that the OTF is often zero in wide ranges. This is the case for both the OTFs of motion blur Eq. (9.7) and defocusing Eq. (9.5). A zero OTF results in an infinite inverse OTF.

Even worse is the situation for reconstruction from focus series. The 3-D OTF discussed in Sections 4.3.5b and 4.3.5c is zero in large areas of the Fourier space. In particular, it is zero along the z axis, which means that all structures in z direction are completely lost and cannot be reconstructed.

Not only the zeroes of the OTF cause problems, but also all ranges in which the OTF is significantly smaller than one. This has to do with the influence of noise. We assume the following simple image formation model in the following discussion:

$$G' = H * G + N \qquad \circ\!\!-\!\!\bullet \qquad \hat{G}' = \hat{H} \cdot \hat{G} + \hat{N}. \tag{9.11}$$

Equation Eq. (9.11) states that the noise is added to the image *after* the image is degraded. With this model, inverse filtering yields according to Eq. (9.8)

$$\hat{G}_R = \hat{H}^{-1} \cdot \hat{G}' = \hat{G} + \hat{H}^{-1} \cdot \hat{N} \tag{9.12}$$

provided that $\hat{H} \neq 0$. This equation states that the restored image is the original image G plus the noise amplified by $\hat{H}^{-1}$. If $\hat{H}$ tends to zero, $\hat{H}^{-1}$ becomes infinite and correspondingly also the noise level. Even worse, the equations state that the signal to noise ratio is not improved at all but stays the same since the noise and the degraded image are multiplied by the same factor.

From this basic fact, we can conclude that inverse filtering does not at all improve the image quality. Even more, it becomes evident that *any* linear technique will not do so. All we can do with linear techniques is to amplify the attenuated structure only up to the point where the noise level remains below an acceptable level.

9.3.5   Model-based Restoration

From the discussion about inverse filtering in Section 9.3.4, we can conclude that only *nonlinear* techniques can gain a significant progress in restoration. Basically, these processes can be improved by the following two techniques:

Additional knowledge. In most cases, we have additional knowledge about the objects to be reconstructed. Mathematically, this means that the solution to the reconstruction is constrained. The difficulty of this approach is twofold. First, it is often difficult to simply express more qualitative constraints in equations. Moreover, it is difficult to devise solutions which meet even simple criteria, for instance, that gray values are positive and not larger than an upper threshold. Often, such constraints lead to complex and nonlinear equations. Second, we cannot advise general solutions with this kind of approach. Every situation requires a specific model. This makes model-based reconstruction techniques very cumbersome.

Distinction between structure and noise. The key restriction of inverse filtering is that it cannot — as a linear filter technique — distinguish noise from image structures and, thus, treats both equally (Section 10.3.10a). Any technique that is able to distinguish noise from real image structures would significantly improve any inverse filter technique. Such techniques are discussed in Sections 10.3.10 and 11.3.3. To the extent that noise is suppressed in the degraded image without losing any significant image information, we can apply inverse filter techniques to lower values of the inverse OTF.

These two techniques carry global information back to local processing. Additional knowledge may bring in completely lost information, while nonlinear noise reduction procedures improve the signal-to-noise ratio.

As a simple example, we discuss the reconstruction of blurred circular objects with known gray values. This situation arises, for instance, in the depth-from-focus technique applied to image gas bubbles submerged by breaking waves at the ocean surface as discussed in Section 1.3.2.

The images taken with this technique can be normalized in such a way that the background is mapped to zero and a sharply imaged bubble to one. Then, the following reconstruction technique can be applied. As long as the blurring is modest, i. e., less than the radius of the bubbles, the intensity at the center of the bubble remains one. If the point spread function is symmetric, the edge of the bubbles is simply given by the position of the value 0.5. Thus, a simple global thresholding operation is sufficient to reconstruct the shape of the bubble. This works because the bubble has no jagged boundary. A jagged boundary would, of course, not be reconstructed but be smoothed by this technique.

If the blurring is more significant, the bubble takes more and more the shape of the PSF. Then, we can still retrieve at least the cross section of the bubble since the sum of all intensities in the blob is still preserved. From this sum, we can compute the original cross section of the bubble by using the knowledge that the intensity in the whole original object was one and that the bubble cross section has the shape of a circular disk.

Although this example is a quite trivial one, it sheds some light onto the principal possibilities of model-based reconstruction techniques. On the one side, they may be far reaching. On the other side, they can be applied to only very specific situations. This is why a more detailed discussion of these techniques is omitted in this general handbook. The general message is, however, clear: linear inverse filter techniques are only of limited use while the model-based techniques are too specific to be considered here. Some further references will be given in Section 9.5.

9.3.6 Radon Transform and Fourier Slice Theorem

In Section 4.4.4, we described the principles of tomographic image formation by a series of projections. With respect to reconstruction, it is important to note that the projection under all the angles ϑ can be regarded as another two-dimensional representation of the image. One coordinate is the position in the projection profile, r, the other the angle ϑ (Fig. 9.3). Consequently, we can regard the parallel projection as a transform which transforms the image into another two-dimensional representation. Reconstruction then just means applying the inverse transform, provided that it exists.

A projection beam is characterized by the angle ϑ and the offset r (Fig. 9.3). The angle ϑ is the angle between the projection plane and the x axis. Furthermore, we assume that we slice the three-dimensional object parallel to the xy plane. Then, the scalar product between a vector $\boldsymbol{x}$ on the projection beam and a unit vector

$$\bar{\boldsymbol{n}} = \left[\begin{array}{c} \cos\vartheta \\ \sin\vartheta \end{array} \right] \tag{9.13}$$

normal to the projection beam is constant and equal to the offset r of the beam

$$\boldsymbol{x}\bar{\boldsymbol{n}} - r = x\cos\vartheta + y\sin\vartheta - r = 0. \tag{9.14}$$

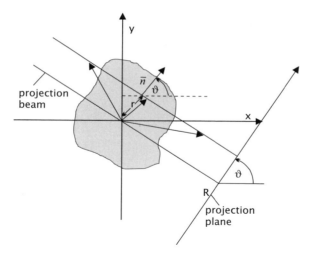

Figure 9.3: *Geometry of a projection beam.*

The projected intensity $P(r, \vartheta)$ is given by integration along the projection beam:

$$P(r, \vartheta) = \int_{\text{path}} g(\boldsymbol{x})\mathrm{d}\boldsymbol{s} = \int_{-\infty}^{\infty} g(\boldsymbol{x})\delta(x_1 \cos \vartheta + x_2 \sin \vartheta - r)\mathrm{d}^2 x. \tag{9.15}$$

The projective transformation of a two-dimensional function $g(\boldsymbol{x})$ onto $P(r, \vartheta)$ is named after the mathematician Radon as the *Radon transform*. To better understand the properties of the Radon transform, we analyze it in the Fourier space. This is easily done by rotating the coordinate system so that the direction of a projection beam coincides with the y' axis. Then the r coordinate in $P(r, \vartheta)$ coincides with the x' coordinate and the Fourier transform of the projection function in the rotated coordinate systems (x', y') and (k_x', k_y'), respectively, is:

$$\hat{P}(k_y', 0) = \int_{-\infty}^{\infty} P(x', 0) \exp(-\mathrm{i}k_x' x')\mathrm{d}x'. \tag{9.16}$$

The angle ϑ is zero in the rotated coordinate system. Replacing $P(x', 0)$ by the definition of the Radon transform Eq. (9.15), we yield

$$\hat{P}(k_x', 0) = \int_{-\infty}^{\infty} \left[\int_{-\infty}^{\infty} g(x', y')\mathrm{d}y_1' \right] \exp(-\mathrm{i}k_x' x')\mathrm{d}x' = \hat{g}(k_x', 0), \tag{9.17}$$

or, with regard to the original coordinate system,

$$\hat{P}(q, \vartheta) = \hat{P}\left[|\boldsymbol{k}|, \arctan(k_2/k_1)\right] = \hat{g}(k\bar{\boldsymbol{n}}), \tag{9.18}$$

where q is the coordinate in the k space in the direction of ϑ. The spectrum of the projection is identical to the spectrum of the original object on a beam normal to the direction of the projection beam. This important result is called the *Fourier slice theorem* or *projection theorem*.

The Fourier slice theorem can also be derived without any computation by regarding the projection as a linear shift-invariant filter operation. Since the projection adds up all

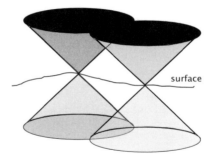

Figure 9.4: *Superposition of the point spread function of two neighboring points on a surface.*

the gray values along the projection beam, the point spread function of the projection operator is a δ line in the direction of the projection beam. In the Fourier domain, this convolution operation corresponds to a multiplication with the transfer function, a δ line normal to the δ line in the space domain (see Appendix B.3). In this way, the projection operator cuts out a slice of the spectrum in the direction normal to the projection beam.

9.4   Procedures

9.4.1   Reconstruction of Depth Maps from Focus Series

The discussion on inverse filtering in Section 9.3.4 has shown that a general 3-D reconstruction from focus series is a hard problem. A much simpler case is the reconstruction of a single opaque surface from focus series. This is a very useful application in the area of optical surface gauging. Steurer et al. [138] developed a simple method to reconstruct a *depth map* from a light microscopic focus series. A depth map is a two-dimensional function which gives the depth of an object point $d(x, y)$ — relative to a reference plane — as a function of the image coordinates (x, y).

With the given restrictions, only one depth value for each image point needs to be found. We can make use of the fact that the three-dimensional point spread function of optical imaging that was discussed in detail in Section 4.3.5 has a distinct maximum on the focal plane because the intensity falls off with the square of the distance from the focal plane. This means that at all locations where distinct image features such as edges, lines, or local extrema are located, we will also obtain a maximum contrast on the focal plane. Figure 9.4 illustrates that the point spread functions of neighboring image points do not influence each other close to the focal plane.

Steurer's method makes use of the fact that a distinct maximum of the point spread function exists in the focal plane. His algorithm includes the following steps:

1. Take a focus series with constant depth differences. If required, to improve the signal-to-noise ratio, digitize and average several images at each depth level.

2. Apply a highpass filter to emphasize small structures. The highpass-filtered images are segmented to obtain a mask for the regions with gray value changes.

3. In the masked regions, search the maximum contrast in all the images of the focus series. The image in which the maximum occurs gives a depth value for the depth map. Sub-pixel accurate height information can be obtained by a regression of the contrast as a function of the depth with a second-order polynomial.

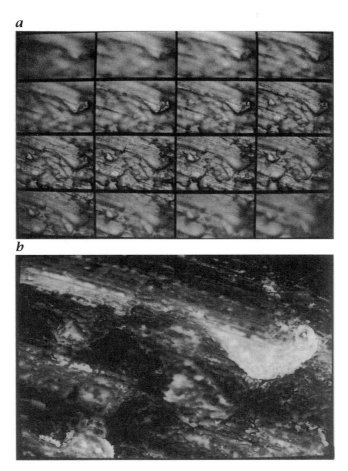

Figure 9.5: *a Focus series with 16 images of a metallic surface taken with depth distances of 2 μm; the focal plane becomes deeper from the left to the right and from top to bottom. **b** Depth map computed from the focus series. Depth is coded by intensity. Objects closer to the observer are shown brighter. From Steurer et al. [138].*

4. Since the depth map will not be dense, interpolate the depth map. Steurer used a region-growing method followed by an adaptive lowpass filtering which is applied only to the interpolated regions in order not to corrupt the directly computed depth values. However, any other technique, especially normalized convolution, is also a valid approach.

This method was successfully used to determine the surface structure of worked metal pieces. Figure 9.5 shows the results of the reconstruction technique. A filing can be seen which sticks out of the surface. Moreover, the surface shows clear traces from the grinding process.

This technique works only if the surface shows fine details. If this is not the case, the confocal illumination technique of Scheuermann et al. [122] can be applied that projects statistical patterns into the focal plane (compare Section 1.5.1 and Fig. 1.22).

a   b   c   d

e   f   g   h

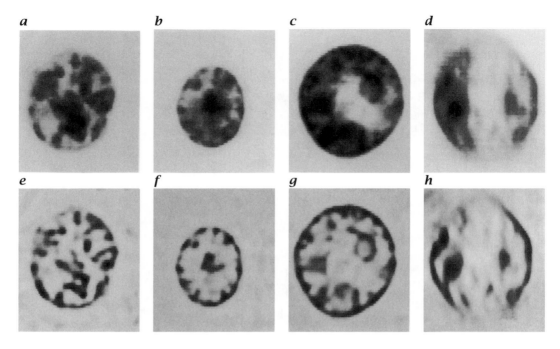

Figure 9.6: *3-D reconstruction of a focus series of a cell nucleus taken with conventional microscopy. Upper row: **a** – **c** selected original images; **d** xz cross section perpendicular to the image plane. Lower row: **e** – **h** reconstructions of the images shown above; courtesy of Dr. Schmitt and Prof. Dr. Komitowski, German Cancer Research Center, Heidelberg.*

9.4.2 3-D Reconstruction by Inverse Filtering

If true three-dimensional objects and not only surfaces are involved, we need to reconstruct the 3-D object function $g(\boldsymbol{x})$ from the 3-D focus series which is blurred by the 3-D point spread function of optical imaging. It is obvious that an exact knowledge of the PSF is essential for a good reconstruction. In Section 4.3.5, we computed the 3-D *PSF* of optical imaging neglecting lens errors and resolution limitation due to diffraction. However, the resolution in high magnification microscopy images is diffraction-limited (Section 4.3.3a).

The diffraction-limited 3-D PSF was computed by Erhardt et al. [27]. The resolution limit basically changes the double-cone close to the focal plane. At the focal plane it does not reduce to a point but to a diffraction disk (Fig. 4.16). As a result, the *OTF* drops off at higher wave numbers in the $k_x k_y$ plane. To a first approximation, we can regard the diffraction-limited resolution as an additional lowpass filter by which the OTF for unlimited resolution is multiplied. This filtering produces the effects on the PSF and OTF described above.

The principal problems with inverse filtering are discussed in Section 9.3.4 in detail. Here, we illustrate some practical approaches. The simplest approach to yield an optimum reconstruction is to limit application of the inverse OTF to the wave number components which are not damped below a critical threshold. This threshold depends on the noise in the images. In this way, the true inverse OTF is replaced by an *effective inverse OTF* which approaches zero again in the wave number regions which cannot be reconstructed.

The results of this reconstruction procedure are shown in Fig. 9.6. A $64 \times 64 \times 64$ focus series has been taken from the nucleus of a cancerous rat liver cell. The resolution in all directions is $0.22\,\mu m$. The images clearly verify the theoretical considerations. The reconstruction considerably improves the resolution in the xy image plane, while the resolution in z direction is clearly worse. Structures that are directed only in z direction could not be reconstructed at all as can be seen from the top and bottom edge of the cell nucleus in Fig. 9.6d and h.

Unconstrained inverse filtering can also be performed in the space domain using an iterative method. Let $\mathcal{H}$ be the blurring operator. We introduce the new operator $\mathcal{H}' = 1 - \mathcal{H}$. Then, the inverse operator

$$\mathcal{H}^{-1} = \frac{1}{1 - \mathcal{H}'} \tag{9.19}$$

can be approximated by the Taylor expansion

$$\mathcal{H}^{-1} = 1 + \mathcal{H}' + \mathcal{H}'^2 + \mathcal{H}'^3 + \ldots + \mathcal{H}'^k, \tag{9.20}$$

or explicitly written for the OTF in the Fourier domain

$$\hat{h}^{-1} = 1 + \hat{h}' + \hat{h}'^2 + \hat{h}'^3 + \ldots + \hat{h}'^k. \tag{9.21}$$

In order to understand how the iteration works, we consider periodic structures. First, we take one which is only slightly attenuated. This means that $\hat{H}$ is only slightly less than one. Thus, $\hat{H}'$ is small and the expansion converges rapidly. The other extreme is a periodic structure that has nearly vanished. Then, $\hat{H}'$ is close to one. Consequently, the amplitude of the periodic structure increases by the same amount with each iteration step (linear convergence). This procedure has the significant advantage that we can stop the iteration when the noise becomes visible.

A direct application of the iteration makes not much sense because the increasing exponents of the convolution masks become larger and thus the computational effort increases from step to step. A more efficient scheme known as *Van Cittert iteration* utilizes Horner's scheme for polynomial computation:

$$G_0 = G', \quad G_{k+1} = G' + (I - H) * G_k. \tag{9.22}$$

In Fourier space, it is easy to examine the convergence of this iteration. From Eq. (9.22)

$$\hat{g}_k(\boldsymbol{k}) = \hat{g}'(\boldsymbol{k}) \sum_{i=0}^{k} (1 - \hat{h}(\boldsymbol{k}))^i. \tag{9.23}$$

This equation constitutes a geometric series with the start value $a_0 = \hat{g}'$ and the factor $q = 1 - \hat{h}$. The series converges only if $|q| = |1 - \hat{h}| < 1$. Then the sum is given by

$$\hat{g}_k(\boldsymbol{k}) = a_0 \frac{1 - q^k}{1 - q} = \hat{g}'(\boldsymbol{k}) \frac{1 - |1 - \hat{h}(\boldsymbol{k})|^k}{\hat{h}(\boldsymbol{k})} \tag{9.24}$$

and converges to the correct value $\hat{g}'/\hat{h}$.

Unfortunately, this condition for convergence is not met for all transfer functions that have negative values. This is, however, the case for motion blurring (Section 9.3.3) and defocusing (Section 9.3.2). Therefore the Van Cittert iteration cannot be applied in these cases.

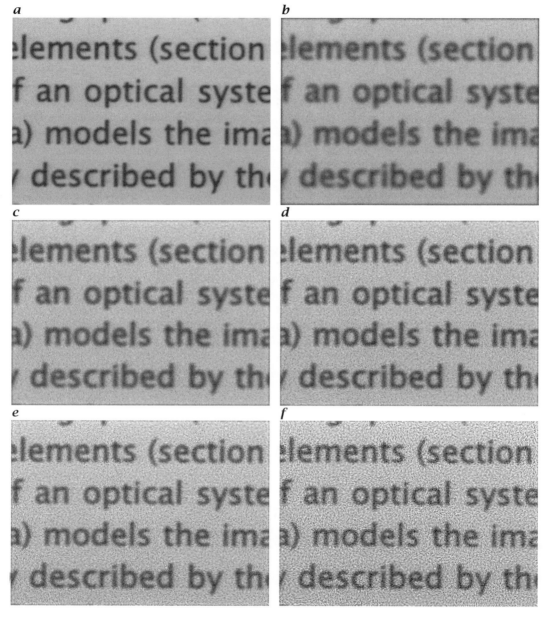

Figure 9.7: *Illustration of iterative inverse filtering: **a** original image, **b** blurred by a 9 × 9 binomial filter and Gaussian noise with a standard deviation of 5 added, **c - f** after first, second, third, and sixth iteration of inverse filtering.*

A slight modification of the iteration process, however, makes it possible to use it also for degradations with partially negative transfer functions. The simple trick is to apply the transfer function twice. The transfer function $\hat{h}^2$ of the cascaded filter $\boldsymbol{H} * \boldsymbol{H}$ is always positive.

The modified iteration scheme is

$$\boldsymbol{G}_0 = \boldsymbol{H} * \boldsymbol{G}', \quad \boldsymbol{G}_{k+1} = \boldsymbol{H} * \boldsymbol{G}' + (\boldsymbol{I} - \boldsymbol{H} * \boldsymbol{H}) * \boldsymbol{G}_k. \tag{9.25}$$

With $a_0 = \hat{h}\hat{g}'$ and $q = 1 - \hat{h}^2$ the iteration converges to the correct value

$$\lim_{k \to \infty} \hat{g}_k(\boldsymbol{k}) = \lim_{k \to \infty} \hat{h}\hat{g}' \frac{1 - |1 - \hat{h}^2|^k}{\hat{h}^2} = \frac{\hat{g}'}{\hat{h}}, \tag{9.26}$$

provided that $|1 - \hat{h}^2| < 1$. This condition is met as long as the absolute value of the transfer function is smaller than 2.

Figure 9.7 illustrates the inverse filtering technique applied to a simulation. The image shown in Fig. 9.7a is blurred by a 9×9 binomial mask. After the blurring, normal distributed noise with a standard deviation of 5 is added in Fig. 9.7b. Figure 9.7c–f shows the result of the inverse filtering after 1, 2, 3, and 6 iteration steps. While the letters become almost as sharp as in the original image, the noise level increases from iteration step to iteration step to an unacceptably high level. The optimum reconstruction result seems to be somewhere between 2 and 3 iteration steps but is still far away from the quality of the original image.

The iterative solution of the inverse problem has the additional benefit that it is quite easy to add further terms that result in better solutions. These extended solutions can be understood most easily if we observe that the solution of the inverse problem minimizes an error functional. If we apply the degradation filter $\mathcal{H}$ to the reconstructed image $\boldsymbol{G}_R$ it should come as close as possible to the degraded image. This means that

$$\|\boldsymbol{G} - \mathcal{H}\boldsymbol{G}_R\| \tag{9.27}$$

should become minimal using an appropriate error norm. If the standard 2-norm is used (*least squares*), we end up with a huge linear equation system that can in principle be solved as described in Appendix B.2. Because of the special structure of the convolution operator $\mathcal{H}$ it can be much more efficiently solved by iterated convolution.

It is now easily possible to add further constraints to the minimization problem. Noise tends to cause strong variations in the signal. Therefore it should be possible to decrease this disturbing effect by demanding that the solution becomes smooth. Such an approach is known as *regularization*. In the simplest case, we use an additional term with the norm of a highpass-filtered image. Then Eq. (9.27) extends to

$$\|\boldsymbol{G} - \mathcal{H}\boldsymbol{G}_R\| + a \|\boldsymbol{C}\boldsymbol{G}_R\|. \tag{9.28}$$

The iterative solution of Eq. (9.28) still has the same form as Eq. (9.25) with just one additional term.

$$\boldsymbol{G}_0 = \boldsymbol{H} * \boldsymbol{G}', \quad \boldsymbol{G}_{k+1} = \boldsymbol{H} * \boldsymbol{G}' + (\boldsymbol{I} - a\boldsymbol{C} * \boldsymbol{C} + \boldsymbol{H} * \boldsymbol{H}) * \boldsymbol{G}_k. \tag{9.29}$$

For a further discussion of such flexible iterative solutions of inverse problems we refer to Jähne [74] and Section 9.5.

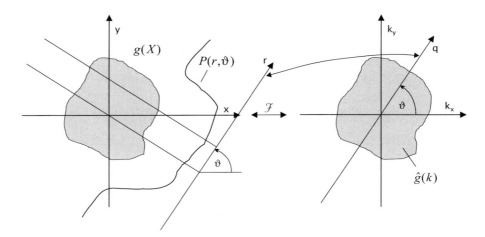

Figure 9.8: *Illustration of the Fourier slice theorem. Each parallel projection yields the spectrum $\hat{g}(\mathbf{k})$ of the object $g(\mathbf{x})$ along a line parallel to the projection plane and through the origin of the Fourier space.*

9.4.3 Filtered Backprojection

9.4.3a Principle. In this section, we turn to reconstruction from projections as in tomographic imaging. The principles of tomographic images were discussed in Section 4.4.4, while the Radon transformation and Fourier slice theorem that give the relations between the image and its projections were treated in Section 9.3.6. Here, we discuss the most important and fastest technique for reconstruction from projection known as *filtered backprojection*. The basic fact established by the Fourier slice theorem says that each projection yields the spectrum on one line in the Fourier space as illustrated in Fig. 9.8.

If the projections include projections from all directions, the obtained slices of the spectrum eventually cover the complete spectrum of the object. Inverse Fourier transform then yields the original object. Filtered backprojection uses this approach with a slight modification. If we just add the spectra of the individual projection beams to obtain the complete spectrum of the object, the spectral density for small wave numbers would be too high since the beams are closer to each other for small radii. Thus, we must correct the spectrum with a suitable weighting factor. In the continuous case, the geometry is very easy. The density of the projection beams goes with $|\mathbf{k}|^{-1}$. Consequently, the spectra of the projection beams must be multiplied by $|\mathbf{k}|$. Thus, filtered backprojection is a two-step process:

Filtering of the projections. In this step, we multiply the spectrum of each projection direction by a suitable weighting function $\hat{w}(|\mathbf{k}|)$. Of course, this operation can also be performed as a convolution with the inverse Fourier transform of $\hat{w}(|\mathbf{k}|)$, $w(r)$. Because of this step, the procedure is called the *filtered* backprojection.

Addition of the backprojections. Each projection gives a slice of the spectrum. Adding up all the filtered spectra yields the complete spectrum. Since the Fourier transform is a linear operation, we can add up the filtered projections in the space domain. In the space domain, each filtered projection contains the part of the object which is constant in the direction of the projection beam. Thus, we can backproject the corresponding filtered 1-D projection along the direction of the projection beam and in this way sum up to the contributions from all projection beams.

9.4.3b Continuous Case. After this illustrative description of the principle of the filtered backprojection algorithm we derive the method for the continuous case. We start
with the Fourier transform of the object and write the inverse Fourier transformation
in polar coordinates (r, ϑ) in order to make use of the Fourier slice theorem

$$g(\boldsymbol{x}) = \int_0^{2\pi} \int_0^\infty q\hat{g}(q, \vartheta) \exp[2\pi iq(x_1 \cos \vartheta + x_2 \sin \vartheta)]\mathrm{d}q\mathrm{d}\theta. \qquad (9.30)$$

In this formula, the spectrum is already multiplied by the wave number q. The integration boundaries are, however, not yet correct to be applied to the Fourier slice
theorem Eq. (9.18). The coordinate, q, should run from $-\infty$ to ∞ and ϑ only from 0 to
π. In Eq. (9.30), we integrate only over half a beam from the origin to infinity. We can
compose a full beam from two half beams at the angles ϑ and $\vartheta + \pi$. Thus, we split
the integral in Eq. (9.30) into two over the angle ranges $0 - \pi$ and $\pi - 2\pi$ and obtain

$$
\begin{aligned}
g(\boldsymbol{x}) &= \int_0^\pi \int_0^\infty q\hat{g}(q, \vartheta) \exp[2\pi iq(x_1 \cos \vartheta + x_2 \sin \vartheta)]\mathrm{d}q\mathrm{d}\vartheta \\
&+ \int_0^\pi \int_0^\infty q\hat{g}(-q, \vartheta') \exp[-2\pi iq(x_1 \cos \vartheta' + x_2 \sin \vartheta')]\mathrm{d}q\mathrm{d}\vartheta' \qquad (9.31)
\end{aligned}
$$

using the following identities: $\vartheta' = \vartheta + \pi$, $\hat{g}(-r, \vartheta) = \hat{g}(r, \vartheta')$, $\cos(\vartheta') = -\cos(\vartheta)$,
$\sin(\vartheta') = -\sin(\vartheta)$. Now we can recompose the two integrals again, if we substitute q
by $-q$ in the second integral and replace $\hat{g}(q, \vartheta)$ by $\hat{P}(q, \vartheta)$ because of the Fourier slice
theorem Eq. (9.18):

$$g(\boldsymbol{x}) = \int_0^\pi \int_{-\infty}^\infty |q|\hat{P}(q, \vartheta) \exp[2\pi iq(x_1 \cos \vartheta + x_2 \sin \vartheta)]\mathrm{d}q\mathrm{d}\vartheta. \qquad (9.32)$$

Equation Eq. (9.32) gives the inverse Radon transform and is the basis for the *filtered
backprojection* algorithm. It can be parted into two steps:

Filtering; inner integral in Eq. (9.32)

$$P' = \mathcal{F}^{-1}(|k|\mathcal{F}P). \qquad (9.33)$$

$\mathcal{F}$ denotes the Fourier transform operator. P' is the projection function P multiplied
in the Fourier space by $|k|$. If we perform this operation as a convolution in the space
domain, we can formally write

$$P' = [\mathcal{F}^{-1}(|k|)] * P. \qquad (9.34)$$

Backprojection; outer integral in Eq. (9.32)

$$g(\boldsymbol{x}) = \int_0^\pi P'(r, \vartheta)\,\mathrm{d}\vartheta. \qquad (9.35)$$

This equation shows how the object is built up by the individual filtered projections
from all directions. Each filtered projection profile $P'(r, \vartheta)$ is backprojected along
the direction of the projection beams to fill the whole image and accumulated to
the backprojected projection profiles from all directions.

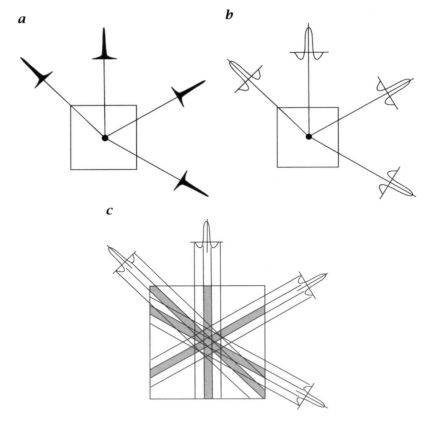

Figure 9.9: *Illustration of the filtered backprojection algorithm with a point object: **a** projections from different directions; **b** filtering of the projection functions; **c** backprojection: adding up the filtered projections.*

9.4.3c Discrete Case.   There are several details which we have not yet discussed but which cause serious problems for the reconstruction in the infinite continuous case, especially:

- The inverse Fourier transform of the weighting function $|\mathbf{k}|$ does not exist, because this function is not square-integrable.
- It is impossible to reconstruct the mean of an object because of the multiplication by $|q| = |\mathbf{k}|$ in the Fourier domain Eq. (9.32) which eliminates $\hat{g}(\mathbf{0})$.

Actually, we never apply the infinite continuous case but only compute using discrete data. Basically, there are three effects which distinguish the idealized reconstruction from the real world:

- The object is of limited size. In practice, the size limit is given by the distance between the radiation source and the detector.
- The resolution of the projection profile is limited by the combined effects of the extent of the radiation source and the resolution of the detector array in the projection plane.
- Finally, we can only take a limited number of projections. This corresponds to a sampling of the angle ϑ in the Radon representation of the image.

We complete the discussion in this section with an illustrative example.

Example 9.1: Projection and Reconstruction of a Point

We can learn a lot about the projection and reconstruction by considering the reconstruction of a simple object, a point, because the Radon transform Eq. (9.15) and its inverse are linear transforms. For a point, the projections from all directions are equal (Fig. 9.9a) and show a sharp maximum in the projection functions $P(r, \vartheta_i)$. In the first step of the filtered backprojection algorithm, P is convolved with the $|k|$ filter. The result is a modified projection function P' which is identical to the point spread function of the $|k|$ filter (Fig. 9.9b).

In a second step, the backprojections are added up in the image. From Fig. 9.9c, we can see that at the position of the point in the image the peaks from all projections add up. At all other positions in the images, the filtered backprojections superimpose each other in a destructive manner, since they show negative and positive values. If the projection directions are sufficiently close to each other, they cancel each other except for the point in the center of the image. Figure 9.9c also demonstrates that an insufficient number of projections leads to star-shaped distortion patterns.

The simple example of the reconstruction of a point from its projections is also useful to show the importance of filtering the projections. Let us imagine what happens when we omit this step. Then, we would add up δ lines as backprojections which rotate around the position of the point. Consequently, we would not obtain a point but a rotational symmetric function which falls off with $|x|^{-1}$. As a result, the reconstructed objects would be considerably blurred. The point spread function of the $|k|$ filter is the $|x|^{-1}$ function.

9.5 Advanced Reference Material

J. C. Russ, 1994. The Image Processing Handbook, 2nd edition, CRC Press, Boca Raton. *Chapter 9 of Russ' handbook discusses reconstruction from tomographic projection. Although superficial in the treatment of the mathematical concepts, a very nice collection of example images is presented that gives a good intuition of what factors determine the quality of the reconstruction.*

J. A. Parker, 1990. Image Reconstruction in Radiology, CRC Press, Boca Raton. *An excellent introduction to the image reconstruction techniques used in medical radiology. The focus of the book is on the mathematical methods being used. Here the book starts with the elementary concepts and covers everything that is required in an intuitive way: vectors, linear systems, complex numbers, linear algebra, eigenfunctions, stochastic processes, least squares estimation techniques, and Fourier transform. The reconstruction techniques themselves are covered only in the last quarter of the book. Excellent for readers with little background in mathematics who want to understand how reconstruction techniques work.*

W. Menke, 1989. Geophysical Data Analysis: Discrete Inverse Theory, Academic Press, San Diego. *Although this book does not deal specifically with problems related to image processing, it is an exceptionally well-presented account of discrete inverse problems.*

R. L. Lagendijk and J. Biemond, 1991. Iterative Identification and Restoration of Images, Kluwer Academic Publishers, Boston. *While in the past the focus was on the restoration techniques themselves, this monograph offers a fresh approach to "estimate the properties of the imperfect imaging system from the observed degraded images". This is an important aspect of image restoration because in practice the point spread function that caused the degradation is often not known. The book gives also a comprehensive overview of iterative restoration techniques.*

Part III

From Images to Features

10 Neighborhoods

10.1 Highlights

Operations in small neighborhoods are the initial processing step to detect objects. The analysis of a small neighborhood is sufficient to judge whether a homogeneous region or an edge is encountered or whether the gray value shows a dominant orientation. Two basic types of neighborhood operations are available:

- "Weighting and accumulation" with convolution or linear shift-invariant (LSI) filter operations (Sections 10.3.3 and 10.3.8)
- "Sorting and selecting" with rank-value filters (Section 10.3.9)

Convolution of a w-dimensional image with a $(2R + 1)^w$-dimensional mask (Sections 10.3.1 and 10.3.3) is given by:

1-D   N-vector
$$g'_n = \sum_{n'=-R}^{R} h_{n'} g_{n-n'}$$

2-D $M \times N$ matrix
$$g'_{mn} = \sum_{m'=-R}^{R} \sum_{n'=-R}^{R} h_{m'n'} g_{m-m',n-n'}$$

3-D $L \times M \times N$ matrix
$$g'_{lmn} = \sum_{l'=-R}^{R} \sum_{m'=-R}^{R} \sum_{n'=-R}^{R} h_{k'm'n'} g_{k-k',m-m',n-n'}$$

The point spread function (PSF, Section 10.3.4) describes the response of the filter to a point, and the transfer function (Section 10.3.5) describes how structures of different scales are modified.

A special class of linear filters are recursive filters (Section 10.3.8). They can easily be "tuned" and computed with only a few multiplications and additions.

Linear filters are limited in the sense that they operate without taking the context into account. In principle, they cannot distinguish objects of interest from noise. They are, however, the basis for a number of nonlinear and adaptive filter strategies such as normalized convolution and steerable filters (Section 10.3.10).

The adequate usage of neighborhood operations requires appropriate filter design techniques. The design criteria for image processing differ significantly from the "classical" design rules in signal processing (Section 10.4.1). Discussed are filter design by windowing techniques (Section 10.4.2), the use of recursive filters for image processing (Section 10.4.3), and filter design by cascading elementary simple filters (Section 10.4.4). A filter set is introduced that constitutes the basis for all filter operations used throughout this handbook. Given the enormous amounts of pixels in images, efficient computation schemes for neighborhood operations are essential (Section 10.4.5). The correct treatment of neighborhood operations at image borders receives attention in Section 10.4.6.

10.2   Task

Pixel operations as discussed in Chapter 7 are not suitable to detect objects of interest in images. The reason lies in the fact that each pixel is treated individually without any consideration of the content of neighboring pixels. Thus, the key for object recognition is the analysis of the *spatial* relation of the gray values. In an image sequence, it is required to study the *temporal* relations as well.

The first step in this direction is the analysis of a small neighborhood. Imagine a simple scene with objects of constant gray values and a dark background. If the gray values do not change in a small neighborhood, we can conclude that we are within the object or the background. If the gray values do change, we are at the edge of an object. In this way, we can recognize areas of constant gray values and edges.

Neighborhood operators generally combine the gray values in a small neighborhood around a pixel and write the result back to this pixel. This procedure is repeated for all pixels of an image. Thus, the result of a neighborhood operation is a new image of the same size. The meaning of the "gray values", however, is now different. If we apply, for instance, a neighborhood operation to detect edges, a bright gray value at a pixel will now indicate that an edge runs across the pixel. If it is dark, we are either in the background or within the object. Therefore, a neighborhood operation converts a gray scale image into a *feature image*. Often, neighborhood operations are also named *filters*. The term *filter* stresses that a certain kind of information has been filtered out of the image and that some information which is contained in the original image is lost.

Neighborhood operations are the key operations to extract features from images. Task list 7 summarizes the tasks to be performed to understand neighborhood operations properly and thus to learn which types of neighborhood operations extract which feature.

The first task is to find out which classes of neighborhood operations are principally possible. We will learn that there are essentially only two classes, *convolution*, also called *linear shift-invariant (LSI) filters* , and *rank-value filters*. The general properties of convolution and rank-value filters will be discussed in Sections 10.3.3 and 10.3.9. An extra section is devoted to recursive filtering, a special type of convolution filters (Section 10.3.8).

Linear filters have apparent limitations. They cannot distinguish whether a gray value structure represents an object feature or noise. Such a distinction is only possible by regarding the context in which the feature is found. This observation led to the development of nonlinear and adaptive filter techniques. At each point in the image it is decided which filter is to be applied. Various strategies for adaptive filtering are discussed in Section 10.3.10.

Filter design techniques for image processing are quite different from those for processing of one-dimensional time series. An important practical goal is to compute neighborhood operations efficiently (Section 10.4.5). This means both fast processing and usage of minimum extra storage space. Another practical problem is the treatment of neighborhood operations at the borders of an image (Section 10.4.6). Depending on the type of neighborhood operation, different strategies are possible to handle neighborhood operators at image borders appropriately.

Task List 7: Neighborhood Operations

Goal	Procedures
Analyze spatial structure (in image sequences spatio-temporal structure) in a small neighborhood	Find out what kind of neighborhood operations are possible
Understand convolution operators (linear shift-invariant, FIR filter)	Find out basic properties
Understand recursive (IIR) filtering	Find out basic properties. Devise strategies to use them appropriately in higher-dimensional signals
Understand rank-value filters	Find out basic properties
Understand adaptive filtering	Find out limitations of linear filtering; devise strategies for adaptive filtering
Compute neighborhoods efficiently (fast and using minimum extra storage)	Develop efficient general control structures and buffering schemes
Recognize difficulties with neighborhood operations at image borders	Devise appropriate strategies for extrapolation of image beyond borders.

Table 10.1: *Number of pixels included in even-sized and odd-sized masks for neighborhood operations in 1-D to 4-D signals; W denotes the dimension.*

Size of Mask	2^W	3^W	4^W	5^W	6^W	7^W	8^W	9^W	13^W	15^W
1-D	2	3	4	5	6	7	8	9	13	15
2-D	4	9	16	25	36	49	64	81	169	225
3-D	8	27	64	125	216	343	512	729	2197	3375
4-D	16	81	256	625	1296	2401	4096	6561	28561	50625

10.3 Concepts

10.3.1 Masks

The first characteristic of a neighborhood operation is the size of the neighborhood, which we call the *window* or *filter mask*. The mask may be rectangular or of any other form. We must also specify the position of the pixel relative to the window which will receive the result of the operation. With regard to symmetry, the most natural choice is to place the result of the operation at the pixel in the center of an odd-sized mask. The smallest size of this type of masks only includes the directly neighboring pixels. In one, two, and three dimensions, the mask includes 3, 9, and 27 pixels, respectively (Fig. 10.1a). At first glance, even-sized masks seem not to be suitable for neighborhood operations since there is no pixel that lies in the center of the mask. With a trick, we can apply them nevertheless, and they turn out to be useful for certain types of neighborhood operations. The result of the neighborhood operation is simply written back to pixels that lie in between the original pixels (Fig. 10.1b). Thus, the resulting image is shifted by half the pixel distance into every direction and the receiving central pixel lies directly in the center of the 1-D, 2-D, or 3-D neighborhoods (Fig. 10.1b). In effect, the resulting image has one pixel less in every direction. It is very important to be aware of this shift by half the pixel distance. Therefore, image features computed by

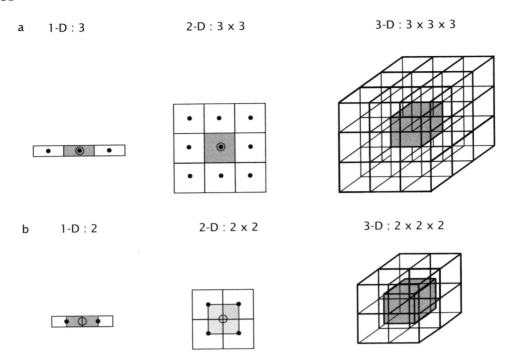

Figure 10.1: *Smallest odd-sized and even-sized masks for neighborhood operations. The result of the operation is written back to the shaded pixel. The center of this pixel is marked by an open circle. For even-sized masks, the results lie on an interleaved grid in the middle between the original pixels.*

even-sized masks should never be combined with original gray values since this would lead to considerable errors.

The number of pixels contained in the masks is considerably increasing with their size (Table 10.1). The higher the dimension the faster the number of pixels with the size of the mask increases. Even small neighborhoods include hundreds or thousands of pixels.

Therefore, it will be a challenging task for 2-D image processing and even more for 3-D and 4-D to develop efficient schemes to compute a neighborhood operation with as few computations as possible. Otherwise, it would not be possible to use them at all. The number of pixels contained in odd and even-sized masks of the dimension W are given by the following formulas:

$$\text{Odd-sized mask}\quad P_o = (2R + 1)^W \qquad \text{Even-sized mask}\quad P_e = (2R)^W. \qquad (10.1)$$

The challenge for efficient computation schemes is to decrease the number of computations from W^{th} order or $O(R^W)$ to a lower order. This means that the number of computations is no longer proportional to R^W but rather to a lower order of the linear size of the mask. The ultimate goal is to achieve computation schemes that increase only linearly with the linear size of the mask ($O(R^1)$) or, even better, do not depend at all on the size of the mask ($O(R^0)$).

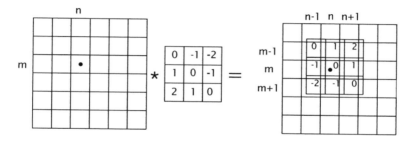

Figure 10.2: *Illustration of the discrete convolution operation with a 3 × 3 filter mask.*

10.3.2 Operators

Throughout this handbook, we use an *operator notation* for image processing operations. It helps us to make complex composite operations easily comprehensible. All operators will be written in calligraphic letters, as $\mathcal{B}, \mathcal{D}, \mathcal{H}, S$. We write

$$G' = \mathcal{H} G \qquad (10.2)$$

for an operator $\mathcal{H}$ which transforms the image G into the image G'. We will systematically reserve special letters for certain operators. Consecutive application is denoted by writing the operators one after the other. The right-most operator is applied first. Consecutive application of the same operator is expressed by an exponent

$$\underbrace{\mathcal{H} \mathcal{H} \ldots \mathcal{H}}_{p \text{ times}} = \mathcal{H}^p. \qquad (10.3)$$

If the operator acts on a single image, the operand, which stands to the right in the equations, will be omitted. In this way we can write operator equations. In Eq. (10.3), we already made use of this notation. Furthermore, we will use braces in the usual way to control the order of execution.

Pixelwise multiplication is denoted by a centered dot (·) in order to distinguish a point operation from successive application of operators. The operator expression $\mathcal{H}'(\mathcal{H} \cdot \mathcal{H})$, for instance, means: apply the operator $\mathcal{H}$ to the image, square the result pixelwise, and then apply the operator $\mathcal{H}'$.

The operator notation has another significant advantage. It does not matter whether the image is represented by an $M \times N$ matrix in the spatial or the Fourier domain. The reason for this *representation-independent notation* lies in the fact that the general properties of operators do not depend on the actual representation. The mathematical foundation is the theory of vector spaces [56]. In this sense, we regard a 2-D (3-D or 4-D) image as an element in a complex-valued $M \times N$ ($L \times M \times N$ or $K \times L \times M \times N$)-dimensional vector space.

10.3.3 Convolution

The most elementary combination of the pixels in a neighborhood is given by an operation which multiplies each pixel in the range of the filter mask with the corresponding weighting factor of the mask, adds up the products, and writes the result to the position of the center pixel (Fig. 10.2). This operation is known as a discrete *convolution* and is defined as:

$$g'_{mn} = \sum_{m'=-R}^{R} \sum_{n'=-R}^{R} h_{m'n'} g_{m-m',n-n'} = \sum_{m'=-R}^{R} \sum_{n'=-R}^{R} h_{-m',-n'} g_{m+m',n+n'}. \qquad (10.4)$$

This equation assumes an odd-sized mask with $(2r + 1) \times (2r + 1)$ coefficients.

Higher-dimensional smoothing masks should generally either have only even or only odd numbers of coefficients in every direction. In this way it is ensured that the images resulting from convolution operations lie either on the original grid or on the interleaved grid.

Further processing must consider the half-pixel shifts in every direction in which the mask is even-sized and never use a mix of images on the interleaved and original grid. If an even-sized mask is applied an even number of times, the resulting image lies on the original grid again. However, care must be taken that even-sized filters switch between a positive and negative shift of half a pixel as illustrated in Example 10.1.

Example 10.1: Elementary even-sized mask

Averaging of neighboring pixels is the simplest case of a filter with an even-sized mask. The run direction of the filters is chosen in such a way that no temporary storage is required. Therefore, these filters can be used in place.

$$g'_m = 1/2(g_m + g_{m-1}) \quad \text{running forward, causing a shift of -1/2} \\ g'_m = 1/2(g_m + g_{m+1}) \quad \text{running backward, causing a shift of 1/2.} \tag{10.5}$$

Equations Eq. (10.4) describes a *discrete convolution* operation. An operator that maps an image onto itself is abbreviated by

$$G' = H * G, \tag{10.6}$$

or, in operator notation (Section 10.3.2),

$$G' = \mathcal{H}\,G. \tag{10.7}$$

In comparison to the continuous convolution, the integral is replaced by a sum over discrete elements (compare Appendices B.3 and B.4).

The negative signs of the indices k and l either for the mask or the image in Eq. (10.4) may cause confusion. It means that we either mirror the mask or the image at its symmetry center before we put the mask over the image. If we want to calculate the result of the convolution at the point (m, n), we center the mirrored mask at this point, perform the convolution, and write the result back to position (m, n) (Fig. 10.2). This operation is performed for all pixels of the image.

Equation Eq. (10.4) indicates that none of the calculated gray values g'_{mn} will flow into the computation at other neighboring pixels. Generally, this means that the result of the convolution operation, i. e., a complete new image, has to be stored in a separate memory area. If we want to perform the filter operation in place, we run into a problem. Let us assume that we perform the convolution line by line and from left to right. Then the gray values at all pixel positions above and to the left of the current pixel are already overwritten by the previously computed results (Fig. 10.3). Consequently, it is required to store the gray values at these positions in an appropriate buffer and to develop strategies to do this efficiently (Section 10.4.5). This problem is common to all kinds of neighborhood operations, not only convolutions.

The convolution mask can — more generally — also be written as a matrix of the same size as the image to which it is applied:

$$g'_{mn} = (H * G)_{mn} = \sum_{m'=0}^{M-1} \sum_{n'=0}^{N-1} h_{m'n'} g_{m-m',n-n'}. \tag{10.8}$$

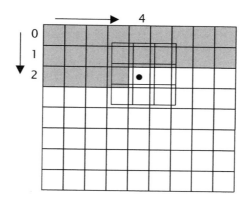

Figure 10.3: Image convolution by scanning the convolution mask line by line over the image. At the shaded pixels the gray values are already replaced by the convolution sum. Thus, the original gray values at the shaded pixels falling within the filter mask need to be stored in an extra buffer.

Both representations are equivalent if we consider the periodicity in the space domain for elements outside of the $[0, M-1] \times [0, N-1]$ interval

$$g_{m+kM,n+lN} = g_{m,n} \quad \forall k, l \in \mathbb{Z}.$$

The restriction of the sum in Eq. (10.4) reflects the fact that the filter mask is zero except for the few points around the center pixel. Thus, this representation is much more practical. For example, the filter mask

$$\begin{bmatrix} 0 & -1 & -2 \\ 1 & 0_{\bullet} & -1 \\ 2 & 1 & 0 \end{bmatrix} \tag{10.9}$$

written as an $M \times N$ matrix reads as

$$\begin{bmatrix} 0_{\bullet} & -1 & 0 & \dots & 0 & 1 \\ 1 & 0 & 0 & \dots & 0 & 2 \\ 0 & 0 & 0 & \dots & 0 & 0 \\ \vdots & \vdots & \vdots & \ddots & \vdots & \vdots \\ 0 & 0 & 0 & \dots & 0 & 0 \\ -1 & -2 & 0 & \dots & 0 & 0 \end{bmatrix}. \tag{10.10}$$

In the following, we will write all filter masks in the comprehensive notation where the filter mask is centered around the point H_{00} as we have introduced discrete convolution at the beginning of this chapter.

10.3.4 Point Spread Function

A convolution operator is completely described by its *point spread function* (PSF) because it is linear and shift-invariant (Section 4.3.5). The PSF is defined as the response of the operator to the discrete delta impulse or point image P which is in one to three

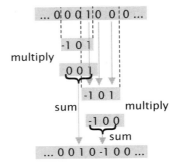

Figure 10.4: *Computation of the PSF of the mask [1 0 -1] by convolution with a delta image. The mask is only shown at the two positions where the accumulation of the product of the coefficients with the values of the image is nonzero. We observe that the inversion of the mask is required for the PSF to be identical to the mask.*

dimensions:

$$p_n = \begin{cases} 1 & m = 0 \\ 0 & \text{otherwise} \end{cases} \quad \text{one dimension,}$$

$$p_{m,n} = \begin{cases} 1 & m = n = 0 \\ 0 & \text{otherwise} \end{cases} \quad \text{two dimensions,} \qquad (10.11)$$

$$p_{l,m,n} = \begin{cases} 1 & l = m = n = 0 \\ 0 & \text{otherwise} \end{cases} \quad \text{three dimensions.}$$

With this definition, the PSF of the convolution operator H is given by

$$\text{PSF} = \sum_{m'=-r}^{r} \sum_{n'=-r}^{r} h_{m'n'} p_{m-m',n-n'} = h_{mn}. \qquad (10.12)$$

For each m and n, this sum contains only one nonzero term at $m - m' = 0$ and $n - n' = 0$. Therefore, the point spread function is identical to the mask of the convolution operator. The following example illustrates this fact with a simple 1-D antisymmetric mask.

Example 10.2: Point spread function of a convolution operator

The PSF of a convolution operator is identical to its mask. To compute the PSF, apply it to a centered impulse image,

[1 0 -1] * [... 0 0 0 1 0 0 0 ...],

mirror the mask, and slide it over the impulse image as depicted in Fig. 10.4. The result is

= [... 0 0 -1 0 1 0 0 ...].

The central importance of the PSF of a convolution operator is directly related to the fact that it is linear and shift-invariant. The *shift-invariance* or homogeneity ensures that the PSF is the same, independent of the position in the image. The linearity tells us that we can decompose any image in a series of shifted point images, apply the convolution operator to each of the point images, and then add up the PSFs multiplied with the gray value at the given position.

The PSF is a very powerful concept commonly used in wide areas of natural sciences. It is useful to describe all kinds of systems that are at least in a good approximation

a

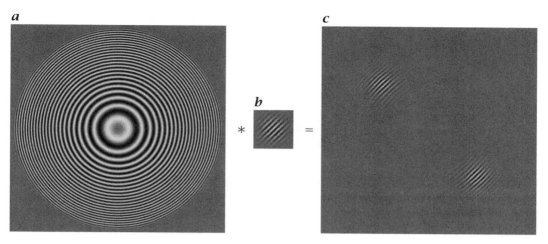

c

b

*   =

Figure 10.5: *Application of a large filter kernel to the ring test pattern. The kernel gives the highest response where the image is most similar to the kernel.* **a** *original test pattern,* **b** *mask,* **c** *filtered pattern.*

linear and homogeneous. In signal processing (time series, electrical signals), it is known as the *impulse response*. In the theory of linear differential equations, it is known as the *Green's function*.

In this way, it is also easy to see that we can judge a convolution operator just by its mask or PSF. The larger the mask is, the more the point will be spread out resulting in coarser resolution of the resulting image. The shape of the mask itself tells us what kinds of features are preserved and which ones are filtered out. A mask, for instance with constant coefficients, will essentially spread out a point to the size of the mask and thus filter out all small scales which are smaller than the size of the mask. As a rule of thumb we can state that an image structure which is most similar to the mask gives the maximum response. This fact is illustrated in Fig. 10.5 by applying a filter with a wavelet-like PSF to the *ring test pattern* (Section 10.4.7a). The reason for this behavior lies in the fact that convolution Eq. (10.8) is very similar to *correlation* which is defined as:

$$(\boldsymbol{H} \star \boldsymbol{G})_{mn} = \sum_{m'=0}^{M-1} \sum_{n'=0}^{N-1} h_{m'n'} g_{m+m',n+n'}. \tag{10.13}$$

The only difference between correlation and convolution is that the mask is not mirrored. For a symmetric filter mask ($h_{-m,-n} = h_{m,n}$) correlation and convolution are identical operations.

10.3.5 Transfer Function

The convolution theorem states that a convolution operation in the spatial domain corresponds to a multiplication in the Fourier domain (Appendix B.4):

$$\boldsymbol{G}' = \boldsymbol{H} * \boldsymbol{G} \circ\!\!\!-\!\!\!\bullet \hat{\boldsymbol{G}}' = \hat{\boldsymbol{H}} \cdot \hat{\boldsymbol{G}}. \tag{10.14}$$

The wavelength-dependent multiplication factor in Fourier space is called the *transfer function* which is the Fourier transform of the convolution mask in the spatial domain. It is easy to understand that convolution in the spatial domain simplifies to a

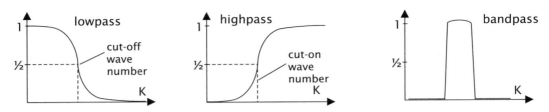

Figure 10.6: *Transfer functions of the three basic classes of 1-D filters: lowpass, highpass, and bandpass.*

multiplication in the Fourier domain. It is obvious that a periodic signal remains a periodic signal under a convolution operation. If we shift the mask by one wavelength, we get the same result. The only two things that can happen to a periodic signal are a change in the amplitude and a shift in the phase. A multiplication by a complex number achieves both changes. Note that the Fourier transform of a real image gives always complex images in the Fourier space (Appendix B.3). Example 10.3 illustrates the *phase shift* and change in the amplitude with a 1-D periodic signal.

Example 10.3: Phase shift and amplitude change of a periodic signal

Let us assume a simple 1-D periodic signal with unit amplitude and the phase φ at the origin (Fig. 10.7). This signal can be described by the real part of a complex exponential $\exp i(kx + \varphi)$ in the Fourier domain. By multiplication of this periodic signal with a complex number

$$a + ib = r \exp(i\Delta\varphi) \quad \text{with} \quad r = \sqrt{a^2 + b^2} \quad \text{and} \quad \varphi = \arctan\frac{b}{a} \qquad (10.15)$$

we get

$$r \exp(i\Delta\varphi) \exp[i(kx + \varphi)] = r \exp[i(kx + \varphi + \Delta\varphi). \qquad (10.16)$$

As a result, the phase of the periodic signal is shifted by $\Delta\varphi$ and the amplitude is multiplied by r.

From these elementary properties of the transfer function, we can immediately conclude that a real transfer function ($b = 0, \Delta\varphi = 0$) results in no phase shifts. In contrast, a pure imaginary transfer function ($a = 0, \Delta\varphi = \pi/2$) causes a 90° phase shift interchanging extremes and zero crossings.

The transfer function is an intuitive description of the convolution operation. It shows by which factor the different scales in an image are multiplied. Figure 10.6 illustrates three classical classes of transfer functions. A *lowpass* is a filter that does not change the coarse structures in an image. It has a value of one at the wave number zero and small wave numbers and then monotonically decreases towards zero for larger wave numbers. A *highpass* filter has the opposite behavior. It is zero for wave number zero meaning that the mean gray value is removed from the image and is then monotonically increasing towards larger wave numbers, leaving predominantly the smallest scales (largest wave number) in the image. A mixed behavior is given by a *bandpass* filter which leaves only a certain range of wave numbers in the image and removes both structures with wave numbers smaller and larger than specified by the bandpass range. This classification scheme originates from signal theory in electrical engineering. We will see that it is less practicable for image processing. Further details are discussed in Section 10.4.1.

Space domain   **Fourier domain**

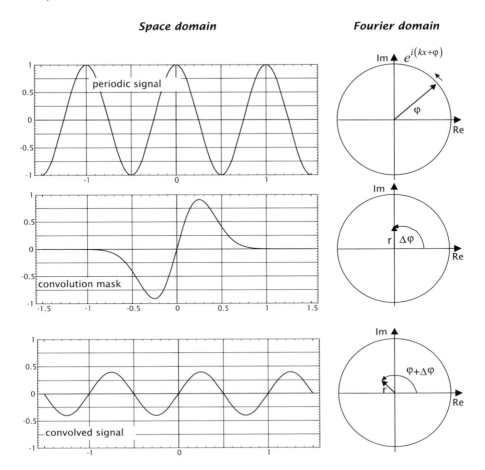

Figure 10.7: *Continuous convolution of a periodic signal with a convolution mask results in a periodic signal of the same wavelength but with a phase shift and amplitude change. Phase shift $\Delta\varphi$ and amplitude attenuation r can be described by multiplication with the complex number $r\exp(\mathrm{i}\Delta\varphi)$. Therefore, convolution reduces to a multiplication with a wavelength-dependent complex number — the transfer function.*

10.3.6 General Properties of Convolution Operators

In this section we discuss the general properties of convolution operators. The theoretical foundations laid down in this section are essential for practical application of this type of operators throughout this handbook.

10.3.6a Linearity. *Linear operators* are defined by the *principle of superposition*. If G and G' are two $M \times N$ images, a and b two complex-valued scalars, and $\mathcal{H}$ is an operator which maps an image onto another image of the same dimension, then the operator is linear if and only if

$$\mathcal{H}(a\boldsymbol{G} + b\boldsymbol{G}') = a\mathcal{H}\boldsymbol{G} + b\mathcal{H}\boldsymbol{G}'. \tag{10.17}$$

We can generalize Eq. (10.17) to the superposition of many input images $\boldsymbol{G}_K$

$$\mathcal{H}\left(\sum_k a_k\boldsymbol{G}_k\right) = \sum_k a_k\mathcal{H}\boldsymbol{G}_k. \tag{10.18}$$

The superposition principle makes linear operators very useful. We can decompose a complex image into simpler components for which we can easily derive the response of the operator and then compose the resulting response from that of the components.

It is especially useful to decompose an image into its individual elements. Formally, this means that we compose the image with the basic images of the chosen representation, which is a series of shifted point images ${}^{mn}P$

$$
{}^{mn}P_{m'n'} = \begin{cases} 1 & m = m', n = n', \\ 0 & \text{otherwise.} \end{cases} \tag{10.19}
$$

Thus we can write any image G as:

$$
G = \sum_{m=0}^{M-1}\sum_{n=0}^{N-1} g_{mn}\,{}^{mn}P. \tag{10.20}
$$

The following example illustrates that convolution is linear, while the nonlinear median filter violates the superimposition principle.

Example 10.4: Demonstration of a linear and a nonlinear neighborhood operator

Linear convolution: same results are obtained when first convolution is performed and results are added, or first components are added and then convolution is performed.

$$
[1\,2\,1] * [\cdots 0\,0\,1\,0\,0\,0\cdots] + [1\,2\,1] * [\cdots 0\,0\,0\,1\,0\,0\cdots]
$$
$$
= [\cdots 0\,1\,2\,1\,0\,0\cdots] + [\cdots 0\,0\,1\,2\,1\,0\cdots]
$$
$$
= [\cdots 0\,1\,3\,3\,1\,0\cdots]
$$

$$
[1\,2\,1] * ([\cdots 0\,0\,1\,0\,0\,0\cdots] + [\cdots 0\,0\,0\,1\,0\,0\cdots])
$$
$$
= [\cdots 0\,1\,3\,3\,1\,0\cdots]
$$

Nonlinear 3-median filter (sort 3 values and select medium (second) value in sorted list).

$$
{}^{3}M[\cdots 0\,0\,1\,0\,0\,0\cdots] + {}^{3}M[\cdots 0\,0\,0\,1\,0\,0\cdots]
$$
$$
= [\cdots 0\,0\,0\,0\,0\,0\cdots] + [\cdots 0\,0\,0\,0\,0\,0\cdots]
$$
$$
= [\cdots 0\,0\,0\,0\,0\,0\cdots]
$$

$$
{}^{3}M([\cdots 0\,0\,1\,0\,0\,0\cdots] + [\cdots 0\,0\,0\,1\,0\,0\cdots])
$$
$$
= {}^{3}M[\cdots 0\,0\,1\,1\,0\,0\cdots]
$$
$$
= [\cdots 0\,0\,1\,1\,0\,0\cdots]
$$

Results are different when first median filtering is performed and then results are added or when first components are added and then median filtering is performed.

10.3.6b Shift Invariance. Another important property of an operator is *shift invariance* or *homogeneity*. It means that the response of the operator does not explicitly depend on the position in the image. If we shift an image, the output image is the same but for the shift applied. We can formulate this property more elegantly if we define a *shift operator* ${}^{m'n'}\!\circlearrowleft$ which is defined as

$$
{}^{m'n'}\!\circlearrowleft g_{mn} = g_{m-m',n-n'}. \tag{10.21}
$$

The superscripts $m'n'$ denote the shifts in y and x directions, respectively. If the shift is not specifically given, the shift operator will be used without superscripts. The shift operator is cyclic. If a part of an image is shifted out at one side, it enters again at the opposite side. Mathematically, this means that the indices are used modulo the size of the corresponding coordinate (see Appendix B.4).

Then, we can define a *shift-invariant* operator in the following way: an operator is shift invariant if and only if it commutes with the shift operator, i. e.,

$$[\mathcal{H}, \circlearrowright] = 0 \quad \text{or} \quad \mathcal{H}\,(\circlearrowright G) = \circlearrowright (\mathcal{H}G), \tag{10.22}$$

where $[\mathcal{H}, \circlearrowright] = \mathcal{H} \circlearrowright - \circlearrowright \mathcal{H}$ is the standard notation of a commutator used in quantum mechanics. The shift operator $\circlearrowright$ itself is a *linear shift-invariant* or, in short, *LSI operator*.

10.3.6c Commutativity.

Commutativity means that the order of cascaded convolution operators can be changed:

$$\mathcal{H}\mathcal{H}' = \mathcal{H}'\mathcal{H}. \tag{10.23}$$

This property is easy to prove in the Fourier domain since there the operators reduce to scalar multiplication which is commutative.

10.3.6d Associativity.

$$\mathcal{H}(\mathcal{H}'G) = (\mathcal{H}\mathcal{H}')G = \mathcal{H}''G \quad \text{with} \quad \mathcal{H}'' = \mathcal{H}\mathcal{H}'. \tag{10.24}$$

Since LSI operations are associative, we can compose a complex operator out of simple operators. Likewise, we can try to decompose a given complex operator into simpler operators. This feature is essential for an effective implementation of convolution operators.

10.3.6e Separability.

A p-dimensional neighborhood operator that can be decomposed into one-dimensional operators is called *separable*. We will denote one-dimensional operators with an index indicating the axis. We are then able to write a separable operator $\mathcal{H}$ in a three-dimensional space as

$$\mathcal{H} = \mathcal{H}_x \mathcal{H}_y \mathcal{H}_z. \tag{10.25}$$

A separable convolution mask is the outer product of the 1-D masks it is composed of (Example 10.5a). A separable p-dimensional mask is easy to recognize. All its rows, columns, planes, etc. are multiples of each other, as can be seen from the masks in Example 10.5a. The higher the dimension of the space the more efficient separable masks are as demonstrated by Example 10.5b.

Example 10.5: Usage of the separability for an efficient computation of convolution

a 5×5 binomial smoothing mask (Section 11.4.2):

$$\begin{bmatrix} 1 & 4 & 6 & 4 & 1 \\ 4 & 16 & 24 & 16 & 4 \\ 6 & 24 & 36 & 24 & 6 \\ 4 & 16 & 24 & 16 & 4 \\ 1 & 4 & 6 & 4 & 1 \end{bmatrix}. \tag{10.26}$$

Direct computation (without further tricks) requires 25 multiplications and 24 additions per pixel. Decomposing this mask into two separable masks

$$\begin{bmatrix} 1 & 4 & 6 & 4 & 1 \\ 4 & 16 & 24 & 16 & 4 \\ 6 & 24 & 36 & 24 & 6 \\ 4 & 16 & 24 & 16 & 4 \\ 1 & 4 & 6 & 4 & 1 \end{bmatrix} = [1\ 4\ 6\ 4\ 1] * \begin{bmatrix} 1 \\ 4 \\ 6 \\ 4 \\ 1 \end{bmatrix} \tag{10.27}$$

reduces the computations to only 10 multiplications and 8 additions.

b Separable three-dimensional $9 \times 9 \times 9$ mask: A direct implementation would cost 729 multiplications and 728 additions per pixel, while a separable mask of the same size only requires 27 multiplications and 24 additions, i. e., about a factor of 30 fewer operations.

10.3.6f Distributivity over Addition.

Since LSI operators are elements of the same vector space on which they can operate, we can define addition of the operators by the addition of the vector elements. We then find that LSI operators distribute over addition

$$\mathcal{H}'G + \mathcal{H}''G = (\mathcal{H}' + \mathcal{H}'')G = \mathcal{H}G. \tag{10.28}$$

Because of this property we can also integrate operator additions into the general operator notation and write $\mathcal{H} = \mathcal{H}' + \mathcal{H}''$.

10.3.6g Eigenfunctions.

Eigenanalysis is one of the most important and widely used mathematical tools throughout natural sciences. Eigenanalysis reveals the fundamental properties of a system. With respect to neighborhood operations, we ask whether special types of image E exist which do not change except for multiplication by a scalar when a convolution operator is applied, i. e.,

$$\mathcal{H}E = \lambda E. \tag{10.29}$$

An image (vector) which meets this condition is called an *eigenimage*, *eigenvector* or *characteristic vector* or, more generally, an *eigenfunction* of the operator, the scaling factor λ the *eigenvalue* or *characteristic value* of the operator.

All linear shift-invariant operators have the same set of eigenimages. We can prove this statement by referring to the *convolution theorem* (see Appendix B.4) which states that convolution is a point-wise multiplication in the Fourier space:

$$G' = H * G \quad \circ\!\!-\!\!\bullet \quad \hat{G}' = \hat{H} \cdot \hat{G}. \tag{10.30}$$

The element-wise multiplication of the two matrices H and G in the Fourier space is denoted by a centered dot to distinguish this operation from matrix multiplication which is denoted without any special sign. Equation Eq. (10.30) tells us that each element of the image representation in the Fourier space $\hat{G}_{uv}$ is multiplied by the complex scalar $\hat{H}_{uv}$. Each point $\hat{G}_{uv}$ in the Fourier space represents a base image, namely the complex exponential ^{uv}W:

$$^{uv}W_{mn} = \exp\left(-\frac{2\pi imu}{M}\right) \exp\left(-\frac{2\pi inv}{N}\right) \tag{10.31}$$

multiplied with the complex-valued scalar $\hat{G}_{uv}$. Since $\hat{G}_{uv}$ is simply multiplied by $\hat{H}_{uv}$, ^{uv}W is an eigenfunction of $\mathcal{H}$. The eigenvalue to the eigenimage ^{uv}W is the corresponding element of the transfer function, $\hat{H}_{uv}$. In conclusion, we can write for the eigenfunctions ^{uv}W of any convolution operator $\mathcal{H}$:

$$\mathcal{H}(^{uv}W) = \hat{H}_{uv}{}^{uv}W. \tag{10.32}$$

10.3.6h Inverse Operators. An inverse convolution operator reverses the action of an operator in such a way that the original image is retained. If the *inverse operator* to $\mathcal{H}$ is denoted by $\mathcal{H}^{-1}$, we can write the condition to be met by the inverse operator in operator notation as

$$\mathcal{H}^{-1}\mathcal{H} = 1, \tag{10.33}$$

where 1 is the *identity* operator. It is obvious that an inverse operator, also denoted as *deconvolution*, does exist only under quite limited conditions. This can be seen immediately by writing Eq. (10.33) in the Fourier domain where convolution reduces to multiplication:

$$\hat{H}^{-1}(\boldsymbol{k})\,\hat{H}(\boldsymbol{k}) = 1 \quad \rightsquigarrow \quad \hat{H}^{-1}(\boldsymbol{k}) = \frac{1}{\hat{H}(\boldsymbol{k})}. \tag{10.34}$$

From this equation we can immediately conclude that an inverse operator only does exist when the transfer function of the operator is unequal to zero for all wave numbers. This is almost never the case. The transfer function of a lowpass filter, for instance, is zero for high wave numbers. Therefore, restoration of images that are degraded by a distortion that can be described by a lowpass filter (such as blurring by defocusing, lens aberrations, or velocity smearing) is not at all a trivial task (Section 9.3.4).

10.3.7 Error Propagation with Filtering

Applying a convolution to an image means the computation of a linear combination of neighboring pixel values. Because of this basic fact, it is easy to compute the influence of a convolution operation on the statistical properties of the image data. In Section 7.3.3c we discussed the important general relation that the *covariance matrix* of the linear combination $\boldsymbol{g}' = \boldsymbol{M}\boldsymbol{g}$ of a random vector $\boldsymbol{g}$ is, according to Eq. (7.34), given by

$$\mathrm{cov}(\boldsymbol{g}') = \boldsymbol{M}\,\mathrm{cov}(\boldsymbol{g})\boldsymbol{M}^T. \tag{10.35}$$

In the special case of a convolution this relation becomes even simpler. First, we consider only 1-D signals. We further assume that the covariance matrix of the signal is homogeneous, i. e., depends only on the distance of the points and not the position itself. Then the variance σ^2 for all elements is equal. Furthermore, the values on the side diagonals are also equal and the covariance matrix takes the simple form

$$\mathrm{cov}(\boldsymbol{g}) = \begin{bmatrix} C_0 & C_1 & C_2 & \cdots & \cdots \\ C_{-1} & C_0 & C_1 & C_2 & \cdots \\ C_{-2} & C_{-1} & C_0 & C_1 & \ddots \\ \vdots & & C_{-2} & C_{-1} & C_0 & \ddots \\ \vdots & \vdots & & \ddots & \ddots & \ddots \end{bmatrix}, \tag{10.36}$$

where the index indicates the distance between the points and $C_0 = \sigma^2$. Generally, the covariances decrease with increasing pixel distance. Often, only a limited number of covariances C_n are unequal to zero. With statistically uncorrelated pixels, the covariance matrix reduces to an identy matrix $\mathrm{cov}(\boldsymbol{g}) = \sigma^2\boldsymbol{I}$.

Because the linear combinations described by $\boldsymbol{M}$ have the special form of a convolution, the matrix has the same form as the homogeneous covariance matrix. For a filter with three coefficients M reduces to

$$
M = \begin{bmatrix}
h_0 & h_{-1} & 0 & 0 & 0 \\
h_1 & h_0 & h_{-1} & 0 & 0 \\
0 & h_1 & h_0 & h_{-1} & 0 \\
0 & 0 & h_1 & h_0 & \ddots \\
0 & 0 & 0 & \ddots & \ddots
\end{bmatrix}.
\tag{10.37}
$$

Thus the matrix multiplications in Eq. (10.35) reduce to convolution operations, apart from edge effects. Introducing the *autocovariance vector*

$$
c = [\ldots, C_{-1}, C_0, C_1, \ldots]^T,
\tag{10.38}
$$

Eq. (10.35) reduces to

$$
c' = {}^-h * c * h,
\tag{10.39}
$$

where ${}^-h$ is the mirrored convolution mask: ${}^-h_n = h_{-n}$. Convolution is a commutative operation (Section 10.3.6c), ${}^-h * h$ can be replaced by a *correlation* operation and the convolution of c with $h \star h$ can also be replaced by a correlation, because the autocorrelation function of a real-valued function is a function of even symmetry. Then Eq. (10.39) becomes

$$
c' = c \star (h \star h).
\tag{10.40}
$$

In the case of uncorrelated data, the autocovariance vector is a *delta function* and the autocovariance vector of the noise of the filtered vector reduces to

$$
c' = \sigma^2 (h \star h).
\tag{10.41}
$$

For a filter with R coefficients, now $2R - 1$ values of the autocovariance vector are unequal to zero. This means that in the filtered signal pixels with a maximal distance of $R - 1$ are now correlated with each other. The variance of the filtered data is given by

$$
\sigma'^2 = C_0 = \sigma^2 (h \star h)_0 = \sigma^2 \sum_{n'=-R}^{R} h_{n'}^2.
\tag{10.42}
$$

Further interesting insight into the statistical properties of the convolution operation is given in the Fourier domain. First we observe that the variance of zero-mean image can also be computed in the Fourier domain (Appendix B.4):

$$
\sigma^2 = \frac{1}{MN} \sum_{m'=0}^{M-1} \sum_{n'=0}^{N-1} |g_{m'n'}|^2 \quad \circ\!\!-\!\!\bullet \quad \sigma^2 = \frac{1}{MN} \sum_{u'=0}^{M-1} \sum_{v'=0}^{N-1} |\hat{g}_{u'v'}|^2 = \int_0^1 \int_0^1 |\hat{h}(\tilde{k})|^2 \, d^2\tilde{k}.
\tag{10.43}
$$

More generally, the *correlation theorem* (Section B.4) states that the *noise spectrum* $|\hat{G}|^2$, i.e., the *power spectrum* of the noise, is the Fourier transform of the *autocovariance* of the signal:

$$
C_{mn} = \frac{1}{MN} \sum_{m'=0}^{M-1} \sum_{n'=0}^{N-1} g_{m'n'} g_{m+m',n+n'} \quad \circ\!\!-\!\!\bullet \quad \hat{c}(k) = |\hat{G}|^2.
\tag{10.44}
$$

Then it is obvious that the autocovariance of a convolved image, C', is given by

$$
C' = C \star (H \star H) \quad \circ\!\!-\!\!\bullet \quad \hat{c}'(k) = \hat{c}(k) \left| \hat{h}(k) \right|^2.
\tag{10.45}
$$

Starting with an image of uncorrelated pixels and homogenous noise with the variance σ^2 and applying a cascade of convolution operators $\boldsymbol{H}_p$, we end up with a noise autocovariance of

$$C = \sigma^2 (\boldsymbol{H}_1 \star \boldsymbol{H}_1) \star (\boldsymbol{H}_2 \star \boldsymbol{H}_2) \ldots \quad \circ\!\!-\!\!\bullet \quad \hat{c}(\boldsymbol{k}) = \sigma^2 \prod_{p=1}^{P} \left| \hat{h}_p(\boldsymbol{k}) \right|^2. \tag{10.46}$$

The variance of the convolved image can be computed in the Fourier domain by

$$\sigma'^2 = \frac{1}{MN} \sum_{u'=0}^{M-1} \sum_{v'=0}^{N-1} |\hat{h}_{u'v'}\hat{g}_{u'v'}|^2 = \int_0^1\!\!\int_0^1 \left| \hat{h}(\tilde{\boldsymbol{k}})\hat{g}(\tilde{\boldsymbol{k}}) \right|^2 \mathrm{d}^2\tilde{k}. \tag{10.47}$$

In the special case of uncorrelated data the noise spectrum is flat (*white noise*) because the autocovariance is a delta function. Therefore the computation of the variance reduces in this case to

$$\sigma'^2 = \sigma^2 \frac{1}{MN} \sum_{u'=0}^{M-1} \sum_{v'=0}^{N-1} |\hat{h}_{u'v'}|^2 = \sigma^2 \int_0^1\!\!\int_0^1 |\hat{h}(\tilde{\boldsymbol{k}})|^2 \mathrm{d}^2\tilde{k}. \tag{10.48}$$

10.3.8 Recursive Convolution

10.3.8a Definition. *Recursive filters* are a special subclass of LSI filters. In contrast to standard convolution filters Eq. (10.8), recursive filters use previously computed output values as inputs for the multiplication with the filter kernel coefficients. For time series (1-D signals), a recursive filter is given by

$$g'_n = h_o g_n + \sum_{n'=1}^{r} h'_n g'_{n-n'}. \tag{10.49}$$

This filter uses the current — yet not filtered — element and r previously filtered elements. By principle, recursive filters work in a certain direction. For time series, this is a quite natural procedure since the current output can only depend on previous outputs. Such a filter proceeds in the time direction and is called a *causal filter*. Recursive filters are the standard approach for the analysis of discrete time series. The theory for 1-D signals is well understood and is handled in all classical text books on linear system theory and discrete time series analysis [108, 115]. The following example shows a simple recursive filter.

Example 10.6: A simple recursive filter

The simplest one-dimensional recursive filter has the general form

$$g'_n = \alpha g'_{n-1} + (1 - \alpha) g_n. \tag{10.50}$$

This filter takes the fraction α from the previously calculated value and the fraction $1 - \alpha$ from the current element. From Eq. (10.50), we can calculate the response of the filter to the *discrete delta function*

$$\delta_k = \begin{cases} 1 & k = 0, \\ 0 & k \neq 0, \end{cases} \tag{10.51}$$

i.e., the PSF or *impulse response* of the filter. Recursively applying Eq. (10.50) to the discrete delta function, we obtain

$$\begin{aligned} g'_{-1} &= 0 \\ g'_0 &= 1 - \alpha \\ g'_1 &= (1 - \alpha)\alpha \\ g'_n &= (1 - \alpha)\alpha^n. \end{aligned} \tag{10.52}$$

Thus, the point spread function is

$$R_n = (1 - \alpha)\alpha^n. \tag{10.53}$$

The transfer function of this infinite impulse response is complex-valued and given by

$$\hat{R}(\tilde{k}) = \frac{1 - \alpha}{1 - \alpha \exp(-i\pi\tilde{k})} \quad |\alpha| < 1. \tag{10.54}$$

The magnitude and phase of the transfer function are (Fig. 10.9)

$$|\hat{R}(k)| = \frac{1 - \alpha}{\sqrt{1 + \alpha^2 - 2\alpha \cos \pi\tilde{k}}} \quad \text{and} \quad \varphi(k) = -\arctan \frac{\alpha \sin \pi\tilde{k}}{1 - \alpha \cos \pi\tilde{k}}.$$

Thus, the recursive filter is a lowpass. It has the disadvantage of a wave number-dependent phase shift and that at the highest wave number the response does not reach zero. The advantage is that the constant α can vary between 0 and 1 to tune the cut-off wave number over a very wide range.

Example 10.6 illustrates some general properties of recursive filters:

- The impulse response can be infinite despite a finite number of coefficients. In contrast, the impulse response of standard convolution is always finite and identical to the filter coefficients (Section 10.3.4). Therefore, filters are sometimes classified into *finite impulse response filters* (*FIR* filters) and *infinite impulse response filters* (*IIR* filters).

- While nonrecursive convolution filters are always stable, recursive filters may be instable. The stability of recursive filters depends on the coefficient α. The filter in the above example is instable for $|\alpha| \geq 1$ as the response to the impulse reponse diverges (see Eq. (10.53) and Fig. 10.8).

- Recursive filters are computationally more efficient than nonrecursive filters. Only a few coefficients are required to obtain an impulse response with large effective widths. The filters in the above example have only two coefficients and yet can be tuned to smooth over wide ranges by setting α to appropriate values (Fig. 10.9a).

- Causal recursive filters show negative phase shifts (delay) that depend nonlinearly on the wave number (Fig. 10.9b).

Application of recursive filters for image processing poses a number of principal problems:

- While for time series a natural run direction exists, this is not the case for images. A recursive filter could run in any spatial direction.

- Image processing requires rather zero-phase (noncausal) filters than phase shifting causal filters (Fig. 10.8) so that the position of edges and features is not shifted during processing.

- While the theory of 1-D recursive filters is well understood, no closed theory for higher dimensional recursive filters is available. This causes fundamental problems for the design of recursive filters and the analysis of the stability of filters.

These three problems force us to apply different approaches for the design of recursive filters for image processing than for time series analysis. Most importantly, strategies must be found that yield filters with symmetrical or antisymmetrical point spread functions to avoid the phase-shift problems (Section 10.4.3).

10.3.8b Recursive Filters and Linear Systems. Recursive filters can be regarded as the discrete counterpart of *analog filters*. Analog filters for electrical signals contain resistors, capacitors, and inductors. Electrical filters actually are only an example

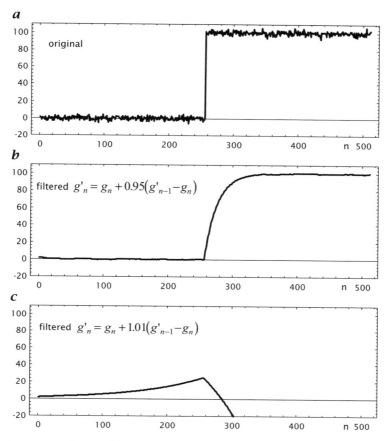

Figure 10.8: *Response of* **b** *a stable (*$\alpha = 0.95$*) and* **c** *instable (*$\alpha = 1.01$*) two-tap recursive filter* $g'_n = g_n + \alpha(g'_{n-1} - g_n)$ *on the 1-D box signal with small random fluctuations shown in* **a** *. The filter is applied from the left to the right.*

of a wide range of systems known as linear systems that can be described by linear differential equations.

Here we discuss the two most important types of linear systems: the relaxation process and the damped harmonic oscillator. Both systems are widely known and applied throughout engineering and natural sciences and thus help us to understand the principal features of recursive filters.

10.3.8c Resistor-Capacitor Circuit; Relaxation Process. We start with the simple resistor-capacitor circuit shown in Fig. 10.10b. The differential equation for this filter can easily be derived from Kirchhoff's current-sum law. The current flowing through the resistor from U_i to U_o must be equal to the current flowing into the capacitor. Since the current flowing into a capacitor is proportional to the temporal derivative of the potential dU_o/dt, we end up with the first order differential equation

$$\frac{U_i - U_o}{R} = C\frac{dU_o}{dt}. \tag{10.55}$$

This equation represents a very important general type of process called a *relaxation process* which is governed by a time constant τ. In our case, the time constant is given

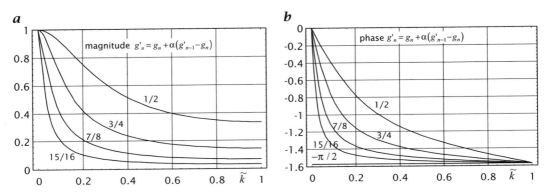

Figure 10.9: a *Magnitude and* **b** *phase of the transfer function of the one-tap recursive lowpass filter* $g'_n = g_n + \alpha(g'_{n-1} - g_n)$ *for values of* α *as indicated.*

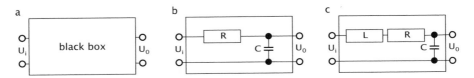

Figure 10.10: *Analog filter for time series:* **a** *black-box model: a signal* U_i *is put into an unknown system; at the output we measure the signal* U_o; **b** *a resistor-capacitor circuit as a simple example for an analog lowpass filter;* **c** *a resistor-capacitor-inductivity circuit as an example for a damped oscillator.*

by $\tau = RC$. Generally, we can write the differential equation of a relaxation process as

$$\tau \frac{dU_o}{dt} + U_o = U_i. \tag{10.56}$$

The impulse response $h(t)$ of this system (in case of an RC-circuit the reaction to a short voltage impulse) and the transfer function are given by

$$h(t) = \begin{cases} 0 & t < 0 \\ (1/\tau)\exp(-t/\tau) & t \geq 0 \end{cases} \qquad \circ\!\!-\!\!\bullet \qquad \frac{1}{1 + i\tau\omega}. \tag{10.57}$$

In case of a continuous function, the impulse response is also known as *Green's function*. Once we know the impulse response of the filter, we can calculate the response to any arbitrary signal by convolution

$$U_o(t) = \int_0^\infty U_i(t - t')h(t')dt', \tag{10.58}$$

since Eq. (10.56) is linear in U. Because the impulse response is zero for $t < 0$ (*causal filter*), the integration limits extend from 0 to ∞ only. A discrete approximation of the analog RC filter can be derived by transforming the differential equation Eq. (10.56) into a finite difference equation, for example

$$U_o(t) = \frac{\tau}{\tau + \Delta t}U_o(t - \Delta t) + \frac{\Delta t}{\tau + \Delta t}U_i(t). \tag{10.59}$$

This equation is equivalent to the simple recursive filter Eq. (10.50). Since the difference equation is only an approximation of the differential equation, discrete and continuous filters are called equivalent when the sampled continuous impulse response is equal to the discrete impulse response. In this way, we can derive the relationship between the constant α and the time constant τ. It is sufficient to sample in time ($t = n\Delta t$) and to compare the exponential terms of Eqs. (10.53) and (10.57). Then from

$$\exp(-n\Delta t/\tau) = \alpha^n = \exp(n \ln \alpha), \tag{10.60}$$

we derive

$$\tau = -\frac{\Delta t}{\ln \alpha} \quad \text{or} \quad \alpha = \exp(-\Delta t/\tau). \tag{10.61}$$

With these equations, we obtain a relationship between a continuous process and its discrete counterpart. Since the discrete samples resemble the analog process exactly, it is not only an approximation. This means that we can exactly simulate analog filters with discrete filters, provided we meet the sampling theorem.

10.3.8d Second-Order Recursive Filter; Damped Harmonic Oscillator. A *second-order* filter relates the current output to the output of the two last samples (two-tap filter):

$$g'_n = 2r\cos(\pi \tilde{k}_0)g'_{n-1} - r^2 g'_{n-2} + g_n \quad 0 < k_0 < 1 \quad |r| < 1. \tag{10.62}$$

In Eq. (10.62), already a special form of the coefficients has been taken in order to simplify further expressions. The impulse response of this filter can be shown according to Oppenheim and Schafer [108] to be

$$h_n = \begin{cases} \dfrac{r^n \sin[\pi \tilde{k}_0 (n + 1)]}{\sin(\pi \tilde{k}_0)} & n \geq 0, \\ 0 & n < 0. \end{cases} \tag{10.63}$$

The transfer function of this asymmetric causal filter is complex-valued:

$$\hat{h}(\tilde{k}) = \frac{1}{1 - r\exp(i\tilde{k}_0)\exp(-\pi i\tilde{k})} \cdot \frac{1}{1 - r\exp(-i\tilde{k}_0)\exp(-\pi i\tilde{k})}, \tag{10.64}$$

with the magnitude

$$|\hat{h}(\tilde{k})|^2 = \frac{1}{r^2 + 1 - 2r\cos[\pi(\tilde{k} - \tilde{k}_0)]} \cdot \frac{1}{r^2 + 1 - 2r\cos[\pi(\tilde{k} + \tilde{k}_0)]}. \tag{10.65}$$

This transfer function is the discrete analogue to the system function of a fundamental physical system, the *damped harmonic oscillator* (Fig. 10.10c). The impulse response describes a sampled damped harmonic oscillation which has been excited at time zero:

$$h(t) = \begin{cases} \exp(-t/\tau)\sin(\omega t) & t \geq 0, \\ 0 & t < 0. \end{cases} \tag{10.66}$$

The transfer function Eq. (10.64) contains the physical meaning of the *resonance curve* for the oscillator (Fig. 10.11). If $r = 1$, the oscillator is undamped, and the transfer function has two *poles* at $\tilde{k} = \pm \tilde{k}_0$. If $r > 1$, the resonator is unstable; even the slightest excitement will cause infinite amplitudes of the oscillation. Only for $r < 1$, the system is stable; the oscillation is damped. Comparing Eqs. (10.63) and (10.66), we can determine the relationship of the eigenfrequency ω and the time constant τ of a real-world oscillator to the parameters of the discrete oscillator, r and $\tilde{k}_0$

$$r = \exp(-\Delta t/\tau) \quad \text{and} \quad \tilde{k}_0 = \omega\Delta t/\pi. \tag{10.67}$$

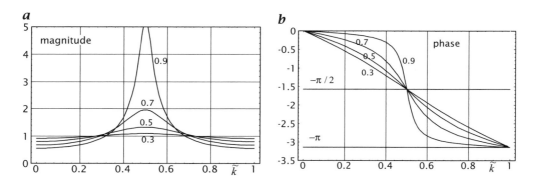

Figure 10.11: *Magnitude and phase shift of the two-tap recursive filter resembling a damped harmonic oscillator:* $g'_n = g_n + 2r\cos(\pi\tilde{k}_0)g'_{n-1} - r^2 g'_{n-2}$. *Shown are curves for* $\tilde{k}_0 = 0.5$ *and values of* r *as indicated.*

With respect to image processing, the second order recursive filter seems to be the basic building block to filter out certain frequencies, respectively wave numbers. The two coefficients r and k_0 allow setting the center wave number and the width of the range that is filtered out of the image (Fig. 10.11).

This simple filter nicely demonstrates the power and the problems of recursive filtering. The filtering is fast (only two multiplications and two additions are required) and can easily be tuned. However, care must be taken that the filter response remains stable. Furthermore, the delay (phase shift) of the filter process makes it impossible to apply these filters directly to image processing tasks.

10.3.8e Linear System Theory and Modeling. The last example of the damped oscillator illustrates that there is a close relationship between discrete filter operations and analog physical systems. Thus, filters may be used to represent a real-world physical process. They model how the corresponding system would respond to a given input signal g. Actually, we have already made use of this equivalence in our discussion of the transfer function of optical imaging in Section 4.3.5. There we found that imaging with a homogeneous optical system is completely described by its point spread function and that the image formation process can be described by convolution.

This generalization is very useful for image processing since we can describe both the image formation and image processing as convolution operations with the same formalism. Moreover, the images observed may originate from a physical process which can be modeled by a linear shift-invariant system. Then, an experiment to find out how the system works can be illustrated using the black-box model (Fig. 10.10a). The black box means that we do not know the composition of the system observed or, physically speaking, the laws which govern it. We can find them out by probing the system with certain signals (input signals) and by watching the response (output signals). If it turns out that the system is linear, it will completely be described by the impulse response. Many biological and medical experiments are performed in this way. Researchers often stimulate them with signals and watch for responses in order to be at least able to make a model. From this model, more detailed research may start to investigate how the observed system functions might be realized. In this way, many properties of biological visual systems have been discovered. But be careful — a model is not the reality! It pictures only the aspect that we probed with the applied signals (see also Section 2.2.5).

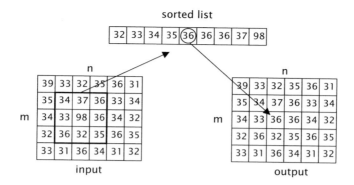

Figure 10.12: *Illustration of the principle of rank value filters with a 3×3 median filter.*

10.3.9 Rank-Value Filters

Rank-value filters are another class of neighborhood operations. They are based on a quite different concept than linear-shift invariant operators. These operators consider all pixels in the neighborhood. It is implicitly assumed that each pixel, distorted or noisy, carries still useful and correct information. Thus, convolution operators are not equipped to handle situations where the value at a pixel carries incorrect information. This situation arises, for instance, when an individual sensor element in a CCD array is defective or a transmission error occurred.

To handle such cases, operations are required that apply selection mechanisms and do not use all pixels in the neighborhood to compute the output of the operator. The simplest of such classes are rank-value filters. They may be characterized by "comparing and selecting".

For this we take all the gray values of the pixels which lie within the filter mask and sort them by ascending gray value. This sorting is common to all rank-value filters. They only differ by the position in the list from which the gray value is picked out and written back to the center pixel. The filter operation which selects the medium value is called the *median filter*. Figure 10.12 illustrates how the median filter works. The filters choosing the minimum and maximum values are denoted as the minimum and maximum filter, respectively.

There are a number of significant differences between linear convolution filters and rank-value filters. First of all, rank-value filters are *nonlinear* filters. Consequently, it is much more difficult to understand their general properties. Since rank-value filters do not perform arithmetic operations but select pixels, we will never run into rounding problems. These filters map a discrete set of gray values onto itself.

10.3.10 Strategies for Adaptive Filtering

10.3.10a Limitations of Linear Filters. Linear filters cannot distinguish between a useful feature and noise. This property can be best demonstrated in the Fourier space (Fig. 10.13). We add white noise to the image. Because of the linearity of the Fourier transform, the Fourier transform of the white noise adds directly to the Fourier transform of the image elevating the Fourier transform of the image. At each wave number, a certain part of the amplitude originates now from the noise, the rest from the image.

Any linear filter operator now works in such a way that the Fourier transform of the image is multiplied with the Fourier transform of the filter. The result is that at each

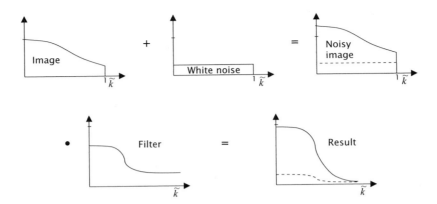

Figure 10.13: *A linear filter does not distinguish between useful features and noise in an image. It reduces the feature and the noise amplitudes equally at each wave number so that the signal-to-noise ratio remains the same.*

wave number the noise level and the image features are attenuated by the same factor. Thus, nothing has improved; the signal-to-noise ratio did not increase but stayed just the same.

From the above considerations, it is obvious that more complex approaches are required than linear filtering. Common to all these approaches is that in one or the other way the filters are made dependent on the context. Therefore a kind of control strategy is an important part of adaptive filtering that tells us which filter or in which way a filter has to be applied at a certain point in the image. Here, we will only outline the general strategies for adaptive filtering. Further details as to how to apply such types of filters for a specific task can be found in the corresponding task-oriented chapters.

10.3.10b Problem-Specific Nonlinear Filters. This approach is the oldest one and works if a certain specific type of distortion is present in an image. A well-known type of this approach is the median filter (Section 10.3.9) which can excellently remove single-distorted pixels with minimum changes to the image features. It is important to stress that this approach, of course, only works for the type of distortion it is designed for. A median filter is excellent for removing a single pixel that has a completely incorrect gray value because of a transmission or data error. It is less well suited, e. g., for white noise.

10.3.10c Pixels With Certainty Measures. So far, we have treated each pixel equally assuming that the information it is carrying is of equal significance. While this seems to be a reasonable first approximation, it is certain that it cannot be generally true. Already during image acquisition, the sensor area may contain bad sensor elements that lead to erroneous gray values at certain positions in the image. Likewise, transmission errors may occur so that individual pixels may carry wrong information. In one way or the other we may attach to each sensor element a certainty measurement.

Once a certainty measurement has been attached to a pixel, it is obvious that the normal convolution operators are no longer a good choice. Instead, the weight we attach to the pixel has to be considered when performing any kind of operation with it. A pixel with suspicious information should only get a low weighting factor in the convolution sum. This kind of approach leads us to what is called *normalized convolution*.

Actually, this type of approach seems to be very natural for a scientist or engineer. He is used to qualifying any data by a measurement error which is then used in any further evaluation of the data. Normalized convolution applies this common principle to image processing.

The power of this approach is related to the fact that we have quite different possibilities to define the certainty measurement. It needs not only be related to a direct measurement error of a single pixel. If we are, for example, interested in computing an estimate of the mean gray value in an object, we could devise a kind of certainty measurement which analyzes neighborhoods and attaches low weighting factors where we may suspect an edge so that these pixels do not contribute much to the mean gray value or feature of the object. In a similar way, we could, for instance, also check how likely the gray value of a certain pixel is if we suspect some distortion by transmission errors or defective pixels. If the certainty measurement of a certain pixel is below a critical threshold, we could replace it by a value interpolated from the surrounding pixels.

10.3.10d Adaptive Filtering. *Adaptive filters* in the narrower sense use a different strategy than normalized convolution. Now, the filter operation itself is made dependent on the neighborhood. Adaptive filtering can best be explained by a classical application, the suppression of noise without significant blurring of image features. The basic idea of adaptive filtering is that in certain neighborhoods we could very well apply a smoothing operation. If, for instance, the neighborhood is flat, we can assume that we are within an object with constant features and thus apply an isotropic smoothing operation to this pixel to reduce the noise level. If an edge has been detected in the neighborhood, we could still apply some smoothing, namely along the edge. In this way, still some noise is removed but the edge is not blurred. With this approach, we need a kind of large filter set of directional smoothing operations. Because of the many filters involved, it appears that adaptive filtering might be a very computational-intensive approach. This is, indeed, the case if either the coefficients of the filter to be applied have to be computed for every single pixel or if a large set of filters has to be used. With the discovery of steerable filters, however, adaptive filtering techniques have become attractive and computationally much more efficient. Steerable filters will be discussed in the following chapter in Section 11.4.7.

10.3.10e Nonlinear Filters by Combining Point Operations and Linear Filter Operations. Normalized convolution and adaptive filtering have one strategy in common. Both use combinations of linear filters and nonlinear point operations such as point-wise multiplication of images. The combination of linear filter operations with nonlinear point operations makes, of course, the whole operation nonlinear.

We may thus speculate that the combination of these two kinds of operations is a very powerful instrument for image processing. We will discover, indeed, that this type of operations does not only include normalized convolution and adaptive filtering but that many advanced image processing techniques discussed in this handbook are of that type. This includes operators to compute local orientation and local wave numbers in images and various operations for texture analysis.

Operators containing combinations of linear filter operators and point operators are very attractive as they can be composed of very simple and elementary operations that are very well understood and for which analytic expressions are available. Thus, these operations in contrast to many other ones can be the subject of a detailed mathematical analysis.

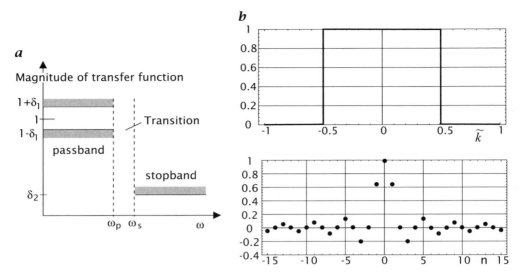

Figure 10.14: *a Design criteria for the analysis of time series demonstrated with a lowpass filter. The magnitude of the transfer function should be one and zero within fixed tolerance levels in the* passband *and* stopband, *respectively. b Transfer function and filter coefficients for an ideal nonrecursive lowpass filter.*

10.4   Procedures

10.4.1   Filter Design Criteria

10.4.1a   Classical Criteria for Filter Design.   Filter design is a well-established area both in continuous-time and discrete-time signal processing. Before we turn to the design techniques themselves, it is helpful to discuss the typical design criteria in order to check whether they are also useful for image processing. Design of filters for time series is centered around the idea of a *passband* and a *stopband* (Fig. 10.14a). In the passband, the transfer function should be one within a tolerance δ_1. In contrast, in the stopband the frequencies should be suppressed as well as possible with a maximum amplitude of δ_2. Finally, the width of the transition region $\omega_s - \omega_p$ determines how fast the filter changes from the passband to the stopband.

These design criteria could, of course, also be applied for higher-dimensional signals. Depending on the application, the shape of the passband could be either circular for isotropic filters or rectangular. Discrete-time signal processing is dominated by causal recursive filters. This type of filters is more efficient than finite impulse response filters, i. e., less filter coefficients are required to meet the same design criteria.

Ideal filters, i. e., filters with a flat passband and stopband and a zero width transition region, can only be gained with a filter with an infinite number of coefficients. This general rule is caused by the reciprocity between the Fourier and space domains. A sharp discontinuity as shown for the ideal lowpass filter in Fig. 10.14b causes the envelope of the filter coefficients to decrease not faster than with $1/n$ from the center coefficient. The narrower the selected wave number range of a filter is, the larger its filter mask is.

10.4.1b   Filter Design Criteria for Image Processing.   Filters used typically for the analysis of time series are not suitable for image processing. Figure 10.15 shows that filtering of a step edge with an ideal lowpass filter leads to undesirable overshooting

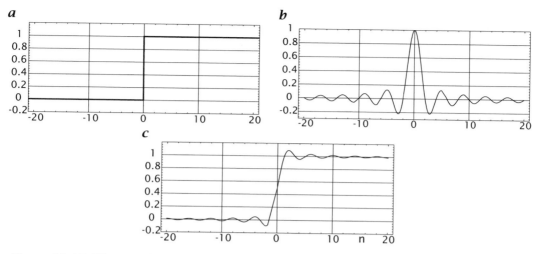

a

b

c

Figure 10.15: *Filtering of a step edge (**a**) with an ideal lowpass filter (**b**) causes overshoots (**c**).*

effects including negative gray values. The causal recursive filter — commonly used to filter time series — causes the edge to shift (Fig. 10.8b). Both effects are not desirable for image processing. The general design criteria for filters to be used for image processing are:

- They should cause no overshooting at step edges.
- The center of gravity or the edges of objects should not be shifted. This condition implies that filters have either an even or odd mask in all directions.
- In order not to bias any directions in images, smoothing filters, edge detectors, and other filters should generally be isotropic.
- Separable filters are preferable because they can be computed much more efficiently than nonseparable filters (see Example 10.5). This requirement is even more important for higher-dimensional images.

10.4.2   Filter Design by Windowing

This approach starts from the observation that the most ideal filters contain discontinuities in their transfer function. This includes lowpass, highpass, bandpass, and bandstop filters (Fig. 10.6). Also, the ideal derivative filter $i\pi\tilde{k}$ has a hidden discontinuity. It appears right at $\tilde{k} = 1$ where the value jumps from π to $-\pi$; the periodic extension of the ideal derivative filter leads to the sawtooth function.

The direct consequence of these discontinuities is that the corresponding non-recursive filter masks, which are the inverse Fourier transforms of the transfer functions, are infinite. The envelope of the filter masks decreases only with the inverse of the distance of the center pixel of the mask. At this point, the idea of windowing enters. We could take the filter coefficients as they are given by the ideal transfer function and limit the number of coefficients in the filter to a finite number by using a window function that is symmetric around the center pixel of the masks. Then, of course, the filter no longer has ideal properties. However, it is very easy to predict the deviations from the ideal behavior. As we multiply the filter mask with the window function, the transfer function — according to the convolution theory — is convolved by the Fourier transform of the window function. Convolution in Fourier space will have no effects

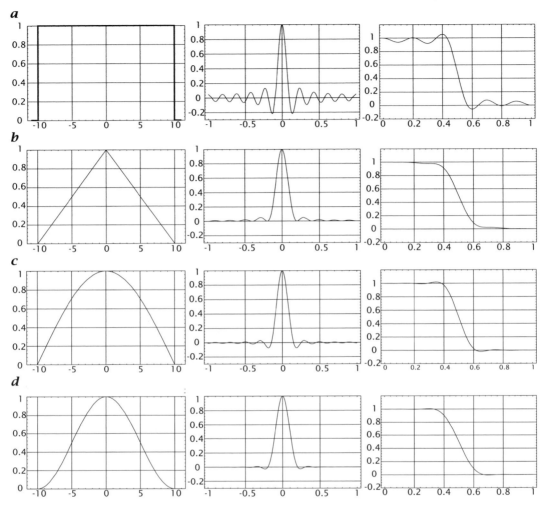

Figure 10.16: *Lowpass filter design by the windowing technique. Shown are the window function (left column), its Fourier transform (middle column), and the convolution of the window Fourier transform with the transfer function of an ideal lowpass with a cutoff wave number $\tilde{k} = 0.5$. The window functions are **a** a rectangle window, **b** a triangle window (Bartlett window), **c** a cosine window (Hanning window), and **d** a cosine squared window.*

where the transfer function is flat or linear-increasing and thus will only change the transfer functions near the discontinuities. Essentially, the discontinuity is blurred by the windowing of the filter mask. From the general properties of the Fourier transform (Appendix B.3), we know that the larger the blurring is the smaller the window is. The exact way in which the discontinuity changes depends on the shape of the window.

Figure 10.16 shows several common window functions, their transfer functions, and the convolution of the transfer functions with a step edge. Figure 10.16a clearly shows that the rectangular window gives the sharpest transition but is not suitable since the transfer functions show considerable oscillations. These oscillations are much less for the cosine (Hanning) and cosine squared window but still not completely removed. The triangle window (Bartlett window) has the advantage that its transfer function is nonnegative, resulting in a monotonous lowpass filter.

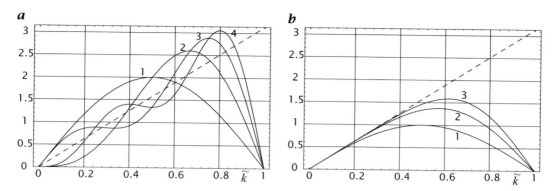

Figure 10.17: *Window technique applied to the design of a 2r + 1 first-order derivative filter. Imaginary part of the transfer function for **a** rectangle window, r = 1, 2, 3, and 4; **b** binomial window, r = 1, 2, and 3; the dashed line is the ideal transfer function, $i\pi\tilde{k}$.*

A window function which is not much mentioned in the design techniques for filters for time series is the Gaussian window. The reason is that the Gaussian window is in principle infinite. However, it decays very rapidly so that it is not a practical problem. Its big advantage is that the shape of the window and its transfer functions are identical. The Fourier transform of the Gaussian window is again a Gaussian window. Moreover, since the Gaussian window is strictly positive, there are no oscillations at all in the transition region. For discrete filters, the Gaussian function is well approximated by the binomial distribution.

Example 10.7: Application of the windowing technique for the design of a derivative filter

The transfer function of an ideal derivative filter is proportional to the wave number:

$$\hat{D} = i\pi\tilde{k}, \quad |\tilde{k}| \le 1. \tag{10.68}$$

An expansion of this transfer function into a Fourier series results in

$$\hat{D}(\tilde{k}) = 2i\left(\sin(\pi\tilde{k}) - \frac{\sin(2\pi\tilde{k})}{2} + \frac{\sin(3\pi\tilde{k})}{3} - \ldots\right) \tag{10.69}$$

corresponding to the following odd filter mask:

$$\ldots -2/5 \quad 1/2 \quad -2/3 \quad 1 \quad -2 \quad 0 \quad 2 \quad -1 \quad 2/3 \quad -1/2 \quad 2/5 \ldots \tag{10.70}$$

The usage of the rectangular window would be equivalent to keeping only a number of elements in the Fourier series. As Fig. 10.17a shows, the oscillations are still large and the slope of the transfer function at low wave numbers flips with the order of the series between 0 and twice the correct value. Thus, the resulting truncated filter is not useful at all.

Much better results can be gained by using a binomial window function.

Window	Resulting mask	
1/4[1 2 1]	1/2[1 0 −1]	
1/16[1 4 6 4 1]	1/12[−1 8 0 −8 1]	(10.71)
1/64[1 6 15 20 15 6 1]	1/60[1 −9 45 0 −45 9 −1]	

The masks in the above equation have been scaled in such a way that the transfer function for small wave numbers is $i\pi\tilde{k}$. These approximations show no oscillations; the deviation from the ideal response increases monotonically (Fig. 10.17b).

10.4.3 Recursive Filters for Image Processing

The filter mask and the point spread function are different for recursive filters. Thus, the real problem is the design of the recursive filter, i. e., the determination of the filter coefficients for a desired point spread function and transfer function. While the theory of one-dimensional recursive filters is standard knowledge in digital signal processing (see, for example, Oppenheim and Schafer [108]), the design of two-dimensional filters is still not adequately understood. The main reasons are the fundamental differences between the mathematics of one- and higher-dimensional z-transforms and polynomials [90].

Because of these differences, we will show here only a simple yet effective strategy to apply recursive filters in digital image processing. It is based on a cascaded operation of simple one-dimensional recursive filters in such a way that the resulting transfer function is either real and even or imaginary and odd. Thus, the filter is either a zero-phase filter suitable for smoothing operations or a derivative filter that shifts the phase by 90° (for further details, see Section 11.4.5).

In the first composition step, we combine causal recursive filters to symmetric filters. We start with a standard one-dimensional causal recursive filter with the transfer function

$$^{+}\hat{A} = a(\tilde{k}) + ib(\tilde{k}). \tag{10.72}$$

The index $+$ denotes the run direction of the filter in a positive coordinate direction. The transfer function of the same filter but running in the opposite direction is (replace $\tilde{k}$ by $-\tilde{k}$ and note that $a(-\tilde{k}) = a(+\tilde{k})$ and $b(-\tilde{k}) = -b(\tilde{k})$) since the transfer function of a real PSF is hermitian (Appendix B.3).

$$^{-}\hat{A} = a(\tilde{k}) - ib(\tilde{k}). \tag{10.73}$$

Thus, only the sign of the imaginary part of the transfer function changes, when the coordinate direction is reversed.

We now have several possibilities to combine these two filters to symmetric filters:

$$
\begin{array}{llll}
\text{Addition} & {}^{e}\hat{A} & = \dfrac{1}{2}\left({}^{+}\hat{A} + {}^{-}\hat{A}\right) & = a(\tilde{k}) \\[2mm]
\text{Subtraction} & {}^{o}\hat{A} & = \dfrac{1}{2}\left({}^{+}\hat{A} - {}^{-}\hat{A}\right) & = ib(\tilde{k}) \\[2mm]
\text{Multiplication} & \hat{A} & = {}^{+}\hat{A}\,{}^{-}\hat{A} & = a^{2}(\tilde{k}) + b^{2}(\tilde{k}).
\end{array}
\tag{10.74}
$$

Addition and multiplication (consecutive application) of the left and right running filter yields filters of even symmetry, while subtraction results in a filter of odd symmetry. This way to cascade recursive filters gives them the same properties as zero- or $\pi/2$-phase shift nonrecursive filters with the additional advantage that they can easily be tuned, and extended point spread functions can be realized with only a few filter coefficients.

Example 10.8: Noncausal even and odd recursive filters from a simple causal lowpass filter

We demonstrate the generation of even and odd filters with the simple two-element lowpass filter we have already discussed in Example 10.6:

$$g'_n = \alpha g'_{n-1} + (1-\alpha)g_n$$

with the point spread function

$$R_n = (1-\alpha)\alpha^n$$

and the transfer function

$$\hat{R}(\tilde{k}) = \frac{1 - \alpha}{1 - \alpha \exp(-i\pi\tilde{k})}.$$

The transfer function parted into the real and imaginary parts reads as

$$^{\pm}\hat{A}(\tilde{k}) = \frac{1 - \alpha}{1 + \alpha^2 - 2\alpha \cos \pi\tilde{k}} \left[(1 - \alpha \cos \pi\tilde{k}) \mp \alpha i \sin \pi\tilde{k} \right].$$

From this transfer function, we can compute the additive ($^{e}\hat{A}$), subtractive ($^{o}\hat{A}$), and multiplicative ($\hat{A}$) application of the filters running in positive and negative direction, $^{+}\hat{A}(k)$, $^{-}\hat{A}(k)$ as

$$^{e}\hat{A}(\tilde{k}) = \frac{1 - \alpha}{1 + \alpha^2 - 2\alpha \cos \pi\tilde{k}} (1 - \alpha \cos \pi\tilde{k}),$$

$$^{o}\hat{A}(\tilde{k}) = \frac{1 - \alpha}{1 + \alpha^2 - 2\alpha \cos \pi\tilde{k}} \alpha \sin \pi\tilde{k},$$

$$\hat{A}(\tilde{k}) = \frac{(1 - \alpha)^2}{1 + \alpha^2 - 2\alpha \cos \pi\tilde{k}}.$$

10.4.4 Design by Filter Cascading

Since the design goal for filters in image processing is not to achieve ideal lowpass, highpass or bandpass filters, cascading of simple filters is a valuable alternative approach. The idea is to start with simple 1-D filters and to build up larger complex filters by applying these filters repeatedly in one direction and then in the other directions in the images. This approach has a number of significant advantages.

1. The properties of the simple filters used as building blocks are known. The transfer function of all these filters can be given analytically in simple equations. As cascading of filters simply means the multiplication of the transfer function, the transfer functions for even complex filters can be computed easily and analyzed analytically.

2. If complex filters can be built up by 1-D filters, that means by separable filters, then this approach does not only work in two dimensions but also in *any* higher dimension. This greatly simplifies and generalizes the filter design process.

3. In most cases, filters designed in this way are computed with a minimum number of arithmetic operations. This means that we not only have found a solution which can be analyzed very well but which can simultaneously also be computed very efficiently.

Table 10.2 and Fig. 10.18 summarize all the simple 1-D filters to build up almost all filters that are used in this handbook. The collection includes nonrecursive and recursive filters with even and odd-sized masks. The three recursive filters include a lowpass and a highpass filter, U and V, and a bandpass (resonance) filter, Q. All recursive filters can be tuned. The U and V filters constitute a mirror filter pair: $\hat{V}(\tilde{k}) = \hat{U}(1 - \tilde{k})$ (Fig. 10.18d and e). All of the filters in this toolbox are either even zero phase filters or odd filters with a 90° phase shift.

Besides cascading, a number of other operations are useful to build up more complex filters (Table 10.3). This includes addition and subtraction of filtered images. An important modification of these filters is to change the step width. Instead of applying the filters with at most three coefficients to the direct neighbors, we could easily also use the second, third, or further neighbors.

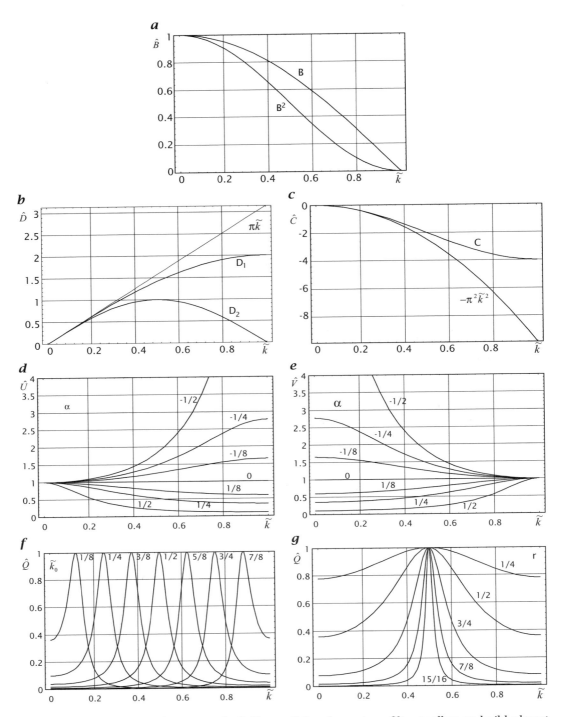

Figure 10.18: *Transfer functions of all filters of the elementary filter toolbox to build almost all filters used in this handbook. The symbols and masks of the filters are indicated (compare Table 10.2).*

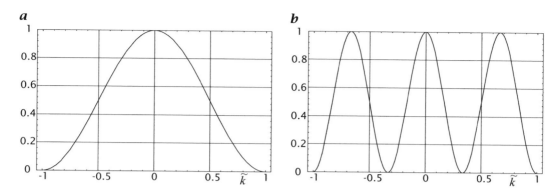

Figure 10.19: *Transfer functions of* **a** $B^2 = 1/4[1\ 2\ 1]$ *and of* **b** $B_3^2 = 1/4[1\ 0\ 0\ 2\ 0\ 0\ 1]$, *a copy of* B^2 *stretched by a factor of three. As an effect of the stretching in the spatial domain, the transfer function shrinks by the same factor and is three times replicated.*

The change in the step width can be applied for nonrecursive and recursive filters. The mask $1/4[1\ 2\ 1]$ stretched by a factor of two, for example, yields the new mask $1/4[1\ 0\ 2\ 0\ 1]$. If the operator B has the transfer function $\hat{B}(\tilde{k})$, then the same filter using only every s^{th} pixel, devoted by B_s, has a transfer function that is shrunk by the factor s:

$$B \circ\!\!-\!\!\bullet \hat{B}(\tilde{k}) \quad \leftrightarrow \quad B_s \circ\!\!-\!\!\bullet \hat{B}(s\tilde{k}). \tag{10.75}$$

This scaling means that the original transfer function occupies only $1/s$ of the whole wave number interval. Because of the periodic repetition, the Fourier domain thus contains s times the transfer function. Figure 10.19 explains this behavior with the elementary smoothing mask B for the step size three. The following examples demonstrate simple applications of the filter tool box technique.

Example 10.9: Simple applications of the filter toolbox technique

    A. Larger smoothing kernels by cascading the binomial mask $\mathcal{B}^4 = \mathcal{B}^2\mathcal{B}^2$
    Mask $1/4[1\ 2\ 1] * 1/4[1\ 2\ 1] = 1/16\ [1\ 4\ 6\ 4\ 1]$
    Transfer function $\cos^4(\pi\tilde{k}/2)$

    B. Larger smoothing kernels by cascading and stretching $\mathcal{B}_1^4\mathcal{B}_2^2$
    Mask $1/16[1\ 4\ 6\ 4\ 1] * 1/4\ [1\ 0\ 2\ 0\ 1] = 1/64\ [1\ 4\ 8\ 12\ 14\ 12\ 8\ 4\ 1]$
    Transfer function $\cos^4(\pi\tilde{k}/2)\cos^2(\pi\tilde{k})$

    C. Bandpass filter by subtraction of smoothing kernels $4(\mathcal{B}_1^2 - \mathcal{B}_1^4)$
    Mask $[1\ 2\ 1] - 1/4\ [1\ 4\ 6\ 4\ 1] = 1/4\ [-1\ 0\ 2\ 0\ -1]$
    Transfer function $\sin^2(\pi\tilde{k})$

10.4.5 Efficient Computation of Neighborhood Operations

10.4.5a In Place Nonrecursive Neighborhood Operations with 2-D masks. In Section 10.3.3 (see also Fig. 10.3), we already discussed that in place computation of nonrecursive neighborhood operations requires intermediate storage. Of course, it would be the easiest way to store the whole image with which a neighborhood operation should be performed. However, often not enough storage area is available to allow for such a simplistic approach. In the following, we will take convolution as an example for a neighborhood operation. All considerations in this section are generally enough that they apply also to any other type of neighborhood operation. Convolution of a 2-D

Table 10.2: *Filter toolbox with simple 1-D filters for cascaded filter operations.*

1. Group: Elementary even-sized masks; filter result is placed at the interleaved grid

Symbol	Mask	Transfer function
$\mathcal{B}$: elementary averaging	$1/2[1\ 1]$	$\cos(\pi\tilde{k}/2)$
$\mathcal{D}_1$: first difference	$[1\ \text{-}1]$	$2i\sin(\pi\tilde{k}/2)$

2. Group: Elementary odd-sized masks

Symbol	Mask	Transfer function
$\mathcal{B}^2 = \mathcal{B}\mathcal{B}$: averaging	$1/4[1\ 2\ 1]$	$\cos^2(\pi\tilde{k}/2)$
$\mathcal{D}_2 = \mathcal{B}\mathcal{D}_1$: first difference	$1/2[1\ 0\ \text{-}1]$	$i\sin\pi\tilde{k}$
$C = \mathcal{D}_1^2$: second difference	$[1\ \text{-}2\ 1]$	$-4\sin^2(\pi\tilde{k}/2)$

3. Group: Elementary recursive filters

Symbol	Recursion	Transfer function
$\pm\mathcal{U}$: lowpass	$g'_n = g_n + \alpha(g'_{n\pm1} - g_n)$	$^{+}\hat{U}^{-}\hat{U} = \dfrac{1}{1 + \beta - \beta\cos\pi\tilde{k}},$ $\beta = 2\alpha/(1-\alpha)^2$
$\pm\mathcal{V}$: highpass	$g'_n = g_n + \alpha(g_{n\pm1} + g_n)$	$^{+}\hat{V}^{-}\hat{V} = \dfrac{1}{1 + \beta + \beta\cos\pi\tilde{k}},$ $\beta = -2\alpha/(1+\alpha)^2$
$\pm\mathcal{Q}$: bandpass	$\begin{aligned} g'_n =\ & (1-r^2)\sin\pi\tilde{k}_0 g_n \\ +\ & 2r\cos\pi\tilde{k}_0 g'_{n-1} \\ -\ & r^2 g'_{n-2} \end{aligned}$	$^{+}\hat{Q}^{-}\hat{Q} =$ $\dfrac{\sin^2\pi\tilde{k}_0}{1 + r^2 - 2r\cos\pi(\tilde{k} - \tilde{k}_0)} \cdot$ $\dfrac{(1-r^2)^2}{1 + r^2 - 2r\cos\pi(\tilde{k} + \tilde{k}_0)}$

image with a 2-D $(2r + 1) \times (2r + 1)$ mask,

$$G'_{mn} = \sum_{m'=-r}^{r} \sum_{n'=-r}^{r} G_{m+m',n+n'} H_{-m',-n'}, \tag{10.76}$$

requires four loops in total. The two inner loops scan through the 2-D mask accumulating the convolution result for a single pixel (indices m' and n'). The two outer loops (indices m and n) are required to move the mask through image rows and image lines.

If the mask moves row by row through the image, it is obvious that an intermediate storage area with a minimum number of lines corresponding to the height of the mask is required. If the mask moves to the next row, it is not required to rewrite all lines. The uppermost line in the buffer can be discarded because it is no longer required and can be overwritten by the lowest row required now. Figure 10.20 explains the details of how the buffers are arranged for the example of a convolution with a 5×5 mask. To compute the convolution for row m, row $m + 2$ is read into one of the line buffers. The five line buffers then contain the rows $m - 2$ through $m + 2$. Thus, all rows are available to compute the convolution with a 5×5 mask for row m. When the

Table 10.3: *Basic operations to build complex filters from a set of simple primitive filters. Assumed are the filters G and $\mathcal{H}$ with the point spread functions G and H and the transfer functions $\hat{G}$ and $\hat{H}$, respectively; a and b are (generally complex-valued) scalar constants. The table lists the operations and the resulting point spread and transfer functions.*

Operation	Point spread function	Transfer function
Cascading $G\mathcal{H}$	Convolution $G * H$	Multiplication $\hat{G} \cdot \hat{H}$
Addition $G + \mathcal{H}$	Addition $G + H$	Addition $\hat{G} + \hat{H}$
Subtraction $G - \mathcal{H}$	Subtraction $G - H$	Subtraction $\hat{G} - \hat{H}$
Multiplication by scalar $a \cdot G$	Multiplication $a \cdot G$	Multiplication $a \cdot \hat{G}$
Stretching G_s	Sampling	Reduction and periodic replication

computation proceeds to row $m + 1$, row $m + 3$ is loaded into one of the line buffers replacing row $m - 2$. The five line buffers now contain the rows $m - 1$ through $m + 3$. Generally, the line buffer number into which a row is written is given by the row number modulo the number of line buffers, i. e., $m \mod (2r + 1)$.

Since a whole row is available, further optimizations are possible. The convolution can be performed in a vectorized way; the accumulation of the multiplication results can be performed with all pixels in a row for one coefficient after the other. This essentially means that the loop with the index n in Eq. (10.76) becomes the innermost loop. It is often advantageous to use an extra accumulator buffer to accumulate the intermediate results. This is, for instance, required when the intermediate results of the convolution sum must be stored with higher precision. For 8-bit, i. e., 1-byte images, it is required to perform the accumulation at least with 16-bit accuracy. After the results with all filter coefficients are accumulated, the buffer is copied back to row m in the image with appropriate scaling.

The procedure described so far for neighborhood operations is also very effective from the point of memory caching. Note that a lot of computations (in our example, 25 multiplications and 24 additions per pixel) are performed fetching data from the 5 line buffers and storing data into one accumulator buffer. These 6 buffers easily fit into the cache of modern CPUs. Proceeding to the next row needs only the replacement of one of these buffers. Thus, most of the data simply remains in the cache.

The vectorization of the convolution operation enables further optimizations. Since the multiplication with one filter coefficient takes place with a whole row, it does not cause much overhead to test the coefficient of the filter. If the coefficient is zero, no operation has to be performed. If it is one, the row is only added to the accumulator buffer and no multiplication is required.

10.4.5b Separable Neighborhood Operations. For separable masks, a simpler approach could be devised as illustrated in Fig. 10.21. The convolution is now a two-step approach. First, the horizontal convolution with a 1-D mask takes place. Row m is read into a line buffer. Again, a vectorized convolution is now possible of the whole line with the 1-D mask. If higher-precision arithmetic is required for the accumulation, an accumulation buffer is again needed; if not, the accumulation could directly take place in row m of the image. After the horizontal convolution, the vertical convolution takes place with exactly the same scheme. The only difference is that now column by column is read into the buffer (Fig. 10.21).

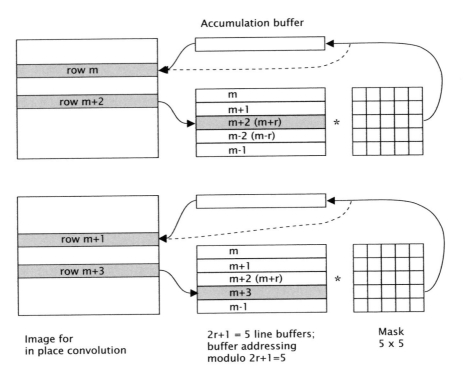

Figure 10.20: *Schematic diagram for the in place computation of the convolution with a 5 × 5 mask using 5 line buffers and an accumulation buffer. The computation of rows m and m + 1 is shown. By using an addressing modulo 5 for the line buffers, only one row (m + 2 or m + 3) needs to be written to the line buffers; all others are still there from the computation of the previous rows.*

This procedure for separable neighborhood operations can easily be extended to higher-dimensional images. For each dimension, a line in the corresponding direction is copied to the buffer, where the convolution takes place and then copied back into the image at the same place. This procedure has the significant advantage that filtering in different directions only requires different copy operations but that the same 1-D neighborhood operation can be used for all directions. It has the disadvantage, though, that a lot of data copying takes place, slowing down neighborhood operations with small masks.

10.4.6 Filtering at Image Borders

The procedures described for efficient computation of the convolution in the previous sections are very helpful also to handle the problems with image borders. If an image is convolved with a $(2r + 1) \times (2r + 1)$ mask, a band of r rows and columns is required around the image in order to perform the convolution up to the pixels at the image edges (Fig. 10.22). Three different strategies are available to extend the image.

1. *Periodic extension of the image.* This is the theoretically correct method, as the Fourier transform expects that the image is periodically repeating through the whole space. If a filter operation is performed in the Fourier domain by directly multiplying the Fourier transformed image with the transfer function of the filter, this periodic

Step 1: horizontal convolution

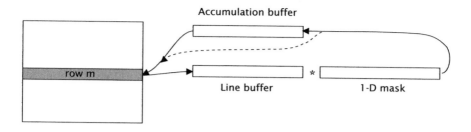

Step 2: vertical convolution

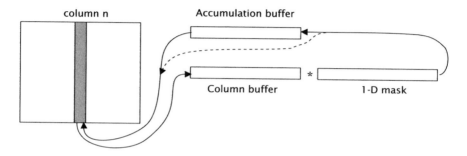

Figure 10.21: *Schematic diagram for the in place computation of a separable convolution operation using a line buffer and an optional accumulation buffer. First the horizontal convolution with the 1-D mask is performed by copying row by row into the line buffer and writing the result of the 1-D convolution operation back to the image. Then, column by column is copied into the column buffer to perform the vertical convolution with the same or another 1-D mask in the same way.*

extension is automatically implied. This mode to perform the convolution is also referred to as *cyclic convolution.*

2. *Zero-extension of the image.* Since we do not know how the images continue outside of the sector which is available, we cannot actually know what kind of values we could expect. Therefore, it seems to be wise to extend the image just by zeros, acknowledging our lack of knowledge. This procedure, of course, has the disadvantage that the image border is seen as a sharp edge which will be detected by any edge detector. Smoothing operators behave more graciously. The image just becomes darker at the borders. The mean gray value of the image is preserved.

3. *Extrapolation techniques.* This technique takes the pixels close to the border of the image and tries to extrapolate the gray values beyond the borders of the image in an appropriate way. The easiest technique is just to set the required pixels outside of the image to the border pixel in the corresponding row or column. A more sophisticated extrapolation technique might use linear extrapolation. Extrapolation techniques avoid that the edge detectors detect an edge at the border of the image and that the image gets darker due to smoothing operations at the edges. It is, however, also far from perfect, as we actually cannot know how the picture is extending beyond its border. In effect, we give too much weight to the border pixels. This also means that the mean gray value of the image is slightly altered.

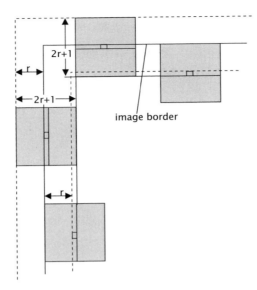

image border

Figure 10.22: *Illustration of the problems of neighborhood operations at the image border: A neighborhood operation with a $(2r + 1) \times (2r + 1)$ mask requires an r-wide band of extra pixels in every direction to compute the values of the border pixel. Within the image, an r-wide border region will be influenced by the way the off-image pixels are chosen.*

It is obvious that none of the techniques, neither periodic extension of the image, zero extension of the image, nor extrapolation techniques, are ideal. Therefore, errors will occur if we detect or analyze any object very close to the border. The only safe way to avoid them is to make sure that the objects of interest are not too close to the border of the image. A safe distance is half of the linear size of the largest mask used to analyze the image (Fig. 10.22).

10.4.7 Test Patterns

The use of computer-generated images is essential to test the accuracy of algorithms. As we have already seen, algorithm-related errors can be separated from sensor-related errors including aliasing due to imperfect sampling in this way. It is easily possible to study the response of the algorithms to well-defined signals. These tests do not substitute experiments with real-world images, but are a very useful preparation in order to detect weaknesses in the algorithms and to optimize them *before* they are applied to real images.

10.4.7a Concentric Ring Test Pattern.

The first test image has been adapted from the work of Granlund's group at Linköping University [80]. They used a pattern with sinusoidal concentric rings. The wave numbers *decreased* with the distance from the center. Such a pattern has the advantage that it includes all local orientations and a wide range of wave numbers. Along a radius, the local orientation (or optical flow) is constant but the wave number is changing gradually. This idea was picked up, but we used concentric rings with wave numbers directly proportional to the distance from the center as shown in Fig. 10.23 generated by

$$g(\boldsymbol{x}) = g_0 \sin\left(\frac{k_m |\boldsymbol{x}|^2}{2r_m}\right) \left[\frac{1}{2} \tanh\left(\frac{r_m - |\boldsymbol{x}|}{w}\right) + \frac{1}{2}\right], \qquad (10.77)$$

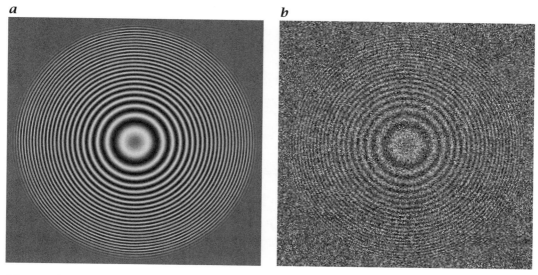

a *b*

Figure 10.23: *Synthetic test image with concentric rings that is suitable to test all types of neighborhood operations. It contains structures in all directions. The wave number is proportional to the distance from the center; the maximum wave number of 0.4 (2.5 samples/wavelength) at the outer edge is close to the limit set by the sample theorem.* **a** *Ring pattern with an amplitude of 100 and an offset of 128.* **b** *Ring pattern with an amplitude of 40 and an offset of 128 superposed by zero mean normal distributed noise with a standard deviation of 40 (the signal-to-noise ratio is one).*

where r_m is the maximum radius of the pattern and k_m the maximum wave number at its edge. In optics, this pattern is known as the *Fresnel zone plate*. Multiplication by the tanh function results in a smooth transition of the ring patterns at the outer edge to avoid aliasing problems. The parameter w controls the width of this transition. This pattern has the advantage that the local wave number is directly given by the Cartesian coordinates with origin at the center of the rings. Consequently, we can directly depict the transfer function of any operation with this test pattern.

For image sequence analysis, this test image has several advantages. First, the velocity is equally distributed in the image, showing all possible values. Thus it is easy to study errors of the algorithms with regard to the velocity or — equivalently — the orientation of the patterns. Second, errors dependent on the scale (wavelength) of the structure become immediately apparent since, as already discussed, the wave number space is directly mapped onto the spatial domain.

10.4.7b Test Image for Edge and Line Detection. The ring test pattern introduced in Section 10.4.7a is not very suitable to test edge and line detectors. Therefore, the test pattern shown in Fig. 10.24 has been designed. It contains curved and straight edges, corners, and various lines of different thickness. In the different quadrants, smoothing filters have been applied to simulate edges of different steepness.

10.5 Advanced Reference Material

A. V. Oppenheim and R. W. Schafer, 1989. Discrete-Time Signal Processing, Prentice-Hall, Inc., Englewood Cliffs. *The classical textbook on digital signal processing.*

a
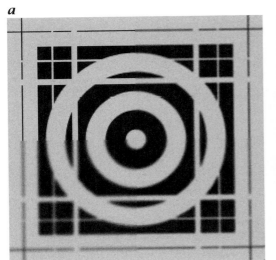

b
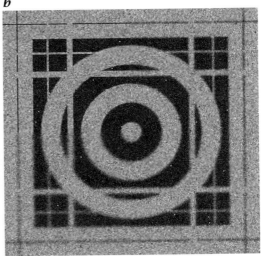

Figure 10.24: Synthetic test image for edge detection containing straight and curved edges, corners, and lines of various widths. **a** Bright areas 220, dark areas 20, **b** bright areas 190, dark areas 60, normal distributed noise with a standard deviation of 20 added. The upper right, lower right, and lower left quadrants in both images are smoothed with B^2, B^4, and B^6 smoothing masks to simulate edges of different steepness.

Gives a thorough treatment of linear system theory. Also a good reference to the standard design techniques for digital filters.

J. G. Proakis and D. G. Manolakis, 1992. Digital Signal Processing, Principles, Algorithms, and Applications, Macmillan Publishing Company, New York. *Another excellent account of classical digital signal processing. Very valuable is the section on filter design techniques.*

J. S. Lim, 1990. Two-Dimensional Signal and Image Processing, Prentice-Hall, Inc., Englewood Cliffs. *Covers the application of classical digital signal processing to image processing.*

I. Pias and A. N. Venetsanopoulos, 1990. Nonlinear Digital Filters, Kluwer Academic Publishers, Boston. *One of the rare monographs on nonlinear digital filter techniques.*

11 Regions

11.1 Highlights

Objects are recognized in images because they consist of connected regions with constant or at least similar gray values. Therefore, one approach to object detection starts with the detection of constant regions. In order to detect these regions and to compute average values, spatial smoothing is required. Spatial averaging is, thus, one of the elementary low-level image processing operations. At first glance, it might appear that this is a rather trivial task about which not much needs to be said. This is not the case, and this chapter is about three major issues related to spatial smoothing.

First, the general properties of smoothing filters are discussed. One of the most important general properties for practical application is what is called isotropy, i. e., that the smoothing is the same in all directions. If this condition is not met, significant errors may result in further processing steps. Another important issue is the technique of weighted averaging (Section 11.3.2). When averages of data are computed, it is a standard technique to take the errors of the data into account as a weighting factor. If this is not done and the errors of the individual data points are different, the average will be biased. In image processing, this technique is hardly used so far. However, with the advent of sophisticated feature extraction algorithms (for example, the structure tensor method to compute local orientation, Chapter 13) which not only deliver the feature but also its uncertainty, weighted averaging techniques become increasingly important. In image processing, they are known as normalized convolution (Section 10.3.10c).

Second, spatial averaging must be performed fast. An analysis of many low-level feature extraction procedures reveals that averaging is an important step and that most computations are spent with spatial averaging. Prominent examples are the computation of local orientation (Chapter 13) and the computation of pyramids (Chapter 14). Four basic techniques will be discussed to speed up spatial averaging: *filter cascading* (Sections 11.4.1 and 11.4.2), *multistep averaging* (Section 11.4.3), *multigrid averaging* (Section 11.4.4), and *recursive averaging* (Section 11.4.5).

Third, how do we average at edges? Just averaging across the boundaries of an object results in useless values. Somehow, the averaging has to be stopped at edges. This insight leads us to controlled averaging (Section 11.3.3). Two techniques will be discussed. First, averaging can be regarded as a diffusion process. Making the diffusion constant dependent on the gray value structure of a neighborhood provides a general solution to controlled averaging. Second, filters that can be steered to average into a certain direction only provide a fast solution to directional averaging (Sections 11.3.4 and 11.4.7).

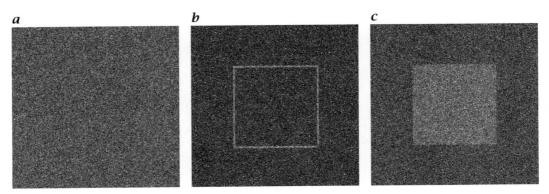

Figure 11.1: *Spatial coherency is required to recognize objects.* **a** *No object can be recognized in an image that contains only randomly distributed gray values.* **b** *One- and* **c** *two-dimensional spatial coherency makes us recognize lines and areas even if a high noise level is present in a 2-D image.*

11.2   Task

Objects may be defined as *regions* of constant gray values. Therefore, the analysis and detection of regions has a central importance as an initial step to detect objects. This approach, of course, implies a simple model of the image content. The objects of interest must indeed be characterized by constant gray values that are clearly different from the background and/or other objects. This assumption is, however, seldom met in real-world applications. The intensities will generally show some variations. These variations may inherently be a feature of the object. They could also be caused by the image formation process. Typical effects are *noise*, a *nonuniform illumination*, or *inhomogeneous background*.

An individual pixel cannot tell us whether it is located in a region. Therefore, it is required to analyze the neighborhood of the pixel. If the pixel shows the same gray value as its neighbors, then we have good reasons to believe that it lies within a region of constant gray values. In an image with totally independent pixel gray values (Fig. 11.1a), nothing can be recognized. *Spatial coherency* either in one (Fig. 11.1b) or two (Fig. 11.1c) dimensions is required in order to recognize lines and regions, respectively. Since we also want to obtain an estimate of the *mean gray value* of the region, averaging within a local neighborhood appears to be a central tool for region detection. In addition, the deviation of the gray values of the pixels in the neighborhood gives a quantitative measure as to how well a region of constant gray values is encountered in the neighborhood.

This approach is very much the same as that used for any type of measurements. A single measurement is meaningless. Only repeated measurements give us a reliable estimate both of the measured quantity and its uncertainty. In image processing, averaging needs not necessarily be performed by taking several images — although this is a very useful procedure. Since spatial information is obtained, spatial averaging offers a new alternative.

Many objects do not show a distinct constant gray value, but we can still recognize them if the pattern they show differs from the background (see Chapter 14). These objects still show a spatially well organized structure. Thus, the concept of spatial coherency is still valid. However, a suitable preprocessing is required to convert the original image into a *feature image*. Such an image contains a suitable parameter that

Task List 8: Regions

Goal	Procedures
Obtain estimates of object features	Perform spatial averaging in images
	Perform spatio-temporal averaging in image sequences
Adapt smoothing to size and shape of objects	Select appropriate class of smoothing filter and degree of smoothing
Speed up smoothing over large areas	Use recursive or multistep filters
Limit smoothing to region not crossing object boundaries	Use advanced smoothing techniques such as inhomogeneous and anisotropic diffusion filters, normalized convolution, or steerable filters

makes the object pattern distinct from that of the background. Then, the feature image can be handled in the same way as a gray scale image for simple objects. When the feature values within the object are constant and different from the background, the model of constant regions can again be applied, now to the feature image.

In complex cases, it is not possible to distinguish objects from the background with just one feature. Then it may be a valid approach to compute more than one feature image from one and the same image. This results in a multicomponent or *vectorial feature image*.

Therefore, the task of averaging must also be applied to vectorial images. The same situation arises when more than one image is taken from a scene as with *color images* or any type of *multispectral images*. In image sequences, averaging is extended into the time coordinate to a spatio-temporal averaging.

This chapter is first about how to perform averaging or smoothing in a proper way. Several classes of averaging filters are available that show different features: *box filters* (Section 11.4.1), *binomial filters* (Sections 11.4.2–11.4.4), and *recursive smoothing filters* (Section 11.4.5).

An important general question for smoothing is how it can be computed efficiently. This question is of special importance if we want to analyze coarse features in the images that require averaging over larger distances and, thus, large smoothing masks (task list 8) or if we apply smoothing to higher-dimensional images such as volumetric images or image sequences. Three elementary techniques will be discussed here: *multistep filters* (Section 11.4.3), *multigrid filters* (Section 11.4.4), and recursive filters (Section 11.4.5).

While averaging works well within regions, it is questionable at the edges of an object. When the filter mask contains pixels from both the object and the background, averaged values have no useful meaning. These values cannot be interpreted as an object-related feature since it depends on the fraction of the object pixels contained in the mask of the smoothing operator. Therefore, smoothing techniques are discussed that stop or at least diminish averaging at discontinuities.

Such an approach is not trivial as it requires the detection of the *edges* before the operation can be applied. Obviously, such advanced smoothing techniques need to analyze the local neighborhoods in more detail and to adapt the smoothing process in one or the other way to the structure of the local neighborhood. We will discuss inhomogeneous and anisotropic diffusion filters (Section 11.4.6), and steerable filters for adaptive filtering (Section 11.4.7).

Table 11.1: *General properties of smoothing neighborhood operators.*

Description	Condition in space domain	Condition in Fourier domain
Zero shift	Symmetrical mask in each coordinate direction $(h_{m,-n} = h_{m,n}, h_{-m,n} = h_{m,n})$	Transfer function is real (zero phase-shift filter): $\Im(\hat{h}(\bar{k})) = 0$
Preserve mean value	$\sum_m \sum_n h_{mn} = 1$	$\hat{h}(0) = 1$
Monotonic smoothing	—	$\hat{h}(\bar{k})$ decreases monotonically
Smoothing independent of direction	Isotropic mask: $h(x) = h(|x|)$	Isotropic transfer function: $\hat{h}(\bar{k}) = \hat{h}(|\bar{k}|)$

11.3 Concepts

11.3.1 General Properties of Averaging Filters

Linear convolution provides the framework for all elementary averaging filters. These filters have some elementary properties in common (Table 11.1).

11.3.1a Preservation of Object Position. With respect to object detection, the most important feature of a smoothing convolution operator is that it must not shift the object position. Any shifts introduced by a preprocessing operator would cause errors in the later estimates of the position and possibly other geometric features of an object. In order to cause no shift, the transfer function of a filter must be real. A filter with this property is known as a *zero-phase filter*. Then, all periodic components of an image experience no phase shift. A real transfer function implies a symmetrical filter mask (Appendices B.3 and B.4):

$$
\begin{aligned}
&\text{1-D} \quad h_{-n} = h_n \\
&\text{2-D} \quad h_{-m,n} = h_{m,n}, \; h_{m,-n} = h_{m,n} \\
&\text{3-D} \quad h_{-l,m,n} = h_{l,m,n}, \; h_{l,-m,n} = h_{l,m,n}, \; h_{l,m,-n} = h_{l,m,n}.
\end{aligned}
\tag{11.1}
$$

For convolution masks with an even number of coefficients, the symmetry relations Eq. (11.1) are valid as well. The resulting smoothed image lies on the *interleaved grid* between the original grid (Section 10.3.1). With respect to this grid, the mask indices are half-integer numbers. The indices for a mask with 2R coefficients run from $-R + 1/2$, $-R + 3/2$, ..., $-1/2$, $1/2$, ..., $R - 1/2$.

11.3.1b Preservation of Mean Value. The mean value must be preserved by a smoothing operator. This condition says that the transfer function for the zero wave number is 1 or, equivalently, that the sum of all coefficients of the mask is 1:

$$
\begin{aligned}
&\text{1-D} \quad \hat{h}(0) = 1, \quad \sum_n h_n = 1 \\
&\text{2-D} \quad \hat{h}(0) = 1, \quad \sum_m \sum_n h_{mn} = 1 \\
&\text{3-D} \quad \hat{h}(0) = 1, \quad \sum_l \sum_m \sum_n h_{lmn} = 1.
\end{aligned}
\tag{11.2}
$$

11.3.1c Nonselective Smoothing. Intuitively, we expect that any smoothing operator attenuates smaller scales stronger than coarser scales. More specifically, a smoothing operator should not completely annul a certain scale while smaller scales still remain in the image. Mathematically speaking, this means that the transfer function decreases monotonically with the wave number:

$$\hat{h}(\tilde{k}_2) \le \hat{h}(\tilde{k}_1) \quad \text{if} \quad \tilde{k}_2 > \tilde{k}_1. \tag{11.3}$$

We may impose a more stringent condition that for the highest wave numbers the transfer function is identical to zero:

1-D $\quad \hat{h}(1) = 0$

2-D $\quad \hat{h}(\tilde{k}_1, 1) = 0 \quad \text{and} \quad \hat{h}(1, \tilde{k}_2) = 0$ (11.4)

3-D $\quad \hat{h}(\tilde{k}_1, \tilde{k}_2, 1) = 0 \quad \text{and} \quad \hat{h}(\tilde{k}_1, 1, \tilde{k}_3) = 0 \quad \text{and} \quad \hat{h}(1, \tilde{k}_2, \tilde{k}_3) = 0.$

Together with the monotony condition Eq. (11.3), this means that the transfer function decreases monotonically from one to zero.

11.3.1d Isotropy. In most applications, the smoothing should be the same in all directions in order not to prefer any direction. Thus, both the filter mask and the transfer function should be isotropic, i.e., depend only on the magnitude of the distance from the center pixel and the magnitude of the wave number, respectively:

$$h(\boldsymbol{x}) = h(|\boldsymbol{x}|) \quad \text{and} \quad \hat{h}(\tilde{\boldsymbol{k}}) = \hat{h}(|\tilde{\boldsymbol{k}}|). \tag{11.5}$$

In discrete space, this condition can — of course — only be met approximately. Therefore, it is an important design goal to construct discrete masks with optimum isotropy.

11.3.1e Symmetric Masks. This property of any smoothing mask deserves further consideration since it forms the base to compute the convolution more efficiently by reducing the number of multiplications. For a 1-D mask we obtain

$$g_n' = h_0 g_n + \sum_{n'=1}^{r} h_{n'}(g_{n-n'} + g_{n+n'}). \tag{11.6}$$

The equations for 2-D and 3-D masks are summarized in ≻ 11.1. For a W-dimensional mask, the number of multiplications is reduced from $(2R + 1)^W$ multiplications to $(R+1)^W$ multiplications. The number of additions remains unchanged at $(2R+1)^W - 1$. Thus, for large R, the number of multiplications is almost reduced by a factor of 2^W; however, the order of the convolution operation does not change and is still $O(R^W)$.

For even-sized averaging masks, the equations corresponding to Eq. (11.6) are considerably simpler. Note that the indices are related to the interleaved output grid. Therefore, the primed coefficient on the original grid takes half-integer values since it is displaced by half a grid constant from the output grid in all directions. For 1-D masks, this leads to

$$g_n' = \sum_{n'=1/2}^{R-1/2} h_{n'}(g_{n-n'} + g_{n+n'}). \tag{11.7}$$

The equations for 2-D and 3-D masks are in ≻ 11.2. For a W-dimensional mask, the number of multiplications is reduced from $(2R)^W$ to R^W, i.e., by a factor of 2^W.

Table 11.2: *Comparison of the number of multiplications (M) and additions (A) required to perform a convolution with symmetric nonseparable and separable masks of various sizes in two and three dimensions. The performance gain is given as the ratio of the total number of additions and multiplications.*

Mask	Nonseparable	Separable	Performance gain
5^2, $R = 2$	9 M, 24 A	6 M, 8 A	2.3
9^2, $R = 4$	25 M, 80 A	10 M, 16 A	4.0
13^2, $R = 6$	49 M, 168 A	14 M, 24 A	5.7
17^2, $R = 8$	81 M, 288 A	18 M, 32 A	7.4
5^3, $R = 2$	27 M, 124 A	9 M, 12 A	7.2
9^3, $R = 4$	125 M, 728 A	15 M, 24 A	21.9
13^3, $R = 6$	343 M, 2196 A	21 M, 36 A	44.5
17^3, $R = 8$	729 M, 4912 A	27 M, 48 A	75.2

The symmetry relations also significantly ease the computation of the transfer functions since only the cosine terms of the complex exponential from the Fourier transform remain in the equations. The transfer function for 1-D symmetric masks with odd number of coefficients is

$$\hat{h}(\tilde{k}) = h_0 + 2 \sum_{n'=1}^{R} h_{n'} \cos(n'\pi\tilde{k}) \tag{11.8}$$

and with even number of coefficients

$$\hat{h}(\tilde{k}) = 2 \sum_{n'=1/2}^{R-1/2} h_{n'} \cos(n'\pi\tilde{k}). \tag{11.9}$$

The equations for 2-D and 3-D masks can be found in the reference part $\succ 11.3$ and 11.4.

11.3.1f Separable Masks. Even more significant savings in the number of computations are gained if the smoothing mask is separable. In this case, the $(2R + 1)^W$-mask can be replaced by a cascaded convolution with W masks of the size $(2R + 1)$ (Section 10.3.6e). Now, only $W(R + 1)$ multiplications and $2WR$ additions are required. The order of the convolution operation with a separable mask is reduced to the order of $O(R)$, i.e., directly proportional to the linear size of the mask. Table 11.2 compares the number of multiplications and additions required for separable and nonseparable masks of various sizes in two and three dimensions.

11.3.1g Standard Deviation. One measure for the degree of smoothing is the standard deviation or, more generally, the second moment. This measure can be applied both in the spatial domain and in the wave number domain. In the spatial domain, the standard deviation is a measure for the distance over which the pixels are averaged. In the Fourier domain, this means the wave number where the attenuation becomes significant.

In a multi-dimensional space, the standard deviation is only a scalar if the filter mask and, thus, also the transfer function are isotropic; otherwise, it is a symmetrical tensor. Since isotropic smoothing filters are most significant in image processing and most smoothing filters are separable, only the equations for the 1-D case are given:

$$\text{Continuous spatial domain} \quad \sigma_x^2 = \int x^2 H(x)\,\mathrm{d}x$$

$$\text{Odd-sized } (2r+1)\text{-mask} \quad \sigma_x^2 = \sum_{m'=-r}^{r} m'^2 H_{m'}$$

$$\text{Even-sized } 2r\text{-mask} \quad \sigma_x^2 = \sum_{m'=-(r-1/2)}^{r-1/2} m'^2 H_{m'} \tag{11.10}$$

$$\text{Wave number domain} \quad \sigma_k^2 = \int k^2 H(k)\,\mathrm{d}k.$$

11.3.1h Noise Suppression. The noise suppression of smoothing filters can be computed directly from the results obtained about error propagation with convolution in Section 10.3.7. According to Eqs. (10.42) and (10.48) the variance of noise for uncorrelated pixels (*white noise*) is given by

$$\frac{\sigma'^2}{\sigma^2} = \sum_{n'=-R}^{R} h_{n'}^2 = \frac{1}{MN} \sum_{u'=0}^{M-1} \sum_{v'=0}^{N-1} |\hat{h}_{u'v'}|^2 = \int_0^1 \int_0^1 |\hat{h}(\tilde{\boldsymbol{k}})|^2 \mathrm{d}^2\tilde{k}. \tag{11.11}$$

The noise variance always reduces because $|\hat{h}(\tilde{\boldsymbol{k}})| \leq 1$. In the case of correlated pixels, the equations for the propagation of the autocovariance function must be used; see Eq. (10.45):

$$\boldsymbol{C}' = \boldsymbol{C} \star (\boldsymbol{H} \star \boldsymbol{H}) \quad \circ\!\!\!-\!\!\!\bullet \quad \hat{c}'(\boldsymbol{k}) = \hat{c}(\boldsymbol{k}) \left|\hat{h}(\boldsymbol{k})\right|^2. \tag{11.12}$$

Example 11.1: Noise suppression for a simple smoothing filter

For the simplest smoothing filter

$$\boldsymbol{B} = [1/2\ 1/2]$$

the variance is, according to Eq. (11.11), reduced to one half:

$$\frac{\sigma'^2}{\sigma^2} = (\boldsymbol{h} \star \boldsymbol{h})_0 = 1/4 + 1/4 = 1/2 \quad \text{or} \quad \frac{\sigma'^2}{\sigma^2} = \int_0^1 \cos^2(\pi\tilde{k}/2) = 1/2.$$

If we apply this filter two times,

$$\boldsymbol{B} = [1/2\ 1/2] * [1/2\ 1/2] = [1/4\ 1/2\ 1/4],$$

the variance reduction is not 1/4, but only 3/8:

$$\frac{\sigma'^2}{\sigma^2} = 1/16 + 1/4 + 1/16 = 3/8 \quad \text{or} \quad \frac{\sigma'^2}{\sigma^2} = \int_0^1 \cos^4(\pi\tilde{k}/2) = 3/8.$$

This lower variance reduction is caused by the fact that the neighboring pixels have become correlated after the application of the first filter. Equation (10.41) yields

$$c_1 = \sigma^2 (\boldsymbol{h} \star \boldsymbol{h})_1 = \sigma^2/4.$$

11.3.2 Weighted Averaging

In Section 10.3.10c, it was discussed that gray values at pixels, just as well as any other experimental data, may be characterized by individual errors that have to be considered in any further processing. As an introduction, we first discuss the averaging of a set of N data g_n with standard deviations σ_n. From elementary statistics, it is known that appropriate averaging requires the weighting of each data point g_n with the inverse of the variance $w_n = 1/\sigma_n^2$. Then, an estimate of the mean value is given by

$$\bar{g} = \sum_{n=1}^{N} g_n/\sigma_n^2 \bigg/ \sum_{n=1}^{N} 1/\sigma_n^2 , \tag{11.13}$$

while the standard deviation of the mean is

$$\sigma_{\bar{g}}^2 = 1 \bigg/ \sum_{n=1}^{N} 1/\sigma_n^2 . \tag{11.14}$$

The weight of an individual data point in Eq. (11.13) for computation of the mean is higher the lower its statistical error is.

The application of weighted averaging to image processing is known as normalized convolution. The averaging is now extended to a local neighborhood. Each pixel enters the convolution sum with a weighting factor associated to it. Thus, normalized convolution requires two images. One is the image to be processed, the other an image with the weighting factors. In analogy to Eqs. (11.13) and (11.14), normalized convolution is defined by

$$G' = \frac{H * (W \cdot G)}{H * W} \tag{11.15}$$

where H is any convolution mask, G the image to be processed, and W the image with the weighting factors. A normalized convolution with the mask H essentially transforms a multicomponent image consisting of the image G and the weighting image W into a new multicomponent image with G' and a new weighting image $W' = H * W$ which can undergo further processing. In this sense, normalized convolution is nothing complicated and special. It is just the adequate consideration of pixels with spatially variable statistical errors. "Standard" convolution can be regarded as a special case of normalized convolution. Then all pixels are assigned to the same weighting factor and it is not required to use a weighting image, since it is and remains a constant.

The flexibility of normalized convolution is given by the choice of the weighting image. The weighting image does not necessarily have to be associated to an error. It can be used to select and/or amplify pixels with certain features. In this way, normalized convolution becomes a versatile nonlinear operator.

11.3.3 Controlled Averaging

The linear averaging filters discussed so far blur edges. Even worse, if the mask of the smoothing operator crosses an object edge it contains pixels from both the object and the background resulting in a meaningless result from the filter. The question, therefore, is whether it is possible to perform an averaging which does not cross object boundaries. Such a procedure can, of course, only be applied if we have already detected the edges. Once we have obtained an estimate for edges, four different strategies can be thought of to avoid smoothing across object boundaries. The discussion in this section is based on the general considerations on adaptive filtering in Section 10.3.10.

11.3.3a   Averaging as Diffusion.   In recent years, a whole new class of image processing operators has been investigated which are known as *diffusion filters*. We will introduce this class of operators by first showing that diffusion is nothing else but a special smoothing convolution kernel. Diffusion is a transport process that tends to level out concentration differences. In physics, diffusion processes govern the transport of heat, matter, and momentum. To apply a diffusion process to an image, we regard the gray value g as the concentration of a chemical species. The elementary law of diffusion states that the flux induced by a concentration difference is against the direction of the concentration gradient and proportional to it:

$$j = -D\nabla g \tag{11.16}$$

where the constant D is known as the *diffusion coefficient*. Using the continuity equation

$$\frac{\partial g}{\partial t} + \nabla j = 0, \tag{11.17}$$

the nonstationary diffusion equation is

$$\frac{\partial g}{\partial t} = \nabla(D\nabla g). \tag{11.18}$$

For the case of a homogeneous diffusion process (D does not depend on the position), the equation reduces to

$$\frac{\partial g}{\partial t} = D\Delta g. \tag{11.19}$$

It is easy to show that the general solution to this equation is equivalent to a convolution with a smoothing mask. To this end, we perform a spatial Fourier transform, which results in

$$\frac{\partial \hat{g}(\mathbf{k})}{\partial t} = -D|\mathbf{k}|^2 \hat{g}(\mathbf{k}) \tag{11.20}$$

reducing the equation to a linear first-order differential equation with the general solution

$$\hat{g}(\mathbf{k}, t) = \exp(-D|\mathbf{k}|^2 t)\hat{g}(\mathbf{k}, 0) \tag{11.21}$$

where $\hat{g}(\mathbf{k}, 0)$ is the Fourier transformed image at time zero. Multiplication of $\hat{g}(\mathbf{k}, 0)$ in the Fourier space with the Gaussian function $\exp(-|\mathbf{k}|^2/(2\sigma_k^2))$ with $\sigma_k^2 = 1/(2Dt)$ as given by Eq. (11.21) is equivalent to a convolution with the same function but of reciprocal width (Appendix B.3). Thus,

$$g(\mathbf{x}, t) = \frac{1}{2\pi\sigma^2(t)} \exp\left(-\frac{|\mathbf{x}|^2}{2\sigma^2(t)}\right) * g(\mathbf{x}, 0) \tag{11.22}$$

with

$$\sigma(t) = \sqrt{2Dt}. \tag{11.23}$$

Equation Eq. (11.22) establishes the equivalence between a diffusion process and convolution with a Gaussian kernel. In the discrete case, the Gaussian kernel can be replaced by binomial filters (Section 11.4.2).

Given the equivalence between convolution and a diffusion process, it is possible to adapt smoothing to the local image structure by making the diffusion constant dependent on it. The two principal possibilities are to make the diffusion process dependent on the position (*inhomogeneous diffusion*) and/or the direction (*anisotropic diffusion*).

11.3.3b Inhomogeneous Diffusion. In order to avoid smoothing of edges, it appears logical to attenuate the diffusion coefficient there. Thus, the diffusion coefficient is made dependent on the strength of the edges as given by the magnitude of the gradient

$$D(g) = D(|\nabla g|). \tag{11.24}$$

Perona and Malik [112] used the following dependency of the diffusion coefficient on the magnitude of the gradient:

$$D = D_0 \frac{\lambda^2}{|\nabla g|^2 + \lambda^2} \tag{11.25}$$

where λ is an adjustable parameter. For low gradients $|\nabla g| \ll \lambda$, D approaches D_0; for high gradients $|\nabla g| \gg \lambda$, D tends to zero.

As simple and straightforward as this idea appears, it is not without problems. Depending on the functionality of D on ∇g, the diffusion process may become unstable resulting even in a steeping of the edges. A safe way to avoid this problem is to use a regularized gradient obtained from a smoothed version of the image as shown by [151].

11.3.3c Anisotropic Diffusion. Inhomogeneous diffusion has one significant disadvantage. It stops diffusion completely and in all directions at edges leaving the edges noisy. Edges are, however, only blurred by diffusion perpendicular to them while diffusion parallel to edges is even advantageous since it stabilizes the edge.

An approach that makes diffusion independent from the direction of edges is known as anisotropic diffusion. With this approach, the flux is no longer parallel to the gradient. Therefore, the diffusion can no longer be described by a scalar diffusion coefficient as in Eq. (11.16). Now, a *diffusion tensor* is required:

$$\boldsymbol{j} = -\boldsymbol{D}\nabla g = -\begin{bmatrix} D_{11} & D_{12} \\ D_{12} & D_{22} \end{bmatrix} \begin{bmatrix} \dfrac{\partial g}{\partial x} \\ \dfrac{\partial g}{\partial y} \end{bmatrix}. \tag{11.26}$$

The properties of the diffusion tensor can best be seen if the symmetric tensor is brought into its principal axis system by a rotation of the coordinate system. Then, Eq. (11.26) reduces to

$$\boldsymbol{j} = -\begin{bmatrix} D_{x'} & 0 \\ 0 & D_{y'} \end{bmatrix} \begin{bmatrix} \dfrac{\partial g}{\partial x'} \\ \dfrac{\partial g}{\partial y'} \end{bmatrix} = -\begin{bmatrix} D_{x'} \dfrac{\partial g}{\partial x'} \\ D_{y'} \dfrac{\partial g}{\partial y'} \end{bmatrix}. \tag{11.27}$$

Now, the diffusion in the two directions of the axes is decoupled. The two coefficients on the diagonal, $D_{x'}$ and $D_{y'}$, are the *eigenvalues* of the diffusion tensor. In analogy to isotropic diffusion, the general solution of the anisotropic diffusion can be written as

$$\hat{g}(\boldsymbol{x}, t) = \frac{1}{2\pi\sigma_{x'}(t)\sigma_{y'}(t)} \exp\left(-\frac{x'^2}{2\sigma_{x'}(t)}\right) * \exp\left(-\frac{y'^2}{2\sigma_{y'}(t)}\right) * g(\boldsymbol{x}, 0), \tag{11.28}$$

in the spatial domain with $\sigma_{x'}(t) = \sqrt{2D_{x'}t}$ and $\sigma_{y'}(t) = \sqrt{2D_{y'}t}$, provided that the diffusion tensor does not depend on the position.

This means that anisotropic diffusion is equivalent to cascaded convolution with two 1-D Gaussian convolution kernels that are steered into the directions of the principal axes of the diffusion tensor. This property makes it not easy to implement anisotropic diffusion. We will discuss practical implementations later in Section 11.4.6. If one of the two eigenvalues of the diffusion tensor is significantly larger than the other, diffusion occurs only in the direction of the corresponding eigenvector. Thus the gray values are smoothed only in this direction. The spatial blurring is — as for any diffusion process — proportional to the square root of the diffusion constant Eq. (11.23).

11.3.4 Steerable Averaging

The idea of *steerable filters* is similar to inhomogeneous and anisotropic diffusion in the sense that the filter is made dependent on the local image structure. However, it is a more general concept. It is not restricted to averaging but can be applied to any type of convolution process. The basic idea of steerable filters is as follows. A steerable filter has some freely adjustable parameters that control the filtering. This could be very different properties such as the degree of smoothing, the direction of smoothing, or both. It is easy to write down a filter mask with adjustable parameters (we have done this, e. g., already in Eqs. (11.22) and (11.28) where the parameter t determines the degree of smoothing). It is, however, not computationally efficient to convolve an image with masks that are different at every pixel. Then, no longer can any advantage be made from the fact that masks are separable.

An alternative approach is to seek a base of a few filters, and to use these filters to compute a set of filtered images. Then, these images are interpolated using parameters that depend on the adjustable parameters. In operator notation this reads as

$$\mathcal{H}(\alpha) = \sum_{p=1}^{P} f_p(\alpha)\mathcal{H}_p \qquad (11.29)$$

where $\mathcal{H}_p$ is the pth filter and $f_p(\alpha)$ a scalar function of the steering parameter α. Two problems must be solved to use steerable filters. First, and most basically, it is not clear that such a filter base H_p exists at all. Second, the relation between the steering parameter(s) α and the interpolation coefficients f_p must be found. If the first problem is solved, we mostly get the solution to the second for free. As a simple example, directional smoothing is discussed in Example 11.2.

Example 11.2: Simple directional smoothing

A directional smoothing filter is to be constructed with the following transfer function

$$\hat{h}_\theta(k,\phi) = f(k)\cos^2(\phi - \theta). \qquad (11.30)$$

In this equation cylinder coordinates (k,ϕ) are used in the Fourier domain. The filter in Eq. (11.30) is a *polar separable* filter with an arbitrary radial function $f(k)$. This radial component provides an isotropic bandpass filtering.

The steerable angular term is given by $\cos^2(\phi - \theta)$. Structures oriented into the direction θ remain in the image, while those perpendicular to θ are completely filtered out. The angular width of the directional filter is $\pm 45°$.

Using elementary trigonometry it can be shown that this filter can be computed from only two base filters in the following way:

$$\hat{h}_\theta(k,\phi) = \frac{1}{2} + \frac{1}{2}[\cos(2\theta)\hat{H}_0(k,\phi) + \sin(2\theta)\hat{H}_{45}(k,\phi)] \qquad (11.31)$$

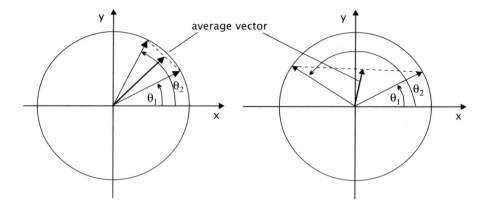

Figure 11.2: *Averaging of an angle by vector addition of unit vectors $\bar{\boldsymbol{n}}_\theta = [\cos\theta, \sin\theta]^T$. The average vector $(\bar{\boldsymbol{n}}_{\theta_1} + \bar{\boldsymbol{n}}_{\theta_2})/2$ points to the correct direction $(\theta_1 + \theta_2)/2$ but its magnitude decreases with the difference angle.*

with the filter base

$$
\begin{aligned}
\hat{h}_0(k, \phi) &= f(k)\cos^2\phi \\
\hat{h}_{45}(k, \phi) &= f(k)\cos^2(\phi - \pi/4) = f(k)\sin^2(\phi).
\end{aligned}
\tag{11.32}
$$

The two base filters are directed into 0° and 45°. The directional filter $\hat{h}_\theta$ can be steered into any direction between -90° and 90°.

Although the equations for this steerable directional smoothing filter are simple, it is not easy to implement the polar separable base filters since they are not (Cartesian) separable and, thus, require quite large convolution kernels. This makes this approach less attractive.

11.3.5 Averaging in Multichannel Images

At first glance, it appears that there is not much special about averaging of multichannel images, just apply the smoothing mask to each of the P channels individually:

$$
\boldsymbol{G}' = \begin{bmatrix} \boldsymbol{G}'_1 \\ \boldsymbol{G}'_2 \\ \vdots \\ \boldsymbol{G}'_p \end{bmatrix} = \boldsymbol{H} * \boldsymbol{G} = \begin{bmatrix} \boldsymbol{H} & * & \boldsymbol{G}_1 \\ \boldsymbol{H} & * & \boldsymbol{G}_2 \\ & \vdots & \\ \boldsymbol{H} & * & \boldsymbol{G}_p \end{bmatrix}.
\tag{11.33}
$$

This simple concept can also be extended to normalized convolution. If to all components the same smoothing kernel is applied, it is sufficient to use one common weighting image that can be appended as the $P + 1$ component of the multicomponent image

$$
\begin{bmatrix} \boldsymbol{G}'_1 \\ \boldsymbol{G}'_2 \\ \vdots \\ \boldsymbol{G}'_P \\ \boldsymbol{W}' \end{bmatrix} = \begin{bmatrix} \boldsymbol{H} * \boldsymbol{W}\boldsymbol{G}_1/\boldsymbol{H} * \boldsymbol{W} \\ \boldsymbol{H} * \boldsymbol{W}\boldsymbol{G}_2/\boldsymbol{H} * \boldsymbol{W} \\ \vdots \\ \boldsymbol{H} * \boldsymbol{W}\boldsymbol{G}_P/\boldsymbol{H} * \boldsymbol{W} \\ \boldsymbol{H} * \boldsymbol{W} \end{bmatrix}.
\tag{11.34}
$$

Figure 11.3: **a** *Color image shown in Plate 16 degraded by addition of normally distributed noise with a standard deviation of 50;* **b** *image shown in* **a** *after smoothing the red, green, and blue channels by a binomial mask with a standard deviation (pixel radius of smoothing) of 3.6;* **c** *smoothing of only the hue channel in the HSI color model with the same filter as in* **b**; **d** *smoothing of only the U and V channels in the YUV color model with the same filter as in* **b**. *(See also Plate 18.)*

A special case of multicomponent images is given when they represent features that can be mapped to angular coordinates. Typically, such features include the direction of an edge or the phase of a periodic signal. Features of this kind are cyclic and cannot be represented well as Cartesian coordinates. They can also not be averaged in this representation. Imagine an angle of +175° and -179°. The mean angle is 178°, since -179° = 360° -179° = 181° is close to 175° and not (175° -179°) / 2 = -2°.

Circular features such as angles are, therefore, better represented as unit vectors in the form $\bar{\boldsymbol{n}}_\theta = [\cos\theta, \sin\theta]^T$. In this representation, they can be averaged correctly as illustrated in Fig. 11.2. The average vector points to the correct direction but its magnitude is generally smaller than 1:

$$(\bar{\boldsymbol{n}}_{\theta_1} + \bar{\boldsymbol{n}}_{\theta_2})/2 = \begin{bmatrix} \cos[(\theta_1 + \theta_2)/2] \\ \sin[(\theta_1 + \theta_2)/2] \end{bmatrix} \cos[(\theta_2 - \theta_1)/2]. \tag{11.35}$$

For an angle difference of 180°, the average vector has zero magnitude. The decrease of the magnitude of the average vector has an intuitive interpretation. The larger the scatter of the angle is, the less is the certainty of the averaged value. Indeed, if all directions are equally probable, the sum vector vanishes, while it grows in length when the scatter is low.

These considerations lead to an extension of the averaging of circular features. We can assign a certainty to the angle by the magnitude of the vector representing it. This is a very attractive form of weighted convolution since it requires — in contrast to normalized convolution (Section 11.3.2) — no time-consuming division. Of course, it works only with features that can adequately be mapped to an angle.

Finally, we consider a measure to characterize the scatter in the direction of vectors. Figure 11.2 illustrates that for low scatter the sum vector is only slightly lower than the sum of the magnitudes of the vector. Thus, we can define an angular coherence measure as

$$c = \frac{|H * G|}{|G|} \tag{11.36}$$

where H is an arbitrary smoothing convolution operator. The operator $|\ |$ means a point operator that computes the magnitude of the vectors at each pixel. This measure is one if all vectors in the neighborhood covered by the convolution operator point into the same direction and zero if they are equally distributed. This definition of a coherence measure works not only in two-dimensional but also in higher-dimensional vector spaces. In one-dimensional vector spaces (scalar images), the coherency measure is, of course, always one.

As a final illustration to multichannel averaging, different types of averaging are applied to a degraded color image in Fig. 11.3. The color image in Fig. 11.3a is degraded by addition of normal distributed noise with a standard deviation of 50. In Fig. 11.3b, the red, green, and blue channels are smoothed by the same binomial mask (Section 11.4.2) with a standard deviation (pixel radius of smoothing) of 3.6. As expected, the noise level is reduced but the edges are blurred. Smoothing of only the hue channel in the HSI color model with the same filter does not blur the edges. But note the false colors in some areas that are generated by averaging a circular feature (the hue) as a scalar. The best results are gained if only the U and V channels in the YUV color system are smoothed with the same filter as in the other cases. This corresponds to the vector averaging discussed above for circular features, since the U and V components are the x and y components of the color vector while the hue is the radial component of the vector. For further details on these color systems we refer to Sections 3.4.7 and 5.5.3 and Plate 16.

11.4 Procedures

11.4.1 Box Filters

11.4.1a Summary of Properties.

Mask and Standard Deviation. Box filters are the simplest smoothing filters one could think of. The mask has constant coefficients. The spatial *variance* of a 1-D box filter with R coefficients is given by

$$\sigma_x^2 = \frac{1}{12}(R^2 - 1). \tag{11.37}$$

This equation is valid for masks with an even and an odd number of coefficients. For the *variance* in the wave number space, no simple equation can be given. The *standard deviation* σ of smoothing increases approximately linearly with the size of the mask. This is different from the binomial filters (Section 11.4.2) where the standard deviation increases only with the square root of the mask size.

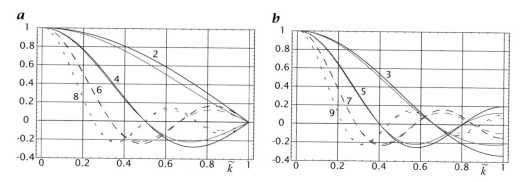

Figure 11.4: Transfer function of discrete 1-D box filters with **a** even and **b** odd number of coefficients as indicated. The gray lines mark the corresponding continuous box filters Eq. (11.40).

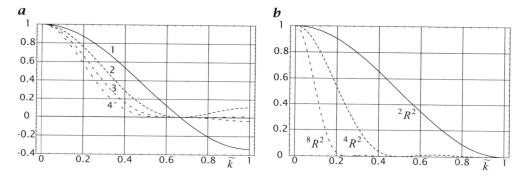

Figure 11.5: Transfer functions of cascaded application of 1-D box filters: **a** repeated application of $^3R = 1/3\,[1\ 1\ 1]$; the number in the graph indicates how often the filter has been applied. **b** Cascaded application of the even-sized masks $^2R^2$, $^4R^2$, $^8R^2$, an efficient scheme for fast smoothing.

Transfer Function. The transfer function can be computed as

$$^R\hat{R}(\tilde{k}) = \frac{\sin(R\pi\tilde{k}/2)}{R\sin(\pi\tilde{k}/2)}. \tag{11.38}$$

For small wave numbers, the transfer function can be expanded in a Taylor series. An expansion up to the fourth order in $\tilde{k}$ yields

$$^R\hat{R} = 1 - \frac{R^2 - 1}{24}(\pi\tilde{k})^2 + \frac{3R^4 - 10R^2 + 7)}{5760}(\pi\tilde{k})^4 + O(\pi\tilde{k}^6). \tag{11.39}$$

Thus, the transfer function decreases for small wave numbers from a unit value proportionally to the wave number squared.

For large R, the transfer function of a discrete R-box filter — corresponding to a continuous box filter with a width of R — slowly converges to the sinc function (Fig. 11.4):

$$^R\hat{R}(\tilde{k}) = \frac{\sin(R\pi\tilde{k}/2)}{R\pi\tilde{k}/2}. \tag{11.40}$$

Therefore, the attenuation of large wave numbers is not steeper than $\propto \tilde{k}^{-1}$. This is only a very weak attenuation of high wave numbers which renders box filters useless for

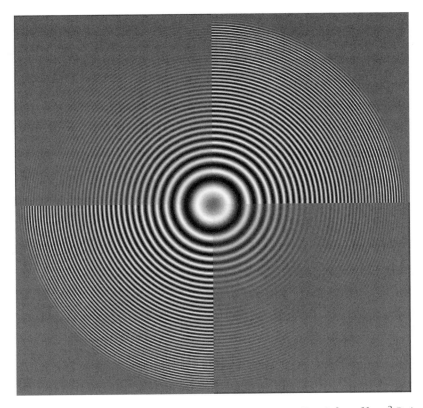

Figure 11.6: *Application of box filters to the ring test pattern:* 3×3 *box filter* $^3\mathcal{R}$ *in the upper left quadrant and* 5×5 *box filter* $^5\mathcal{R}$ *in the lower right quadrant. Notice the nonmonotonic smoothing and the contrast inversion in some areas. They can best be observed at the edges to the unfiltered parts of the test pattern.*

many applications. Cascaded application of box filters, however, significantly improves this limiting attenuation. If a box filter is applied n times, the asymptotic attenuation goes $\propto \tilde{k}^{-n}$ and the secondary peaks diminish. An even number of applications of the same mask results in a nonnegative transfer function (Fig. 11.5).

With the exception of the ^2R box filter, the transfer function does not decrease monotonically. It rather shows significant oscillations. This implies the following disadvantages. First, certain wave numbers are eliminated:

$$^{2R}\hat{r}(\tilde{k}) = 0 \quad \forall \tilde{k} = \frac{n}{R/2}, \quad 1 \leq n \leq R. \tag{11.41}$$

Note that the transfer function for the highest wave number $\tilde{k} = 1$ vanishes only for even-sized box filters. Second, the transfer function becomes negative in certain wave number ranges. This means a $180°$ phase shift and a contrast inversion (Fig. 11.6).

Noise Suppression. Because the box filter simply computes the average of R pixels ("running mean"), the variance of the averaged pixel reduces to (see Section 11.3.1h)

$$\frac{\sigma'}{\sigma} = \frac{1}{\sqrt{R}} \tag{11.42}$$

and the autocovariance is given by

$$C_n = \begin{cases} \dfrac{\sigma^2}{R^2}\,(R - |n|) & -R < n < R \\ 0 & \text{otherwise} \end{cases} \qquad \circ\!\!-\!\!\bullet \qquad \hat{c}(\tilde{k}) = \sigma^2 \frac{\sin^2(R\pi\tilde{k}/2)}{R^2 \sin^2(\pi\tilde{k}/2)}, \qquad (11.43)$$

provided that the input pixels are statistically independent (*white noise*).

Separability. The box filter is separable. Higher-dimensional box filters result from a cascaded application of the 1-D box filter in all directions:

$$\mathbf{R} = \mathbf{R}_y\mathbf{R}_x, \quad \mathbf{R} = \mathbf{R}_z\mathbf{R}_y\mathbf{R}_x, \quad \text{or} \quad \mathbf{R} = \mathbf{R}_t\mathbf{R}_y\mathbf{R}_x. \qquad (11.44)$$

Response to a Step Edge. A box filter converts a step edge into a ramp edge.

Isotropy. A box filter is isotropic only for small wave numbers Eq. (11.39). Generally, it is strongly nonisotropic (see below and example implementations).

Efficient computation $O(R^0)$**.** The box filter with $2R + 1$ coefficients can be computed very efficiently as a recursive filter according to the following equation:

$$g'_n = g'_{n-1} + \frac{1}{2R + 1}(g_{n+R} - g_{n-R-1}). \qquad (11.45)$$

This recursion can be understood by comparing the computations for the convolution at neighboring pixels. When the box mask is moved one position to the right, it contains the same weighting factor for all pixels except for the last and the first pixel. Thus, we can simply take the result of the previous convolution, (g'_{n-1}), subtract the first pixel that just moved out of the mask, (g_{n-R-1}), and add the gray value at the pixel that just came into the mask, (g_{n+R}). In this way, the computation of a box filter does not depend on its size; the number of computations is of $O(R^0)$. Only one addition, one subtraction, and one multiplication are required to compute it.

11.4.1b 1-D Box Filters $^R\mathcal{B}$.

Advantages. Fast computation ($O(R^0)$)) with only two additions and one division Eq. (11.45). Since an integer division cannot be performed efficiently on most hardware platforms, a significantly more efficient integer implementation is possible with 2^q coefficients. The division in Eq. (11.45) can then be replaced by a shift operation.

Disadvantages. Nonmonotonic transfer function eliminating certain wave numbers Eq. (11.41). Negative parts of transfer functions result in contrast inversion (Figs. 11.4 and 11.6). Step edges become ramp edges which are perceived by the human eye as double contours since it is also sensitive to discontinuities in gray scale gradients.

Conclusion. Not recommended

11.4.1c 2-D Box Filters $^R\mathcal{R}_y\,^R\mathcal{R}_x$.

Advantages. Fast computation ($O(R^0)$)) with a separable recursive scheme applied first in x and then in y direction (Sections 11.4.1a and 11.4.1d) results in only four additions and two multiplications.

Disadvantages. To all the disadvantages of 1-D box filters (Section 11.4.1b) comes the significant anisotropy of the transfer function. Structures in axes directions are attenuated much less than in the directions of the diagonals (Fig. 11.7). Application of these filters to the ring test pattern demonstrates the elimination of certain wave numbers and the contrast inversion (180° phase shift, Fig. 11.6). Only the smallest filter, $^2R_y{}^2R_x$, has a nonnegative transfer function.

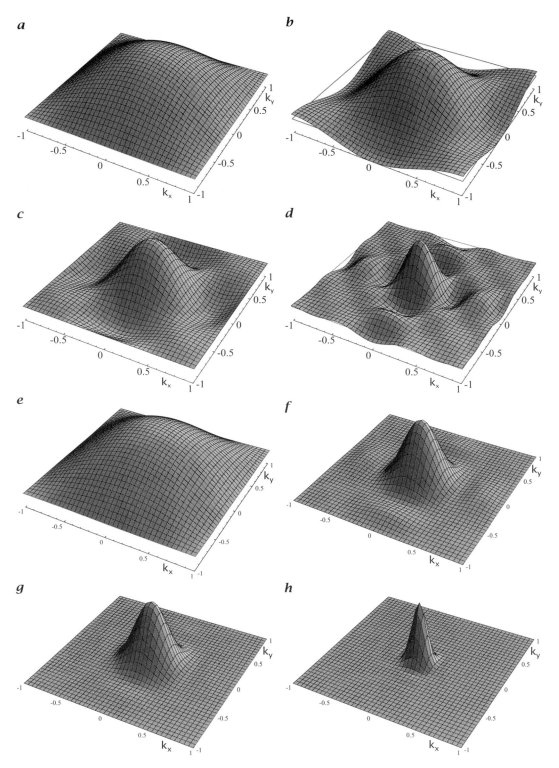

a   *b*

c   *d*

e   *f*

g   *h*

Figure 11.7: *Transfer functions of some 2-D box filters and cascaded 2-D box filters in a pseudo 3-D plot. The wave number 0 is located in the center of the plot.* $\boldsymbol{a}$ $^2R_y{}^2R_x$; $\boldsymbol{b}$ $^3R_y{}^3R_x$; $\boldsymbol{c}$ $^4R_y{}^4R_x$; $\boldsymbol{d}$ $^7R_y{}^7R_x$; $\boldsymbol{e}$ $^2R^2$; $\boldsymbol{f}$ $^4R^2$; $\boldsymbol{g}$ $^4R^2{}^2R^2$; $\boldsymbol{h}$ $^8R^2{}^4R^2{}^2R^2$.

Table 11.3: *Standard deviation, white noise suppression factor, effective mask size, and computational effort (number of required multiplications N_M and additions N_A per pixel) for some cascaded box filters.*

Filter	Standard deviation	Noise suppression	Effective mask size	N_M	N_A
${}^{2}R^{2}$	$\sqrt{1/2} \approx 0.7$	$\sqrt{3/8} \approx 0.61$	3	2	4
${}^{4}R^{2}$	$\sqrt{5/2} \approx 1.58$	$\sqrt{11/64} \approx 0.41$	7	2	4
${}^{4}R^{2}\,{}^{2}R^{2}$	$\sqrt{3} \approx 1.73$	≈ 0.397	9	4	8
${}^{8}R^{2}\,{}^{4}R^{2}\,{}^{2}R^{2}$	$\sqrt{27/2} \approx 3.67$	≈ 0.273	23	6	12
${}^{2}R^{2}$	$\sqrt{1/2} \approx 0.7$	$3/8 = 0.375$	3×3	4	8
${}^{4}R^{2}$	$\sqrt{5/2} \approx 1.58$	$11/64 \approx 0.172$	7×7	4	8
${}^{4}R^{2}\,{}^{2}R^{2}$	$\sqrt{3} \approx 1.73$	≈ 0.158	9×9	8	16
${}^{8}R^{2}\,{}^{4}R^{2}\,{}^{2}R^{2}$	$\sqrt{27/2} \approx 3.67$	≈ 0.075	23×23	12	24
${}^{2}R^{2}$	$\sqrt{1/2} \approx 0.7$	≈ 0.230	$3 \times 3 \times 3$	6	12
${}^{4}R^{2}$	$\sqrt{5/2} \approx 1.58$	≈ 0.071	$7 \times 7 \times 7$	6	12
${}^{4}R^{2}\,{}^{2}R^{2}$	$\sqrt{3} \approx 1.73$	≈ 0.063	$9 \times 9 \times 9$	12	24
${}^{8}R^{2}\,{}^{4}R^{2}\,{}^{2}R^{2}$	$\sqrt{27/2} \approx 3.67$	≈ 0.0205	$23 \times 23 \times 23$	18	36

Conclusion. Not recommended

11.4.1d 3-D Box Filters ${}^{R}\mathcal{R}_{z}{}^{R}\mathcal{R}_{y}{}^{R}\mathcal{R}_{x}$.

Advantages. Fast computation ($O(R^{0})$) with separable 1-D masks resulting in only six additions and three multiplications.

Disadvantages. Same as for 1-D and 2-D box filters. The anisotropy is even more pronounced than cross-sections of the 3-D transfer function in planes along two axes, along an area-diagonal, and along a space-diagonal.

Conclusion. Not recommended. The only excuse to use these filters is their computational efficiency.

11.4.1e Cascaded Box Filters.

The previous sections clearly showed that box filters are very fast to compute but have too many disadvantages to be really useful. This overall negative conclusion changes, however, when box filters are applied multiple times.

Advantages. Cascading box filters avoid or at least diminish many of the disadvantages of box filters as demonstrated in Figs. 11.5 and 11.7. Since individual box filters can be computed independently of their size, cascading them still remains independent of the size. The computational effort can be balanced against the remaining anisotropy and other distortions of the filter. Table 11.3 summarizes some properties for some example implementations, including the standard deviation, the noise suppression factor and the number of multiplications and additions per pixel required to apply the cascaded filter.

Disadvantages. None

Conclusion. Only recommended way of using box filters.

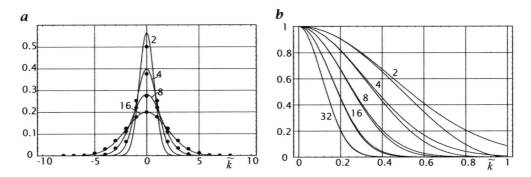

Figure 11.8: *One-dimensional binomial filters $\mathcal{B}^R$ for R = 2 ,4, 8, 16, and 32.* **a** *Masks shown as dots. The curves are the equivalent Gaussian masks with the same standard deviation σ.* **b** *Transfer functions; the light gray curves show the transfer function of the equivalent Gaussian mask. The numbers in the two graphs indicate the exponent R of the mask.*

11.4.2 Binomial Filters

11.4.2a Summary of Properties.

Mask and Standard Deviation. *Binomial filters* are built by cascading the simplest and most elementary smoothing mask we can think of. In the one-dimensional case this is

$$^2R = B = \frac{1}{2}[1\ 1], \tag{11.46}$$

taking the mean value of the two neighboring pixels. We can cascade this mask R times. This corresponds to the filter mask

$$\frac{1}{2^R}\underbrace{[1\ 1] * [1\ 1] * \ldots * [1\ 1]}_{R\ \text{times}}, \tag{11.47}$$

or written as an operator equation

$$\mathcal{B}_x^R = \underbrace{\mathcal{B}_x\mathcal{B}_x\ldots\mathcal{B}_x}_{R\ \text{times}}. \tag{11.48}$$

Cascaded application of the $B_x = 1/2\,[1\ 1]$ filter alternates the resulting images between the interleaved and original grid. Only odd-sized masks should be applied if the resulting smoothed image should lie on the same grid.

The coefficients of the mask are the values of the *binomial distribution* (Fig. 11.8a). The iterative composition of the mask by consecutive convolution with the $1/2\,[1\ 1]$ mask is equivalent to the computation scheme of *Pascal's triangle*:

R	f		σ^2
0	1	1	0
1	1/2	1 1	1/4
2	1/4	1 2 1	1/2
3	1/8	1 3 3 1	3/4
4	1/16	1 4 6 4 1	1
5	1/32	1 5 10 10 5 1	5/4
6	1/64	1 6 15 20 15 6 1	3/2
7	1/128	1 7 21 35 35 21 7 1	7/4
8	1/256	1 8 28 56 70 56 28 8 1	2

$$(11.49)$$

where R denotes the order of the binomial, f the scaling factor 2^{-R}, and σ^2 the variance, i. e., the effective width of the mask. The standard deviation σ of the binomial mask, B^R, is given by Papoulis [110]

$$\sigma^2 = \frac{R}{4}. \tag{11.50}$$

The standard deviation increases only with the square root of the mask size. For high R the mask size ($R + 1$ coefficients) is therefore much larger than the standard deviation $\sqrt{R}/2$.

Two- and higher-dimensional binomial filters can be composed by cascading filters along the corresponding axes:

$$\mathcal{B}^R = \mathcal{B}_y^R \mathcal{B}_x^R, \quad \mathcal{B}^R = \mathcal{B}_t^R \mathcal{B}_y^R \mathcal{B}_x^R, \quad \text{or} \quad \mathcal{B}^R = \mathcal{B}_z^R \mathcal{B}_y^R \mathcal{B}_x^R. \tag{11.51}$$

The smallest odd-sized mask ($R = 2$) of this kind is a (3×3)-binomial filter in 2-D:

$$B = B_y B_x = \frac{1}{4} \begin{bmatrix} 1 & 2 & 1 \end{bmatrix} * \frac{1}{4} \begin{bmatrix} 1 \\ 2 \\ 1 \end{bmatrix} = \frac{1}{16} \begin{bmatrix} 1 & 2 & 1 \\ 2 & 4 & 2 \\ 1 & 2 & 1 \end{bmatrix} \tag{11.52}$$

and in 3-D:

$$B = B_z B_y B_x = \frac{1}{64} \left[\begin{bmatrix} 1 & 2 & 1 \\ 2 & 4 & 2 \\ 1 & 2 & 1 \end{bmatrix}, \begin{bmatrix} 2 & 4 & 2 \\ 4 & 8 & 4 \\ 2 & 4 & 2 \end{bmatrix}, \begin{bmatrix} 1 & 2 & 1 \\ 2 & 4 & 2 \\ 1 & 2 & 1 \end{bmatrix} \right]. \tag{11.53}$$

Transfer Function. The transfer function of the mask $\mathcal{B}$ is according to Eq. (11.9)

$$\hat{b}(\tilde{k}) = \cos(\pi \tilde{k}/2). \tag{11.54}$$

Therefore the transfer function of $\mathcal{B}^R$ is given as the Rth power of the transfer function of $\mathcal{B}$.

$$\hat{b}^R(\tilde{k}) = \cos^R(\pi \tilde{k}/2) = 1 - \frac{R}{8}(\pi \tilde{k})^2 + \left(\frac{3R^2 - 2R}{384} \right)(\pi \tilde{k})^4 + O(\tilde{k}^6), \tag{11.55}$$

where $\tilde{k}$ is the wave number normalized to the Nyquist wave number. The transfer function decreases monotonically and approaches zero at the largest wave number $\tilde{k} = 1$.

For larger masks, both the transfer function and the filter masks quickly approach the Gaussian or normal distribution (Fig. 11.8):

$$b^R(x) \longrightarrow \exp(-2x^2/R), \quad \hat{b}^R(\tilde{k}) \longrightarrow \exp(-R\pi^2 \tilde{k}^2/8). \tag{11.56}$$

The transfer function of the 2-D binomial filter $\mathcal{B}^R$ with $(R + 1) \times (R + 1)$ coefficients is easily derived from the transfer functions of the 1-D filters Eq. (11.55):

$$\hat{b}^R(\tilde{\boldsymbol{k}}) = B_y^R B_x^R = \cos^R(\pi \tilde{k}_y/2) \cos^R(\pi \tilde{k}_x/2), \tag{11.57}$$

and correspondingly for the 3-D filter:

$$\hat{b}^R(\tilde{\boldsymbol{k}}) = B_z^R B_y^R B_x^R = \cos^R(\pi \tilde{k}_z/2) \cos^R(\pi \tilde{k}_y/2) \cos^R(\pi \tilde{k}_x/2). \tag{11.58}$$

Noise Suppression. Because of the simple expressions for the transfer function of binomial filters (Eq. (11.55)), the noise suppression factors for uncorrelated pixels (see Section 11.3.1h) can be computed analytically to

$$\frac{\sigma'}{\sigma} = \frac{\sqrt{2R!}}{2^R R!} = \left(\frac{\Gamma(R+1/2)}{\sqrt{\pi}\Gamma(R+1)}\right)^{1/2} \approx \left(\frac{1}{R\pi}\right)^{1/4}\left(1 - \frac{1}{16R}\right) \qquad (11.59)$$

and the autocovariance is given by

$$C_n = \begin{cases} \dfrac{\sigma^2 2R!}{4^R (R+n)!(R-n)!} & -R \le n \le R \\ 0 & \text{otherwise} \end{cases} \quad \circ\!\!-\!\!\bullet \quad \hat{c}(\tilde{k}) = \sigma^2 \cos^{2R}(\pi\tilde{k}/2), \quad (11.60)$$

provided that the input pixels are statistically independent (*white noise*).

Isotropy. Generally, the transfer function Eq. (11.57) is not isotropic. A Taylor expansion in $\tilde{k}$

$$\hat{b}^R(\tilde{k}) \approx 1 - \frac{R}{8}(\pi\tilde{k})^2 + \frac{2R^2 - R}{256}(\pi\tilde{k})^4 - \frac{R\cos 4\phi}{768}(\pi\tilde{k})^4 \qquad (11.61)$$

shows that only the second-order term is isotropic. Equation Eq. (11.61) is written in polar coordinates $k = [k,\phi]^T$. The fourth and higher-order terms are anisotropic and increase the transfer function in the directions of the diagonals. With increasing r, the anisotropy becomes less, because the isotropic term goes with R^2 while the anisotropic term goes only with R. Since binomial filters are computed by cascaded convolutions, it is also very easy to achieve elliptically shaped transfer functions to make up for different grid spacing on rectangular grids:

$$\mathcal{B}^{k,m,n} = \mathcal{B}_t^k \mathcal{B}_y^m \mathcal{B}_x^n. \qquad (11.62)$$

Monotonic transfer function. The transfer function decreases monotonically and is zero for the highest wave number for *all* binomial masks.

Efficient computation. In contrast to box filters (Section 11.4.1a), no algorithm is available that does not depend on the mask size (not of order $0(R^0)$). It is only possible to replace all multiplication operations by shift operations. This constitutes a significant advantage for integer arithmetic. If we decompose the binomial mask into the elementary smoothing mask $B = 1/2 [1\ 1]$ and apply this mask in all W directions R times each, we only need WR additions. For example, the computation of a 13×13 binomial filter ($R = 12$) requires only 24 additions and some shift operations compared to 169 multiplications and 168 additions needed for the direct approach.

11.4.2b   1-D Binomial Filters $\mathcal{B}^R$.

Advantages. Nonnegative transfer function decreases monotonically towards higher wave numbers, see Eq. (11.54), and is zero for the highest wave number $\tilde{k} = 1$ (Fig. 11.8).

Disadvantages. The computation of binomial filters is less efficient (not order $0(R^0)$) than for box filters (Section 11.4.1a), but can be computed in integer arithmetic with only addition and shift operations. The standard deviation of smoothing increases only with the square root of the mask size ($\sigma = \sqrt{R}/2$) and the noise suppression factor is only proportional to $R^{-1/4}$.

Conclusion. Recommended to use for smoothing only over small distances, otherwise computationally inefficient.

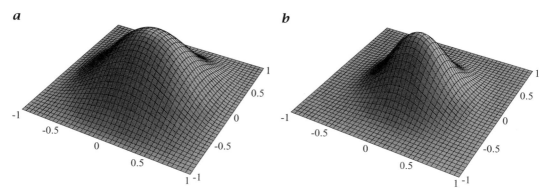

a   *b*

Figure 11.9: *Pseudo 3-D plot of the transfer function of two-dimensional binomial masks:* a 3×3, $\mathcal{B}^2$; b 5×5, $\mathcal{B}^4$.

11.4.2c 2-D Binomial Filters $\mathcal{B}^R$ $\mathcal{B}_x^R$.

Advantages. Nonnegative transfer function decreasing monotonically towards higher wave numbers Eq. (11.54) and being zero for the highest wave number $\tilde{k} = 1$ (Fig. 11.9). Although the 2-D binomial masks are not exactly isotropic, the deviations are negligible for most applications.

Disadvantages. Less efficient computation than for box filters (Section 11.4.1a), but can be computed in integer arithmetic with only addition and shift operations. Standard deviation of smoothing increases only with the square root of the mask size ($\sigma = \sqrt{R}/2$).

Conclusion. Recommended to use for smoothing only over small distances, otherwise computationally inefficient.

11.4.2d 3-D Binomial Filters $\mathcal{B}_z^R$ $\mathcal{B}_y^R$ $\mathcal{B}_x^R$.

Advantages. Nonnegative transfer function decreasing monotonically towards higher wave numbers Eq. (11.54) and being zero for the highest wave number $\tilde{k} = 1$. Although the 3-D binomial masks are not exactly isotropic, the deviations are negligible for most applications.

Disadvantages. Less efficient computation than for box filters (Section 11.4.1a), but can be computed in integer arithmetic with only addition and shift operations. Standard deviation of smoothing increases only with the square root of the mask size ($\sigma = \sqrt{R}/2$).

Conclusion. Recommended to use for smoothing only over small distances, otherwise computationally inefficient.

11.4.3 Cascaded Multistep Filters

11.4.3a Principle. Despite the efficient implementation of binomial smoothing filters $\mathcal{B}^R$ by cascaded convolution with $\mathcal{B}$, the number of computations increases dramatically for smoothing masks with low cutoff wave numbers, because the standard deviation of the filters is proportional to the square root of R according to Eq. (11.50): $\sigma = \sqrt{R/4}$. Let us consider a smoothing operation over a circle with a radius of about only 1.73 pixels, corresponding to a variance $\sigma^2 = 3$. According to Eq. (11.50) we need to apply $\mathcal{B}^{12}$ which — even in an efficient separable implementation — requires 24 (36) additions and 2 (3) shift operations for each pixel in a 2-D (3-D) image. If we want to smooth over the doubled distance ($\sigma^2 = 12$, radius ≈ 3.5, $\mathcal{B}^{48}$) the number of additions quadruples to 96 (144) per pixel in 2-D (3-D) space.

The problem originates from the small distance of the pixels averaged in the elementary $B = 1/2\,[1\ 1]$ mask. In order to overcome this problem, we may use the same elementary averaging process but with more distant pixels and increase the standard deviation for smoothing correspondingly. In two dimensions, the following masks could be applied along diagonals (σ $\sqrt{2}$ times larger):

$$B_{x+y} = \frac{1}{4}\begin{bmatrix} 1 & 0 & 0 \\ 0 & 2 & 0 \\ 0 & 0 & 1 \end{bmatrix}, \quad B_{x-y} = \frac{1}{4}\begin{bmatrix} 0 & 0 & 1 \\ 0 & 2 & 0 \\ 1 & 0 & 0 \end{bmatrix} \tag{11.63}$$

and along axes (σ two times larger):

$$B_{2x} = \frac{1}{4}[1\ 0\ 2\ 0\ 1], \quad B_{2y} = \frac{1}{4}\begin{bmatrix} 1 \\ 0 \\ 2 \\ 0 \\ 1 \end{bmatrix}. \tag{11.64}$$

In three dimensions, the multistep masks along the axes are:

$$B_{2x} = \frac{1}{4}[1\ 0\ 2\ 0\ 1], \quad B_{2y} = \frac{1}{4}\begin{bmatrix} 1 \\ 0 \\ 2 \\ 0 \\ 1 \end{bmatrix}, \quad B_{2z} = \frac{1}{4}[[1],[0],[2],[0],[1]]. \tag{11.65}$$

The subscripts in these elementary masks denote the number of steps along the coordinate axes between two pixels to be averaged. B_{x+y} averages the gray values at two neighbored pixels in the direction of the main diagonal. B_{2x} takes the mean at two pixels which are two grid points distant in x direction. The standard deviation of these filters is proportional to the distance of the pixels. The most efficient implementations are multistep masks along the axes. They have the additional advantage that the algorithms do not depend on the dimension of the images.

The problem with these filters is that they perform a subsampling. Consequently, they are no longer good smoothing filters for larger wave numbers. If we take, for example, the symmetric 2-D $\mathcal{B}_2^2 = \mathcal{B}_{2x}^2\mathcal{B}_{2y}^2$ filter, we effectively work on a grid which is twice as large in the spatial domain. Hence, the reciprocal grid in the wave number is half the size, and we see the periodic replication of the transfer function. The zero lines of the transfer function show the reciprocal grid for the corresponding subsample grids. For convolution with two neighboring pixels in the direction of the two diagonals, the reciprocal grid is turned by 45°. The grid constant of the reciprocal grid is a factor of $\sqrt{2}$ smaller than that of the original grid.

Used individually, these filters are not of much help. But we can use them in cascade, starting with directly neighboring pixels and increasing the distance recursively. This filter technique is called multistep smoothing. Then, the zero lines of the transfer functions, which lie differently for each pixel distance, efficiently force the transfer function close to zero for large wave number ranges.

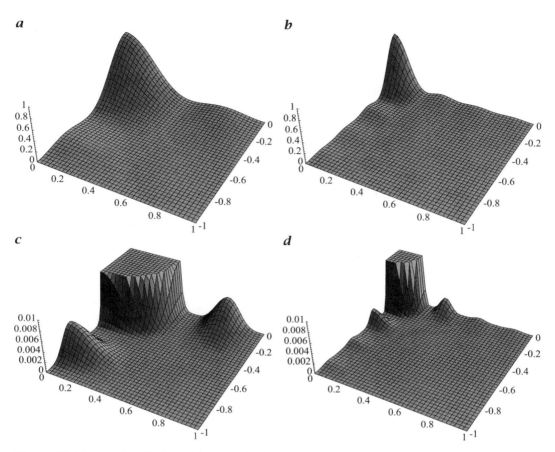

Figure 11.10: *Pseudo 3-D plot of the transfer functions of 2-D cascaded multistep binomial filters:* **a** $\mathcal{B}_2^2\mathcal{B}_1^2$, **b** $\mathcal{B}_4^2\mathcal{B}_2^2\mathcal{B}_1^2$, **c** $\mathcal{B}_2^4\mathcal{B}_1^4$, **d** $\mathcal{B}_4^2\mathcal{B}_2^4\mathcal{B}_1^4$. *Only a quarter of the wave number space is shown. In* **c** *and* **d** *only a range of very low values from 0 to 0.01 for the transfer function is shown to depict the residual inhomogeneities and anisotropies.*

11.4.3b Efficient Implementation. Cascaded multistep binomial filtering leads to a significant performance increase for large-scale smoothing. For normal separable binomial filtering, the number of computations is of the order $0(\sigma^2)$. For multistep binomial filtering it reduces to a order logarithmic in σ ($0(\mathrm{ld}\sigma)$) if a cascade of filter operations with step doubling is performed:

$$\underbrace{B_{2^{S-1}x}^R B_{8x}^R B_{4x}^R B_{2x}^R B_x^R}_{S \text{ times.}} \qquad (11.66)$$

Such a mask has the standard deviation

$$\sigma^2 = \underbrace{R/4 + R + 4R + \ldots + 4^{S-1}R}_{S \text{ times}} = \frac{R}{12}(4^S - 1). \qquad (11.67)$$

Thus, for S steps only RS additions are required while the standard deviation grows exponentially with $\sigma \approx 2^S \sqrt{R/12}$.

With the parameter R, we can adjust the degree of isotropy and the degree of residual inhomogeneities in the transfer function. A very efficient implementation is given by using R = 2 (B^2 = 1/4[1 2 1] in each direction). However, the residual peaks at high wave numbers with maximal amplitudes up to 0.08 still constitute some significant disturbances (Fig. 11.10a and b).

With the next larger odd-sized masks (R = 4, B^4 = 1/16[1 4 6 4 1] in each direction) these residual bumps at high wave numbers are suppressed well below 0.005 (Fig. 11.10c and d). This is about the relative resolution of 8-bit images and should therefore be sufficient for most applications. With still larger masks, they could be suppressed even further.

11.4.4 Cascaded Multigrid Filters

The multistep cascaded filter approach can be even further enhanced by converting it into a multiresolution technique. The idea of *multigrid smoothing* is very simple. When a larger-step mask is involved, this operation can be applied on a correspondingly coarser grid. This means that the last operation before using the larger-step mask needs to compute the convolution only at the grid points used by the following coarser grid operator. This sampling procedure is denoted by the sampling operator $\Downarrow$. An additional index specifies the sampling direction and step width. $\Downarrow_{2x}$ means that the preceding operation is performed only at every second pixel in x direction. Therefore, $\mathcal{B}_x \Downarrow_{2x} \mathcal{B}_x^2$ means: Apply $\mathcal{B}_x^2$ to every second pixel in horizontal direction resulting in a sampled row of half length; then apply the $\mathcal{B}_x$ filter mask to this filtered and sampled row.

Multigrid smoothing makes the number of computations essentially independent of the standard deviation of the smoothing mask. Recursively applied, the operator sequence is

$$\underbrace{\Downarrow_{2x} \mathcal{O} \dots \Downarrow_{2x} \mathcal{O} \Downarrow_{2x} \mathcal{O}}_{S \text{ times}}.$$

If the $\Downarrow_{2x} \mathcal{O}$ operator takes n operations, the sequence takes

$$n(1 + 1/2 + 1/4 + \dots) < 2n$$

steps. Thus, smoothing to any degree takes less than two times the operations for smoothing for the first step! As for multistep binomial filters, the standard deviation grows by a factor of two. Also — as long as the sample theorem is met — the transfer functions of the filters are the same as for the multistep filters that were discussed in Section 11.4.3.

11.4.5 Recursive Smoothing

The principle of recursive convolution has already been discussed in Section 10.3.8. In Section 10.4.1, we saw that zero-phase recursive filters which are required for smoothing can be easily designed by letting the filter run in the forward and backward directions. The most efficient way for smoothing is to cascade the forward and backward running filter,

$$\mathcal{A}_x = {}^-\mathcal{A}_x {}^+\mathcal{A}_x. \tag{11.68}$$

For 2-D images this means that four runs of the same filter are required, two in horizontal and two in vertical direction,

$$\mathcal{A} = {}^-\mathcal{A}_y {}^+\mathcal{A}_y {}^-\mathcal{A}_x {}^+\mathcal{A}_x. \tag{11.69}$$

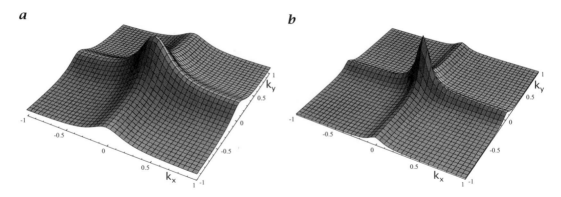

Figure 11.11: *Pseudo 3-D plot of the transfer function of recursive smoothing filters according to Eq. (11.71) for* **a** $\alpha = 1/2$, **b** $\alpha = 3/4$.

The simplest recursive filter has already been discussed in Section 10.4.3. The most efficient and numerically stable implementation

$$g'_m = g_m + \alpha(g'_{m-1} - g_m) \tag{11.70}$$

requires only one multiplication, one addition, and one subtraction for each run. The zero-phase transfer function of this simple lowpass filter for consecutive runs in backward and forward direction is given by:

$$\hat{A}(\tilde{k}) = \frac{1-\alpha}{1+\alpha^2 - 2\alpha\cos\pi\tilde{k}}(1-\alpha\cos\pi\tilde{k}). \tag{11.71}$$

A Taylor expansion of the transfer function for small $\tilde{k}$ yields:

$$\hat{A}(\tilde{k}) = 1 - \frac{a(1+a)}{(1-a)^2}(\pi\tilde{k})^2 + \frac{a(1+a)(1+10a+a^2)}{24(1-a)^4}(\pi\tilde{k})^4 + 0(\pi\tilde{k})^6. \tag{11.72}$$

Recursive smoothing has the following general features:

- It can be computed very efficiently. The number of computations does not depend on the standard deviation of smoothing.
- Separable recursive smoothing is not isotropic. Especially disturbing is the feature that structures in the direction of the axes are much less diminished than in all other directions (Fig. 11.11).

Given the significant disadvantage of anisotropic filtering, it is preferable instead of recursive smoothing filters to use the efficient cascaded multistep or multigrid filters that are based on binomial filters. These filters can also be controlled in a flexible way by the filter being cascaded and by the number of filters in the cascade (Sections 11.4.3 and 11.4.4).

11.4.6 Inhomogeneous and Anisotropic Diffusion

In this and the following sections we discuss two approaches for smoothing operations that depend on the spatial gray value structure and thus smooth regions without blurring edges. The basic principles of these techniques were discussed in Section 11.3.3.

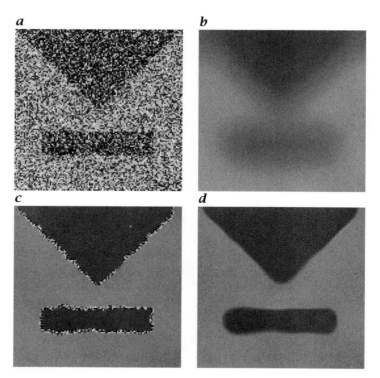

Figure 11.12: *a Original, smoothed by b linear diffusion, c inhomogeneous but isotropic diffusion, and d anisotropic diffusion. From Weickert [152].*

Nonlinear diffusion diminishes the blurring by decreasing the diffusion coefficient at edges. The simplest approach is inhomogeneous diffusion. The diffusion coefficient decreases with increasing magnitude of the gradient. Weickert [152] used

$$D = 1 - \exp\left(-\frac{c_m}{(|\nabla(B^r * g)(\boldsymbol{x})|/\lambda)^m}\right). \tag{11.73}$$

This equation implies that for small magnitudes of the gradient the diffusion coefficient is constant. At a certain threshold of the magnitude of the gradient, the diffusion coefficient quickly decreases towards zero. The higher the exponent m is, the steeper the transition is. With the values used by Weickert [152], $m = 4$ and $c_4 = 3.31488$, the diffusion coefficient falls from 1 at $|\nabla g|/\lambda = 1$ to about 0.15 at $|\nabla g|/\lambda = 2$. Note that a regularized gradient has been chosen in Eq. (11.73), because the gradient is not computed from the image $g(\boldsymbol{x})$ directly, but from the image smoothed with a binomial smoothing mask. A properly chosen regularized gradient stabilizes the inhomogeneous smoothing process and avoids instabilities and steepening of the edges.

A simple implementation of inhomogeneous diffusion controls binomial smoothing using the operator $\mathcal{I} + \mathcal{D} \cdot (\mathcal{B} - \mathcal{I})$. The operator $\mathcal{D}$ computes the diffusion coefficient according to Eq. (11.73) and results in a control image which is one in constant regions and drops towards zero at edges. Weickert [152] did not use this simple explicit method but a more sophisticated implicit solution scheme which, however, requires the solution of large linear equation systems.

For *anisotropic diffusion*, the diffusion constant needs to be replaced by a *diffusion tensor* as discussed in Section 11.3.3c. Then, the eigenvalue of the diffusion tensor

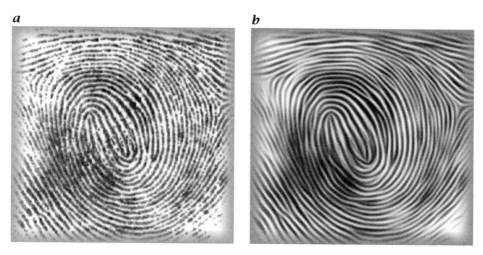

Figure 11.13: *a Original finger print; b smoothed by coherence-enhancing anisotropic diffusion. From Weickert [151].*

perpendicular to the edge is set to the value given by Eq. (11.73), while the other eigen-value parallel to the edge is set constant to one. In this way, only smoothing across the edge but not along the edge is hindered:

$$
\begin{aligned}
D_{x'} &= 1 - \exp\left(-\frac{c_m}{(|\nabla(B^r * g)(\boldsymbol{x})|/\lambda)^m}\right) \\
D_{y'} &= 1.
\end{aligned}
\tag{11.74}
$$

The different approaches to nonlinear smoothing are compared in Fig. 11.12 with a noisy test image containing a triangle and a rectangle. Standard smoothing, i.e., linear diffusion, significantly suppresses the noise but results in a significant blurring of the edges (Fig. 11.12b). Inhomogeneous diffusion does not lead to a blurring of the edges although the constant regions are perfectly smoothed (Fig. 11.12c). However, the edges remain noisy since no smoothing is performed at all at the edges. This disadvantage is avoided by anisotropic diffusion since this technique still performs smoothing along the edges (Fig. 11.12d). The smoothing along edges has, however, the disadvantage that the corners at the edges are now blurred as with linear diffusion. This did not happen with inhomogeneous diffusion (Fig. 11.12c).

Anisotropic diffusion can also be used to enhance directed structures in images. Weickert [152] called this *coherence-enhancing diffusion*. Now smoothing occurs *only* along edges. Smoothing is not performed in regions of constant gray values and in regions where no directed gray values are present. In this case the diffusion process is controlled by the *structure tensor* which will be discussed in Section 13.3.3. Further details of this approach can be found in Weickert [152]. Here we will only show two applications. A classical application is the enhancement of finger print images (Fig. 11.13a). Coherence-enhancing diffusion connects disrupted lines and removes noise in regions with directed structures (Fig. 11.13b). This technique is also useful to detect defects in fabrics. Figure 11.14 shows a sequence of increasingly smoothed images where more and more coarser structures are removed leaving finally only the defect in the image.

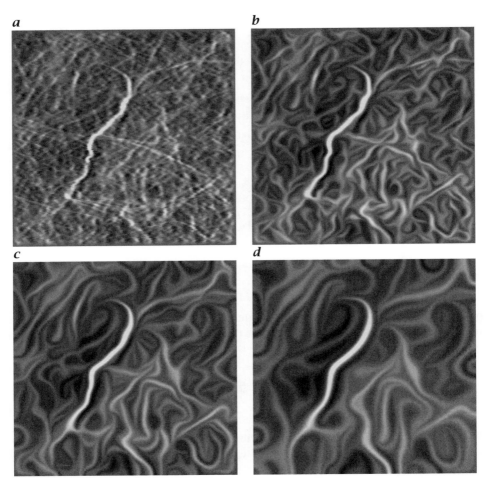

a   *b*

c   *d*

Figure 11.14: *a Original image of a fabric with a defect; b - d sequence of increasingly smoothed images of a using coherence-enhancing anisotropic diffusion. From Weickert [151].*

11.4.7 Steerable Directional Smoothing

Steerable averaging and a simple example for directional smoothing based on polar separable filters was discussed in Section 11.3.4. While this concept is elegant, it is hard to implement. In this section, we introduce a simple and fast technique to approximate polar separable filters for a steerable directional smoothing filter. The general transfer function of the polar separable directional smoothing filter was according to Eq. (11.30)

$$\hat{H}_\theta(k, \phi) = f(k) \cos^2(\phi - \theta). \tag{11.75}$$

A steerable filter of this type required only two base filters, one in the direction of 0° and one in the direction of 45°:

$$\hat{H}_\theta(k, \phi) = \frac{1}{2} + \frac{1}{2}[\cos(2\theta)\hat{H}_0(k, \phi) + \sin(2\theta)\hat{H}_{\pi/4}(k, \phi)]. \tag{11.76}$$

Using separable filters, a polar separable directional filter can be approximated only in a narrow wave number range. Thus $f(k)$ in Eq. (11.75) must be a bandpass filter. The

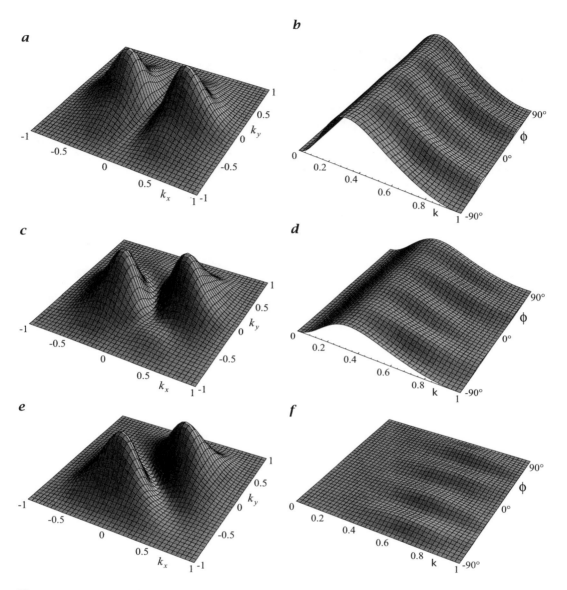

Figure 11.15: *Transfer function of the steerable directional smoothing filter steered in **a** 0°, **c** 22.5°, and **e** 45°. Isotropy of the transfer function as a function of the wave number and the steering angle for angle difference between the direction of the filter and the direction of the filtered structure of **b** 0°(maximum response), **d** 45°, and **f** 90°(minimum response).*

following filter set turns out to be a good approximation. It uses only binomial filters along axes ($\mathcal{B}_x$ and $\mathcal{B}_y$) and diagonals ($\mathcal{B}_{x-y}$ and $\mathcal{B}_{x+y}$) with equal variance:

$$
\begin{aligned}
\mathcal{H}_0 &= (1 - \mathcal{B}_x^{2R}\mathcal{B}_y^{2R} - (\mathcal{B}_x^{2R} - \mathcal{B}_y^{2R}))/2 \\
\mathcal{H}_{\pi/4} &= (1 - \mathcal{B}_x^{2R}\mathcal{B}_y^{2R} - (\mathcal{B}_{x+y}^{R} - \mathcal{B}_{x-y}^{R}))/2
\end{aligned}
\tag{11.77}
$$

with the transfer function (Fig. 11.15a,e)

a   **b**

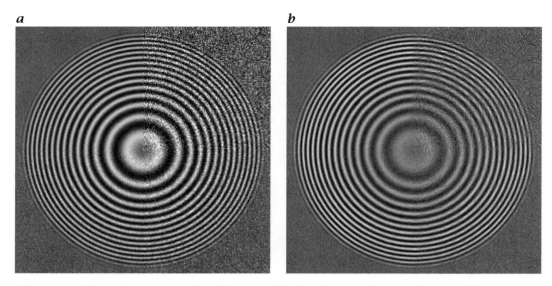

Figure 11.16: a *Ring test image with an amplitude of 100 superposed by zero mean normal distributed noise with a standard deviation of 10, 20, and 40 in three quadrants. **b** Image **a** after two iterations of steerable directional smoothing.*

$$2\hat{h}_0 = 1 - \cos(\pi\tilde{k}_x/2)^{2R}\cos(\pi\tilde{k}_y/2)^{2R} - \cos(\pi\tilde{k}_x/2)^{2R} + \cos(\pi\tilde{k}_y/2)^{2R}$$

$$2\hat{h}_{\pi/4} = 1 - \cos(\pi\tilde{k}_x/2)^{2R}\cos(\pi\tilde{k}_y/2)^{2R} - \cos(\pi(\tilde{k}_x + \tilde{k}_y)/2)^{R} + \cos(\pi(\tilde{k}_x - \tilde{k}_y)/2)^{R}.$$

For small wave numbers the transfer function of the filter steered in the direction θ agrees with the required form Eq. (11.75):

$$\hat{h}_\theta \approx \frac{R}{2}\cos^2(\phi - \theta)(\pi\tilde{k})^2. \tag{11.78}$$

The transfer function does not depend — except for high wave numbers, where the transfer function is low — on the steering angle. Figure 11.15b, d, and f shows this for the maximum response ($\theta = \phi$), $\theta = \phi + \pi/4$, and the minimum response ($\theta = \phi + \pi/2$). Thus it is not surprising that this filter can well be used to remove noise in images with directed gray value structures when this filter is steered to smooth in the direction of constant gray values (Fig. 11.16).

11.5 Advanced Reference Material

Equations for symmetric masks with odd-numbered size in 2-D and 3-D

11.1

2-D symmetric $(2R+1)^2$-mask

$$
\begin{aligned}
g'_{mn} =\; & h_{00}g_{nm} \\
& + \sum_{n'=1}^{R} h_{0n'}(g_{m,n-n'} + g_{m,n+n'}) + \sum_{m'=1}^{R} h_{m'0}(g_{m-m',n} + g_{m+m',n}) \\
& + \sum_{m'=1}^{R}\sum_{n'=1}^{R} h_{m'n'}(g_{m-m',n-n'} + g_{m-m',n+n'} + g_{m+m',n-n'} + g_{m+m',n+n'})
\end{aligned}
\tag{11.79}
$$

3-D symmetric $(2R+1)^3$-mask

$$
\begin{aligned}
g'_{lmn} =\; & h_{000}g_{lnm} \\
& + \sum_{n'=1}^{R} h_{00n'}(g_{l,m,n-n'} + g_{l,m,n+n'}) \\
& + \sum_{m'=1}^{R} h_{0m'0}(g_{l,m-m',n} + g_{l,m+m',n}) \\
& + \sum_{l'=1}^{R} h_{l'00}(g_{l-l',m,n} + g_{l+l',m,n}) \\
& + \sum_{m'=1}^{R}\sum_{n'=1}^{R} h_{0m'n'}(g_{l,m-m',n-n'} + g_{l,m-m',n+n'} + g_{l,m+m',n-n'} + g_{l,m+m',n+n'}) \\
& + \sum_{l'=1}^{R}\sum_{n'=1}^{R} H_{l'0n'}(g_{l-l',m,n-n'} + g_{l-l',m,n+n'} + g_{l+l',m,n-n'} + g_{l+l',m,n+n'}) \\
& + \sum_{l'=1}^{R}\sum_{m'=1}^{R} H_{l'm'0}(g_{l-l',m-m',n} + g_{l-l',m+m',n} + g_{l+l',m-m',n} + g_{l+l',m+m',n}) \\
& + \sum_{l'=1}^{R}\sum_{m'=1}^{R}\sum_{n'=1}^{R} h_{l'm'n'}(g_{l-l',m-m',n-n'} + g_{l-l',m-+m',n+n'} \\
& \qquad + g_{l-l',m+m',n-n'} + g_{l-l',m+m',n+n'} \\
& \qquad + g_{l+l',m-m',n-n'} + g_{l+l',m-m',n+n'} + g_{l+l',m+m',n-n'} + g_{l+l',m+m',n+n'})
\end{aligned}
\tag{11.80}
$$

Equations for symmetric masks with even-numbered size in 2-D and 3-D

11.2

2-D symmetric $(2R)^2$-mask

$$
\begin{aligned}
g'_{m,n} =\; & \sum_{m'=1/2}^{R-1/2}\sum_{n'=1/2}^{R-1/2} h_{m'n'} \cdot \\
& (g_{m-m',n-n'} + g_{m-m',n+n'} + g_{m+m',n-n'} + g_{m+m',n+n'})
\end{aligned}
\tag{11.81}
$$

3-D symmetric $(2R)^3$-mask

$$
\begin{aligned}
g'_{lmn} = & \sum_{l'=1/2}^{R-1/2} \sum_{m'=1/2}^{R-1/2} \sum_{n'=1/2}^{R-1/2} h_{l'm'n'} \cdot \\
& \cdot (g_{l-l',m-m',n-n'} + g_{l-l',m-+m',n+n'} \\
& + g_{l-l',m+m',n-n'} + g_{l-l',m+m',n+n'} \\
& + g_{l+l',m-m',n-n'} + g_{l+l',m-m',n+n'} \\
& + g_{l+l',m+m',n-n'} + g_{l+l',m+m',n+n'})
\end{aligned}
\tag{11.82}
$$

11.3 Transfer functions for symmetric 2-D masks with even and odd number of coefficients

2-D symmetric $(2R+1)^3$-mask

$$
\begin{aligned}
\hat{g}(\tilde{\boldsymbol{k}}) = & \; h_{00} \\
& + 2\sum_{n'=1}^{R} h_{0n'}\cos(n'\pi\tilde{k}_1) + \sum_{m'=1}^{R} h_{m'0}\cos(m'\pi\tilde{k}_2) \\
& + 4\sum_{m'=1}^{R}\sum_{n'=1}^{R} h_{m'n'}\cos(n'\pi\tilde{k}_1)\cos(m'\pi\tilde{k}_2)
\end{aligned}
\tag{11.83}
$$

2-D symmetric $(2R)^2$-mask

$$
\hat{g}(\tilde{\boldsymbol{k}}) = 4\sum_{m'=1/2}^{R-1/2}\sum_{n'=1/2}^{R-1/2} h_{m'n'}\cos(n'\pi\tilde{k}_1)\cos(m'\pi\tilde{k}_2)
\tag{11.84}
$$

11.4 Transfer functions for symmetric 3-D masks with even and odd number of coefficients

3-D symmetric $(2R+1)^3$-mask

$$
\begin{aligned}
\hat{g}(\tilde{\boldsymbol{k}}) = & \; h_{000} \\
& + 2\sum_{n'=1}^{R} h_{00n'}\cos(n'\pi\tilde{k}_1) + 2\sum_{m'=1}^{R} h_{0m'0}\cos(m'\pi\tilde{k}_2) + 2\sum_{l'=1}^{R} h_{l'00}\cos(l'\pi\tilde{k}_3) \\
& + 4\sum_{m'=1}^{R}\sum_{n'=1}^{R} h_{0m'n'}\cos(m'\pi\tilde{k}_2)\cos(n'\pi\tilde{k}_1) \\
& + 4\sum_{l'=1}^{R}\sum_{n'=1}^{R} h_{l'0n'}\cos(l'\pi\tilde{k}_3)\cos(n'\pi\tilde{k}_1) \\
& + 4\sum_{l'=1}^{R}\sum_{m'=1}^{R} h_{l'm'0}\cos(l'\pi\tilde{k}_3)\cos(m'\pi\tilde{k}_2) \\
& + 8\sum_{l'=1}^{R}\sum_{m'=1}^{R}\sum_{n'=1}^{R} h_{k'm'n'}\cos(l'\pi\tilde{k}_3)\cos(m'\pi\tilde{k}_2)\cos(n'\pi\tilde{k}_1)
\end{aligned}
\tag{11.85}
$$

3-D symmetric $(2r)^3$-mask

$$
\hat{g}(\tilde{\boldsymbol{k}}) = 8\sum_{l'=1/2}^{R-1/2}\sum_{m'=1/2}^{R-1/2}\sum_{n'=1/2}^{R-1/2} H_{k'm'n'}\cos(l'\pi\tilde{k}_3)\cos(m'\pi\tilde{k}_2)\cos(n'\pi\tilde{k}_1)
\tag{11.86}
$$

Reference to advanced topics

T. Lindeberg, 1994. Scale-Space Theory in Computer Vision, Kluwer Academic Publishers, Dordrecht. *Covers scale space theory as the theory behind diffusion filters. Limited, however, to linear diffusion.*

J. Weickert, 1998. Anisotropic Diffusion and Image Processing, Teubner, Stuttgart. *An excellent account of the mathematical foundation of inhomogeneous and anisotropic diffusion filters for image processing. Discusses also a number of applications.*

W. T. Freeman and E. H. Adelson, 1991. The design and use of steerable filters, IEEE Trans. PAMI, 13, 891–906. *Groundbreaking paper on steerable filters.*

12   Edges and Lines

12.1   Highlights

The detection of edges offers an alternative way to find objects in images. The knowledge of the boundary of an object is sufficient to describe its shape and, thus, equivalent to the detection of regions of constant values.

The shape of edges in images depends on many parameters: the geometrical and optical properties of the object, the illumination conditions, the point spread function of the image formation process, and the noise level in the images. Only if all these processes are modeled adequately can we develop an optimal edge detector from which we can also expect a subpixel-accurate estimate of the position of an edge.

The base of edge detection is differentiation. In discrete images, differentiation has to be approximated by differences between neighboring pixels. Therefore, any difference scheme is only an approximation and, thus, associated with errors both in the estimation of the steepness and orientation of an edge. This elementary fact is often not considered adequately and one of the most common sources of errors in low-level image processing.

It will be shown that there are much better alternatives to classical operators such as popular and widely used operators, Roberts and Sobel, to name only two of them.

12.2   Task

Edge detection requires neighborhood operators that are sensitive to changes and suppress areas of constant gray values. In this way, a feature image is formed in which those parts of the image appear bright where changes occur while all other parts remain dark.

Mathematically speaking, an ideal edge is a discontinuity of the spatial gray value function $g(x)$. It is obvious that this is only an abstraction which often does not meet the reality. Real edges deviate in one or the other aspect from an ideal edge because of the requirement to meet the sampling theorem (Section 6.3.2), the illumination conditions (Section 3.4.3), or the properties of the objects. Thus, the first task of edge detection is to find out the properties of the edge contained in the image to be analyzed (task list 9). Only if we can formulate a model of the edges will it be possible to determine how accurate and under which conditions it will be possible to detect an edge and to find an optimum edge detection. A simple edge model is discussed in Section 12.3.1.

One important special case of discontinuities is lines. While edges are the border between two regions with different gray values, lines are a kind of second-order or double discontinuity. On both sides of a line, the regions may have the same gray

391

Task List 9: Edges and Lines

Task	Procedures
Determine type of edges (discontinuity) in images: edges, lines, surfaces	Analyze optical properties of observed objects; analyze illumination conditions; analyze point spread function of image formation process
Select an edge detector on criteria such as maximum errors for edge direction and edge magnitude	Choose from various approximations of first and second-order derivation filters and from different regularization schemes
Determine "edge strength"	Find an objective confidence measure to identify an edge

values. Only the line itself has a different gray value but has — in the limit of abstraction — no area.

Edge detection is always based on differentiation in one or the other form. In discrete images, differentiation is replaced by discrete differences which are only an approximation to differentiation. The errors associated with these approximations require careful consideration. They cause effects that are not expected in the first place. The two most serious errors are: edge detection becomes anisotropic, i. e., edges are detected not equally well in all directions, and significant errors occur in the estimate of the direction of the edges.

While it is obvious what edges are in scalar images, different measures are possible in multicomponent or vectorial images (Section 12.3.4). An edge might be a feature that shows up in only one component or in all. Likewise, edge detection becomes more complex in higher-dimensional images. In three dimensions, for example, volumetric regions are separated by surfaces and edges are rather discontinuities in the orientation of surfaces.

Another important question is the reliability of the edge estimates. We do not only want to find an edge, but also know how significant it is. Thus, we need a measure for *edge strength*. Closely related to this issue is the question of optimum edge detection. Once edge detectors do not only deliver edges but also an objective confidence measure, different edge detectors can be compared to each other and optimization of edge detection becomes possible. An important figure of merit for edge detectors in two and higher-dimensional signals is their isotropy; this means their ability to detect edges equally independent of the direction of the edge.

12.3 Concepts

12.3.1 Edge Models

Edge detection cannot be performed without a model. The simplest type of model is objects with constant gray values and a uniform background and step edges between these two regions (Fig. 12.1a). The limited resolution of the imaging system (Section 6.3.2a) blurs the edge; a nonideal sensor and sensor electronics add noise (Fig. 12.1b). Under the assumption that the point spread function (PSF) is symmetric, the convolution of the step edge with the PSF still gives a symmetric edge curve. The position of the edge can be identified by one of the following three properties.

1. Position of the mean value of the object and background gray values $g = (g_0 + g_b)/2$.

2. Position of the steepest slope of the edge.

3. Position of zero curvature.

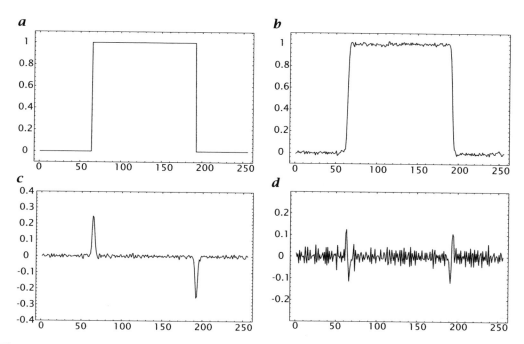

Figure 12.1: *Idealized model for a 1-D edge between objects and background of uniform radiance.* **a** *Original sharp transition (step edge) between background and object.* **b** *Blurred, noisy edge generated by an image formation system with limited resolution and a noisy sensor.* **c** *First derivative;* **d** *second derivative.*

Any of these criteria would allow determining the position of the edge with arbitrary accuracy for such an ideal edge even in sampled images as already discussed in Section 6.3.5. In reality, however, the accuracy is limited by two effects.

Systematic errors. Systematic errors are caused by deviations from the model assumptions. In our simple example model, this could be an asymmetric point spread function of the imaging system that would result in an asymmetric edge or an additional curvature in the background (Fig. 12.1d and e). Example 12.1 discusses the systematic error caused by a curved background in more detail.

Statistical error. Statistical errors add uncertainty by random derivations. If this noise is additive to the signal and not correlated to it, the edge position is not biased. Noise, however, limits the accuracy with which any quantity can be determined (Section 7.3.1). Given a well-defined mathematical concept (*statistics*), statistical errors are much easier to handle. It is much more difficult to identify systematic errors. For a discussion of statistical and systematic errors, see also Sections 2.2.3 and 2.2.5.

Example 12.1: Edge on a curved background

In this example, we investigate the systematic error introduced by a curved background on the determination of an ideal 1-D edge. The following model is used:

$$g(x) = \frac{1}{\sqrt{2\pi}\sigma} \int\limits_{-\infty}^{x} \exp(-x'^2/(2\sigma^2))\mathrm{d}x' + a_0 + a_1 x + a_2 x^2.$$

Table 12.1: *Pairs of constant features with their corresponding discontinuity in multidimensional images.*

Dimension	Constant features and corresponding discontinuities
1	constant region ⇔ edge
2	constant region ⇔ edge, edge ⇔ corner
3	constant region ⇔ surface, surface ⇔ edge, edge ⇔ corner

As the result of a convolution of a step edge with $\exp(-x'^2/(2\sigma^2))$, the edge has a width of 2σ. The background has been modeled by a quadratic spatial variation (second-order polynomial). The first and second derivative of this model edge are

$$\frac{\partial g}{\partial x} = \frac{1}{\sqrt{2\pi}\sigma}\exp(-x^2/(2\sigma^2)) + a_1 + 2a_2 x$$

and

$$\frac{\partial^2 g}{\partial x^2} = -\frac{x}{\sqrt{2\pi}\sigma^3}\exp(-x^2/(2\sigma^2)) + 2a_2.$$

The edge position can now be found by setting the last equation to zero corresponding to a zero crossing in the second derivative or an extreme in the first derivative. Generally, this condition results in a transcendental equation that is not readily solved. If, however, the shift in the edge position is small, the exponential term can be set to one and we obtain

$$x_e \approx 2\sqrt{2\pi}a_2\sigma^3.$$

This result states that a shift in the edge position by a background modeled with a parabolic spatial variation is only caused by the quadratic term a_2. A linear change in the background radiance (a_1) has no influence on the edge position. The error increases with the cube of the width of the edge transition σ^3. Consequently, the position of blurred edges (large σ) is very sensitive to superposition by spatially varying background radiance.

The edge models, as discussed so far, were limited to one-dimensional signals. In higher dimensions, the relation between discontinuities and regions of constant gray values becomes more complex. In two-dimensional images, a second form of discontinuity exists. The edge may end, or show, a discontinuity in its slope, a *corner*.

In three dimensions, the dual quantity to a region is not an edge but a closed surface completely surrounding the object. An edge is now rather a discontinuity in the surface slope. A summary of the dual quantities in one to three dimensions is given in Table 12.1.

12.3.2 Principal Methods for Edge Detection

Edge detection requires filter operations which emphasize the changes in gray values and suppress areas with constant gray values. Figure 12.1 illustrates that derivative operators are suitable for such an operation. The first derivative (Fig. 12.1c) shows an extreme at the edge while the second derivative (Fig. 12.1d) crosses zero where the edge has its steepest ascent. Both criteria can be used to detect edges. It can be noticed, however, that the noise level in the profile of the second derivative is significantly higher than in the first derivative.

An n-th order derivative operator corresponds to multiplication by $(ik)^n$ in the Fourier space (Appendix B.3). Thus, the transfer function of any derivative operator is zero for a wave number of zero. This means that areas of constant gray values are set to zero. Depending on the order of the derivative filter, high wave numbers are strongly enhanced by the multiplication with $(ik)^n$. The first-order derivative directly measures the *steepness* of an edge, while the second-order derivative computes the *curvature*.

12.3.2a The Gradient Vector. In two and higher dimensions, derivations can be taken in all coordinate directions. For first-order derivation, the derivations in all directions are commonly collected in a vector

$$\nabla g = \left[\frac{\partial g}{\partial x_1}, \frac{\partial g}{\partial x_2}, ..., \frac{\partial g}{\partial x_W} \right]^T \tag{12.1}$$

known as the *gradient vector*. This vector is perpendicular to the direction of edges (see Section 13.3.1a for a proof). As a vector quantity, the magnitude of the gradient

$$|\nabla g| = \left(\sum_{w=1}^{W} \left(\frac{\partial g}{\partial x_w} \right)^2 \right)^{1/2} \tag{12.2}$$

is a measure of the steepness of the edge independent of its direction.

The location of an edge is at the maximum of the magnitude of the gradient. The direction of an edge is directly given by the direction of the gradient vector. Thus, the direction of the *normal* to the edge, i. e., the direction in which the values change most (direction of steepest ascent or steepest descent) and not the direction of the edge, is specified. In two dimensions, the angle θ of the edge to the x axis can then be expressed as

$$\theta = \arctan \left(\frac{\partial g}{\partial y} \Big/ \frac{\partial g}{\partial x} \right). \tag{12.3}$$

Since this expression is limited to the two-dimensional case, it is more appropriate to seek an expression that is valid for arbitrary dimensions. Such an expression is given if we form a unit vector from the gradient vector by dividing it by its magnitude,

$$\bar{n} = \frac{\nabla g(x)}{|\nabla g(x)|}. \tag{12.4}$$

This *direction vector* has the same direction as the gradient vector but shows the advantage that from its components the direction can be read directly. Each component of the direction vector contains the cosine of the angle between the gradient vector and the corresponding axis:

$$\bar{n}_w = \left(\frac{\nabla g(x)}{|\nabla g(x)|} \right)_w = \frac{\nabla g(x) \bar{e}_w}{|\nabla g(x)|} = \cos(\angle(\nabla g(x), \bar{e}_w)), \tag{12.5}$$

where $\bar{e}_w$ is a unit vector in the direction of the coordinate axis w.

The usage of first-order derivative operators is illustrated in Fig. 12.2 by applying them to a test image that contains edges and lines in various directions (Fig. 12.2). The horizontal and vertical derivatives miss horizontal and vertical edges (Fig. 12.2b and c), respectively, while the magnitude of the gradient shows all edges in equal strength independent of their direction (Fig. 12.2d). Lines appear as double contours. For the horizontal and vertical derivatives, one contour has a negative and the other a positive value (Fig. 12.2b and c). In the magnitude of the gradient image (Fig. 12.2d), both contours show positive values. Thus first-order derivatives are less suitable for line detection.

Figure 12.2: *First-order and second-order derivatives of a test image shown in **a**, **b** first-order derivative in horizontal (x) direction, **c** first-order derivative in vertical (y) direction, **d** magnitude of the gradient, **e** second-order derivative in horizontal direction ($\partial^2/\partial x^2$), **f** second-order derivative in vertical direction ($\partial^2/\partial y^2$), **g** second-order cross derivative ($\partial^2/\partial x\partial y$), **h** Laplacian operator ($\partial^2/\partial x^2 + \partial^2/\partial y^2$).*

12.3.2b The Laplacian Operator. For second-order derivative operators, the addition of the derivation in all directions

$$\Delta g = \sum_{w=1}^{W} \frac{\partial^2 g}{\partial x_w^2}, \tag{12.6}$$

known as the *Laplacian operator*, results in an isotropic edge detector. In contrast to the nonlinear operations required to compute the magnitude of the gradient, this is a linear operator. The isotropy of the Laplacian operator can also be seen in the Fourier space:

$$\mathcal{F}\left(\Delta g(\boldsymbol{x})\right) = \sum_{w=1}^{W} - k_w^2 \hat{g}(\boldsymbol{k}) = -k^2 \hat{g}(\boldsymbol{k}). \tag{12.7}$$

A comparison between Eqs. (12.2) and (12.6) shows that it is much easier to achieve an isotropic operator with second-order derivation. This explains why the Laplacian operator has become popular in image processing. Edges are zero-crossings in Laplace-filtered images.

The second-order derivative shows each discontinuity as a transition from a dark to a bright stripe (Fig. 12.2e and f). The cross-derivative $(\partial^2 / \partial x \partial y)$ (Fig. 12.2g) highlights corners but also all edges that are not either horizontal or vertical, while the Laplacian operator (Fig. 12.2h) as the magnitude of the gradient is an isotropic operator. The Laplacian operator seems to be a better line detector than first-order derivative filters. The position of the line can be characterized by an extreme curvature and, consequently, shows up with a distinct maximum or minimum in the Laplace-filtered image (Fig. 12.2h).

12.3.2c Edge Coefficient. The edge operators, such as the gradient vector (Section 12.3.2a) and the Laplacian operator (Section 12.3.2b), extract candidates for edges. The magnitude of the gradient, e. g., has a maximal value at an edge. The question remains, however, how large this value must be as to provide a reliable edge estimate. To answer this question, we must consider both systematic and statistical errors.

We first investigate statistical errors. If a certain noise level is present in an image, the magnitude of the gradient will not be zero even in a region with a mean constant gray value. Thus, an edge is reliably detected only when the magnitude of the gradient is above a certain threshold which is related to the noise level in the image. Therefore the certainty of the edge detection can be expressed by the following *edge coefficient*

$$c_e = \frac{|\nabla g|^2}{|\nabla g|^2 + \gamma^2 \sigma_n^2}, \tag{12.8}$$

where σ_n^2 and γ are the variance of the noise and a factor that controls the confidence level for the edge coefficient. As long as the magnitude of the gradient is lower than the standard deviation of the noise, the edge coefficient is low. When $|\nabla g| = \gamma \sigma_n$, the edge coefficient is one half. The addition of squares in Eq. (12.8) causes a rapid transition of the edge coefficient from 0 to 1 once the magnitude of the gradient becomes larger than the standard deviation of the noise. When the magnitude of the gradient is 3 and 10 times larger, the edge coefficient is 9/10 and 100/101, respectively.

12.3.3 General Properties

In this section, some general properties of derivative filters are discussed. This section forms the bridge between the general discussion of derivative filters in Section 12.3.2 and the practical implementation in Section 12.4.

12.3.3a Preservation of Object Position. With respect to object detection, the most important feature of a derivative convolution operator is that it must not shift the object position. For a smoothing filter, this constraint required a real transfer function and a symmetrical convolution mask (Section 11.3.1a). For a first-order derivative filter, a real transfer function makes no sense, since extreme values should be mapped onto zero crossings and the steepest slopes to extreme values. This property requires a 90° phase shift. Therefore, the transfer function of a first-order derivative filter must be imaginary. An imaginary transfer function implies an antisymmetrical filter mask (Appendices B.3 and B.4). An antisymmetrical convolution mask is defined as

$$h_{-m} = -h_m. \tag{12.9}$$

For a convolution mask with an odd number of coefficients, this implies that the central coefficient is zero. For convolution masks with an even number of coefficients, the symmetry relations Eq. (12.9) are valid as well. However, the mask indices have to be taken as fractional numbers (compare Section 11.3.1a). The indices for a mask with $2R$ coefficients are $R' + 1/2$ with $R' \in \mathbb{Z}$, running from $-R$ to $R - 1$. The image resulting from the convolution lies on the *interleaved grid* (Section 10.3.1).

A second-order derivative filter detects curvature. Extremes in gray values should coincide with extremes in the curvature. Consequently, a second-order derivative filter should be symmetric and all properties for symmetric filters that have been discussed for smoothing filters also apply to these filters (Section 11.3.1).

12.3.3b No Response to Mean Value. A derivative filter of any order must not show any response to constant values or an offset in a signal. This condition implies that the sum of the coefficients must be zero and that the transfer function is zero for a zero wave number:

$$
\begin{aligned}
&\text{1-D} \quad \hat{h}(0) = 0, \quad \sum_m h_m = 0 \\
&\text{2-D} \quad \hat{h}(\mathbf{0}) = 0, \quad \sum_m \sum_n h_{mn} = 0 \\
&\text{3-D} \quad \hat{h}(\mathbf{0}) = 0, \quad \sum_l \sum_m \sum_n h_{lmn} = 0.
\end{aligned}
\tag{12.10}
$$

A second-order derivative filter should also not respond to a constant slope. This condition implies no further constraints since it directly follows from the even symmetry of the filter and the zero sum condition Eq. (12.10).

12.3.3c Nonselective Derivation. Intuitively, we expect that any derivative operator amplifies smaller scales stronger than coarser scales. Mathematically speaking, this means that the transfer function should increase monotonically with the wave number:

$$\hat{h}(\tilde{k}_2) \geq \hat{h}(\tilde{k}_1) \quad \text{if} \quad \tilde{k}_2 > \tilde{k}_1. \tag{12.11}$$

12.3.3d Noise sensitivity. Another important issue for edge detection in real and thus noisy images is the sensitivity of the edge operator to noise in the image. In contrast to smoothing filters (Section 11.3.1h), the noise can be amplified by derivative filters because the magnitude of the transfer function can be larger than one. Nevertheless, Eqs. (10.42) and (10.48) can be used to compute the noise amplification for uncorrelated pixels (*white noise*):

$$\frac{\sigma'^2}{\sigma^2} = \sum_{n'=-R}^{R} h_{n'}^2 = \frac{1}{MN} \sum_{u'=0}^{M-1} \sum_{v'=0}^{N-1} |\hat{h}_{u'v'}|^2 = \int_0^1 \int_0^1 |\hat{h}(\tilde{k})|^2 d^2 \tilde{k}. \tag{12.12}$$

In the case of correlated pixels, the equations for the propagation of the autocovariance function must be used, see Eq. (10.45).

12.3.3e Symmetrical Masks. The symmetry properties deserve some further consideration since they form the base to compute the convolution more efficiently by reducing the number of multiplications and simplify the computations of the transfer functions. For a one-dimensional $2R + 1$ mask of odd symmetry (antisymmetrical mask), we obtain

$$g'_n = \sum_{n'=1}^{R} h_{n'} (g_{n-n'} - g_{n+n'}) \tag{12.13}$$

and for a $2R$ mask

$$g'_n = \sum_{n'=1/2}^{R-1/2} h_{n'} (g_{n-n'} + g_{n+n'}). \tag{12.14}$$

The number of multiplications is reduced to R multiplications. The number of additions is still $2R$.

The symmetry relations also significantly ease the computation of the transfer functions since only the sine terms of the complex exponential from the Fourier transform remain in the equations. The transfer function for a 1-D symmetric mask with an odd number of coefficients is

$$\hat{g}(\tilde{k}) = 2i \sum_{n'=1}^{R} h_{n'} \sin(n' \pi \tilde{k}) \tag{12.15}$$

and with an even number of coefficients

$$\hat{g}(\tilde{k}) = 2i \sum_{n'=1/2}^{R-1/2} h_{n'} \sin(n' \pi \tilde{k}). \tag{12.16}$$

12.3.4 Edges in Multichannel Images

It is significantly more difficult to analyze edges in multichannel images. The principal difficulty lies in the fact that the different channels may contain conflicting information about edges. In channel A, the gradient can point to a different direction than in channel B. The simple addition of the gradients in all channels

$$\sum_{p=1}^{P} \nabla g_p(x) \tag{12.17}$$

is no useful quantity. It may happen that the gradients in two channels point to opposite directions and, thus, cancel each other. Then, the sum of the gradient over all channels could be zero, although the individual channels have nonzero gradients.

Thus, a more suitable measure of the total edge strength is the sum of the squared magnitude of gradients in all channels:

$$\sum_{p=1}^{P} |\nabla g_p|^2 = \sum_{p=1}^{P} \sum_{w=1}^{W} \left(\frac{\partial g_p}{\partial x_w} \right)^2. \tag{12.18}$$

While this expression gives a useful estimate of the overall edge strength, it still does not handle the problem of conflicting edge directions adequately. This means that we cannot distinguish noise-related edges from coherent edge information. This distinction can be made by using the following symmetric $P \times P$ matrix S:

$$S = J^T J \tag{12.19}$$

where *J* is known as the *Jacobi matrix*

$$
J = \begin{bmatrix}
\dfrac{\partial g_1}{\partial x_1} & \dfrac{\partial g_1}{\partial x_2} & \cdots & \dfrac{\partial g_1}{\partial x_W} \\[2ex]
\dfrac{\partial g_2}{\partial x_1} & \dfrac{\partial g_2}{\partial x_2} & \cdots & \dfrac{\partial g_2}{\partial x_W} \\[2ex]
\vdots & & \ddots & \vdots \\[2ex]
\dfrac{\partial g_P}{\partial x_1} & \dfrac{\partial g_P}{\partial x_2} & \cdots & \dfrac{\partial g_P}{\partial x_W}
\end{bmatrix}.
\tag{12.20}
$$

The elements of the matrix *S* are

$$
S_{kl} = \sum_{p=1}^{P} \frac{\partial g_p}{\partial x_k} \frac{\partial g_p}{\partial x_l}.
\tag{12.21}
$$

The importance of this matrix can most easily be seen if we rotate the coordinate system in such a way that *S* becomes diagonal. Then, we can write

$$
S = \begin{bmatrix}
\displaystyle\sum_P \left(\dfrac{\partial g_p}{\partial x_1'}\right)^2 & 0 & \cdots & 0 \\[3ex]
0 & \displaystyle\sum_P \left(\dfrac{\partial g_p}{\partial x_2'}\right)^2 & \ddots & 0 \\[3ex]
0 & \ddots & \ddots & 0 \\[3ex]
0 & \cdots & \cdots & \displaystyle\sum_P \left(\dfrac{\partial g_p}{\partial x_W'}\right)^2
\end{bmatrix}.
\tag{12.22}
$$

In the case of an ideal edge, only one of the diagonal terms of the matrix will be nonzero. This is the direction perpendicular to the discontinuity. In all other directions it will be zero. Thus, *S* is a matrix of rank one in this case.

In contrast, if the edges in the different channels point randomly to all directions, all diagonal terms will be nonzero and equal. In this way, it is principally possible to distinguish random changes by noise from coherent edges. The trace of the matrix *S*

$$
\mathrm{trace}(S) = \sum_{w=1}^{W} S_{ww} = \sum_{w=1}^{W} \sum_{p=1}^{P} \left(\frac{\partial g_p}{\partial x_w}\right)^2
\tag{12.23}
$$

gives a measure of the edge strength that is independent of the orientation of the edge since the trace of a symmetric matrix is invariant under a rotation of the coordinate system.

Figure 12.3 illustrates edge detection in multichannel images with the simple and common example of an RGB color image. The first three images (Fig. 12.3a, b, and c) are the *edge coefficient*, as defined in Eq. (12.8) in Section 12.3.2c, for the red, green, and blue components of the color image shown in Plate 16 (a gray scale version of this image is shown in Fig. 5.27). We notice that the edge between the black letters "ADAPTIVE FILTER THEORY" and the red background becomes visible only in the red channel, while the edges between the white letters "SECOND EDITION" and the red background appear only in the green channel.

In Fig. 12.3d, the edge coefficients of the three channels are composed to a color image. The color of the edge indicates in which channels the edge appears. If the edge

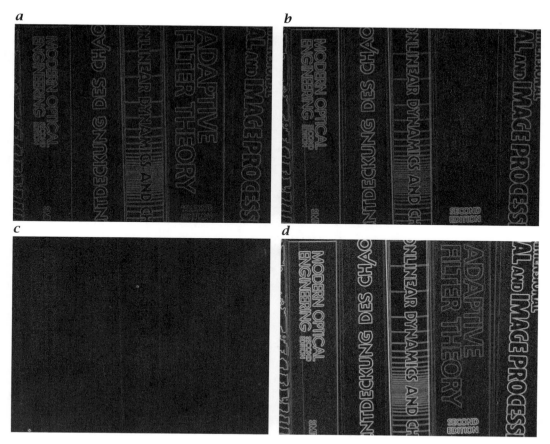

Figure 12.3: *Edges in color images a, b, and c; edge coefficient computed from the red, green, and blue channnels, respectively; d edge coefficients of the red, green, and blue channels combined. (See also Plate 19.)*

is white, it occurs in all channels. If it is red, it is predominantly in the red channel, and if it is yellow, it is visible in both the red and green channels. In this way, multiple features such as in a color image give more possibilities to detect edges and thus to separate objects than a monochrome image.

12.3.5 Regularized Edge Detection

Detecting edges within a local neighborhood has serious limitations in noisy images. This is demonstrated in Fig. 12.4. While it is easily possible to see the edge in the image (Fig. 12.4a), we cannot recognize it from a profile or derivative of the profile perpendicular to the edge (Fig. 12.4b). This is, however, how local edge detectors work.

Obviously, local operators miss an important point. This may be called spatial coherency. An edge is not an isolated event at a single point in the image but extended in the direction along the edge. If in the simple example shown in Fig. 12.4 we average the profile in vertical direction, the edge becomes visible (Fig. 12.4c).

Techniques that stabilize the detection of features in noisy images are called *regularization*. They are, of course, generally not so simple as in the example shown in Fig. 12.4. But the potential is significant. If we assume that a regularization technique

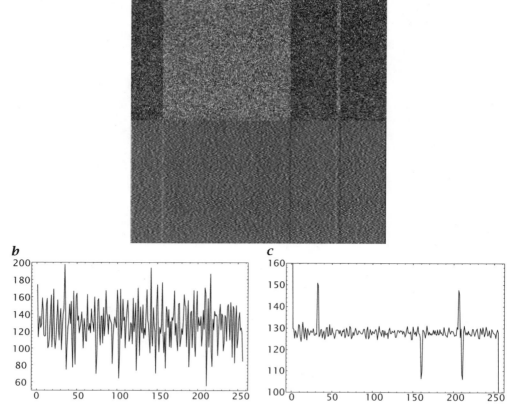

Figure 12.4: *Illustration how bad local edge detectors work. The edge is "seen" by the human visual system in the image but not in the profile of the first-order derivative. **a** Noisy image with two vertical edges and a vertical line; the lower half of the image shows the first-order horizontal derivative. **b** Profile of the first derivative perpendicular to the edges. **c** Averaging of the profile in vertical direction makes the edges visible.*

effectively averages over 100 pixels along an edge, the signal-to-noise ratio would be improved by a factor of 10 without blurring the edge.

The simplest regularization technique is averaging. Since averaging and differentiation are both convolution operators, it does not matter whether we first perform spatial averaging and then the derivation or vice versa. Spatial averaging works better the higher the dimension of a signal is. This is related to the fact that the number of pixels included in a neighborhood for a certain smoothing distance increases with σ^W. If we, for example, average isotropically over a distance of 4 pixels, the averaged volume includes 4, 12.6, and 33.5 pixels in 1, 2, and 3 dimensions, respectively. Simple averaging is adequate as long as the edges are straight. Corners, however, are smoothed. The smoothing must be, of course, the same for all components of the gradient. Otherwise, the smoothing would be part of the differentiation biasing the derivative operator. It is especially not a valid approach to smooth only in the direction perpendicular to the direction into which the derivative operator is applied. This would result in a new type of derivative operator.

This train of thoughts leads therefore to the following general approach for regularized gradients in a W-dimensional image:

$$\mathcal{B} \begin{bmatrix} \mathcal{D}_1 \\ \mathcal{D}_2 \\ \vdots \\ \mathcal{D}_W \end{bmatrix} G = \begin{bmatrix} \mathcal{D}_1 \\ \mathcal{D}_2 \\ \vdots \\ \mathcal{D}_W \end{bmatrix} (\mathcal{B}G). \tag{12.24}$$

This equation shows that the smoothing of the gradient can also be regarded as applying the gradient to a new smoothed version of the image. Smoothing could first be interpreted as if it causes an error in the magnitude of the gradient. This is, of course, true with respect to the original image, but the computed gradient is correct for the regularized version of the image.

This generalized approach to regularized derivative operators gives a lot of freedom in designing filters. The transfer function of a gradient operator can take the following general form

$$\hat{\boldsymbol{d}}(\boldsymbol{k}) = 2\pi\mathrm{i}\boldsymbol{k}b(\boldsymbol{k}) = \begin{bmatrix} 2\pi\mathrm{i}k_x \\ 2\pi\mathrm{i}k_y \\ \vdots \\ 2\pi\mathrm{i}k_W \end{bmatrix} b(\boldsymbol{k}), \tag{12.25}$$

where $b(\boldsymbol{k})$ is an arbitrary smoothing function common to *all* components of the gradient. This approach can also be extended to higher-order derivative filters:

$$\hat{d}_{p,q}(\boldsymbol{k}) = (2\pi\mathrm{i})^{P+q} k_x^p k_y^q b(\boldsymbol{k}). \tag{12.26}$$

Again $b(\boldsymbol{k})$ is an arbitrary smoothing function that must be common to *all* derivative operators used together.

More complex regularization methods include the nonlinear techniques discussed in Chapter 11: normalized convolution (Section 11.3.2), anisotropic diffusion (Section 11.3.3c), and adaptive averaging (Section 11.3.4).

12.4 Procedures

12.4.1 First-Order Derivation

In this section, various approximations to first-order derivative filters will be discussed.

12.4.1a First-Order Discrete Differences. On a discrete grid, a derivative operator can only be approximated. In case of the first partial derivative in x direction, one of the following approximations may be used:

$$\frac{\partial f(x)}{\partial x} \approx \frac{f(x + \Delta x/2) - f(x - \Delta x/2)}{\Delta x}$$
$$\approx \frac{f(x + \Delta x) - f(x - \Delta x)}{2\Delta x}. \tag{12.27}$$

Both approximations are formulated symmetrically around the point at which the derivative is to be computed, and correspond to the filter masks

$$\boldsymbol{D}_{1x} = [1 \ {-1}]$$
$$\boldsymbol{D}_{2x} = 1/2 \, [1 \ 0 \ {-1}]. \tag{12.28}$$

a

b

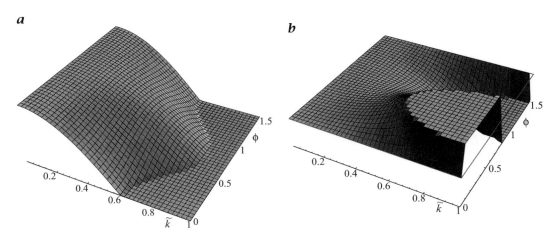

Figure 12.5: *Error in the gradient vector based on the difference filters D_{2x} and D_{2y} shown in a pseudo 3-D plot. The parameters are the magnitude of the wave number (0–1) and the angle to the x axis, $0 - \pi/2$ (90°). **a** Error in the magnitude; range -50% – 0%; **b** error in the direction; range ± 0.2 radians ($\pm 11.5°$).*

The two-element mask puts the difference at the interleaved grid with points in between the original grid points. The transfer function of $\boldsymbol{D}_{1x}$ is given by

$$\hat{D}_{1x} = \exp(\mathrm{i}\pi\tilde{k}_x/2)\left[1 - \exp(-\mathrm{i}\pi\tilde{k}_x)\right] = 2\mathrm{i}\sin(\pi\tilde{k}_x/2), \tag{12.29}$$

where the first term results from the shift by half a grid point. The transfer function of the difference operator $\boldsymbol{D}_{2x}$ reduces to

$$\hat{D}_{2x} = \frac{1}{2}\left[\exp(\mathrm{i}\pi\tilde{k}) - \exp(-\mathrm{i}\pi\tilde{k})\right] = \mathrm{i}\sin(\pi\tilde{k}_x). \tag{12.30}$$

These simple difference filters are only poor approximations for an edge detector. This is demonstrated in Fig. 12.5 showing the errors in the magnitude and orientation of the gradient vectors as a function of the magnitude of the wave number and the direction from 0–90° to the x axis. The magnitude of the gradient is quickly decreasing from the correct value. A Taylor expansion in $\tilde{k}$ yields for the relative error in the magnitude

$$\frac{\Delta|\nabla g|}{|\nabla g|} \approx -\frac{(\pi\tilde{k})^2}{24}(3 + \cos 4\phi). \tag{12.31}$$

The decrease is also anisotropic; it is slower in the diagonal directions. The errors in the direction of the gradient are also large. While in the direction of the axes and diagonals the error is zero, in the directions in between it reaches values already of about $\pm 10°$ at $\tilde{k} = 0.5$. A Taylor expansion in $\tilde{k}$ gives for the angle error $\Delta\phi$:

$$\Delta\phi \approx \frac{(\pi\tilde{k})^2}{24}\sin 4\phi. \tag{12.32}$$

12.4.1b Higher-order approximations by Taylor expansion. We start with the fact that a first-order derivative operator has a filter kernel with odd symmetry. Therefore, a mask with $2R + 1$ coefficients must generally be of the form

$${}^{R}\boldsymbol{D} = 1/2\,[d_R \;\ldots\; d_2\;d_1\;0\;-d_1\;-d_2\;\ldots\;-d_R] \tag{12.33}$$

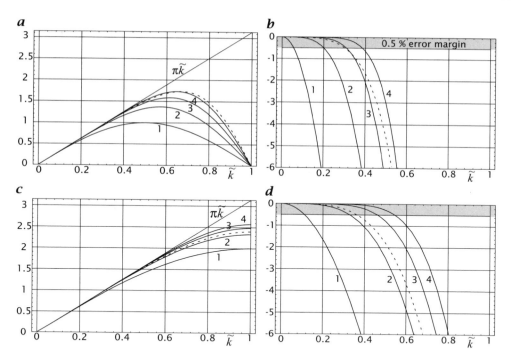

Figure 12.6: *a and c Transfer functions of a family of optimized antisymmetric derivative operators for a given filter length with odd $(2R + 1)$ or even $(2R)$ number of coefficients for $R = 1, 2, 3, 4$, according to Eqs. (12.34) and (12.36); b and d same but shown as the relative deviation from the ideal derivative filter $i\pi\tilde{k}$. Also shown as a dashed curve is the derivative filter based on cubic B-spline interpolation according to Eq. (12.42) and, as a thin line (in a and c), the ideal transfer function. The shaded area in b and d indicates the 0.5% error margin.*

and has the transfer function

$$^{R}\hat{d}(\tilde{k}) = i \sum_{u=1}^{R} d_u \sin(u\pi\tilde{k}). \tag{12.34}$$

For a given mask size r, the coefficients d_u must now be chosen in such a way that Eq. (12.34) approximates the ideal derivative operator $i\pi\tilde{k}$ in an optimum way. One way to perform this task is the Taylor series method. The sine function in Eq. (12.34) is expanded in a Taylor series of order $2r - 1$:

$$\sin(u\pi\tilde{k}) = \sum_{r'=1}^{R} (-1)^{r'-1} \frac{(u\pi\tilde{k})^{2r'-1}}{(2r'-1)!} + 0(\tilde{k}^{2r'+1}). \tag{12.35}$$

For each sin term in Eq. (12.34) we obtain R coefficients for the powers in $(u\pi\tilde{k})^{2r'-1}$. Because R terms are available, we can set R conditions for the R coefficients d_u. To approximate $i\pi\tilde{k}$, for an optimum derivative operator we choose $d_1 = 1$; all other coefficients should vanish. These conditions lead to the following linear equation system:

$$\begin{bmatrix} 1 & 2 & 3 & \cdots & R \\ 1 & 8 & 27 & \cdots & R^3 \\ 1 & 32 & 243 & \cdots & R^5 \\ \vdots & \vdots & \vdots & \ddots & \vdots \\ 1 & 2^{2R-1} & 3^{2R-1} & \cdots & R^{2R-1} \end{bmatrix} \begin{bmatrix} d_1 \\ d_2 \\ d_3 \\ \vdots \\ d_R \end{bmatrix} = \begin{bmatrix} 1 \\ 0 \\ 0 \\ \vdots \\ 0 \end{bmatrix}. \tag{12.36}$$

Table 12.2: *Derivative filters optimized with the Taylor series method. The upper and lower halves of the table contain masks with odd and even number of coefficients, respectively; the corresponding transfer functions are shown in Fig. 12.6. The table also includes the noise amplification factor σ'/σ according to Eq. (12.12).*

R	σ'/σ	Mask	Transfer function
1	0.707	1/2 [1 0 −1]	$i\pi\tilde{k} - i(\pi\tilde{k})^3/6 + O(\tilde{k}^5)$
2	0.950	1/12 [−1 8 0 −8 1]	$i\pi\tilde{k} - i(\pi\tilde{k})^5/30 + O(\tilde{k}^7)$
3	1.082	1/60 [1 −9 45 0 −45 9 −1]	$i\pi\tilde{k} - i(\pi\tilde{k})^7/140 + O(\tilde{k}^9)$
4	1.167	1/840 [−3 32 −168 672 0 −672 168 −32 3]	$i\pi\tilde{k} - i(\pi\tilde{k})^9/630 + O(\tilde{k}^{11})$
1	1.414	[1 −1]	$i\pi\tilde{k} - i(\pi\tilde{k})^3/48 + O(\tilde{k}^5)$
2	1.592	1/24 [−1 27 −27 1]	$i\pi\tilde{k} - 3i(\pi\tilde{k})^5/640 + O(\tilde{k}^7)$
3	1.660	1/1920 [9 −125 2250 −2250 125 −9]	$i\pi\tilde{k} - 5i(\pi\tilde{k})^7/7168 + O(\tilde{k}^9)$

Figure 12.6a shows how the transfer function of these optimized kernels converges monotonically to the ideal derivative operator with increasing R. Convergence is very slow, but the transfer functions do not show any ripples. The deviation from the ideal derivative filter is increasing monotonically (Fig. 12.6b). Example 12.2 illustrates the computations for a filter with 5 coefficients. Figure 12.6 also shows optimized even-sized filters. These masks give much better approximations. A significant advantage is that the transfer function increases monotonically in contrast to the odd-sized masks (Fig. 12.6a and b) where the transfer function decreases towards zero at $\tilde{k} = 1$. However, even-sized masks are of limited value in two and higher dimensions because the derivatives for each direction are put at a different interleaved grid. The coefficients for both types of filters and the equations for the Taylor expansions of the transfer function in $\tilde{k}$ are summarized in Table 12.2.

Example 12.2: A simple better approximation for a first-order derivative filter

As a simple example of the general mathematical formalism, we consider the case $r = 2$. If we expand the transfer function in Eq. (12.34) to the third order in $\tilde{k}_x$, we obtain

$$^{(2)}\hat{d}(\tilde{k}) = \begin{array}{ll} d_1\pi\tilde{k} & - \quad d_1/6(\pi\tilde{k})^3 \\ + \quad 2d_2\pi\tilde{k} & - \quad 8d_2/6(\pi\tilde{k})^3 \end{array}$$

or

$$^{(2)}\hat{d}(\tilde{k}) = (d_1 + 2d_2)\pi\tilde{k} - 1/6(d_1 + 8d_2)(\pi\tilde{k})^3.$$

Since the factor of the $\tilde{k}^3$ term should vanish and the factor for the $\tilde{k}$ term should be equal to one, we have two equations with the two unknowns d_1 and d_2. The solution is $d_1 = 4/3$ and $d_2 = -1/6$. According to Eq. (12.33), we yield the filter mask

$$^{(2)}\hat{d}(\tilde{k}) = \frac{1}{12}[-1\ 8\ 0\ -8\ 1] \tag{12.37}$$

and a transfer function

$$^{(2)}\hat{d}(\tilde{k}) = 4i/3\sin\pi\tilde{k} - i/6\sin 2\pi\tilde{k} = i\pi\tilde{k} - i\frac{(\pi\tilde{k})^5}{30} + O(\tilde{k}^7). \tag{12.38}$$

12.4.1c Differentiation of a Spline Representation. The cubic B-spline transform that is discussed in Section 8.3.2c for interpolation yields a continuous representation of an image by a cubic spline function that interpolates the values at the discrete grid points and that is continuous in its first and second derivative:

$$g_3(x) = \sum_n c_n \beta_3(x - n), \tag{12.39}$$

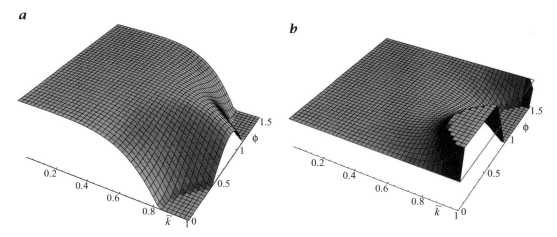

Figure 12.7: *Error in the gradient vector based on the cubic B-spline based derivative filter according to Eq. (12.42) shown in a pseudo 3-D plot. The parameters are the magnitude of the wave number (0–1) and the angle to the x axis, 0–$\pi/2$ (90°).* ***a*** *Error in the magnitude; range -50% – 0%;* ***b*** *error in the direction; range ± 0.2 radians (± 11.5°).*

where $\beta_3(x)$ is the cubic B-spline function defined in Eq. (8.31). Once the B-spline transform is given, the spatial derivative can be computed by taking the spatial derivation of $g_3(x)$ at all the grid points which is given by

$$\frac{\partial g_3(x)}{\partial x} = \sum_n c_n \frac{\partial \beta_3(x-n)}{\partial x}. \tag{12.40}$$

The evaluation of this function at grid points is simple since we only need to compute the derivative at grid points. From Fig. 8.7a it can be seen that the cubic B-spline function covers at most 5 grid points. The maximum occurs at the central grid point. Therefore, the derivative at this point is zero. It is also zero at the two outer grid points. Thus, the partial derivative is only unequal to zero at the neighbors to the central point. The computation of the partial derivative operator thus reduces to

$$\frac{\partial g_3(x)}{\partial x}\bigg|_m = (c_{m+1} - c_{m-1})/2. \tag{12.41}$$

In conclusion, we find that the computation of the first-order derivative based on cubic B-spline transformation is indeed an efficient solution. It is also superior to the Taylor series method concerning the approximation of an ideal derivative operator (Fig. 12.6). The transfer function is given by

$$\hat{d}(\tilde{k}) = i\frac{\sin(\pi\tilde{k})}{2/3 + 1/3\cos(\pi\tilde{k})} = i\pi\tilde{k} - i\frac{\pi^5\tilde{k}^5}{180} + O(\tilde{k}^7) \tag{12.42}$$

and shown in Fig. 12.6. A comparison with Eq. (12.38) shows that the fifth-order term is 6 times lower.

The errors in the magnitude and direction of a gradient vector based on the B-spline derivative filter are shown in Fig. 12.7. They are considerably less than for the simple difference filters (Fig. 12.5). This can be seen more quantitatively from Taylor expansions for the relative errors in the magnitude of the gradient

$$\frac{\Delta|\nabla g|}{|\nabla g|} \approx -\frac{(\pi\tilde{k})^4}{1440}(5 + 3\cos 4\phi) \tag{12.43}$$

a

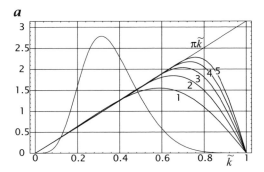

b

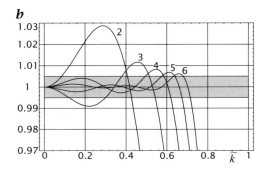

Figure 12.8: a *Transfer functions of a family of least-squares optimized antisymmetric derivative operators for a given filter length with an odd number of coefficients $2R + 1$ for $R = 2, 3, 4, 5, 6$. Shown as thinner curves are also the ideal transfer function and the weighting function of the wave numbers used for the optimization.* **b** *Same but shown as the relative deviation from the ideal derivative filter $i\pi\tilde{k}$. The shaded area indicates the $\pm 0.5\%$ error margin.*

Table 12.3: *Coefficients of antisymmetric derivative filters optimized with a weighted least-squares optimization method. The last column shows the relative weighted standard deviation σ_e of the ideal transfer function in percent. Since the filter mask is antisymmetric, the table gives only R coefficients of the $2R + 1$ coefficients starting from the center of the mask. For the corresponding transfer functions, see Fig. 12.8.*

R	Mask coefficients	σ_e [%]
2	0.74038, -0.12019	4.00
3	0.833812, -0.229945, 0.0420264	1.19
4	0.88464, -0.298974, 0.0949175, -0.0178608	0.46
5	0.914685, -0.346228, 0.138704, -0.0453905, 0.0086445	0.21
6	0.934465, -0.378736, 0.173894, -0.0727275, 0.0239629, -0.00459622	0.11

and the angle error

$$\Delta\phi \approx \frac{(\pi\tilde{k})^4}{720} \sin 4\phi. \qquad (12.44)$$

For comparison see Eqs. (12.31) and (12.32).

12.4.1d Least-Squares First-Order Derivation. In this section, first-order derivative filters are discussed that have been optimized using the least-squares optimization technique introduced in Section 8.3.2d. Two classes of filters are presented, nonrecursive filters and filters with a recursive correction filter which yields significantly better results. Recursive filters cannot, however, be used in all cases since they must run in forward and backward direction. Therefore, they are only efficient when the whole image is kept in the memory.

For the nonrecursive family of filters, we start with a slightly different transfer function as for the Taylor expansion method in Section 12.4.1b:

$$^R\hat{d}(\tilde{k}) = i\sin(\pi\tilde{k}) + i\sum_{u=2}^{R} d_u \left[\sin(u\pi\tilde{k}) - u\sin(\pi\tilde{k}) \right]. \qquad (12.45)$$

This equation yields the additional constraint that the transfer function must be equal to $i\pi\tilde{k}$ for small wave numbers. This constraint reduces the degree of freedom by one,

a
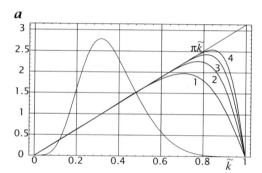

b
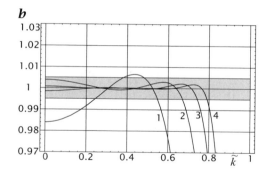

Figure 12.9: *a Transfer functions of a family of least-squares optimized antisymmetric derivative operators with recursive correction for a given filter length with an odd number of coefficients $2R + 1$ for $R = 1, 2, 3, 4$. Shown as thinner curves are also the ideal transfer function and the weighting function of the wave numbers used for the optimization. b Same but shown as the relative deviation from the ideal derivative filter $i\pi\tilde{k}$. The shaded area in b indicates the $\pm 0.5\%$ error margin.*

Table 12.4: *Coefficients of antisymmetric derivative filters with the recursive U correction filter $(g'_n = g_n + \alpha(g'_{n\pm 1} - g_n)$, see Table 10.2) optimized with a weighted least-squares optimization method. The last column shows the relative weighted standard deviation σ_e of the ideal transfer function in percent. Since the filter mask is antisymmetric, the table gives only R coefficients of the $2R + 1$ coefficients starting from the center of the mask. For the corresponding transfer functions, see Fig. 12.9.*

R	α	Mask coefficients	σ_e [%]
1	-0.33456	0.491996	0.74
2	-0.483831	0.44085, 0.0305481	0.134
3	-0.575841	0.416642, 0.0478413, -0.00435846	0.038
4	-0.639506	0.403594, 0.058262, -0.00805505, 0.00110454	0.014

and for a filter with R coefficients, only $R - 1$ can be varied. The first coefficient d_1 is then given by

$$d_1 = 1 - \sum_{u=2}^{R} u d_u. \tag{12.46}$$

The resulting transfer functions are shown in Fig. 12.8 and the filter coefficients in Table 12.3. These filters are considerably better than those designed with the Taylor series technique or the filters based on the cubic B-spline interpolation (Fig. 12.6). The relative deviations from the ideal derivative filter are less than 1% for a wave number up to 0.6 for the least-squares optimized filters, while this error margin is held only up to a wave number of 0.4 for the Taylor series filter.

Even better results are obtained if a recursive correction filter of type $\mathcal{U}$ (Table 10.2) is applied in the same way as it has been applied for the recursive implementation of the optimized interpolation filters (Section 8.3.2d). The transfer functions and filter coefficients of the recursive implementation are shown in Fig. 12.9 and Table 12.4, respectively.

Finally, the errors in the magnitude and direction of the gradient computed from a filter with four coefficients for both filter families are shown in Figs. 12.10 and 12.11. The plots are drawn in a polar coordinate system. The two axes are the magnitude of

a   b

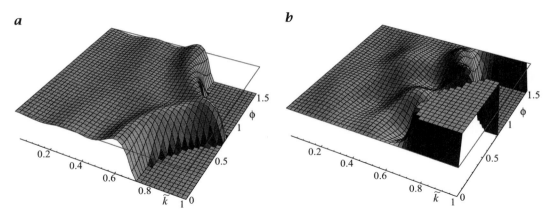

Figure 12.10: *Error in the gradient vector based on the optimized derivative filter according to Eq. (12.45) with R = 4 shown in a pseudo 3-D plot. The parameters are the magnitude of the wave number (0 – 1) and the angle to the x axis, (0 – π/2) (90°). **a** Error in the magnitude; range ±3 %; **b** error in the direction; range ±0.005 radians (±0.29°).*

a   b

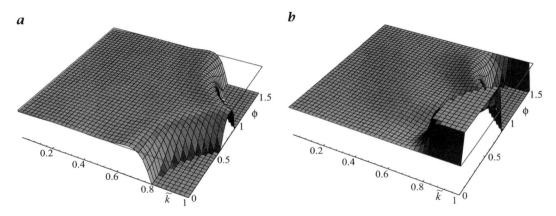

Figure 12.11: *Error in the gradient vector based on the optimized recursive derivative filter with R = 4 shown in a pseudo 3-D plot. The parameters are the magnitude of the wave number (0 – 1) and the angle to the x axis, 0 – π/2 (90°). **a** Error in the magnitude; range ±3 %; **b** error in the direction; range ±0.005 radians (±0.29°).*

the wave number and the angle to the x axis between 0 and 90°. In such a type of plot it is easier to evaluate the anisotropy.

A comparison of Figs. 12.10 and 12.11 with Figs. 12.5 and 12.7 and Figs. 12.14 and 12.15 shows that the derivative filters optimized by the weighted least squares approach perform significantly better. Even for the nonrecursive filter, the deviation from the correct values of the magnitude of the gradient is less than 1%. The deviation in the direction is less than 0.1° for wave numbers less than 0.6.

12.4.2 Second-Order Derivation

With second-order derivatives, we can easily form an isotropic linear operator, the *Laplacian operator*. We can directly derive second-order derivative operators by a two-fold application of first-order operators

$$D_2 = D_1^2 = D_1 D_1 \qquad (12.47)$$

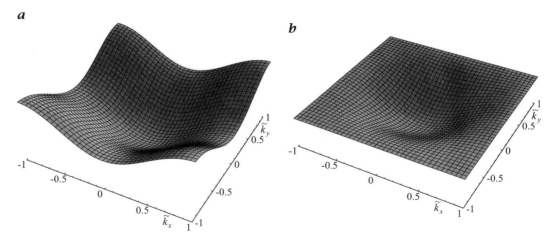

Figure 12.12: *Transfer functions of discrete Laplacian operators: a) $\hat{L}$ Eq. (12.50); b) $\hat{L}'$ Eq. (12.54).*

with the transfer function

$$\hat{d}_2(\tilde{k}) = -4\sin^2(\pi\tilde{k}/2). \tag{12.48}$$

This results in the mask

$$[1\ -1] * [1\ -1] = [1\ -2\ 1]. \tag{12.49}$$

The discrete Laplacian operator $\mathcal{L} = \mathcal{D}_x^2 + \mathcal{D}_y^2$ has the filter mask

$$\boldsymbol{L} = \begin{bmatrix} 1 & -2 & 1 \end{bmatrix} + \begin{bmatrix} 1 \\ -2 \\ 1 \end{bmatrix} = \begin{bmatrix} 0 & 1 & 0 \\ 1 & -4 & 1 \\ 0 & 1 & 0 \end{bmatrix} \tag{12.50}$$

and the transfer function

$$\hat{l}(\tilde{\boldsymbol{k}}) = -4\sin^2(\pi\tilde{k}_x/2) - 4\sin^2(\pi\tilde{k}_y/2). \tag{12.51}$$

As in other discrete approximations of operators, the Laplacian operator is only isotropic for small wave numbers (Fig. 12.12a):

$$\hat{l}(\tilde{\boldsymbol{k}}) \approx -(\pi\tilde{k})^2 + \frac{3}{48}(\pi\tilde{k})^4 + \frac{1}{48}\cos(4\phi)(\pi\tilde{k})^4 + O(\tilde{k}^6). \tag{12.52}$$

Because of the quadratic increase of the transfer function with the wave number, Laplace operators are much more noise sensitive than first order difference filters. The noise amplification factor $\sigma'\sigma$ according to Eq. (12.12) is $\sqrt{6} \approx 2.45$ and $2\sqrt{5} \approx 4.47$ for the 1-D and 2D Laplacian operator, respectively, whereas the noise amplification factor of the simple symmetric difference operator is only $\sqrt{2}/2 \approx 0.71$ (Table 12.2).

There are many other ways to construct a discrete approximation for the Laplacian operator. One possibility is the use of binomial masks. With Eq. (11.57), we can approximate all binomial masks for sufficiently small wave numbers by

$$\hat{b}^{2R}(\tilde{\boldsymbol{k}}) \approx 1 - R\frac{\pi^2}{4}|\tilde{\boldsymbol{k}}|^2 + O(\tilde{k}^4). \tag{12.53}$$

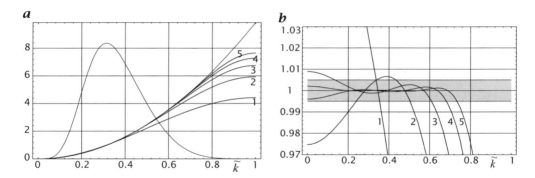

Figure 12.13: *a Transfer functions of a family of least-squares optimized second-order derivative operators for a given filter length with an odd number of coefficients $2R + 1$ for $R=1$–5. The thin curve peaking at a wave number of about 0.3 shows the weighting function of the wave numbers used for the optimization. b Same but shown as the relative deviation from the ideal filter $(\pi\tilde{k})^2$. The shaded area in b indicates the $\pm0.5\%$ error margin.*

Table 12.5: *Coefficients of second-order derivative filters optimized with a weighted least-squares optimization method. The last column shows the relative weighted standard deviation σ_e of the ideal transfer function in percent. Since the masks are symmetric, of the $2R + 1$ coefficients only $R + 1$ coefficients are included in the table starting with the center coefficient. For the corresponding transfer functions, see Fig. 12.13.*

R	Mask coefficients	σ_e [%]
1	-2.20914, 1.10457	5.653
2	-2.71081, 1.48229, -0.126882	0.86
3	-2.92373, 1.65895, -0.224751, 0.0276655	0.21
4	-3.03578, 1.75838, -0.291985, 0.0597665, -0.00827	0.066
5	-3.10308, 1.81996, -0.338852, 0.088077, -0.0206659, 0.00301915	0.025

From this equation, we can conclude that any operator $\mathcal{B}^n - \mathcal{I}$ constitutes a Laplacian operator for small wave numbers. For example,

$$\boldsymbol{L}' = 4(\boldsymbol{B}^2 - \boldsymbol{I}) = \frac{1}{4}\left[\begin{bmatrix} 1 & 2 & 1 \\ 2 & 4 & 2 \\ 1 & 2 & 1 \end{bmatrix} - \begin{bmatrix} 0 & 0 & 0 \\ 0 & 16 & 0 \\ 0 & 0 & 0 \end{bmatrix}\right] = \frac{1}{4}\begin{bmatrix} 1 & 2 & 1 \\ 2 & -12 & 2 \\ 1 & 2 & 1 \end{bmatrix} \quad (12.54)$$

with the transfer function

$$\hat{l}'(\tilde{\boldsymbol{k}}) = 4\cos^2(\pi\tilde{k}_x/2)\cos^2(\pi\tilde{k}_y/2) - 4, \quad (12.55)$$

which can be approximated for small wave numbers by

$$\hat{l}'(\tilde{\boldsymbol{k}}) \approx -(\pi\tilde{k})^2 + \frac{3}{32}(\pi\tilde{k})^4 - \frac{1}{96}\cos(4\phi)(\pi\tilde{k})^4 + O(\tilde{k}^6). \quad (12.56)$$

For large wave numbers, the transfer functions of both Laplacian operators show considerable deviations from an ideal Laplacian, $-(\pi\tilde{k})^2$ (Fig. 12.12). $\mathcal{L}'$ is by a factor of two less anisotropic than $\mathcal{L}$.

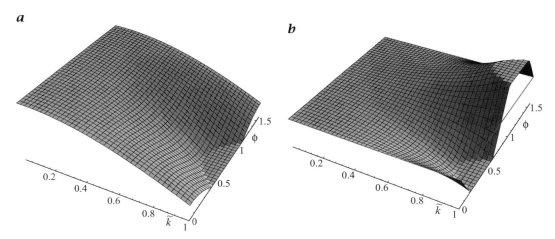

Figure 12.14: *Error in the gradient vector based on the Roberts edge detector shown in a pseudo 3-D plot. The parameters are the magnitude of the wave number (0–1) and the angle to the x axis, 0-π/2 (90°).* **a** *Error in the magnitude; range -50% – 0%;* **b** *error in the direction; range ± 0.2 radians (± 11.5°).*

12.4.2a Least-Squares Second-Order Derivation. For second-order derivatives only the nonrecursive family of least-squares optimized filters are presented. A comparison between Tables 12.3 and 12.5 shows that these filters approximate the second-order derivatives much better than the first-order derivative filters. The filters have been optimized with the constraint that the sum of the coefficients is zero. The transfer function and filter coefficients of these filters are shown in Fig. 12.13 and Table 12.5, respectively.

12.4.3 Regularized Edge Detectors

Edge detection is noise sensitive as it has, for instance, been demonstrated in Section 12.3.5 and Fig. 12.4. Better estimates can be gained by spatially averaging the gradients based on the principles discussed in Section 12.3.5. We will also show that many popular operators, such as the Sobel operator, the derivative of Gaussian, and others are only special cases of the general class of regularized edge detectors — and often have a quite poor performance.

12.4.3a Classical Edge Detectors. The Roberts operator uses the smallest possible difference filters to compute the gradient:

$$D_{x-y} = \begin{bmatrix} 1 & 0 \\ 0 & -1 \end{bmatrix} \quad \text{and} \quad D_{x+y} = \begin{bmatrix} 0 & 1 \\ -1 & 0 \end{bmatrix}. \tag{12.57}$$

The difference filters in diagonal direction result in a gradient vector rotated by 45°.

The Sobel operator uses difference filters that average in the direction perpendicular to the differentiation:

$$D_x B_y^2 = \begin{bmatrix} 1 & 0 & -1 \\ 2 & 0 & -2 \\ 1 & 0 & -1 \end{bmatrix} \quad \text{and} \quad D_y B_x^2 = \begin{bmatrix} 1 & 2 & 1 \\ 0 & 0 & 0 \\ -1 & -2 & -1 \end{bmatrix}. \tag{12.58}$$

The errors in the magnitude and direction obtained with these filters are shown in Figs. 12.14 and 12.15. Both show only a marginal improvement as compared to

a

b

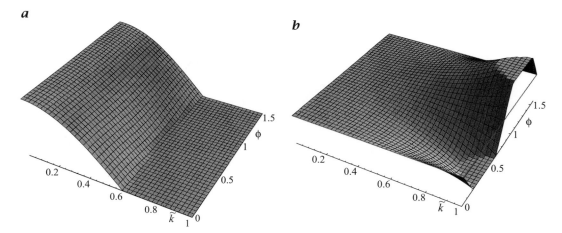

Figure 12.15: *Error in the gradient vector based on the Sobel edge detector shown in a pseudo 3-D plot. The parameters are the magnitude of the wave number (0–1) and the angle to the x axis, 0–π/2 (90°). a Error in the magnitude; range -50 % – 0%; b error in the direction; range ± 0.2 radians (± 11.5°).*

the gradient computation based on the simple difference operator $D_2 = 1/2[1\ 0\ -1]$ (Fig. 12.5). For the Sobel operator, we provide a more quantitative analysis. A Taylor expansion in the wave number yields the following approximations

$$\frac{\Delta|\boldsymbol{\nabla}g|}{|\boldsymbol{\nabla}g|} \approx -\frac{(\pi\tilde{k})^2}{48}(9 - \cos 4\phi) \tag{12.59}$$

for the error of the magnitude and

$$\Delta\phi \approx -\frac{(\pi\tilde{k})^2}{48}\sin 4\phi \tag{12.60}$$

for the direction of the gradient. A comparison with the corresponding equations for the simple difference filter Eqs. (12.31) and (12.32) shows that both the anisotropy and the angle error of the Sobel operator are smaller by a factor of two. The error increases, however, still with the square of the wave number. Already the B-spline based gradient is a significant improvement over the Sobel and Roberts operator (see Fig. 12.7 and equations Eqs. (12.43) and (12.44)). The error in the direction of the Sobel gradient is up to 5° at a wave number of 0.5. With optimized derivative filters, it is not difficult to achieve accuracies in the gradient direction well below 0.1° (Figs. 12.10 and 12.11).

12.4.3b Derivatives of Gaussian.   The derivative of a Gaussian smoothing filter is a well-known general class of regularized derivative filters. It was, e. g., used by Canny [15] for optimal edge detection and is also known as the *Canny edge detector*. On a discrete lattice this class of operators is best approximated by a derivative of a binomial operator (Section 11.4.2) as

$$^{(B,R)}\mathcal{D}_w = \mathcal{D}_{2w}\mathcal{B}^R \tag{12.61}$$

with nonsquare $(2R + 3) \times (2R + 1)$ masks and the transfer function

$$^{(B,R)}\hat{d}_w(\tilde{\boldsymbol{k}}) = \mathrm{i}\sin(\pi\tilde{k}_w)\cos^{2R}(\pi\tilde{k}_x/2)\cos^{2R}(\pi\tilde{k}_y/2). \tag{12.62}$$

Surprisingly, this filter turns out to be not optimal, because its anisotropy is the same as for the simple symmetric difference filter. This can be seen immediately for the direction of the gradient. Eq. (12.3) to compute the angle of the gradient vector includes the quotient of the derivative filters in y and x direction. The smoothing term is the same for both directions and thus cancels out. The remaining terms are the same as for the symmetric difference filter and thus lead to the same angle error at the same wave number.

In the same way, larger Sobel-type quadratic difference operators of the type

$$^R S_w = \mathcal{D}_w \mathcal{B}_w^{R-1} \prod_{w' \neq w} \mathcal{B}_{w'}^R \qquad (12.63)$$

with a $(2R+1)^W$ W-dimensional mask and the transfer function

$$^R \hat{S}_d(\tilde{k}) = \mathrm{i} \tan(\pi \tilde{k}_d / 2) \prod_{w=1}^{W} \cos^{2R}(\pi \tilde{k}_d / 2) \qquad (12.64)$$

show the same anisotropy at the same wave number as the 3×3 Sobel operator.

12.4.4 LoG and DoG Filter

As discussed in Section 12.4.2, Laplace filters tend to enhance the noise level in images considerably, because the transfer function is proportional to the wave number squared. Thus, a better edge detector is again found by first smoothing the image and then applying the Laplacian filter. This leads to a kind of regularized edge detection and to a class of filters called *Laplace of Gaussian* filters (*LoG* for short) or *Marr-Hildreth operator* [99]:

$$\hat{l}(\tilde{k})\hat{b}^R(\tilde{k}) = -4 \left[\sin^2(\pi \tilde{k}_x / 2) + \sin^2(\pi \tilde{k}_y / 2) \right] \cos^p(\pi \tilde{k}_x / 2) \cos^R(\pi \tilde{k}_y / 2). \quad (12.65)$$

For small wave numbers, the transfer function can be approximated by

$$\hat{l}\hat{b}^R \approx -(\pi \tilde{k})^2 + \left[\frac{1}{16} + \frac{1}{8}R + \frac{1}{48} \cos(4\phi) \right] (\pi \tilde{k})^4. \qquad (12.66)$$

The discussion in Section 12.4.2 showed that a Laplace filter can be even better approximated by operators of the type $\mathcal{B}^R - 1$. If additional smoothing is applied, this approximation for the Laplacian filter leads to the *difference of Gaussian* type of Laplace filter, or *DoG* filters:

$$4(\mathcal{B}^S - 1)\mathcal{B}^R = 4(\mathcal{B}^{R+S} - \mathcal{B}^R). \qquad (12.67)$$

The DoG filter $4(\mathcal{B}^{R+2} - \mathcal{B}^R)$ has the transfer function

$$
\begin{aligned}
4(\hat{b}^{R+2} - \hat{b}^R)(\mathbf{k}) &= 4 \cos^{R+2}(\pi \tilde{k}_x / 2) \cos^{R+2}(\pi \tilde{k}_y / 2) \\
&\quad - 4 \cos^R(\pi \tilde{k}_x / 2) \cos^R(\pi \tilde{k}_y / 2),
\end{aligned}
\qquad (12.68)
$$

which can be approximated for small wave numbers by

$$4(\hat{b}^{R+2} - \hat{b}^R)(\tilde{k}, \phi) \approx -(\pi \tilde{k})^2 + \left[\frac{3}{32} + \frac{1}{8}R - \frac{1}{96} \cos(4\phi) \right] (\pi \tilde{k})^4. \qquad (12.69)$$

The DoG filter is significantly more isotropic. A filter with even less deviation in the isotropy can be obtained by comparing Eqs. (12.66) and (12.69). The anisotropic $\cos 4\phi$ terms have different signs. Thus they can easily be compensated by a mix of LoG and DoG operators of the form $2/3\mathrm{DoG} + 1/3\mathrm{LoG}$, which corresponds to the operator $(8/3\mathcal{B}^2 - 8/3\mathit{1} - 1/3\mathcal{L})\mathcal{B}^R$.

DoG and LoG filter operators have some importance for the human visual system [98].

a

b

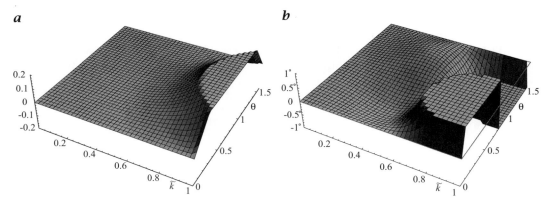

Figure 12.16: *a Anisotropy of the magnitude and **b** error in the direction of the gradient based on the optimized Sobel edge detector Eq. (12.71). Parameters are the magnitude of the wave number (0 to 1) and the angle to the x axis (0 to $\pi/2$).*

12.4.5 Optimized Regularized Edge Detectors

In the previous section it was shown that classical regularized derivative filters obviously are not optimized for optimal edge detection in the sense of a minimum error in the estimate of the direction of the edge or a minimum anisotropy. Obviously, one can do better. A comparison of Eqs. (12.32) and (12.60) shows that these two filters have angle errors in opposite directions. Thus it appears that the Sobel operator performs too much smoothing in cross direction, while the symmetric difference operator does no smoothing in cross direction at all. Consequently, we may expect that a combination of both operators may result in a much lower error.

Indeed, we have one degree of freedom in the design of a 3×3 derivative filter. Any filter of the type

$$
\begin{bmatrix}
\alpha & 0 & -\alpha \\
1-2\alpha & 0 & -1+2\alpha \\
\alpha & 0 & -\alpha
\end{bmatrix}
\tag{12.70}
$$

is a cross-smoothing symmetric difference filter. We can choose the parameter α so that the angle error of the gradient becomes minimal.

Jähne et al. [71] used a nonlinear optimization technique and showed that the operators

$$
1/4\boldsymbol{D}_{2x}(3\boldsymbol{B}_y^2 + \boldsymbol{I}) = \frac{1}{32}
\begin{bmatrix}
3 & 0 & -3 \\
10 & 0 & -10 \\
3 & 0 & -3
\end{bmatrix},
$$

$$
1/4\boldsymbol{D}_{2y}(3\boldsymbol{B}_x^2 + \boldsymbol{I}) = \frac{1}{32}
\begin{bmatrix}
3 & 10 & 3 \\
0 & 0 & 0 \\
-3 & -10 & -3
\end{bmatrix}
\tag{12.71}
$$

have a minimum angle error for the direction of the gradient (Fig. 12.16). The residual error is well below 0.5° and ten times less than for the Sobel operator (Fig. 12.15). Similar optimizations are possible for larger-sized regularized derivative filters.

12.5 Advanced Reference Material

Reference to advanced topics

12.1

B. Jähne, H. Scharr, and S. Körgel, 1999. Principles of filter design, Computer Vision and Applications, Volume 2, Signal Processing and Pattern Recognition, edited by B. Jähne and H. Haußecker and P. Geißler, Academic Press, San Diego, pp. 125–151. *Details principles of filter design with respect to image processing.*

M. Unser, A. Aldroubi, and M. Eden, 1993. B-spline signal processing: Part I — Theory, Part II— efficient design and applications, IEEE Trans. Signal Proc. 41, 821–848. *Image processing based on B-splines, including B-spline based derivative operators. Discusses also a number of applications.*

13 Orientation and Velocity

13.1 Highlights

Image processing becomes more complex if we turn to objects with textures. Then, more advanced techniques than averaging (Chapter 11) and edge detection (Chapter 12) are required. However, the basic idea for object recognition remains the same. An object is still defined as a region with homogeneous features. Only the feature has become more complex and is no longer as simple as the gray value itself. The detection problem can be reduced to smoothing and edge detection techniques if operators are applied that describe patterns by one or more figures and, thus, convert an image into a *feature image*.

In this chapter, the concept of orientation is introduced. It allows a mathematically well-founded description of local neighborhoods. The simplest case is given when the gray values change only in one direction (Section 13.3.1). Then, the neighborhood is characterized by the direction in which the gray value changes. Since the direction is only defined in an angle range of 180°, we speak of *orientation* to distinguish this feature from the direction of a gradient that is defined in the full angle range of 360°.

Such an oriented neighborhood is known as a *simple neighborhood*, *local orientation*, or *linear symmetry*. These concepts can also be applied to higher-dimensional signals. The question, in how many directions do the gray values change, results in a classification of local structures that can be applied to simple objects with constant gray values, textures, and image sequences (Section 13.3.2). In a 2-D image with simple objects, constant areas, edges, and corners can be distinguished, in a volumetric image constant areas, surfaces, edges, and corners. Since a constant velocity constitutes also oriented gray value structures, motion determination can be treated with the same concepts and is, thus, also handled in this chapter.

Based on these simple ideas, it turns out that the mathematically correct first-order presentation of a local structure for W-dimensional data is a symmetric $W \times W$ tensor, the *structure tensor* whose components contain averaged products of derivatives in all directions (Section 13.3.3). An eigenvalue analysis of this tensor differentiates the different classes of neighborhoods.

The tensor has to be computed and analyzed at each pixel. That sounds computationally very expensive and prohibitive to use this technique. Indeed, the first implementations to compute the structure tensor with a set of directional quadrature filters (Section 13.4.1) required thousands of operations per pixel. Meanwhile, much more efficient algorithms have been found. The tensor can be computed as combinations of simple point operations, derivatives, and smoothing filters that require only a few computations per pixel (Section 13.4.2).

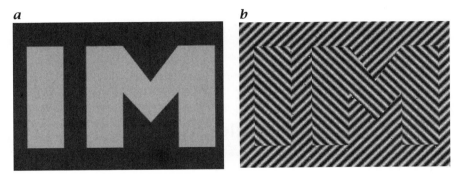

Figure 13.1: *An object can be recognized because it differs from the background **a** in the gray value or **b** in the orientation of a pattern.*

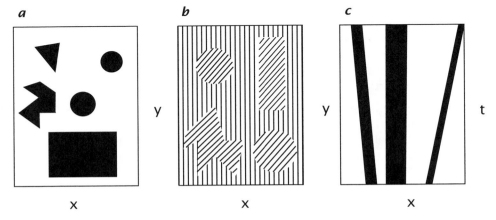

Figure 13.2: *Three different interpretations of local structures in 2-D images: **a** edge between uniform object and background; **b** orientation of pattern; **c** orientation in a 2-D space-time image indicating the velocity of 1-D objects.*

The beauty of the tensor method for the analysis of local neighborhoods is that its performance can be analyzed analytically. This pays off in one of the most robust techniques known in image processing. It can be proved that the technique is exact if it is implemented correctly and that noise causes no systematic error in the determination of the orientation.

13.2 Task

Objects and patterns in images are detected and recognized because of spatial brightness changes. As demonstrated in Fig. 13.1, the human visual system does not only detect luminance differences but also different orientation of a pattern from the background. To perform this task with a digital image processing system, we need an operator which determines the orientation of the pattern. After such an operation, we can distinguish differently oriented patterns in the same way we can distinguish gray values.

The meaning of local structures may be quite different as illustrated in Fig. 13.2 for 2-D imagery:

Task List 10: Orientation

Task	Procedures
Determine type of local neighborhood structure	Compute the structure tensor and perform an eigenvalue analysis
Determine orientation of edges in 2-D images	Compute the orientation and the certainty of the orientation from the structure tensor
Determine orientation of texture	Same as above
Determine velocity in space-time images	Same as above

- In the simplest case, the observed scene consists of objects and a background with uniform intensity (Fig. 13.2a). Then, a gray value change in a local neighborhood indicates that an edge of an object is encountered and the analysis of orientation obtains the orientation of the edges.

- In Fig. 13.2b, the objects differ from the background by the orientation of the *texture*. Now, the local spatial structure does not indicate an edge but characterizes the texture of the objects.

- In image sequences, the local structure in the space-time domain is determined by the motion as illustrated by Fig. 13.2c for a two-dimensional space-time image.

Although the three examples refer to entirely different image data, they have in common that the local structure is characterized by an orientation, i. e., the gray values locally change only in one direction. In this sense, the concept of orientation is an extension of the concept of edges. In Section 13.3.1, it is analyzed how an oriented local neighborhood can be described mathematically. This is the base for a more general classification of neighborhoods that is discussed in Section 13.3.2.

This classification can be achieved by computing a symmetrical tensor, known as the *structure tensor*, from the spatial derivatives in the image and to perform an eigenvalue analysis (Section 13.4.2). From the tensor we can compute all other information about the local neighborhood. The same techniques can be applied to determine the presence and orientation of edges, the orientation of texture, and the velocity in image sequences. Thus, the task list for the orientation analysis in task list 10 is rather short.

13.3 Concepts

13.3.1 Simple Neighborhoods

13.3.1a Definition. For a mathematical description of orientation, we use continuous functions. Then, it is much easier to formulate the concept of local orientation. As long as the corresponding discrete image meets the sampling theorem, all the results derived from continuous functions remain valid, since the sampled image is an exact representation of the continuous gray value function.

A local neighborhood with *ideal* local orientation is characterized by the fact that the gray value only changes in one direction; in the other directions it is constant. Since the curves of constant gray values are lines, local orientation is also denoted as *linear symmetry* or a *simple neighborhood* [6, 45]. If we orient the coordinate system along these two principal directions, we can write the gray values as a one-dimensional function of only one coordinate. Generally, we will denote the direction of local orientation with a unit vector $\bar{k}_0$ which is perpendicular to the lines of constant gray values. Then,

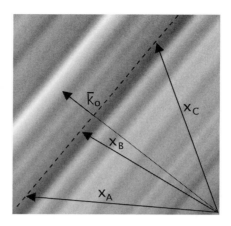

Figure 13.3: *Illustration of a linear symmetric or simple neighborhood, $g(x\bar{k}_0)$. The gray values depend only on a coordinate in the direction of the unit vector $\bar{k}_0$.*

we can write for a local neighborhood with an ideal local orientation:

$$g(x) = g(x\bar{k}_0). \tag{13.1}$$

We can easily verify that this representation is correct by computing the gradient

$$\nabla g(x\bar{k}_0) = \left[\begin{array}{c} \dfrac{\partial g(x\bar{k}_0)}{\partial x_1} \\[3mm] \dfrac{\partial g(x\bar{k}_0)}{\partial x_2} \end{array} \right] = \left[\begin{array}{c} \bar{k}_{0x} g'(x\bar{k}_0) \\[2mm] \bar{k}_{0y} g'(x\bar{k}_0) \end{array} \right] = \bar{k}_0 g'(x\bar{k}_0) \tag{13.2}$$

The gradient vector lies in the direction of $\bar{k}_0$. With g' we denote the derivative of g with respect to the scalar variable $x\bar{k}_0$. Equation Eq. (13.1) is also valid for image data with more than two dimensions. The projection of the vector x onto the unit vector $\bar{k}_0$ makes the gray values only dependent on a single coordinate in the direction of $\bar{k}_0$ as illustrated in Fig. 13.3.

13.3.1b   Representation in Fourier Space. A simple neighborhood has also a special form in Fourier space. From the very fact that a simple neighborhood is constant in all directions except $\bar{k}_0$, we infer that the Fourier transform must be confined to a line. The direction of the line is given by $\bar{k}_0$:

$$g(x\bar{k}_0) \quad \circ\!\!-\!\!\bullet \quad \hat{g}(k\bar{k}_0)\delta(k - \bar{k}_0(k\bar{k}_0)), \tag{13.3}$$

where k denotes the coordinate in the direction of $\bar{k}_0$. The argument in the δ function is only zero when k is parallel to $\bar{k}_0$. Equation Eq. (13.3) assumes that the whole image is a simple neighborhood. A restriction to a local neighborhood is equivalent to a multiplication of $g(x\bar{k}_0)$ with a window function $w(x - x_0)$ in Fourier space. This means a convolution with the Fourier transform of the window function, $\hat{w}$, in Fourier space

$$w(x - x_0) \cdot g(x\bar{k}_0) \quad \circ\!\!-\!\!\bullet \quad \exp(-2\pi i k x_0)\hat{w}(k) * \hat{g}(k\bar{k}_0)\delta(k - \bar{k}_0(k\bar{k}_0)). \tag{13.4}$$

The limitation to a local neighborhood, thus, "blurs" the line in Fourier space to a "sausage-like" shape. Its thickness is inversely proportional to the size of the window

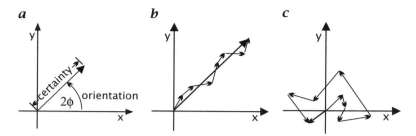

Figure 13.4: *Representation of local orientation as a vector:* **a** *the orientation vector;* **b** *averaging of orientation vectors from a region with homogeneous orientation;* **c** *same for a region with randomly distributed orientation.*

because of the reciprocity of scales between the two domains. From this elementary relation, we can conclude that accuracy of the orientation is directly related to the ratio of the window size to the wavelength of the smallest structures in the window.

13.3.1c Orientation versus Direction.   For an appropriate representation of simple neighborhoods, it is important to distinguish *orientation* from *direction*. The direction (of a gradient vector, for example) covers the full angle range of 2π (360°). Two gradient vectors that point in opposite directions, i.e., they differ by 180°, mean something different. The gradient always points into the direction into which the gray values are increasing. With respect to a bright object on a dark background, this means that the gradient at the edge is pointing towards the object. If the object is less bright than the background, the gradient vectors would all point away from the object. To describe the direction of a local neighborhood, however, an angle range of 180° is sufficient. If a pattern is rotated by 180°, it still has the same direction. Thus, the direction of a simple neighbor is different from the direction of a gradient. While for the edge of an object, gradients pointing in opposite directions are conflicting and inconsistent, for the direction of a simple neighborhood it is consistent information.

In order to distinguish the two types of "directions", we will speak of *orientation* in all cases where only an angle range of 180° is required. Orientation is, of course, still a *cyclic* quantity. Increasing the orientation beyond 180° flips it back to 0°. This means that an appropriate representation of orientation requires an angle doubling to be properly represented as a cyclic quantity.

13.3.1d Vector Representation of Local Orientation.   A simple neighborhood is not well represented by a scalar quantity because of its cyclic nature. Moreover, it seems to be useful to add a certainty measure that describes how well the neighborhood approximates a simple neighborhood. Both pieces of information are best represented by a vector in the following way. The magnitude of the vector is set proportionally to the certainty measure and the direction of the vector is set to the doubled orientation angle 2ϕ (Fig. 13.4a).

The vector representation of orientation has two significant advantages. First, it is more suitable for further processing than a separate representation of orientation in two scalar quantities. We demonstrate this feature with averaging. When vectors are summed up, they are chained up und the resulting vector sum is the vector from the starting point of the first vector to the end point of the last vector (Fig. 13.4b). The weight of an individual vector in the sum is given by its length. In this way, the certainty of the orientation measurement is adequately taken into account. The sum vector also reflects the coherency of the orientation estimate in the right way. In case of

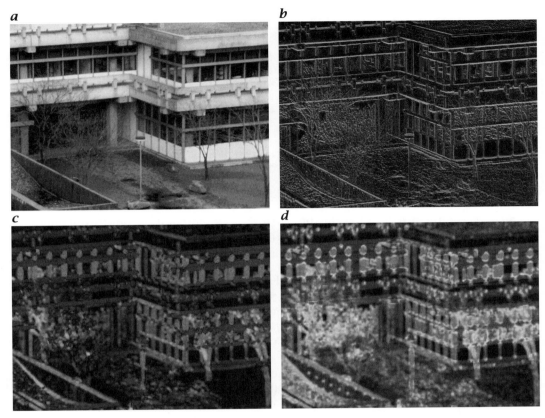

Figure 13.5: *Different possibilities for color coding of edge orientation: a original image, b color coding of the gradient: the magnitude and direction of the gradient is mapped onto the intensity and hue, respectively. Edges in opposite direction are shown in complementary colors. c Representation of the orientation vector as a color image (Section 13.3.1d). d Representation of the structure tensor as a color image (Section 13.4.2d). (See also Plate 20.)*

a region which shows a homogeneous orientation, the vectors line up to a large vector (Fig. 13.4b), i. e., a certain orientation estimate. However, in a region with randomly distributed orientation the resulting vector remains small, indicating that no significant local orientation is present (Fig. 13.4c).

Second, it is difficult to display orientation as a gray scale image. While orientation is a cyclic quantity, the gray scale representation shows an unnatural jump between the smallest and the largest angle. This jump dominates the appearance of the orientation images and, thus, gives no good impression of the orientation distribution. The orientation vector, however, can well be represented as a color image (Fig. 13.5c). It appears natural to map the certainty measure onto the luminance and the orientation angle as the hue of the color. Our attention is then drawn to the bright parts in the images where we can distinguish the colors well. The darker a color is the more difficult it gets to distinguish the different colors visually. In this way, our visual impression coincides with the certainty of the orientation. This color coding is used throughout this handbook to represent orientation. For further images see Plates 21 and 25.

13.3.2 Classification of Local Structures

The discussion on simple neighborhoods suggests that local structures can be classified based on the number of directions in which the gray values change. This classification is especially useful in higher-dimensional images. In a simple neighborhood, the gray values keep constant in all directions perpendicular to this direction, i. e., in a hyperplane normal to $\bar{k}_0$. Other types of local structure are possible. Table 13.1 summarizes the different classes and identifies their meaning for three types of image data: images with a) simple objects showing constant gray values, b) textured objects, and c) image sequences, i. e., space-time images.

The complexity of local structure increases with the number of dimensions. In three dimensions, for example, four types of local neighborhood are possible. We discuss their meaning for simple objects, texture, and image sequences in the following.

13.3.2a Simple Objects. *Simple objects* mean objects with constant features. Then, the four types of local neighborhood have the following meaning in three dimensions:

1. Constant gray value: neighborhood within an object (or background).
2. Gray values change in one direction: neighborhood includes the *surface* of an object. The gray values change only in the direction normal to the surface. Thus, $\bar{k}_0$ represents the surface normal.
3. Gray values change in two directions: the neighborhood includes an *edge* (discontinuity of the surface orientation) of an object. The vector $\bar{k}_0$ represents the direction of the edge.
4. Gray values change in all directions: the neighborhood includes a *corner* of an object.

13.3.2b Texture.

1. The gray value is constant; the object has no texture but is uniform.
2. The gray value is a function of a scalar variable, the inner product between x and the vector $\bar{k}_0$ pointing into the direction of the gray value changes. The texture appears "layered", as if layers with different gray values have been put onto each other.
3. The gray value is constant only in one direction and varies in the other two directions. In this case, k_0 gives the direction in which the gray values do not change. Such a type of 3-D texture can be generated by 'extruding' any 2-D gray value function. A typical example is a fur.
4. The gray values vary in all directions; there is no preferred direction.

13.3.2c Image Sequence. An image sequence can be represented as a three dimensional image to which a third coordinate, the time, has been added. It may be represented as an *image cube* as shown in Fig. 13.6. At each visible face of the cube we map a slice in the corresponding direction. The front face shows the last image of the sequence, the top face a xt slice and the right face a yt slice. The slices were taken at depths marked by the white lines on the front face. From the image cube it is obvious that constant motion in this space-time or spatio-temporal image constitutes gray value structures of constant orientation. At the beginning of the scene, cars are lined up at a red traffic light. Thus, the direction in which the gray values do not change is parallel to the time axis. Then, the traffic light turns green and one car after the other is accelerating until they drive along with constant speed. This can be seen in the yt

Figure 13.6: *A 3-D image sequence of a traffic scene in the Hanauer Landstraße, Frankfurt/Main, represented as an image cube. The time axis is running towards the viewer. The front face shows the last image of the sequence, the left face a yt slice and the top face an xt slice. The positions of the space-time slices are marked by the white lines on the front face (from Jähne [72]).*

slice by the turning orientation of the gray value structures. In order to derive a relation between the orientation of the gray value structures and the velocity, a coordinate system moving along with the object is defined

$$\boldsymbol{x}' = \boldsymbol{x} - \boldsymbol{u}t, \tag{13.5}$$

where $\boldsymbol{u}$ is the velocity vector of the object. If $\bar{\boldsymbol{e}}_0$ is a unit vector pointing in the direction of constant gray values, i.e., into the time axis of the coordinate system moving with the object, the velocity vector can be computed from $\bar{\boldsymbol{e}}_0$ using Eq. (13.5) with $\boldsymbol{x}' = \boldsymbol{0}$ by

$$\boldsymbol{u} = \frac{1}{\bar{e}_{0t}} \left[\begin{array}{c} \bar{e}_{0x} \\ \bar{e}_{0y} \end{array} \right]. \tag{13.6}$$

The equation relates the local spatio-temporal orientation, expressed by the unit vector $\bar{\boldsymbol{e}}_0$ pointing in the direction of constant gray values, to the local velocity $\boldsymbol{u}$. Thus, motion can be determined by analyzing orientation. Once we know how to compute the orientation, we can also determine the velocity. A detailed treatment of motion analysis is given by Jähne [72] and Jähne [73]. Here, we restrict the further discussion of motion analysis to the classification of the local structure as we did for simple objects and textures, and we will discuss some examples of motion analysis in Section 13.4.3.

As for simple objects and texture four classes of local structures can be distinguished in a space-time image:

1. The gray value is constant. No motion can be detected.
2. Gray values change only in one direction. A spatial structure with linear symmetry, e. g., an edge, is moving with constant speed. Physicists would speak of a planar nondispersive wave. In this case, we can only infer the velocity normal to the direction of constant gray values. This case is well known as the *aperture problem*.

Table 13.1: Classification of local structures in multidimensional image data. The first column contains the number of orthogonal directions in which the gray values change. The cases marked by a star refer to linear symmetry.

Rank	Object type	Texture	Spatio-temporal image
		One-dimensional image $g(x)$	
0	Interior	$g(x)$=c, none	—
1	Edge	Asymmetric	—
		Two-dimensional image $g(x, y)$ or $g(x, t)$	
0	Interior	$g(\boldsymbol{x}) = c$, none	$g(x, t) = c$ No motion detectable
1*	Edge	$g(\boldsymbol{x}\bar{\boldsymbol{k}}_0)$ linear symmetry	$g(x - u_0 t)$ Constant velocity
2	Corner	Asymmetric	Complex motion
		Three-dimensional image $g(x, y, z)$ or $g(x, y, t)$	
0	Interior	$g(\boldsymbol{x}) = c$ Constant $g(\boldsymbol{x}) = c$	$g(x, y, t) = c$ No motion detectable
1*	Surface	$g(\boldsymbol{x}\bar{\boldsymbol{k}}_0)$ Layered	$g(\boldsymbol{x}\boldsymbol{k}_0 - \omega t)$ Normal velocity $u_\perp = \omega/\lvert\boldsymbol{k}_0\rvert$
2	Edge	$g(\boldsymbol{x} - \bar{\boldsymbol{k}}_0(\boldsymbol{x}\bar{\boldsymbol{k}}_0))$ Extruded	$g(\boldsymbol{x} - \boldsymbol{u}_0 t)$ Constant velocity, moving "corner"
3	Corner	Asymmetric	Complex motion
		Four-dimensional spatio-temporal image $g(x, y, z, t)$	
0	Interior	—	$g(x, y, z, t) = c$ No motion detectable
1*	Surface	—	$g(\boldsymbol{x}\boldsymbol{k}_0 - \omega t)$ Normal velocity $u_\perp = \omega/\lvert\boldsymbol{k}_0\rvert$
2	Edge	—	Two components $\perp$ edge detectable
3	Corner	—	$g(\boldsymbol{x} - \boldsymbol{u}_0 t)$ Constant velocity
4	Asymmetric	—	Complex motion

3. The spatial structure does not show linear symmetry but has distributed orientations. A gray value corner or extreme moving with a constant speed is an example of such a structure. The gray values remain, however, constant in the direction of motion. In this case, both components of the velocity vector can be determined.
4. The local structure does not show any symmetry at all. Motion is not constant. It is important to note that this class includes many cases. There might be a) a spatial change in the motion field, for example, a discontinuity as at the edge of a moving object, b) a temporal change in the motion (accelerated or decelerated motion) as for a ball reflected at a wall, or c) a motion superimposition of transparent objects. The important issue here is how abrupt the changes in the motion field must be in order for it to belong to this class. This question will be studied in detail later, since it is important for the practical usefulness of this class.

In summary, we conclude that the same concepts for local structure can be used to analyze boundaries of uniform objects and texture in multidimensional images and motion in space-time images. Thus, operators for local structure are a central tool for low-level image processing for a wide class of image processing tasks.

13.3.3 First-Order Tensor Representation

The vectorial representation discussed in Section 13.3.1d is incomplete. While it is suitable to represent the orientation of simple neighborhoods, it cannot distinguish between neighborhoods with constant values and distributed orientation, e. g., uncorrelated noise. Both cases result in an orientation vector with zero magnitude. A first-order description of a local neighborhood requires a more complex description. It should be able to distinguish all cases classified in Section 13.3.2 and determine a unique orientation (given by a unit vector $\bar{k}_0$).

A suitable representation can be introduced by the following optimization strategy to determine the local orientation. The optimum orientation is defined as the orientation which shows the least deviations from the directions of gradients. A suitable measure for the deviation must treat gradients pointing to opposite directions equally. The squared scalar product between the gradient vector and the unit vector representing the local orientation $\bar{k}_0$ meets this criterion:

$$(\nabla g \bar{k}_0)^2. \tag{13.7}$$

This quantity is maximal when ∇g and $\bar{k}_0$ are parallel or antiparallel, and zero if they are perpendicular to each other. Thus, the following expression is minimized in a local neighborhood

$$\int w(\boldsymbol{x} - \boldsymbol{x}') \left(\nabla g(\boldsymbol{x}')\bar{k}_0\right)^2 \mathrm{d}^W x' \to \text{maximum} \tag{13.8}$$

where the window function determines the range around a point $\boldsymbol{x}$ about which the orientation is averaged. The maximization problem must be solved for each point $\boldsymbol{x}$. Equation Eq. (13.8) can be rewritten in the following way

$$\bar{k}_0^T J \bar{k}_0 \to \text{maximum} \tag{13.9}$$

with

$$J = \int w(\boldsymbol{x} - \boldsymbol{x}') \left(\nabla g(\boldsymbol{x}') \otimes \nabla g(\boldsymbol{x}')\right) \mathrm{d}^W x'.$$

The components of this symmetric tensor are

$$J_{pq}(\boldsymbol{x}) = \int\limits_{-\infty}^{\infty} w(\boldsymbol{x} - \boldsymbol{x}') \left(\frac{\partial g(\boldsymbol{x}')}{\partial x'_p} \frac{\partial g(\boldsymbol{x}')}{\partial x'_q}\right) \mathrm{d}^W x'. \tag{13.10}$$

These equations indicate that a tensor, denoted as the *structure tensor*, is an adequate first-order representation of a local neighborhood. The term first-order has a double meaning. First, first-order derivatives are involved. Second, only simple neighborhoods can be described in the sense that we can analyze in which direction(s) the gray values change.

The complexity of Eqs. (13.9) and (13.10) somewhat hides their simple meaning. The tensor is symmetric. By a rotation of the coordinate system, it can be brought into a diagonal form. Then, Eq. (13.9) reduces to

$$J = \begin{bmatrix} k_{0x'}, k_{0y'} \end{bmatrix} \begin{bmatrix} J_{x'x'} & 0 \\ 0 & J_{y'y'} \end{bmatrix} \begin{bmatrix} k_{0x'} \\ k_{0y'} \end{bmatrix} \to \text{maximum}. \tag{13.11}$$

Without loss of generality, we assume that $J_{x'x'} > J_{y'y'}$. Then, it is obvious that the unit vector $\bar{k}_0' = [1\ 0]$ maximizes Eq. (13.11). The maximum value is $J_{x'x'}$. A unit vector $\bar{k}_0' = [\cos\phi\ \sin\phi]$ in an arbitrary direction ϕ gives the value

$$J = J_{x'x'}\cos^2\phi + J_{y'y'}\sin^2\phi.$$

In conclusion, this approach does not only yield a tensor representation for the local neighborhood, but also shows the way to determine the orientation. Essentially, we have to solve what is known as an *eigenvalue problem*. We must determine the *eigenvalues* and *eigenvectors* of **J**. The eigenvector to the maximum eigenvalue then gives the orientation of the local neighborhoods as discussed in Section 13.3.2. We will detail the two-dimensional and three-dimensional case here. Algorithms to compute the eigenvector and eigenvalue and, thus, determine the local orientation are discussed in Section 13.4.2.

13.3.3a Two-Dimensional Eigenvalue Analysis. The eigenvalues (diagonal term of the diagonalized tensor in Eq. (13.11)) are abbreviated with J_1 and J_2. Then, we can distinguish the following cases:

$J_1 = J_2 = 0$ Both eigenvalues are zero. This means that the mean squared magnitude of the gradient ($J_1 + J_2$) is zero. The local neighborhood has constant values.

$J_1 > 0, J_2 = 0$ One eigenvalue is zero. This means that the values do not change in the direction of the corresponding eigenvector. The local neighborhood is a simple neighborhood with ideal orientation.

$J_1 > 0, J_2 > 0$ Both eigenvalues are unequal to zero. The gray values change in all directions. In the special case of $J_1 = J_2$, we speak of an isotropic gray value structure since it changes equally in all directions.

13.3.3b Three-Dimensional Eigenvalue Analysis. The classification of the eigenvalues in three dimensions is similar to the two-dimensional case.

$J_1 = J_2 = J_3 = 0$ The gray values do not change in any direction; constant neighborhood.

$J_1 > 0, J_2 = J_3 = 0$ The gray values change only in one direction. This direction is given by the eigenvector to the nonzero eigenvalue. The neighborhood includes a boundary between two objects or a layered texture. In a space-time image, it means a constant motion of a spatially oriented pattern ("planar wave").

$J_1 > 0, J_2 > 0, J_3 = 0$ The gray values change in two directions and are constant in a third. The eigenvector to the zero eigenvalue gives the direction of the constant gray values. The neighborhood includes an edge of an object or an "extruded" type of texture. In a space-time image, this condition indicates the constant motion of a spatially distributed pattern. In this condition, both components of the velocity vector can be determined.

$J_1 > 0, J_2 > 0, J_3 > 0$ The gray values change in all directions. The neighborhood includes a corner of an object or a nonoriented texture. In a space-time image, this case is a strong indicator for either an abrupt motion change or an occlusion.

The eigenvalue analysis directly reflects the classes discussed in Section 13.3.2. In practice, it will not be checked whether the eigenvalues are zero but below a critical threshold that is determined by the noise level in the image.

13.4 Procedures

13.4.1 Set of Directional Quadrature Filters

13.4.1a Introductionary Remarks. The technique described in this section initiated the field of local structure analysis and coined the term local orientation. The general idea of this approach is to use a set of directionally selective filters and to compare the filter results. If we get a clear maximum for one of the filters but only little response from the other filters, the local neighborhood contains a pattern oriented in the corresponding direction. If several of the filters give a comparable response, the neighborhood contains a distribution of oriented patterns.

Although the concept seems to be straightforward, a number of tricky problems need to be solved. These get more complex in higher dimensional spaces. Which properties have to be met by the directional filters in order to ensure an exact determination of local orientation, if at all possible? For computational efficiency, we need to use a minimum number of filters to interpolate the angle of the local orientation. But what is the minimum number of filters? The concepts introduced in this section are based on the work of Granlund [44], Knutsson [80], and Knutsson et al. [81] on local orientation in spatial images.

13.4.1b Polar Separable Quadrature Filters. First we will discuss the selection of appropriate directional filters. The design goal is filters that give an exact orientation estimate. We can easily see that simple filters cannot be expected to yield such a result. A simple first-order derivative operator, for example, would not give any response at local minima and maxima of the gray values and thus would not allow determination of local orientation in these places. There is a special class of operators, called *quadrature filters*, which perform better. They can be constructed in the following way. Imagine we have a certain directional filter with the transfer function $\hat{h}(\boldsymbol{k})$. We calculate the transfer function of this filter and then rotate the phase of the transfer function by 90°. By this action, the wave number components in the two filter responses differ by a shift of a quarter of a wavelength for every wave number. Where one filter response shows zero crossings, the other shows extremes. If we now square and add the two filter responses, we actually obtain an estimate of the *spectral density*, or, physically speaking, the *energy* of the corresponding periodic image structure. We can best demonstrate this property by applying the two filters to a periodic structure $a\cos(\boldsymbol{kx})$. We assume that the first filter does not cause a phase shift, but the second causes a phase shift of 90°. Then,

$$\begin{aligned} g_1(\boldsymbol{x}) &= \hat{h}(\boldsymbol{k})a\cos(\boldsymbol{kx}) \\ g_2(\boldsymbol{x}) &= \hat{h}(\boldsymbol{k})a\sin(\boldsymbol{kx}). \end{aligned}$$

Squaring and adding the filter results, we get a constant *phase-independent* response of $\hat{h}^2 a^2$. We automatically obtain a pair of quadrature filters if we choose an even real and an odd imaginary transfer function with the same magnitude.

Returning to the selection of an appropriate set of directional filters, we can further state that they should have the same shape, except that they point in different directions. This condition is met if the transfer function can be separated into an angular and a wave number part. Such a filter is called *polar separable* and may conveniently be expressed in polar coordinates

$$\hat{h}(q, \phi) = \hat{r}(q)\hat{a}(\phi), \tag{13.12}$$

a   b

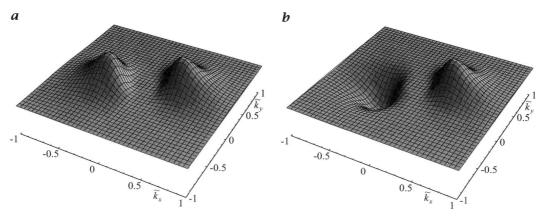

Figure 13.7: *Transfer function for **a** the even and **b** the odd part of the directional quadrature filter according to Eq. (13.13) with l = 2 and B = 2 in 22.5° direction. The 3-D plot covers all wave numbers normalized to the Nyquist wave number.*

with $q^2 = k_1^2 + k_2^2$ and $\tan \phi = k_2/k_1$. Knutsson [80] suggested the following family of directional quadrature filters:

$$\hat{r}(q) = \exp\left[-\frac{\ln^2(q/q_0)}{(B/2)^2 \log 2}\right]$$

$$\hat{a}_e(\phi) = \cos^{2l}(\phi - \phi_k) \tag{13.13}$$

$$\hat{a}_o(\phi) = \mathrm{i}\cos^{2l}(\phi - \phi_k)\, \mathrm{signum}\,[\cos(\phi - \phi_k)].$$

q denotes the magnitude of the wave number; q_0 and ϕ_k are the peak wave number and the direction of the filter, respectively. The indices e and o indicate the even and odd component of the quadrature pair. The constant B determines the half-width of the wave number in the number of octaves and l the angular resolution of the filter. In a logarithmic wave number scale, the filter has the shape of a Gaussian function. Figure 13.7 shows the transfer function of the even and odd parts of this quadrature filter.

A set of directional filters is obtained by a suitable choice of different ϕ_k:

$$\phi_k = \frac{\pi k}{K} \quad k = 0, 1, \cdots, K - 1. \tag{13.14}$$

Knutsson used four filters with 45° increments in the directions 22.5°, 67.5°, 112.5°, and 157.5°. These directions have the advantage that only *one* filter kernel has to be designed. The kernels for the filter in the other directions are obtained by mirroring the kernels at the axes and diagonals. These filters have been designed in the wave number space. The filter coefficients are obtained by inverse Fourier transform.

If we choose a reasonably small filter mask, we will cut off a number of nonzero filter coefficients. This causes deviations from the ideal transfer function. Therefore, Knutsson modified the filter kernel coefficient using an optimization procedure in such a way that it approaches the ideal transfer function as close as possible. It turned out that at least a 15×15 filter mask is necessary to get a good approximation of the anticipated transfer function.

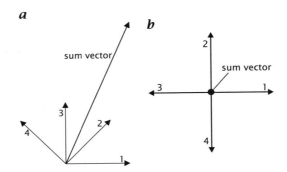

Figure 13.8: *Computation of local orientation by vector addition of the four filter responses 1 to 4. Shown is an example where the neighborhood is isotropic concerning orientation: all filter responses are equal. The angles of the vectors are equal to the filter directions in **a** and twice the filter directions in **b**.*

13.4.1c Computation of Orientation Vector. The local orientation can be computed from the responses of the four filters by vector addition if we represent them as an orientation vector: the magnitude of the vector corresponds to the filter response, while the direction is given by doubling the angle of the filter direction. Using a representation with complex numbers for the orientation vector, we can write the filter in operator notation as

$$\mathcal{Q}_{\phi_k} = \sqrt{{}^e\mathcal{H}_{\phi_k} \cdot {}^e\mathcal{H}_{\phi_k} + {}^o\mathcal{H}_{\phi_k} \cdot {}^o\mathcal{H}_{\phi_k}} \exp(2i\phi_k), \qquad (13.15)$$

where ${}^e\mathcal{H}_{\phi_k}$ and ${}^o\mathcal{H}_{\phi_k}$ are the even and odd component of the directional quadrature filter $\hat{h}(q, \phi)$, respectively Eq. (13.12).

Figure 13.8 explains the necessity of the angle doubling with an example where the responses from all four filters are equal. In this case the neighborhood contains structures in all directions. Consequently, we do not observe local orientation and the vector sum of all filter responses should vanish. This happens if we double the orientation angle (Fig. 13.8b), but not if we omit this step (Fig. 13.8a). Using Eq. (13.15), the complex operator to determine the orientation vector can be written as

$$^q\mathcal{O} = \sum_{k=0}^{K-1} \mathcal{Q}_{\phi_k}. \qquad (13.16)$$

The subscript q for $\mathcal{O}$ denotes that the quadrature filter set method is used to determine the orientation vector.

13.4.1d Evaluation. The serious drawback of the quadrature filter technique is its high computational effort. The image has to be convolved with four complex 15×15 nonseparable filter masks. This results in a total of 1800 multiplications and 1792 additions per pixel just to compute the four filter responses. The number of computations to determine the orientation vector requires much less computations so that the filtering itself is really the bottleneck. This means that in higher dimensions when, for instance, 3-D $15 \times 15 \times 15$ nonseparable kernels are required, the computational costs are even higher. However, Karlholm [77] recently published an efficient algorithm to compute the set of directional quadrature filters that reduces the number of computations by about a factor of 30.

13.4.2 2-D Tensor Method

13.4.2a Computation of the Tensor Components. The structure tensor discussed in Section 13.3.3 can be computed in a straightforward way as a combination of *linear convolution* and *nonlinear point operations*. The partial derivatives are approximated by discrete derivative operators. The integration corresponds to the convolution of a smoothing filter which has the shape of the window function. If we denote the discrete partial derivative operator with respect to the coordinate p by $\mathcal{D}_p$ and the (isotropic) smoothing operator by $\mathcal{B}$, the local structure tensor of a gray value image can be computed with the following structure tensor operator

$$\mathcal{J}_{pq} = \mathcal{B}(\mathcal{D}_p \cdot \mathcal{D}_q). \tag{13.17}$$

The equation is written in an operator notation. Pixelwise multiplication is denoted by $\cdot$ in order to distinguish from successive application of convolution operators. Equation Eq. (13.17) means verbatim: the pq component of the tensor is computed by convolving the image independently with $\mathcal{D}_p$ and $\mathcal{D}_q$, multiplying the two images pixelwise, and smoothing the resulting image with $\mathcal{B}$. This algorithm works with images of any dimension $W \geq 2$. In a W-dimensional image, the symmetric structure tensor has $W(W+1)/2$ components. These components are best stored in a component image with $W(W+1)/2$ components. These figures clearly show that the $W(W+1)/2$ convolutions for smoothing are the most costly operation. For an efficient implementation it is thus most important to speed up the smoothing.

13.4.2b Computation of the Orientation Vector. With the simple convolution and point operations discussed in the previous section, we computed the components of the structure tensor. In this section, we solve the eigenvalue problem to determine the orientation vector. In two dimensions, we can readily solve the eigenvalue problem. The orientation angle can be determined by rotating the structure tensor to the principal axes coordinate system:

$$\begin{bmatrix} J_x & 0 \\ 0 & J_y \end{bmatrix} = \begin{bmatrix} \cos\phi & -\sin\phi \\ \sin\phi & \cos\phi \end{bmatrix} \begin{bmatrix} J_{xx} & J_{xy} \\ J_{xy} & J_{yy} \end{bmatrix} \begin{bmatrix} \cos\phi & \sin\phi \\ -\sin\phi & \cos\phi \end{bmatrix}.$$

Using the trigonometric identities $\sin 2\phi = 2\sin\phi\cos\phi$ and $\cos 2\phi = \cos^2\phi - \sin^2\phi$, the matrix multiplications result in

$$\begin{bmatrix} J_x & 0 \\ 0 & J_y \end{bmatrix} = \begin{bmatrix} \cos\phi & -\sin\phi \\ \sin\phi & \cos\phi \end{bmatrix} \begin{bmatrix} J_{xx}\cos\phi - J_{xy}\sin\phi & J_{xx}\sin\phi + J_{xy}\cos\phi \\ -J_{yy}\sin\phi + J_{xy}\cos\phi & J_{yy}\cos\phi + J_{xy}\sin\phi \end{bmatrix}$$

$$= \begin{bmatrix} J_{xx}\cos^2\phi + J_{yy}\sin^2\phi - J_{xy}\sin 2\phi & 1/2(J_{xx}-J_{yy})\sin 2\phi + J_{xy}\cos 2\phi \\ 1/2(J_{xx}-J_{yy})\sin 2\phi + J_{xy}\cos 2\phi & J_{yy}\sin^2\phi + J_{yy}\cos^2\phi + J_{xy}\sin 2\phi \end{bmatrix}.$$

Now we can compare the matrix coefficients on the left and right side of the equation. Because the matrices are symmetric, we have three equations with three unknowns, ϕ, J_1, and J_2. Although the equation system is nonlinear, it can readily be solved for ϕ. A comparison of the off-diagonal elements on both sides of the equation

$$1/2(J_{xx} - J_{yy})\sin 2\phi + J_{xy}\cos 2\phi = 0 \tag{13.18}$$

yields the orientation angle as

$$\tan 2\phi = \frac{2J_{xy}}{J_{yy} - J_{xx}}. \tag{13.19}$$

Without any presumptions, we obtained the anticipated angle doubling for orientation. Since $\tan 2\phi$ is gained from a quotient, we can regard the dividend as the y and the divisor as the x component of a vector and we can form the vectorial orientation operator $\mathcal{O}$, as it has been introduced by Granlund [44]

$$\mathcal{O} = \left[\begin{array}{c} \mathcal{J}_{yy} - \mathcal{J}_{xx} \\ 2\mathcal{J}_{xy} \end{array} \right]. \tag{13.20}$$

The phase of this vector gives the orientation angle (or its tangent the optical flow in xt images) and the magnitude a certainty measure for local orientation (or constant velocity in xt images) as discussed in Section 13.3.1d.

The result of Eq. (13.20) is remarkable in the respect that the computation of the components of the orientation vector from the components of the orientation tensor requires just one subtraction and one multiplication by two. Since these components of the orientation vector are used for further processing, this was all that was required to solve the eigenvalue problem in two dimensions.

13.4.2c Orientation Coherency. The orientation vector reduces local structure to local orientation. From three independent components of the tensor representation only two are used. When we fail to observe an orientated structure in a neighborhood, we do not know whether no gray value variations or a distributed orientation is encountered. This information is included in the missing component of the operator, $\mathcal{J}_{xx} + \mathcal{J}_{yy}$, which gives the mean square gradient. Consequently, a well-equipped structure (or motion) operator needs to include also the third component. A suitable linear combination is

$$S = \left[\begin{array}{c} \mathcal{J}_{yy} - \mathcal{J}_{xx} \\ 2\mathcal{J}_{xy} \\ \mathcal{J}_{xx} + \mathcal{J}_{yy} \end{array} \right]. \tag{13.21}$$

This structure operator contains the two components of the orientation vector and, as an additional component, the mean square magnitude of the gradient which is a rotation invariant parameter. Comparing the latter with the squared magnitude of the orientation vector a constant gray value area and an isotropic gray value structure without preferred orientation can be distinguished. In the first case, both squared magnitudes are zero, in the second only that of the orientation vector. In the case of a perfectly oriented pattern (or constant optical flow), both squared magnitudes are equal. Thus the ratio of them seems to be a good *coherency measure* for local orientation (or constant 1D velocity):

$$C = \frac{(\mathcal{J}_{yy} - \mathcal{J}_{xx}) \cdot (\mathcal{J}_{yy} - \mathcal{J}_{xx}) + 4\mathcal{J}_{xy} \cdot \mathcal{J}_{xy}}{(\mathcal{J}_{xx} + \mathcal{J}_{yy}) \cdot (\mathcal{J}_{xx} + \mathcal{J}_{yy})} = \left(\frac{J_x - J_y}{J_x + J_y} \right)^2. \tag{13.22}$$

The coherency ranges from 0 to 1. For ideal local orientation ($J_x = 0, J_y > 0$) it is one; for an isotropic gray value structure ($J_x = J_y > 0$) it is zero.

13.4.2d Color Coding of the Structure Tensor. In Section 13.3.1d we discussed a color representation of the orientation vector. The question is whether it is also possible to represent the structure tensor in an adequate way as a color image. As is evident from Eq. (13.21), a symmetric 2-D tensor has three independent pieces of information, which fit well to the three degrees of freedom available to represent color, for example, luminance, hue, and saturation.

A color represention of the structure tensor requires only two slight modifications as compared to that for the orientation vector (Fig. 13.5d). First, instead of the length of the orientation vector, the squared magnitude of the gradient is mapped onto the luminance. Secondly, the coherency measure Eq. (13.22) is used as the saturation. In the color representation for the orientation vector, the saturation is always one. The mapping of the hue remains the same.

In practice, a slight modification of this color representation is useful. The squared magnitude of the gradient shows variations too large to be displayed in the narrow dynamic range of a display screen with only 256 luminance levels. Therefore, a suitable normalization is required. The basic idea of this normalization is to compare the squared magnitude of the gradient with the noise level. Once the gradient is well above the noise level it is regarded as a significant piece of information. This train of thought suggests the following normalization for the luminance I:

$$I = \frac{\mathcal{J}_{xx} + \mathcal{J}_{yy}}{(\mathcal{J}_{xx} + \mathcal{J}_{yy}) + \gamma \sigma_n^2}, \tag{13.23}$$

where σ_n is an estimate of the standard deviation of the noise level. This normalization provides a rapid transition of the luminance from one, when the magnitude of the gradient is larger than σ_n, to zero when the gradient is smaller than σ_n. The factor γ is used to optimize the display. This normalization has been used throughout this handbook. A similar normalization is also used for the color display of orientation vectors. A comparison of the color coding of the orientation vector and the structure tensor is shown in Fig. 13.5. For further color images with this representation, see Plates 21 and 25.

13.4.2e Theoretical Performance and Systematic Errors. Before we discuss a fast and accurate implementation of the 2-D structure tensor method with discrete images, the theoretical performance is discussed. It is one of the most significant advantages of the structure tensor method that its performance can be studied analytically. The basic idea is to use the general expressions for simple neighborhoods given in Section 13.3.1 and to modify them with all kinds of disturbances such as noise and to study the systematical and statistical errors in the orientation angle caused by these disturbances. We give only the results here. The proofs and further details can be found in *Jähne* [1993c].

Exact orientation estimate in a simple neighborhood. This result came somewhat as a surprise. It was long believed that, if in a local neighborhood higher-order derivatives are significant, the orientation estimate is biased. The analytical studies show, however, that this is not the case. The orientation estimate is exact for *any* spatial variation of the gray values, provided it is a simple neighborhood.

Unbiased orientation estimate in noisy images. No bias is introduced into the orientation estimate if normal distributed noise is added to a simple neighborhood.

Curved Patterns. The first deviation from a simple neighborhood is curved patterns. The analytical studies show two important results: First, there is a small bias in the orientation angle, but it is negligibly small. Secondly, the coherency is not sensitive to *gradual* changes in the orientation; it is only slightly reduced (compare also the results with the ring test pattern in Fig. 13.9).

Corners. At corners (in image sequences this corresponds to a motion discontinuity), i. e., sudden changes in the local orientation, the coherency is significantly decreased. In the simple case that the edge strength (magnitude of the gradient) is the same at both sides of the corner, the coherency Eq. (13.22) is simply given by $c = \cos \alpha$, where alpha is the angle between the edges. If the edges at the corners

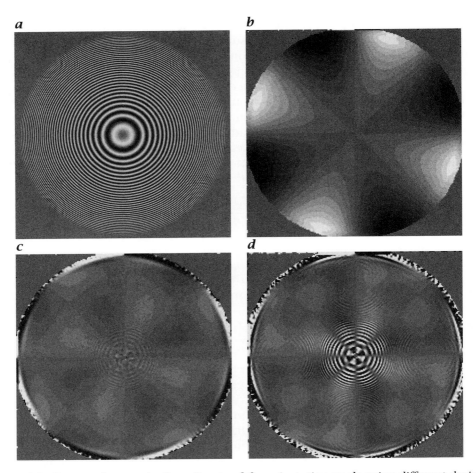

Figure 13.9: *Systematic error in the estimate of the orientation angle using different derivative filters: **a** ring test pattern, the maximal normalized wave number is 0.35, **b** angle error map with the standard 1/2[1 0 -1] derivative filter, range ±16°, distance of 2°; **c** same but using an optimized recursive derivative filter with 2 nonrecursive coefficients, range ±0.4°, distance of 0.05°; **d** same but using an optimized recursive derivative filter with 4 nonrecursive coefficients, range ±0.04°, distance of 0.005°.*

are perpendicular to each other, the coherency becomes zero. Thus, the structure tensor method can also be used to identify corners.

13.4.2f Accurate Implementation. We first turn to the question of an accurate implementation and look at the results of an experimental error analysis of the orientation estimate using the ring test pattern (Fig. 13.9a). The straightforward implementation of the algorithm using the standard derivative filter mask 1/2[1 0 -1] results in a surprisingly high error map (Fig. 13.9b) with a maximum error in the orientation angle of more than 20°. The error depends both on the wave number and the orientation of the local structure (or velocity in space-time images). For orientation angles in axes and diagonal directions, the error vanishes. The high error and the structure of the error map results from the transfer function of the derivative filter. The transfer function shows significant deviation from the transfer function for an ideal derivative filter for high wave numbers (Section 12.4.1). According to Eq. (13.19), the orientation angle depends on the ratios of derivatives. Along the axes, one of the derivatives is zero and, thus, no error occurs. Along the diagonals, the derivatives in x and y direction are the same. Consequently, the error in both cancels in the ratio.

The error in the orientation angle can significantly be suppressed if better derivative filters as discussed in detail in Section 12.4.1d are taken (Fig. 13.9c and d). Even for the smallest optimized filter with two nonrecursive and two recursive coefficients (Fig. 13.9c), the residual error is only about 0.1°. For a filter with four nonrecursive coefficients (Fig. 13.9c), the error in the angle estimate is no longer governed by the filter but by numerical inaccuracies. The little extra effort in optimizing the derivative filters thus pays off in an accurate orientation estimate. A residual angle error of less than 0.1° is sufficient for almost all applications. For motion analysis an error of 0.1° in the orientation means an error in the velocity that corresponds to a shift of 0.0017 pixel between consecutive images.

13.4.2g Fast Implementation. In Section 13.4.2a we found that the smoothing operations consume the largest number of operations to compute the components of the structure tensor. Therefore, a fast implementation must, in the first place, apply a fast smoothing algorithm. A fast algorithm can be established based on the general observation that higher-order features always show a lower resolution than the features it is computed from. This means that the structure tensor can be stored on a coarser grid and thus in a smaller image. A convenient and appropriate subsampling rate is to reduce the scale by a factor of two by storing only every second pixel in every second row.

This procedure leads us in a natural way to multigrid data structures that are discussed in detail in Chapter 14. More specifically, we can take the fast algorithms used to smooth images for the computation of the Gaussian pyramid (Section 14.3.4b). Storing higher-order features on coarser scales has another significant advantage. Any subsequent processing is sped up simply by the fact that much fewer pixels have to be processed. A linear scale reduction by a factor of two results in a reduction in the number of pixels — and thus a corresponding speed up of the computations — by a factor of 4 and 8 in two and three dimensions, respectively.

13.4.2h Statistical Errors. An important property of any image processing algorithm is its *robustness*. This term means how sensitive an algorithm is against noise. Two questions are of importance. How large is the error of the estimated features in a noisy image? And, do we obtain a useful result at all? To answer these questions, the laws of statistics are used to study error propagation. In this framework, noise makes the estimates only uncertain but not erroneous, but the mean — if we make a sufficient number of estimates — is still correct.

Worse things can happen. In noisy images an algorithm can completely fail and give results that are either biased, i. e., the mean shows a significant deviation from the correct value, or the algorithm becomes even unstable delivering random results.

In Section 13.4.2e we have already discussed that the estimate of the structure tensor and the local orientation is not biased by noise. Figure 13.10 demonstrates that the estimate of orientation is also a remarkable robust algorithm. Even when the signal-to-noise ratio is one, the orientation estimate is still correct. Figure 13.11 gives a more quantitative account of the error in the orientation analysis. Even for a signal-to-noise ratio of 2.5, the standard deviation of the estimate of the orientation angle is only 1.5°.

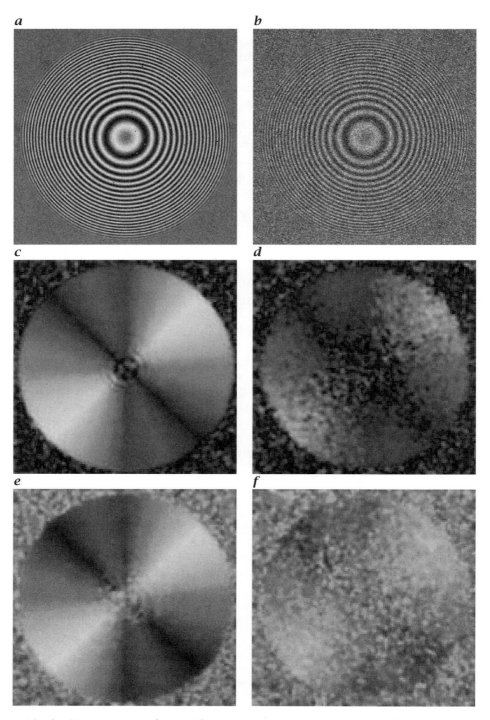

Figure 13.10: *Orientation analysis with a noisy ring test pattern. **a** and **b** ring pattern with amplitude 100 and 50, standard deviation of normal distributed noise 10 and 50, respectively; **c** and **d** color representation of the orientation vector; **e** and **f** color representation of the structure tensor. (See also Plate 21.)*

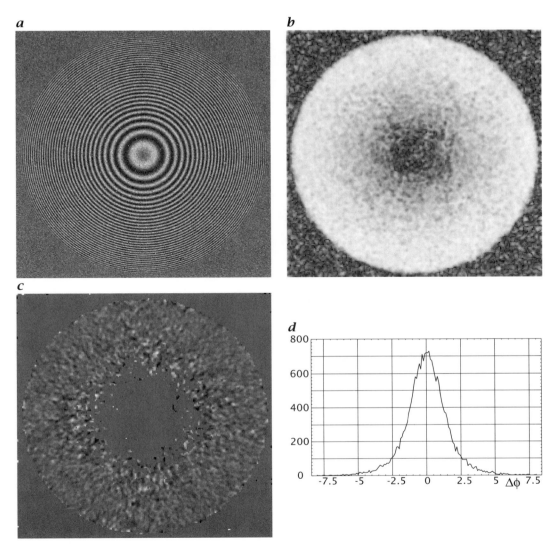

Figure 13.11: *Analysis of the error in the estimate of the orientation angle using different derivative filters:* **a** *noisy ring test pattern with a signal-to-noise ratio of 2.5;* **b** *coherency of the orientation estimate;* **c** *map of the angle error* $\Delta\phi$*; the error is only shown in the areas where the coherency was higher than 0.8;* **d** *histogram of c.*

13.4.3 Motion Analysis in Space-Time Images

In the next chapter (Chapter 14), we discuss in detail the application of the structure tensor for texture analysis. Motion analysis is another excellent application for local structure analysis and orientation estimates. The basic analysis in Sections 13.2 and 13.3.2 (see also Fig. 13.2) showed that these techniques can likewise be applied to simple objects, textures, and motion. Therefore, this chapter closes with an application of motion analysis in space-time images which demonstrate the power of this approach. Motion analysis requires a three-dimensional analysis of the local structure which is not treated in detail in this handbook but in Jähne [73].

Figure 13.12: *Plant leaf growth studies: images of young ricinus leaves taken in the visible (**a** and **c**, left) and near infrared wavelength range (800 – 900 nm) (**b** and **c**, right); **d** example image of a sequence taken to measure leaf growth; color representation of **e** the orientation vector and **f** the structure tensor of the image shown in **d**. (See also Plate 22.)*

Figure 13.13: 2-D velocity determination of plant leaf growth: **a** masked horizontal velocity; **b** masked vertical velocity; **c** interpolated horizontal velocity; **d** interpolated vertical velocity; **e** and **f** area dilation (divergence of the velocity vector field). (See also Plate 23.)

In many sciences, dynamical processes are of essential importance. Image sequence analysis gives a direct insight into the spatial distribution of these processes which is not possible with conventional techniques. Figures 13.12–13.13 show the analysis of leaf growth with exactly the techniques described in this section. The analysis goes two steps ahead. Normalized convolution (Section 11.3.2) is used to compute continuous velocity fields (Fig. 13.13a-d). For growth, not the velocity field itself is of interest but the divergence of the vector field. This quantity is shown in Fig. 13.13e and f, nicely demonstrating the change of the growth regions with time.

13.5 Advanced Reference Material

13.1 | **References for orientation and motion analysis**

G. Granlund and H. Knutsson, 1995. Signal Processing for Computer Vision, Kluwer Academic Publishers, Dordrecht. *This monograph is written by the pioneers of vector and tensor-based image signal processing and gives an excellent overview of these techniques.*

A. Singh, 1991. Optic Flow Computation: A Unified Perspective, IEEE Computer Society Press, Los Alamitos, CA.

D. J. Fleet, 1992. Measurement of Image Velocity, Kluwer Academic Publisher, Dordrecht. *Phase-based determination of optical flow.*

H. Haußecker and H. Spies, 2000. Motion, in Computer Vision and Applications, edited by B. Jähne and H. Haußecker, Academic Press, San Diego, pp. 347-395. *Recent review of motion analysis.*

B. Jähne, 1993. Spatio-Temporal Image Processing, Theory and Scientific Applications, Lecture Notes in Computer Science, 751 Springer-Verlag, Berlin. *Application of the local structure analysis to image sequences.*

B. Jähne, ed., 2004. Image Sequence Analysis to Investigate Dynamic Processes, Lecture Notes in Computer Science, Springer-Verlag, Berlin. *Estimation of parameters of dynamic processes from image sequences: theory and applicatons.*

14 Scale and Texture

14.1 Highlights

Texture, i. e., pattern in images, is one of the features that is ubiquitous but makes image processing a difficult task. Some say this is where image processing really starts since objects with uniform gray scales are not difficult to detect and to analyze.

Anybody who has some experience in texture analysis knows about the frustration in performing this task. On the one side, it is tedious and time consuming to analyze texture. It requires a lot of adjustments for each individual case just to end up with a solution that constitutes an acceptable solution. On the other side, the human and other biological visual systems master this task fast and with ease.

This tells us two things. First, it *is* possible to recognize patterns in images and to differentiate them. Second, we do not yet know how to do it right. Therefore, it is not surprising that a confusing wide range of empirical and semi-empirical approaches exists. All of them work in some cases. All of them fail in most other cases.

We do not want to confront the reader with this mess. This is not to state that the revolutionary new texture analysis technique is introduced in this chapter. We rather want to go some humble first steps that are related to the general approach to image processing taken in this handbook. We will focus on a quadruple of basic texture parameters: mean, variance, orientation, and scale. The beauty of this approach is that it can be applied in a hierarchical manner and, thus, is able to analyze also complex hierarchical textures. Once a higher-level feature such as orientation or a measure for the scale has been determined, the elementary operators mean and variance can be applied again.

One of the most important characteristics of texture is the scale parameter. Texture consists of small elementary patterns that are repeated periodically or quasi-periodically. A complex texture may be built up with a complex hierarchy of such repeating patterns on different scales.

This is why in this chapter data structures are also handled that — in contrast to the Fourier transform — allow analyzing periodic structures but which still preserve spatial resolution. These multiscale data structures are known as pyramids and we will discuss the Gaussian pyramid (Section 14.3.4), the Laplacian pyramid (Section 14.3.5), and directio-pyramidal decomposition (Section 14.3.6).

The term phase is also of elementary importance since it is the key to determining scale parameters such as the local wave number (Sections 14.3.7 and 14.4.3).

Task List 11: Scale and Texture

Task	Procedures
Check your application to find out what type of texture parameters are required	Choose between rotation variant/invariant features and scale variant/invariant features
Determine simple texture parameters	Choose from variance and higher-order statistical parameters, local orientation, characteristic scale, and combinations
Determine phase	Use quadrature filter pairs; use Hilbert filter
Determine local wave numbers	Compute phase gradients

14.2 Task

Real-world objects often have no uniform surface but show distinct patterns (Fig. 14.1). Our visual system is capable of recognizing and distinguishing such patterns with ease, but it is difficult to describe the differences precisely. A pattern which characterizes objects is called a *texture* in image processing. Actually, textures demonstrate the difference between an artificial world of objects whose surfaces are only characterized by the color and reflectivity properties to that of real-world imagery.

The task in this chapter is to systematically investigate operators to extract suitable features from textures. These operators should be able to describe even complex patterns with a few, but characteristic, figures. We thereby reduce the texture recognition problem to the known task of distinguishing scalar and vectorial features.

We can separate texture parameters into two classes. Texture parameters may or may not be *rotation* and *scale invariant*. This classification is motivated by the task we have to perform. Imagine, on the one hand, a typical industrial or scientific application in which we want to recognize objects which are randomly orientated in the image. We are not interested in the orientation of the objects but in the distinction from other objects. Therefore, texture parameters which depend on the orientation are of no interest. On the other hand, we might still use them but only if the objects have a characteristic shape which then allows us to determine their orientation.

We can use similar arguments for scale-invariant features. If the objects of interest are located at different distances from the camera, the texture parameter used to recognize them should also be scale-invariant. Otherwise, the recognition of the object will depend on the distance. If the texture changes its characteristics with the scale, invariant texture features may not exist at all. Then, the use of textures to characterize objects at different distances does not work.

In the examples above, we were interested in the objects themselves but not in their position in space. The orientation of surfaces is, however, a key feature for another image processing task, the reconstruction of a three-dimensional scene from a two-dimensional image. If we know that the surface of an object shows a uniform texture, we can analyze the orientation and scales of the texture to determine the orientation of the surface in space. Then, the characteristic scales and orientations of the texture are critical parameters.

Texture analysis is one of those areas in image processing which still lacks fundamental knowledge. Consequently, the literature contains many different empirical and semiempirical approaches to texture analysis. These approaches are not presented here. As in the whole handbook, we rather take a well-founded approach. It is not known whether the approach presented here is optimal but it is straightforward in the sense that only four fundamental texture parameters are used: *mean* and *variance*, *orientation* and *scale*. The first two parameters are rotation and scale independent, the

Figure 14.1: *Examples of natural textures. (See also Plate 24.)*

latter two determine orientation and scale. With this set of parameters, the important separation in scale and rotation invariant and variant parameters is already made and, thus, significantly simplifies texture analysis. This is a strong hint on the elementary nature of these parameters.

The power of this approach lies in the simplicity and orthogonality of the parameter set. The parameters can be applied hierarchically. We can determine the mean and variance of the orientation or scale to refine the description of texture at larger scales.

Texture analysis is combined in this chapter with the analysis of scale as can be seen from task list 11. The important concept of phase is introduced in Section 14.3.2. It is not only of fundamental importance to understand the nature of textures but also a key parameter to compute "texture energy" and local wave numbers (Section 14.4.3).

It is an important part of the analysis of texture to analyze the scales, but this task appears also elsewhere. An appropriate treatment of scales requires a partition of an image into different scales. This task requires new data structures. Pyramids are one of them which can be computed efficiently and also speed up the computations for larger scales significantly (Sections 14.3.4 and 14.3.5).

14.3 Concepts

14.3.1 What Is Texture?

As an introduction to texture, we discuss a collection of synthetic images shown in Figs. 14.2-14.4. In each of the images, the object (text "IMAGE" (Fig. 14.2a)) differs from the background only by a certain texture parameter. Below the original image, the orientation vector (left image) and the structure tensor (right image) are shown in a color representation using the techniques described in Sections 13.3.1d and 13.4.2d. For color figures, see Plates 25-26.

In Fig. 14.4, object and background have the same mean gray value. The pattern is also the same and consists of normal distributed noise. It differs only in the variance. This is sufficient to recognize the letters although it is not as easy as with gray values (Fig. 14.2a). It is surprisingly easy to recognize an object which has the same oriented pattern as the background but where the pattern is oriented in a different direction (Fig. 14.2d and g). Astonishingly, it is even easier to recognize the text with the coarser pattern in Fig. 14.2g. The letters appear to be floating in front of the striped background. In both cases, we see sharp boundary lines between the letters and the background. The determination of orientation gives — as expected — a clear separation with the fine patterns (Fig. 14.2e and f), but fails with the coarse pattern (Fig. 14.2h and i), because the pattern is too coarse for an orientation analysis in a small neighborhood. The orientation can, however, be computed at a coarser scale. Then, the pattern shrinks in size and comes into the reach of local operators. Results from such an analysis, blown up to the original size again, are shown in Fig. 14.2j and k. These examples shed some light already on the importance of multiscale image processing that is detailed in Sections 14.3.4 and 14.3.5.

The distinction of patterns by orientation works also if the patterns themselves are not oriented as demonstrated in Fig. 14.3d and g. The orientation analysis in this case gives no oriented patterns and, thus, appears as white disoriented areas in Fig. 14.3f. Interestingly, if because of a shorter scale, one orientation is slightly dominant (as in Fig. 14.3g), this orientation is picked out and clearly distinguishes background from the letters (Fig. 14.3).

From orientation, we turn to scale. Figure 14.4d and g demonstrates that an object can be recognized even if it only differs by the scale of the pattern. Mean and variance are the same in the background and in the objects. This time, the recognition task becomes harder if the structures are coarser (Fig. 14.4g).

The example images demonstrate that the basic parameters mean, variance, orientation, and scale are a solid fundament for texture analysis. Obviously, the human visual system can use any of these parameters alone to distinguish texture. Texture is, of course, more than these parameters. If we find, for instance, that a texture is

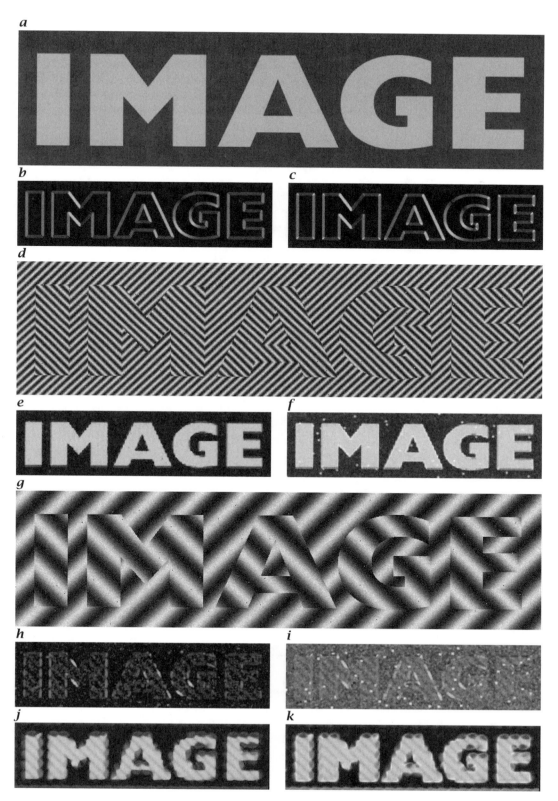

Figure 14.2: *Analysis of textures with the structure tensor technique; for details, see text. (See also Plate 25.)*

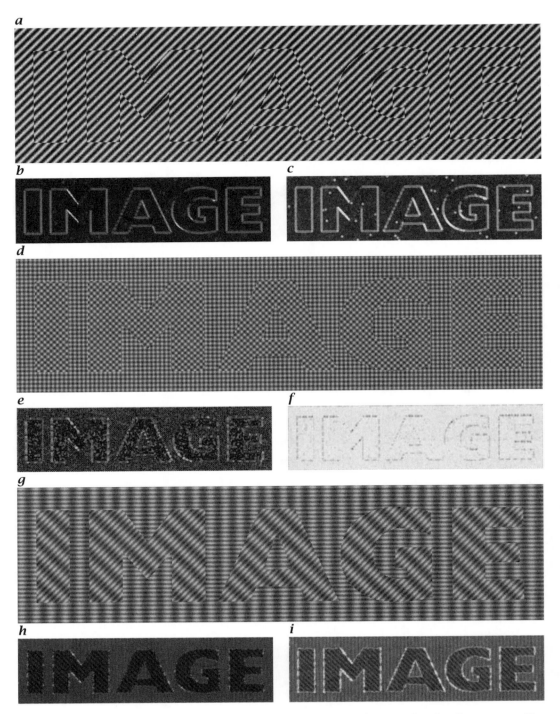

Figure 14.3: *Analysis of textures with the structure tensor technique; for details, see text. (See also Plate 26.)*

Figure 14.4: *Analysis of textures with the structure tensor technique; for details, see text. (See also Plate 27.)*

oriented in a certain direction, we restrict a general spatial function $g(\boldsymbol{x})$ to a function $g(\boldsymbol{x}\bar{\boldsymbol{k}}_0)$, where $\bar{\boldsymbol{k}}_0$ is a unit vector in the orientation into which the textures change (Section 13.3.1a). Although this is a significant reduction (essentially from a 2-D to a 1-D function), we have not yet described the spatial variation and a large number of textures fall into the class of oriented patterns.

Fortunately, the problem is not too bad because of an important general feature of textures. The elementary pattern of a texture is small. This pattern is then repeated periodically or quasi-periodically in space like a pattern on a wall paper. Thus, it is sufficient to describe the small elementary pattern and the repetition rules. The latter gives the characteristic scale of the texture.

Texture analysis in this respect can be compared to the analysis of the structure of crystalline solids. A solid state physicist must find out the repetition pattern and the distribution of atoms in the elementary cell. Texture analysis is, however, complicated by the fact that both the patterns and periodic repetition may show significant random fluctuation.

14.3.2   The Wave Number Domain

Fourier transform, i. e., decomposition of an image into periodic structures, is an extremely helpful tool to understanding image formation, digitization, and processing. Since texture means spatial changes of gray values and the scale at which these changes occur is a key parameter of texture, it appears that the Fourier transform plays also an important role in the analysis of texture.

As outlined in Appendix B.4, the discrete Fourier transform (DFT) can be regarded as a coordinate transformation in a finite-dimensional vector space. Therefore, the image information is completely preserved. We can perform the inverse transformation to obtain the original image. In Fourier space, we observe the image just from another "point of view". Each point in the Fourier domain represents a periodic component of the image with the discrete wave numbers u and v (number of wavelengths in x and y direction). The complex numbers carry two pieces of information: the *amplitude* and the *phase*, i. e., relative position of a periodic structure.

What information do these two quantities mean in images? Is the phase or the amplitude of more importance? At first glance, we might be tempted to give the amplitude more importance because if there is no amplitude there is no information. In order to answer this question, we perform a simple experiment. Figure 14.5a and b shows two images. One shows book backs, the other tree rings. Both images are Fourier transformed and then the phase and amplitude are interchanged as illustrated in Fig. 14.5c and d. The result of this interchange is surprising. The phase determines the content of an image for both images. Both images look patchy but the significant information such as the letters or tree rings is preserved.

The amplitude alone implies only *that* such a periodic structure is contained in the image but not *where*. We can also illustrate this important fact with the shift theorem (see Appendix B.3). Shifting of an object in the spatial domain leads to a shift of the phase in the wave number domain. If we do not know the phase of its Fourier components, we know neither what the object looks like nor where it is located.

From these considerations, we can draw important conclusions for texture analysis.

1. The power spectrum, i. e., the squared amplitudes of the Fourier transformed image, $|\hat{g}(\boldsymbol{k})|^2$, contains very little information. The phase is lost and the periodic components spread out over the whole image.

a **b**

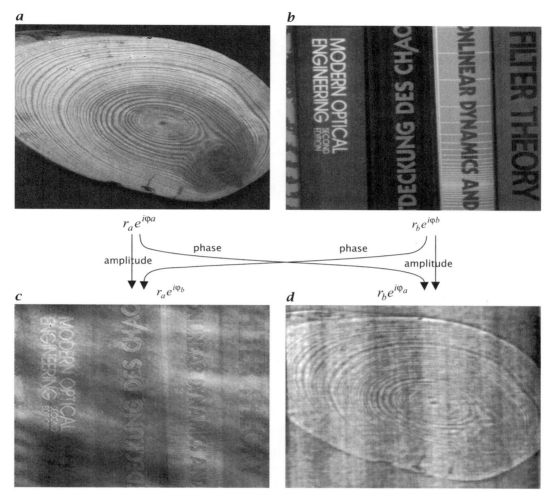

Figure 14.5: *Illustration of the importance of phase and amplitude in Fourier space for the image content: **a**, **b** two original images; **c** composite image using the phase from image **b** and the amplitude from image **a**; **d** composite image using the phase from image **a** and the amplitude from image **b**.*

2. We need a separation into periodic structures as provided by the Fourier transform but with different features: it must preserve spatial resolution and thus retain the phase information.

14.3.3 Hierarchy of Scales

The problem of scales will be illuminated in this section from another perspective: the hierarchical nature of texture. Neighborhood operators (Chapter 10) can only extract local features that are significantly smaller than the image. More complex features such as local orientation (Chapter 13) require already larger neighborhoods (Section 12.4.1). Larger neighborhoods can show a larger set of features which requires more complex operations to reveal them. If we extrapolate this approach by analyzing larger scales in the image with larger filter kernels, we inevitably run into a dead end. The computation of the more complex operators will become so tedious that they are no longer

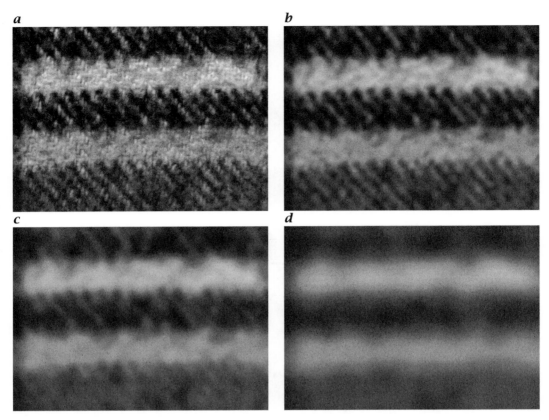

a
b
c
d

Figure 14.6: *Illustration of the hierarchical nature of texture. Shown is a piece of woven fabric at different resolution levels: **a** original image; **b**, **c**, **d** increasingly smoothed images (second, third, and fourth level of the Gaussian pyramid). At different levels, the texture appears different.*

useful. Figure 14.6 shows a hierarchically organized texture that demonstrates the need to analyze texture on different scales. The sequence of images in Fig. 14.6 has been generated by smoothing the original image to a larger and larger extent. At the finest resolution in Fig. 14.6a, the texture of the individual threads running in vertical direction is visible. When this structure is blurred away, the next coarser scale is diagonal stripes (Fig. 14.6b). At even larger scales (Fig. 14.6d), only horizontal stripes remain in the image.

We can also run into problems related to scales from another side. Imagine that we intend to calculate the first horizontal derivative of an image which shows only slowly varying gray values, i. e., large-scale features. Then, only small gray value differences exist between neighboring pixels. Because of the limited resolution of the gray values and noise, the result of the operation will contain significant inaccuracies. The problem is caused by a *scale mismatch*: the gray values only vary on a large scale, while the derivative operator operates on a much finer scale.

From these two observations, we can conclude that we need new data structures to separate the information in the image into different "scale channels". The fine structures which are contained in an image must still be represented on a fine grid, whereas the large scales can be represented on a much coarser grid. On the coarse grid, the large scales come within the range of effective neighborhood operators with small kernels. This approach leads to a *multigrid* or *multiscale* data structure.

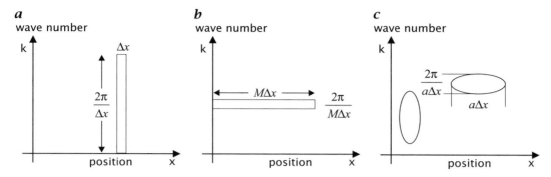

Figure 14.7: *Illustration of the interdependence of resolution in the spatial and wave number domain for a 1-D signal:* **a** *representation in the space domain;* **b** *representation in the wave number domain;* **c** *windowed Fourier transform with two different widths of the window function.*

If we represent a 1-D signal on a grid in the spatial domain, we do not know anything about the scales. We know the position with an accuracy of the grid constant, Δx, but the local wave number at this position may be any out of the range of possible wave numbers from 0 to $2\pi/\Delta x$ (Fig. 14.7a). The other extreme is given by the representation of the 1-D signal in the wave number domain. Each pixel in this domain represents one wave number with the highest possible resolution given, $2\pi/(M\Delta x)$, but any positional information is lost since these periodic structures are spread over the whole spatial extension, $M\Delta x$ (Fig. 14.7b). The same type of considerations is valid for each coordinate in higher-dimensional data.

We can conclude that the representation of an image in either the spatial or wave number domain describes limiting cases. Between these extremes, many other possibilities exist to represent image data. These data structures manifest the fact that the image is separated into different ranges of wave numbers but still preserves some spatial resolution.

Indeed, it is easy to construct such a data structure. All we have to do is to apply a window to image data and take the Fourier transform of these windowed data. The width of the window function gives the spatial resolution. The wave number resolution can be inferred from the fact that the multiplication with the window function in the spatial domain corresponds to a convolution with the Fourier transformed window function in the wave number domain. If a Gaussian window function is used with a standard deviation of σ_x, the Fourier transformed window is a Gaussian function with the standard derivation of $\sigma_k = 1/\sigma_x$.

The product of the resolutions in the spatial and wave number domains is, thus, interrelated and cannot be lower than 1 for Gaussian windows. This result is known as the classical *uncertainty relation*. It states that $\sigma_x \sigma_k \geq 1$. The Gaussian function is optimal in the sense that it reaches the lower limit. Figure 14.7c illustrates the combined space/wave number resolution of the windowed Fourier transform. The windowed Fourier transform gives us the first hint that we can generate data structures which decompose an image into different scales but still preserve some spatial resolution. In the next two sections, the Gaussian (Section 14.3.4) and the Laplacian pyramids (Section 14.3.5) are discussed as examples of such data structures.

Figure 14.8: The first five planes of the Gaussian pyramid.

14.3.4 Gaussian Pyramid

14.3.4a Principle. In this section, we discuss the transformation of an image which is represented on a grid in the spatial domain into a multigrid data structure. From what we have learned in Section 6.3.2, it is obvious that we cannot just *subsample* the image by taking, for example, every second pixel in every second line. If we did so, we would disregard the *sampling theorem* (Section 6.3.2c). For example, a structure which is sampled three times per wavelength in the original image would only be sampled one and a half times in the subsampled image and, thus, appear as an aliased pattern. Consequently, we must ensure that all wave numbers which are sampled less than four times per wavelength are suppressed by an appropriate smoothing filter to ensure a properly subsampled image. Generally, the requirement for the smoothing filter can be formulated as

$$\hat{B}(\tilde{\boldsymbol{k}}) = 0 \quad \forall \tilde{k}_p \geq \frac{1}{S_p}, \tag{14.1}$$

where S_p is the subsampling rate in the pth coordinate and $\tilde{\boldsymbol{k}}$ the wave number normalized to the maximum possible wave number on the current grid.

If we repeat the smoothing and subsampling operations iteratively, we obtain a series of images called the *Gaussian pyramid*. From level to level, the resolution decreases by a factor of two; the size of the images is decreasing correspondingly. Consequently, we can think of the series of images as being arranged in the form of a pyramid.

The pyramid does not require much storage space. Generally, if we consider the formation of a pyramid from a W-dimensional image with a subsampling factor of two and M pixels in each coordinate direction, the total number of pixels is given by

$$M^W \left(1 + \frac{1}{2^W} + \frac{1}{2^{2W}} + \ldots \right) < M^W \frac{2^W}{2^W - 1}. \tag{14.2}$$

For a two-dimensional image, the whole pyramid only needs one third more storage space than the original image. Likewise, the computation of the pyramid is equally effective. The *same* smoothing filter is applied to each level of the pyramid. Thus, the computation of the *whole* pyramid only needs four thirds of the operations necessary for the first level.

The pyramid brings large scales into the range of local neighborhood operations with small kernels. Moreover, these operations are performed efficiently. Once the pyramid has been computed, we can perform neighborhood operations on large scales

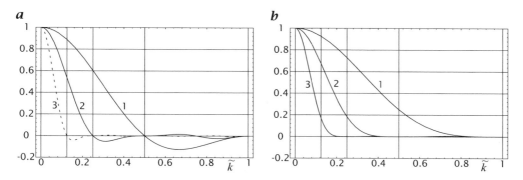

Figure 14.9: *Transfer functions of the smoothing filters used by Burt [1984] to generate the Gaussian pyramid: **a** 1/8 [1 2 2 2 1], **b** 1/16 [1 4 6 4 1]. The maximum wave numbers of 0.5, 0.25, and 0.125 for planes 1, 2, and 3 are marked by vertical lines. Both filters are shown in a cascaded application to generate the planes one to three of the pyramid. Both introduce significant aliasing, **a** even a contrast inversion (negative transfer function).*

in the upper levels of the pyramid — because of the smaller image sizes — much more efficiently than for finer scales. The Gaussian pyramid consists of a series of lowpass filtered images in which the cut-off wave numbers decrease by a factor of two (an octave) from level to level. Thus, only more and more coarse details remain in the image (Fig. 14.8). Only a few levels of the pyramid are necessary to span a wide range of wave numbers. From a 512×512 image, we can usefully compute a seven-level pyramid. The smallest image is then 8×8.

Originally, the Gaussian pyramid was suggested by Burt and Adelson [13] and Burt [12]. Example 14.1 discusses their approach.

Example 14.1: The original Gaussian pyramid from Burt

Burt used an even, *separable, symmetric* 5×5 smoothing mask $G = G_x G_y$ with the one-dimensional filter kernel

$$G = [\gamma/2 \quad \beta/2 \quad \alpha \quad \beta/2 \quad \gamma/2]^T \tag{14.3}$$

and the transfer function

$$\hat{G} = \alpha + \beta \cos(\pi \tilde{k}) + \gamma \cos(2\pi \tilde{k}). \tag{14.4}$$

Burt and Adelson tried to infer the proper coefficients α, β, and γ from the following principles:

Normalization. A proper smoothing mask requires preservation of the mean gray values, i.e., $\hat{G}(0) = 1$. From Eq. (14.4) we obtain

$$\alpha + \beta + \gamma = 1. \tag{14.5}$$

Equal contribution. All points should equally contribute to the next higher level. Each even point is one time central (factor α) and two times edge point (factor $\gamma/2$); each odd point is weighted two times by $\beta/2$. Hence,

$$\alpha + \gamma = \beta. \tag{14.6}$$

Adequate smoothing. A useful smoothing mask should make the highest wave number vanish, i.e., $\hat{G}_{x,y}(1) = 0$. This gives the same condition as Eq. (14.6).

Equations Eqs. (14.5) and (14.6) leave one degree of freedom in the choice of the filter coefficients. Subtracting both equations yields

$$\begin{aligned} \beta &= 1/2 \\ \alpha + \gamma &= 1/2. \end{aligned} \qquad (14.7)$$

Masks which meet these requirements are the binomial filter $\mathcal{B}^4$ (Section 11.4.2, $\alpha = 6/16$)

$$\mathcal{B}^4 = 1/16\,[1\ 4\ 6\ 4\ 1]$$

and a cascaded box filter ($\alpha = 1/4$)

$$^4R^2R = 1/8\,[1\ 2\ 2\ 2\ 1] = 1/4\,[1\ 1\ 1\ 1] * 1/2\,[1\ 1].$$

The transfer functions of these filters for cascaded applications to compute levels 1, 2, and 3 of the Gaussian pyramid are shown in Fig. 14.9. Both filters do not meet the subsampling requirement Eq. (14.1) and, thus, cause significant aliasing. Although this aliasing may not be visible, it causes errors in the orientation estimate for positions of edges, etc. Thus, better filters are needed for quantitative image processing.

The lowest level of the Gaussian pyramid consists of the original image $G^{(0)}$. We denote the level of the pyramid with a braced superscript. This image will be smoothed with the operator $\mathcal{B}^{(0)}$. The braced superscript again denotes the level on which the operator is applied. We obtain the first level of the Gaussian pyramid if we apply a *subsampling operator* $\Downarrow^{(0)}$ onto the smoothed image which picks out every second pixel in every second line

$$G^{(1)} = (\Downarrow \mathcal{B})^{(0)} G^{(0)}. \qquad (14.8)$$

The same operations are performed with the new image $G^{(1)}$ and all subsequent levels of the pyramid. Generally,

$$G^{(p+1)} = (\Downarrow \mathcal{B})^{(p)} G^{(p)} \qquad (14.9)$$

and

$$G^{(P)} = \left(\prod_{p=0}^{P-1} (\Downarrow \mathcal{B})^{(p)} \right) G^{(0)}. \qquad (14.10)$$

With the operator product, we have to be careful with the order of the indices, since the operators are not commutative. Thus, the indices increase from right to left.

14.3.4b Fast and Accurate Filters. Example 14.1 showed that the filters used originally for the design of the Gaussian pyramid were inadequate. Better filters can easily be designed. The basic trick for a fast and accurate filtering is to split the filtering in a prefiltering on the original resolution level and a postfiltering at the reduced level:

$$G^{(1)} = \mathcal{B}^{(1)} (\Downarrow \mathcal{B})^{(0)} G^{(0)}. \qquad (14.11)$$

The postfiltering considerably reduces the number of computations since it is applied to a coarser grid and, thus, performs a more efficient smoothing. If binomial filters are used which are close to zero towards the maximum wave numbers, the postfiltering can also efficiently remove residual aliasing of wave numbers just above the maximum wave number which are aliased into wave numbers just below the maximum wave number. The two example implementations in Fig. 14.10 show the transfer functions of the filters

$$(\mathcal{B}^2)^{(1)} (\Downarrow \mathcal{B}^4)^{(0)} \qquad \text{and} \qquad (\mathcal{B}^4)^{(1)} (\Downarrow \mathcal{B}^4)^{(0)} \qquad (14.12)$$

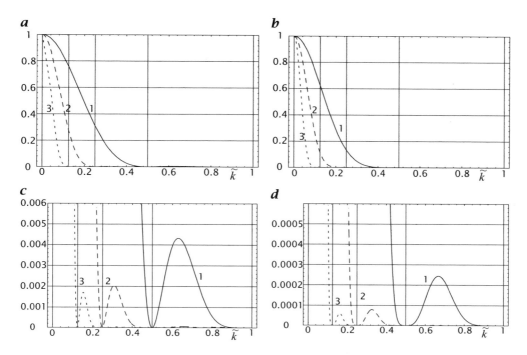

Figure 14.10: *Optimized fast transfer functions of the smoothing filters to generate the Gaussian pyramid:* a $(\mathcal{B}^2)^{(1)}(\Downarrow \mathcal{B}^4)^{(0)}$, b $(\mathcal{B}^4)^{(1)}(\Downarrow \mathcal{B}^4)^{(0)}$. *The maximum wave numbers of 0.5, 0.25, and 0.125 for planes 1, 2, and 3 are marked by vertical lines. Both filters are shown in cascaded applications to generate the planes one to three of the pyramid.* c *and* d *are a blown up section of* a *and* b, *respectively, to show the residual aliasing caused by these filters.*

with the filter masks

$$B^2 = 1/4\,[1\ 2\ 1] * 1/4 \begin{bmatrix} 1 \\ 2 \\ 1 \end{bmatrix} \quad \text{and} \quad B^4 = 1/16\,[1\ 4\ 6\ 4\ 1] * 1/16 \begin{bmatrix} 1 \\ 4 \\ 6 \\ 4 \\ 1 \end{bmatrix} \quad (14.13)$$

at three levels of the Gaussian pyramid. The maximum residual peaks beyond the maximal wave numbers are less than 0.5 % and 0.03 %, respectively (Fig. 14.10c and d). This suppression of aliasing is considerably better than for Burt's original filters (Fig. 14.9) and is sufficient for almost all applications. The number of computations increases only marginally, since the postfiltering on the next level of the pyramid must be performed only with a quarter of the pixels.

14.3.5 Laplacian Pyramid

The Gaussian pyramid is not optimal with respect to the scale matching criterion discussed in Section 14.3.3. It requires that at a certain resolution level only those structures are contained that can just be resolved while larger structures should be contained at corresponding coarser resolution levels.

Such a pyramid, called the *Laplacian pyramid*, is formed by subtracting the images between consecutive levels in the Gaussian pyramid. In this way, only the fine

0

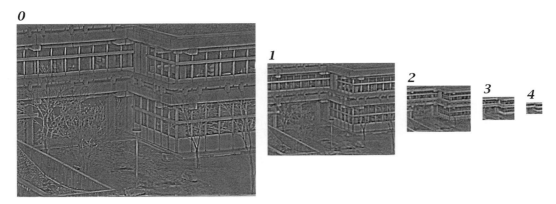

Figure 14.11: The first five planes of the Laplacian pyramid.

scales, removed by the smoothing operation to compute the next level of the Gaussian pyramid, remain in the finer level. The Laplacian pyramid is an effective scheme for a *bandpass decomposition* of the image. The center wave number is halved from level to level. The last image of the Laplacian pyramid is a lowpass-filtered image containing only the coarsest structures.

In order to perform the subtraction operation, we first must expand the image from the higher level to the double size so that both images show the same resolution. This operation is performed by an *expansion operator*, $\Uparrow$, which is the inverse of the reduction operator $\Downarrow$. Essentially, the expansion operator puts the known image points from the coarser image onto the even pixel positions at the even rows and interpolates the missing other points from the given points with the techniques discussed in detail in Section 8.3.2. Thus, the first level of the Laplacian pyramid is formed by the following operation:

$$\boldsymbol{L}^{(0)} = \boldsymbol{G}^{(0)} - \Uparrow^{(1)} \boldsymbol{G}^{(1)} = \left[\boldsymbol{\mathcal{I}}^{(0)} - \Uparrow^{(1)} (\Downarrow \mathcal{B})^{(0)} \right] \boldsymbol{G}^{(0)}. \qquad (14.14)$$

In a similar manner, we obtain the higher levels of the Laplacian pyramid

$$\boldsymbol{L}^{(P)} = \boldsymbol{G}^{(P)} - \Uparrow^{(P+1)} \boldsymbol{G}^{(P+1)} = \left[\boldsymbol{\mathcal{I}}^{(P+1)} - \Uparrow^{(P+1)} (\Downarrow \mathcal{B})^{(P)} \right] \left(\prod_{p=0}^{P-1} (\Downarrow \mathcal{B})^{(p)} \right) \boldsymbol{G}^{(0)}. \qquad (14.15)$$

If we add up the images at the different levels of the Laplacian pyramid starting with the highest level, we obtain the Gaussian pyramid and, thus, also the original image again. Reconstruction of the original image from its representation in a Laplacian image starts at the highest level. There, we have a smoothed image, $\boldsymbol{G}^{(P+1)}$. From Eq. (14.15), we see that we obtain the next lower level of the Gaussian pyramid by expanding $\boldsymbol{G}^{(P+1)}$ and adding $\boldsymbol{L}^{(P)}$:

$$\boldsymbol{G}^{(P)} = \boldsymbol{L}^{(P)} + \Uparrow^{(P+1)} \boldsymbol{G}^{(P+1)}. \qquad (14.16)$$

We continue this operation until we obtain the original image at the lowest level

$$\boldsymbol{G} = \boldsymbol{G}^{(0)} = \boldsymbol{L}^{(0)} + \sum_{p=0}^{P} \left(\prod_{q=1}^{p} \Uparrow^{(q)} \right) \boldsymbol{L}^{(p)} + \left(\prod_{q=1}^{P+1} \Uparrow^{(q)} \right) \boldsymbol{G}^{(P+1)}. \qquad (14.17)$$

The reconstruction of the original image is exact except for round-off errors, even if the expansion operator is not perfect because it is used both for the generation of

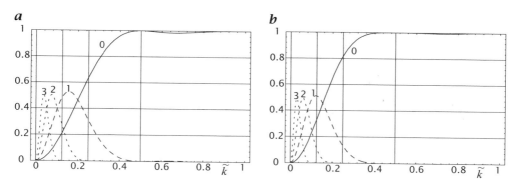

Figure 14.12: *Radial cross section of the transfer functions for the first four levels of the Laplacian pyramid using the isotropic smoothing filters (Section 14.3.4b):* **a** $(\mathcal{B}^2)^{(1)} (\Downarrow \mathcal{B}^4)^{(0)}$, **b** $(\mathcal{B}^4)^{(1)} (\Downarrow \mathcal{B}^4)^{(0)}$.

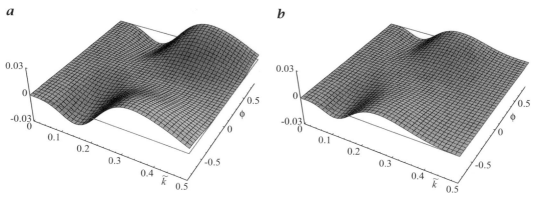

Figure 14.13: *Residual anisotropy of the Laplacian pyramid shown in a polar pseudo 3-D plot. The wave number range is from 0 to 0.5, the angle from -45° to 45° to the x axis. The following smoothing filters were used to form the pyramid (Section 14.3.4b):* **a** $(\mathcal{B}^2)^{(1)} (\Downarrow \mathcal{B}^4)^{(0)}$, **b** $(\mathcal{B}^4)^{(1)} (\Downarrow \mathcal{B}^4)^{(0)}$.

the Laplacian pyramid and the reconstruction. Thus, the errors cancel out, but not, however, residual aliasing if improper smoothing kernels $\mathcal{B}$ are applied.

Figure 14.12 shows a radial cross section of the transfer functions for the first four levels of the Laplacian pyramid for the two types of smoothing filters introduced in Section 14.3.4b. The binomial smoothing filters result in a rather wide bandpass decomposition in the Laplacian pyramid. The maximum of the transfer function in one plane is just about 0.5. Each wave number is distributed over about three levels.

The use of different smoothing filters gives control to optimize the Laplacian pyramid for a given image processing task. By varying the cut-off wave number of the smoothing filter, we can adjust where the peak of the wave number occurs in the pyramid.

14.3.6 Directio-Pyramidal Decomposition

The Laplacian pyramid decomposes an image in logarithmic wave number intervals one octave (factor 2) distant. A useful extension is the additional decomposition into different directions. Such a decomposition is known as *directio-pyramidal decomposition* [66]. This decomposition should have the same properties as the Laplacian pyramid:

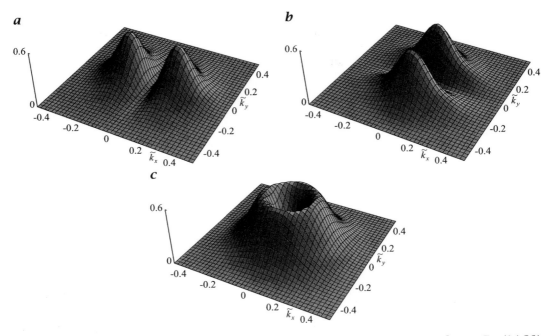

a **b** **c**

Figure 14.14: *Pseudo 3-D plots of the directio-pyramidal decomposition according to Eq. (14.20). a Directional bandpass filter in x direction, b directional bandpass filter in y direction, c isotropic bandpass filter, sum of a and b.*

the addition of all partial images should give the original. This implies that each level of the Laplacian pyramid must be decomposed into several directional components. An efficient scheme for decomposition into two directional components is as follows. The smoothing is performed by separable smoothing filters

$$\boldsymbol{G}^{(1)} = \Downarrow_2 \mathcal{B}_x \mathcal{B}_y \boldsymbol{G}^{(0)}. \tag{14.18}$$

The first level of the Laplacian pyramid then is

$$\mathcal{L} = \boldsymbol{G}^{(0)} - \Uparrow_2 \boldsymbol{G}^{(1)}. \tag{14.19}$$

Then, the two directional components are given by

$$\begin{aligned} \mathcal{L}_1 &= 1/2(\boldsymbol{G}^{(0)} - \Uparrow_2 \boldsymbol{G}^{(1)} - (\mathcal{B}_x - \mathcal{B}_y)\boldsymbol{G}^{(0)}) \\ \mathcal{L}_2 &= 1/2(\boldsymbol{G}^{(0)} - \Uparrow_2 \boldsymbol{G}^{(1)} + (\mathcal{B}_x - \mathcal{B}_y)\boldsymbol{G}^{(0)}). \end{aligned} \tag{14.20}$$

From Eq. (14.20) it is evident that the two directional components $\mathcal{L}_1$ and $\mathcal{L}_2$ add up to the isotropic Laplacian pyramid $\mathcal{L}$. The scheme requires minimal additional computations as compared to the computation of the isotropic Laplacian pyramid. Only one more convolution (of $\boldsymbol{G}^{(0)}$ with $\mathcal{B}_y$) and three more additions are required.

The results from these directional decompositions are surprisingly good. The transfer functions are shown in Fig. 14.14 while example images can be found in Fig. 14.15.

14.3.7 Phase and Local Wave Number

14.3.7a Local Wave Number. In Chapter 14.3.2 we recognized the importance of the phase for image signals in the Fourier domain. We saw that the phase is essential to

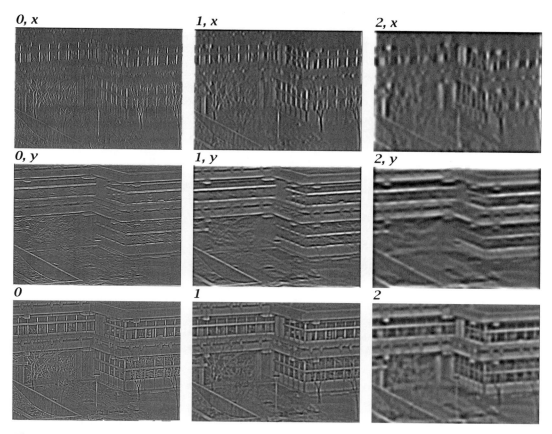

Figure 14.15: *Directio-pyramidal decomposition of an example image as indicated. Shown are the first three planes blown up to the original size.*

locate image features. Here we will learn that the phase is of even greater importance. With techniques such as the Gaussian and Laplacian pyramid it is possible to separate the content of the image into different scales.

We have, however, so far no technique that allows us to estimate the scale of a local feature directly, i. e., how fast it is changing spatially. The spatial change of a periodic signal such as a wave is measured in physics by a term known as the wave number. This quantity tells us how many periods of a signal fit into a unit length. For not strictly periodic signals it should still be possible to determine a *local wave number* as a measure for a characteristic local scale.

As an introduction we discuss a simple example and consider the one-dimensional periodic signal

$$g(x) = g_0 \cos(kx). \tag{14.21}$$

The argument of the cosine function is known as the phase ϕ of the periodic signal:

$$\phi(x) = kx. \tag{14.22}$$

This equation shows that the phase is a linear function of the position and the wave number. Thus we obtain the wave number of the periodic signal by computing the first-order spatial derivative of the phase signal

$$\frac{\partial \phi(x)}{\partial x} = k. \tag{14.23}$$

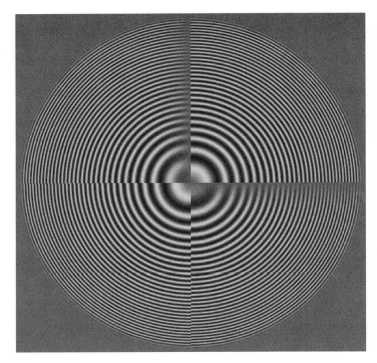

Figure 14.16: *Application of the Hilbert transform to the ring test pattern: upper left quadrant: in the horizontal direction; lower right quadrant: in the vertical direction.*

The question now arises, how the concept of phase and local wave number can be extended from periodic signals to general signals. The key to solve this problem will be two transforms known as the *Hilbert transform* and the *Riesz transform*.

14.3.7b Hilbert Transform and Analytic Signal. In order to explain the principle of computing the phase of a signal, we first stick to the example of the simple periodic signal from the previous section. We suppose that a transform is available to delay the signal by a phase of 90°. This transform would convert the $g(x) = g_0 \cos(kx)$ signal into a $g'(x) = -g_0 \sin(kx)$ signal as illustrated in Fig. 14.16. Using both signals, the phase of $g(x)$ can be computed readily by

$$\phi(g(x)) = \arctan\left(\frac{-g'(x)}{g(x)}\right). \tag{14.24}$$

As only the ratio of $g'(x)$ and $g(x)$ goes into Eq. (14.24), the phase is independent of the amplitude. If we take the signs of the two functions $g'(x)$ and $g(x)$ into account, the phase can be computed over the full range of 360°.

Thus all we need to determine the phase of a signal is a linear operator that shifts the phase by 90°. Such an operator is known as the *Hilbert filter H* and has the transfer function

$$\hat{h}(k) = \begin{cases} \text{i} & k > 0 \\ 0 & k = 0 \\ -\text{i} & k < 0 \end{cases}. \tag{14.25}$$

The magnitude of the transfer function is one as the amplitude remains unchanged. As the Hilbert filter has a purely imaginary transfer function, it must be of odd symmetry to generate a real-valued signal. Therefore positive wave numbers are shifted by 90° ($\pi/2$) and negative wave numbers by $-90°$ ($-\pi/2$).

A special situation occurs at the wave number zero where the transfer function is set to zero. This exception has the following reason. A signal with wave number zero is a constant. It can be regarded as a cosine function with infinite wave number sampled at the phase zero. Consequently, the Hilbert filtered signal is the corresponding sine function at phase zero, which is zero.

Because of the discontinuity of the transfer function of the Hilbert filter at the origin, its point spread function is of infinite extent:

$$h(x) = -\frac{1}{\pi x}. \tag{14.26}$$

The convolution with Eq. (14.26) can be written as

$$g_h(x) = \frac{1}{\pi} \int\limits_{-\infty}^{\infty} \frac{g(x')}{x' - x} dx'. \tag{14.27}$$

This integral transform is known as the *Hilbert transform* [95]. A real-valued signal and its Hilbert transform can be combined into a complex-valued signal by

$$g_a = g - ig_h. \tag{14.28}$$

This complex-valued signal is denoted as the *analytic function* or *analytic signal*. According to Eq. (14.28) the analytic filter has the point spread function and transfer function

$$a(x) = 1 + \frac{i}{\pi x} \quad \circ\!\!-\!\!\bullet \quad \hat{a}(k) = \begin{cases} 2 & k > 0 \\ 1 & k = 0 \\ 0 & k < 0 \end{cases}. \tag{14.29}$$

Thus all negative wave numbers are suppressed. Although the transfer function of the analytic filter is real, it results in a complex signal because it is asymmetric.

The analytic signal can be regarded as just another representation of a real signal with two important properties. The magnitude of the analytic signal gives the *local amplitude* and the argument the *local phase*:

$$|\mathcal{A}|^2 = \mathcal{I} \cdot \mathcal{I} + \mathcal{H} \cdot \mathcal{H} \quad \text{and} \quad \arg(\mathcal{A}) = \arctan\left(\frac{-\mathcal{H}}{\mathcal{I}}\right). \tag{14.30}$$

We used $\mathcal{A}$ and $\mathcal{H}$ for the analytic and Hilbert operators, respectively.

The original signal and its Hilbert transform can be obtained from the analytic signal using Eq. (14.28) by

$$\begin{aligned} g(x) &= (g_a(x) + g_a^*(x))/2 \\ g_h(x) &= i(g_a(x) - g_a^*(x))/2. \end{aligned} \tag{14.31}$$

The concept of the analytic signal also makes it possible to extend the ideas of local phase into multiple dimensions. The transfer function of the analytic operator uses only the positive wave numbers, i.e., only half of the Fourier space. If we extend this partitioning to higher dimensions, we have more than one choice to partition the Fourier space into two half spaces. Instead of the wave number, we can take the scalar product between the wave number vector k and a unit vector $\bar{n}$ and suppress the half space for which the scalar product $k\bar{n}$ is negative:

$$\hat{a}(k) = \begin{cases} 2 & k\bar{n} > 0 \\ 1 & k\bar{n} = 0 \\ 0 & k\bar{n} < 0 \end{cases}. \tag{14.32}$$

The unit vector $\bar{\boldsymbol{n}}$ gives the direction in which the Hilbert filter is to be applied.

This extension of the Hilbert transform to higher dimensions has serious limitations. For one-dimensional signals, discrete Hilbert filters do not work well for small wave numbers (Section 14.4.2b and Fig. 14.20). In multiple dimensions this means that a Hilbert filter does not work well if $\tilde{\boldsymbol{k}}\bar{\boldsymbol{n}} \ll 1$. Thus the spectral density of the image should be zero close to the separation plane, in order to avoid errors. This fact makes the application of Hilbert filters and thus the determination of the local phase in higher-dimensional signals significantly more complex. It is not sufficient to use bandpass filtered images, e. g., a Laplace pyramid (Section 14.3.5). In addition, a decomposition into directional components as discussed in Section 14.3.6 is required.

14.3.7c Riesz Transform and Monogenic Signal. The extension of the Hilbert transform from a 1-D signal to higher-dimensional signals is not satisfactory because it can only be applied to directionally filtered signals. For wave numbers close to the separation plane, the Hilbert transform does not work as can be seen from Fig. 14.16. What is really required is an isotropic extension of the Hilbert transform. It is obvious that no scalar-valued transform for a multidimensional signal can be both isotropic and of odd symmetry.

A vector-valued extension of the analytic signal meets both requirements. It is known as the *monogenic signal* and was introduced to image processing by Felsberg and Sommer [30]. The monogenic signal is constructed from the original signal and its *Riesz transform*. The transfer function of the Riesz transform is given by

$$\hat{\boldsymbol{h}}(\boldsymbol{k}) = \mathrm{i}\frac{\boldsymbol{k}}{|\boldsymbol{k}|}. \tag{14.33}$$

The magnitude of the vector $\boldsymbol{h}$ is one for all values of $\boldsymbol{k}$. Thus the Riesz transform is isotropic and it is also of odd symmetry because

$$\hat{\boldsymbol{h}}(-\boldsymbol{k}) = -\hat{\boldsymbol{h}}(\boldsymbol{k}). \tag{14.34}$$

The Riesz transform can be applied to a signal of any dimension. For a 1-D signal it reduces to the Hilbert transform.

For a 2-D signal the transfer function of the Riesz transform can be written using polar coordinates as

$$\hat{\boldsymbol{h}}(\boldsymbol{k}) = \mathrm{i}\left[\frac{k\cos\theta}{|\boldsymbol{k}|}, \frac{k\sin\theta}{|\boldsymbol{k}|}\right]^T. \tag{14.35}$$

The convolution mask or PSF of the Riesz transform is given by

$$\boldsymbol{h}(\boldsymbol{x}) = -\frac{\boldsymbol{x}}{2\pi\,|\boldsymbol{x}|^3}. \tag{14.36}$$

The original signal and the signal convolved by the Riesz transform can be combined for a 2-D signal to the 3-D monogenic signal as

$$\boldsymbol{g}_m(\boldsymbol{x}) = [g, p, q]^T \quad \text{with} \quad p = h_1 * g, q = h_2 * g. \tag{14.37}$$

The local amplitude a of the monogenic signal is given as the norm of the vector of the monogenic signal as in the case of the analytic signal (Eq. (14.30)):

$$a = |\boldsymbol{g}_m|^2 = g^2 + p^2 + p^2. \tag{14.38}$$

In addition, the monogenic signal gives an estimate of both the *local phase* ϕ and the *local orientation* θ by the following relations:

$$g = a\cos\phi, \quad p = a\sin\phi\cos\theta, \quad q = a\sin\phi\sin\theta. \tag{14.39}$$

14.4   Procedures

14.4.1   Texture Energy

All parameters derived from first-order statistics of the gray values for individual pixels are independent of the orientation of the objects. In Section 7.3.1, we discussed that the gray value distribution of random variables is described by the mean, variance, and higher moments.

To be suitable for texture analysis, the estimate of these parameters has to be adapted to a local neighborhood. In the simplest case, we can select a window W and compute the parameters only from the pixels contained in this window. The *variance operation*, for example, is then given by

$$V_{mn} = \frac{1}{P-1} \sum_{k,l \in W} \left(G_{m-k,n-l} - \overline{G}_{mn} \right)^2. \tag{14.40}$$

The sum runs over the P image points of the window. The expression $\overline{G}_{mn}$ denotes the mean of the gray values at the point (m, n), computed from the pixels in the same window W:

$$\overline{G}_{mn} = \frac{1}{P} \sum_{k,l \in W} G_{m-k,n-l}. \tag{14.41}$$

It is important to note that the variance operator is nonlinear. However, it resembles the general form of a neighborhood operation — a convolution. Combining Eqs. (14.40) and (14.41) shows that it is a combination of linear convolution and nonlinear point operations

$$V_{mn} = \frac{1}{P-1} \left[\sum_{k,l \in W} G_{m-k,n-l}^2 - \left(\frac{1}{P} \sum_{k,l \in W} G_{mn} \right)^2 \right], \tag{14.42}$$

or, in operator notation,

$$\mathcal{V} = \mathcal{R}(\mathcal{I} \cdot \mathcal{I}) - (\mathcal{R} \cdot \mathcal{R}). \tag{14.43}$$

The operator $\mathcal{R}$ denotes a smoothing over all the image points with a box filter of the size of the window W. The operator $\mathcal{I}$ is the identity operator. Therefore, the operator $\mathcal{I} \cdot \mathcal{I}$ performs a nonlinear point operation, namely the squaring of the gray values at each pixel. In the end, the variance operator does nothing else but to subtract the square of a smoothed gray value from the smoothed squared gray values. From discussions on smoothing in Section 11.4, we know that a box filter is not an appropriate smoothing filter. Thus, we obtain a better variance operator if we replace the box filter $\mathcal{R}$ with a binomial filter $\mathcal{B}$ or another isotropic smoothing filter

$$\mathcal{V} = \mathcal{B}(\mathcal{I} \cdot \mathcal{I}) - (\mathcal{B} \cdot \mathcal{B}). \tag{14.44}$$

The variance operator is isotropic. It is also scale independent if the window is larger than the largest scales in the textures and if no fine scales of the texture disappear because the objects are located further away from the camera.

The application of the variance operator Eq. (14.44) to two example images is shown in Fig. 14.17. In Fig. 14.17a, the variance operator turns out as an isotropic edge detector, since the original image contains areas with more or less uniform gray values. The other example in Fig. 14.17 shows the variance images from a textured surface. The variance operator distinguishes the brighter horizontal stripes with less variations in the gray values from the more structured stripes in-between (Fig. 14.17b).

a   b

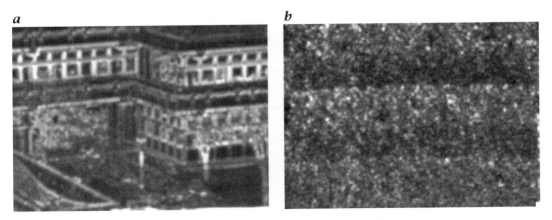

Figure 14.17: *Variance operator according to Eq. (14.44) applied to **a** Fig. 14.8 and **b** Fig. 14.6a*

Besides the variance, we could also use the higher moments of the gray value distribution as defined in Section 7.3.1 for a more detailed description. The significance of this approach may be illustrated with examples of two quite different gray value distributions, a normal and a bimodal distribution

$$p(g) = \frac{1}{\sqrt{2\pi}\sigma} \exp\left(-\frac{g - \overline{g}}{2\sigma^2}\right), \quad p'(g) = \frac{1}{2}\left(\delta(\overline{g} + \sigma) + \delta(\overline{g} - \sigma)\right).$$

Both distributions show the same mean and variance but differ in higher-order moments.

The variance operator takes a very simple form with a Laplacian pyramid, since the mean gray value — except for the coarsest level — is zero:

$$\mathcal{V} = \mathcal{B}(\mathcal{L}^{(p)} \cdot \mathcal{L}^{(p)}). \tag{14.45}$$

The variance operator applied to the Laplacian pyramid essentially is a local energy estimator. This feature can be illustrated if we apply the variance operator to a periodic structure with the amplitude a:

$$g(\boldsymbol{x}) = a \cos(\boldsymbol{kx}). \tag{14.46}$$

After squaring, we obtain two terms:

$$g(\boldsymbol{x}) = a^2 \cos^2(\boldsymbol{kx}) = \frac{a^2}{2} + \frac{a^2}{2} \cos(2\boldsymbol{kx}), \tag{14.47}$$

one with a double wave number and another with a constant term. The squaring, thus, results in a wave number doubling. If the smoothing with the variance operator in Eq. (14.45) extends over more than half a wave period of the structure, the term with the double wave number (half the wavelength) is efficiently filtered out. Thus, only the constant term with the amplitude squared remains in the variance image. In a physical system, the squared amplitude is related to the energy.

Figure 14.18 demonstrates that the variance of the texture depends on the scale. The figure shows the variance operator applied to the first four levels of the Laplacian pyramid. In the first two levels, the variance of the texture is dominated by the small-scale texture. The two different textures are best separated on level 2 of the Laplacian pyramid. At level 3 of the Laplacian pyramid, the character of the variance image

0   1

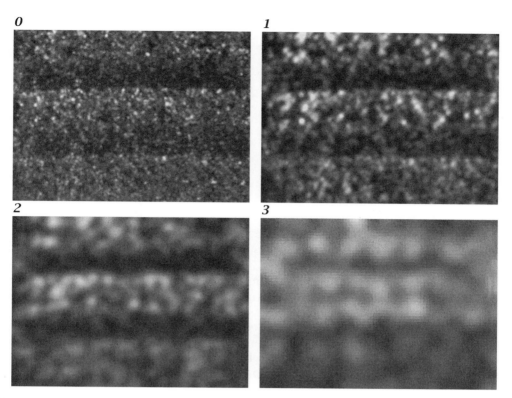

2   3

Figure 14.18: Variance operator applied to different levels (as indicated) of the Laplacian pyramid generated from the image shown in Fig. 14.6a.

completely changes. The resolution is now too coarse. Thus, the two textured regions appear now rather as regions of constant but different gray value. Consequently, the bright areas in the variance image now mark the edges between the two regions.

14.4.2   Phase Determination

In this section, we discuss two techniques to determine the phase of bandpassed images. The quadrature pair filter method can be applied to original images since it filters a selectable wave number range from an image. In contrast, application of the Hilbert filter requires a directionally and bandpass-filtered image.

14.4.2a   Quadrature Pair Filters.   Quadrature filters were discussed in Section 13.4.1. Their main feature is two filters with equal magnitude of the transfer function but a phase that differs by $90°$. Because of this feature, the phase can directly be computed from quadrature filters. Normally, one filter does not cause any phase shift and has, thus, a real transfer function and a symmetrical mask while the other with $90°$ phase shift has a pure imaginary transfer function and a antisymmetrical mask. Because of their symmetry features, they are denoted as the even filter, ^{+}Q, and the odd filter, ^{-}Q. From this filter pair, the phase is given by

$$\phi(\boldsymbol{x}) = \arctan\frac{q_-(\boldsymbol{x})}{q_+(\boldsymbol{x})}, \tag{14.48}$$

where $q_+(\boldsymbol{x})$ and $q_-(\boldsymbol{x})$ denote the signals filtered by ^{+}Q and ^{-}Q, respectively.

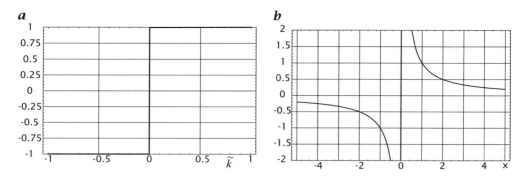

Figure 14.19: *The 1-D Hilbert filter:* **a** *imaginary part of the transfer function (the real part is zero);* **b** *point spread function.*

14.4.2b Hilbert Filters. Another technique to determine the phase is a *Hilbert filter*. A Hilbert filter performs what is known as a *Hilbert transform*. It does not change the amplitude of the different spectral components but just shifts their phase by $\pi/2$. Therefore, the magnitude of the transfer function of the Hilbert filter is one. Because of the $\pi/2$ phase shift, its transfer function is purely imaginary, of odd symmetry jumping from i to -i at $k = 0$ (Fig. 14.19). Because of these properties, the point spread function (convolution kernel) is also odd and decreases only with x^{-1}. Consequently, the convolution mask of the Hilbert filter is infinite in principle and belongs to the class of filters such as lowpass filters with sharp cut-off which cannot be implemented with a small kernel.

The approach to design an effective implementation of a Hilbert filter is guided by the following two criteria:

- The filter should precisely shift the phase by $\pi/2$. This requirement comes from the fact that we cannot afford an error in the phase since it includes the position information. A wave-number dependent phase shift would cause wave-number dependent errors for further processing. The requirement of a phase shift of $\pi/2$ is met by any convolution kernel of odd symmetry.

- The requirements for the magnitude of the transfer function can be less restricted if applied to a bandpassed signal, e.g., the Laplacian pyramid. Then, the Hilbert filter must only show a magnitude of one in the passband range of the bandpass filter used. This feature allows for an effective design of a Hilbert filter adapted to a certain wave number range.

Optimized Hilbert filters are generated with the same techniques used for interpolation filters (Section 8.3.2d) and first-order and second-order derivative filters (Sections 12.4.1d and 12.4.2a). Because of symmetry reasons, the following formulation is used:

$$\hat{H}(\tilde{k}) = \sum_{r=1}^{R} h_r \sin\left((2r-1)\pi\tilde{k}\right) \tag{14.49}$$

for nonrecursive filters and

$$\hat{H}(\tilde{k}) = \frac{\sum_{r=1}^{R} h_r \sin\left((2r-1)\pi\tilde{k}\right)}{(1+\alpha) + \alpha\cos 2\pi\tilde{k}} \tag{14.50}$$

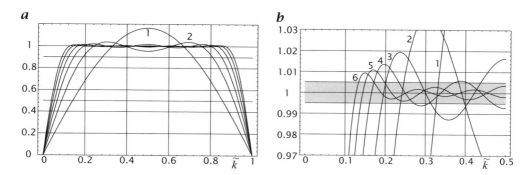

Figure 14.20: *a Transfer function of a family of least-squares optimized Hilbert operators for a given number of filter coefficients R = 1 − 6. The same weighting function of the wave numbers is used as for the optimization of derivative filters in Section 12.4.1d. b Small sector of a; the shaded area indicates the ± 0.5 % error margin.*

Table 14.1: *Coefficients of Hilbert filters optimized with a weighted least-squares optimization method with R = 1 − 6 coefficients. The same weighting function of the wave numbers is used as for the optimization of derivative filters in Section 12.4.1d. For the corresponding transfer functions, see Fig. 14.20.*

R	Mask coefficients	σ_e [%]
1	0.581678	15.2
2	0.595595, 0.118259	4.59
3	0.614034, 0.148466, 0.0426724	1.84
4	0.62078, 0.168256, 0.0629796, 0.0190886	0.87
5	0.625148, 0.179386, 0.0780877, 0.0317812, 0.00969851	0.46
6	0.627811, 0.186828, 0.0882246, 0.0424971, 0.0177172, 0.00538212	0.26

for recursive filters. The recursive correction filter is based on the $\mathcal{V}_2$ filter with double step width of the filter toolbox summarized in Table 10.2. This approach leads to a filter mask with alternating zeros

$$[\cdots -h_2 \ 0 \ -h_1 \ 0 \ h_1 \ 0 \ h_2 \cdots]. \tag{14.51}$$

The filter coefficients obtained by the optimization procedure are summarized in Tables 14.1 and 14.2 and the transfer functions are plotted in Figs. 14.20 and 14.21. The results show that the recursive implementation is significantly better and that Hilbert filters with a few coefficients can be generated for wide bandwidths ($\tilde{k} \in [0.1, 0.9]$) with less than 0.5 % amplitude error and no phase error.

14.4.3 Local Wave Number

14.4.3a Principle. The key to compute the local wave number is the phase signal as discussed in Section 14.4.2. According to Eq. (14.48), the phase can be computed from a Hilbert pair $q_+(\boldsymbol{x})$, $q_-(\boldsymbol{x})$:

$$\phi(\boldsymbol{x}) = \arctan \frac{q_-(\boldsymbol{x})}{q_+(\boldsymbol{x})}. \tag{14.52}$$

Intuitively, it is clear that the phase changes more rapidly in space for small scales than for large scales. Thus, the phase gradient must be related to the wave number. Indeed,

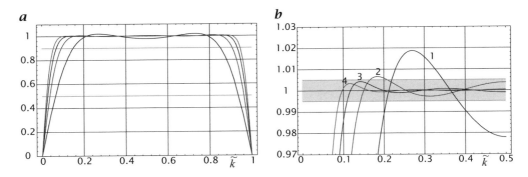

Figure 14.21: a *Transfer function of a family of least-squares optimized Hilbert operators with recursive correction for R = 1, 2, 3, 4 coefficients. The same weighting function of the wave numbers is used as for the optimization of derivative filters in Section 12.4.1d.* **b** *Small sector of* **a**; *the shaded area indicates the ± 0.5 % error margin.*

Table 14.2: *Coefficients of Hilbert filters with a recursive $\mathcal{V}_2$ correction filter ($g'_m = g_m + \alpha(g'_{m\pm2} + g_m)$, see Table 10.2) optimized with a weighted least-squares optimization method. The last column shows the relative weighted standard deviation σ_e of the ideal transfer function in percent. For the corresponding transfer functions, see Fig. 14.21.*

R	α	Mask coefficients	σ_e [%]
1	0.250151	0.488928	2.07
2	0.474813	0.450185, -0.0516525	0.36
3	0.586885	0.438317, -0.065951, -0.00477828	0.109
4	0.656726	0.433351, -0.0726725, -0.0069692, -0.00113187	0.043

for a periodic structure with a wave number k, the phase is

$$\phi(x) = kx, \tag{14.53}$$

from which we immediately obtain

$$k = \nabla\phi(x). \tag{14.54}$$

In generalization to arbitrary signals, we can use Eq. (14.54) as the definition of a *local wave number*.

14.4.3b Phase Gradient. Direct computation of the partial derivatives from the phase signal Eq. (14.52) is not advisable because of the inherent discontinuities in the phase signal (restriction to the main interval $[-\pi, \pi[$). As pointed out by Fleet and Jepson [33], this problem can be avoided by computing the phase gradient from gradients of $q_+(x)$ and $q_-(x)$. The result is

$$\nabla\phi(x) = \frac{q_+(x)\,\nabla q_-(x) - q_-(x)\,\nabla q_+(x)}{q_+^2(x) + q_-^2(x)}. \tag{14.55}$$

This formulation of the phase gradient also eliminates the need for using trigonometric functions to compute the phase signal and is, therefore, also significantly faster. Figure 14.22 demonstrates the computation of the local wave number for the ring test pattern. Since a Hilbert filter in x direction is applied (Fig. 14.22b), the determination of

Figure 14.22: *Computation of the local wave number demonstrated with the ring test pattern:* **a** *ring test pattern,* **b** *filtered with a Hilbert filter in x direction,* **c** *x component of the local wave number,* **d** *y component of the local wave number,* **e** *magnitude of the local wave number,* **f** *local amplitude.*

the phase gradient will be incorrect for all wave numbers that have a small component in x direction. Therefore all areas have been masked out in Fig. 14.22c–e where the amplitude of the periodic structures (Fig. 14.22f) is below 0.9 times the correct value. With this limitation, the local wave number is computed correctly.

Correct local structures in all directions can be obtained if first a directio-pyramidal decomposition of the image is performed as described in Section 14.3.6. Then, Hilbert filters in x and y direction are applied to the horizontal and vertical structures, respectively.

Alternatively, the local wave number for signals with arbitrary directions can also be calculated using the *Riesz transform* (Section 14.3.7c). In magnitude of the local wave number is given from the signals g, p, and q as

$$k = \frac{g\left(\dfrac{\partial p}{\partial x} + \dfrac{\partial q}{\partial y}\right) - p\dfrac{\partial g}{\partial x} - q\dfrac{\partial g}{\partial y}}{g^2 + p^2 + q^2}. \tag{14.56}$$

14.5 Advanced Reference Material

14.1 **References to advanced topics**

A. R. Rao, 1990. A Taxonomy for Texture Description and Identification, Springer, New York. *One of the few monographs about texture analysis.*

J. Bigün and J. M. H. du Buf, 1994. N-folded symmetries by complex moments in Gabor spaces and their application to unsupervised texture analysis, *IEEE Trans. PAMI, 16,* 80–87. *This paper extends the tensor method for orientation analysis to higher-order symmetries.*

D. J. Fleet, 1992. Measurement of Image Velocity, Kluwer Academic Publisher, Dordrecht. *Includes a detailed investigation on the determination of phase and local wave number.*

G. Sommer, 2001. Geometric Computing with Clifford Algebras, Springer, Berlin. *Includes mathematical background to the monogenic signal and the Riesz transform.*

Part IV

From Features to Objects

15 Segmentation

15.1 Highlights

Segmentation is a decision process. Based on the preceding image processing tasks to extract suitable features, we have to decide whether or not a pixel belongs to an object. It is obvious that with this decision, a significant amount of information is lost. From the segmented image, there is no way to infer the original image content.

Traditional approaches to segment images include three classes of techniques: pixel-based, region-based, or edge-based segmentation techniques.

Pixel-based techniques are stupid in the sense that they do not consider the spatial context but only decide solely on the base of the gray value or features at individual pixels. However, when the right features have been extracted, the corresponding feature extraction operator already performs this job adequately, and segmentation reduces to a rather trivial task.

Region-based and edge-based are dual techniques; one approach focuses on continuity, the other one discontinuity. Each of these techniques misses half of the available information. It is, however, not easy to merge region-based and edge-based techniques. One possible way is to control the averaging process by the edge content of the image in order to automatically average the features in the region but still to preserve the discontinuities. We will discuss inhomogeneous and anisotropic diffusion that work in this way.

A handicap of all these techniques is that only local information is used. One way to introduce global information is through models. Hough transformation techniques are well-known techniques of this kind that can be applied to straight lines, circles, and any other type of curves that can be expressed by a small set of parameters. The trouble with the Hough transformation is, however, that the technique is computational intensive. Here we introduce a new type of fast Hough transformation that is based on the analysis of local orientation.

15.2 Task

All image processing operations discussed in the Chapters 10–14 aimed at a better recognition of objects of interest, i. e., to find suitable local features which allow us to distinguish them from other objects and from the background. The next step is to check each individual pixel whether it belongs to an object of interest or not. This operation is called *segmentation* and produces a *binary image*. A pixel has the value one if it belongs to the object; otherwise it is zero. Segmentation is the operation at the threshold between *low-level* and *high-level image processing*. With segmentation, we leave the pixel-oriented image data since we decide which pixels belong to an object. Only at this step, we gain knowledge that objects are contained in the image.

475

a

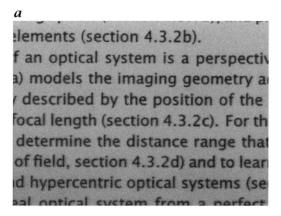

elements (section 4.3.2b).
f an optical system is a perspectiv
a) models the imaging geometry a
y described by the position of the
focal length (section 4.3.2c). For th
determine the distance range that
of field, section 4.3.2d) and to lear
d hypercentric optical systems (se
al optical system from a perfect

b

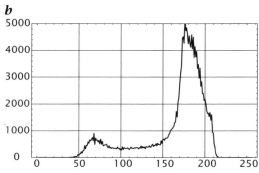

Figure 15.1: *Illustration of segmentation: **a** original image; **b** histogram.*

This establishes the base for further processing. Now, we can analyze the shape of objects (Chapter 16) and use this information to distinguish various classes of objects (Chapter 17).

In this chapter, we discuss several types of segmentation methods. Basically we can think of three concepts for segmentation. Pixel-based methods (Section 15.3.1) only use the gray values of the individual pixels. Edge-based methods (Section 15.3.3) detect edges and then try to follow them. Finally, region-based methods (Section 15.3.2) analyze the gray values in larger areas.

15.3   Concepts

15.3.1   Pixel-Based Segmentation

Point-based or *pixel-based segmentation* is conceptually the simplest approach we can take for segmentation. We may argue that it is the only valid approach. Why? For two reasons.

First, instead of trying a complex segmentation procedure, we should rather first use the whole palette of techniques we have discussed so far in this handbook to extract those features that characterize an object in a unique way *before* we apply a segmentation procedure. It is always better to solve the problem at its root. If, for instance, an image is unevenly illuminated, the first thing to do is to optimize the illumination of the scene. If this is not possible, the next step would be to identify the unevenness of the illumination system and to use corresponding image processing techniques to correct it. One possible technique has been discussed in Section 7.4.2.

Second, if segmentation errors remain in the binary image, it is possible to correct them with morphological operators that are discussed in Section 16.4.1.

If we have found a good feature to separate the object from the background, the histogram of the gray values — or more generally feature values — shows a *bimodal distribution* with two distinct maxima. Ideally, a zone will exist between the two maxima where no features exist. Then the histogram is zero in this range and we can place a threshold anywhere in this range yielding a perfect separation between object and background.

This ideal situation almost never occurs even in cases where the human visual system can still recognize objects very well. If, for instance, the objects to be segmented are rather small as the letters in Fig. 15.1a, the image contains a lot of gray values between the gray value of the object and the background because the edges with intermediate gray values occupy a large fraction of the image. This effect fills the histogram in between the gray values for object and background (Fig. 15.1b). Mis-correspondences do not occur, as long as the intermediate gray values result from the border region between the object and background and not from overlapping distributions of the object and background. If the camera response is linear, the optimum threshold will be the mean of the object and background gray values.

If the probability density functions for the gray values of the object and the background overlap, mis-correspondences are unavoidable. Some object pixels will be recognized as background and vice versa. If we know the probability distribution for the object and the background pixels, we can use a statistical analysis of the decision process to find an optimum threshold with the minimum number of erroneous correspondences [121]. The gray value distributions of the object and the background can be estimated by local histograms that only include areas from either the object or the background.

In less favorable circumstances, the histogram might not show a minimum at all, in which case an adequate threshold does not exist. This situation occurs, for example, in a scene with uneven illumination even if the object clearly juts out of the background (Fig. 7.7a and b). The literature is full of concepts to handle such cases. We can, for example, compute histograms from smaller areas and use these histograms to find local thresholds. But again, as discussed at the beginning of this section, this problem should, if possible, be solved at its root.

15.3.2 Region-Based Segmentation

Region-based methods focus our attention on an important aspect of the segmentation process we miss with point-based techniques. These techniques classify a pixel as an object pixel judging solely on its gray value independent of the context. This means that any isolated points or small areas could be classified as being object pixels, despite the fact that an important characteristic of an object is its *connectivity*.

In this section we will not discuss such standard techniques as spilt-and-merge or region-growing techniques. Interested readers are referred to Rosenfeld and Kak [121] or Jain [75]. Here we point out one of the central problems of the segmentation process.

If we do not use the original image but a feature image for the segmentation process, the feature does not represent a single pixel but a small neighborhood depending on the mask sizes of the operators used. At the edges of the objects, however, where the mask size includes pixels from both the object and the background, no useful feature values can be computed because features from the object and the background are mixed up. The correct procedure would be to limit the mask size at the edge to points of either the object or the background. But how can this be achieved if we can only distinguish the object and the background after computation of the feature?

Obviously, this problem cannot be solved in one step, but only iteratively using a procedure in which feature computation and segmentation are performed alternately. Principally, we proceed as follows. In the first step, we compute the features disregarding any object boundaries. Then we perform a preliminary segmentation and compute the features again, now using the segmentation results to limit the masks of the neighborhood operations at the object edges to either the object or the background pixels, depending on the location of the center pixel. To improve the results, we can repeat both steps until the procedure converges to a stable result.

15.3.3 Edge-Based Segmentation

Even with a perfect illumination, pixel-based segmentation may easily result in a bias of the size of the segmented object. Figure 15.3 illustrates how the size of the objects depends on the threshold level. In the segmented images shown in Fig. 15.3b-d, it can be observed that the size of the letters even increases from the top to the bottom of the image because of a vertically slanted illumination. The size variation results from the fact that the gray values at the edge of an object change only gradually from the background to the object value. No bias in the size occurs if we take the mean of the object and the background gray values as the threshold. However, this approach is only possible if all objects show the same gray values. In the case of differently bright objects and a black background, a bias in the size of the objects is unavoidable. Darker objects will become too small, brighter objects too large.

An *edge-based segmentation* approach can be used to avoid a bias in the size of the segmented object. The position of an edge is given by an extreme of the first-order derivative or a zero crossing in the second-order derivative (see Section 12.3.2).

Edge-based segmentation is a sequential method. The image is scanned line by line for, e. g., maxima in the gradient. When a maximum is encountered, a *tracing algorithm* tries to follow the maximum of the gradient around the object until it reaches the starting point again. Then, the next maximum in the gradient is searched. This procedure is repeated until the complete image is scanned. As region-based segmentation, edge-based segmentation takes into account that an object is characterized by adjacent pixels. Contour-following algorithms are discussed in Jain [75].

15.3.4 Model-Based Segmentation

15.3.4a Introduction. In Section 12.3.5, we discussed that local operators are not sufficient to detect objects. Figure 12.4 demonstrates that it is not possible to detect an edge while the human visual system can still recognize it. Figure 12.4c gives a hint already how to solve such hard problems. If we utilize the knowledge that the edges are vertical and extended over the whole image, we can average the profiles in the vertical direction and then detect the edges.

Thus any model-based segmentation technique tries to include both local information directly derived from the gray values given in the image or features estimated from the image and some model constraints that incorporate global knowledge. Model-based segmentation techniques can either be contour-based or region-based or both.

The most common approach to contour-based models are *snakes*, introduced by Kass et al. [78]. With a snake, local and global information is balanced in the following way. The closeness of the contour to the image data is measured by a distance measure between the points on the contour and features like the *edge coefficient* (Section 12.3.2c) that represent elements of the contour derived from the image data. The desired contour should match the position of these features as close as possible. Therefore the distance measure should be minimal. On the other hand, certain features of the contour such as smoothness or known shape properties should also be preserved. The deviation from the desired contour shape is measured with a second distance. Both distance measurements are then minimized together to achieve an optimal balance between the similarity to the image data and the desired contour shape.

In a similar way, it is possible to devise region-based models [123]. Here a model is set up that measures the difference between the measured features and model features in different image regions. The image regions are then selected in such a way that the

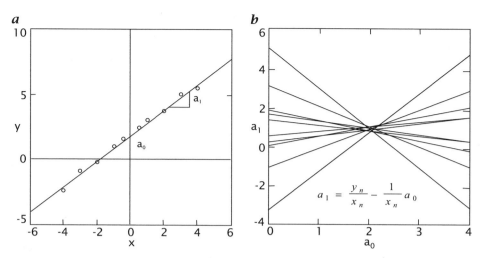

Figure 15.2: *Illustration of the Hough transform with the example of a straight-line fit: the Hough transform maps the (x, y) data space onto the (a_0, a_1) model space:* **a** *data space;* **b** *model space.*

global difference in the whole image becomes minimal. In this way both contour-based and region-based segmentation can be formulated as a variational approach.

15.3.4b Model Spaces. In this section we discuss a very simple model to learn how global constraints can help with edge-based segmentation. A strong but still rather general model constraint that is useful for edges is the assumption that edges are straight lines. For all points on a line, the following condition must be met:

$$y_n = a_0 + a_1 x_n \qquad (15.1)$$

where a_0 and a_1 are the offset and slope of the line. We can read Eq. (15.1) also as a condition for the parameters a_0 and a_1:

$$a_1 = \frac{y_n}{x_n} - \frac{1}{x_n} a_0. \qquad (15.2)$$

This is again the equation for a line in a new space spanned by the parameters a_0 and a_1. In this space, the line has the offset y_n/x_n and a slope of $-1/x_n$. With one point given, we have already no longer a free choice of a_0 and a_1 but the parameters must meet Eq. (15.2).

The space spanned by the model parameters a_0 and a_1 is called the *model space*. Each point reduces the model space to a line. Thus, we can draw a line in the model space for each point in the data space as illustrated in Fig. 15.2. If all points lie on a straight line in the data space, all lines in the model space meet in one point which gives the parameters a_0 and a_1 of the lines. In this way, a line is mapped onto a point. This transformation from the data space to the model space via a model equation is called the *Hough transform*. It is a versatile instrument to detect lines even if they are disrupted or incomplete.

The drawback of the Hough transform method for line detection is the high computational effort. For each point in the image, we must draw a line in the parameter space, i. e., increment a large number of points. Thus, it is worthwhile to investigate faster algorithms. In Section 15.4.3, we discuss a fast technique based on local orientation.

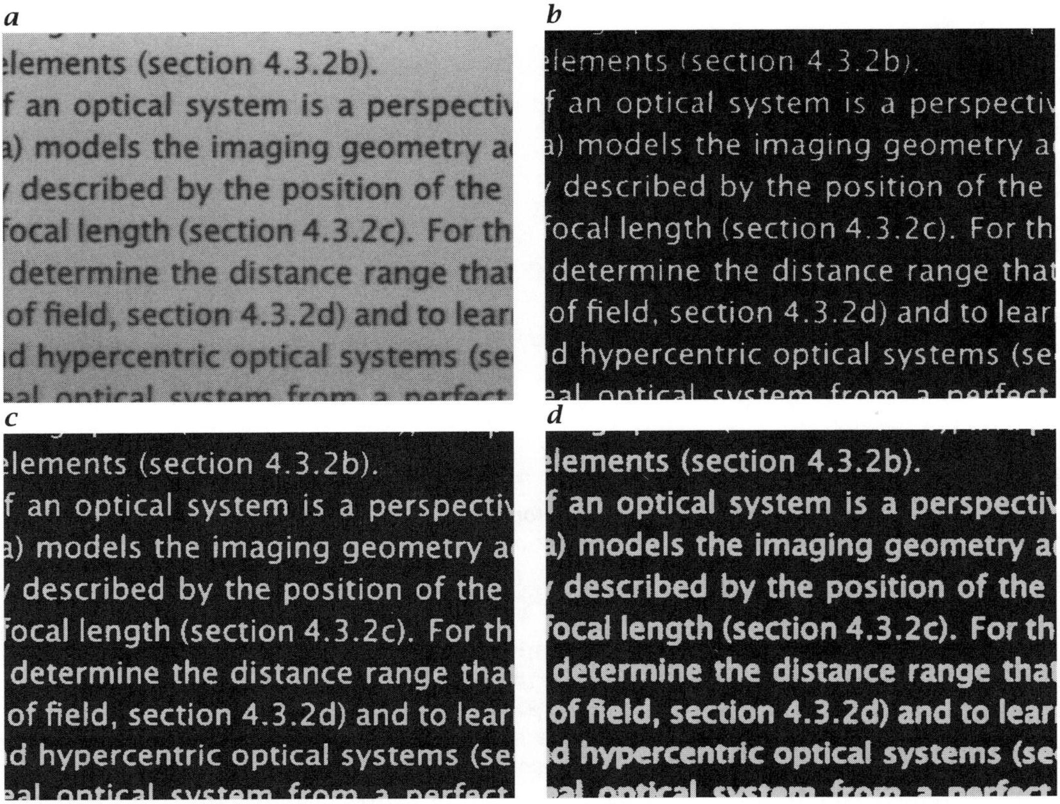

a ... *b* ... *c* ... *d*

*Figure 15.3: The size of the segmented objects depends on the segmentation threshold: **a** original image (same as in Fig. 15.1a, for histogram, see Fig. 15.1b); **b** – **d** segmented images with thresholds of 90, 120, and 150, respectively.*

15.4   Procedures

15.4.1   Global Thresholding

This technique requires homogeneous object and background intensities. If this condition is not met, serious segmentation errors occur. Figure 15.3 illustrates the bias in the size of objects caused by a slanted background illumination. Such a bias is acceptable for some applications, for instance, just the recognition of typeset letters. However, it is a serious flaw for any application where object sizes and areas are to be measured.

A more serious effect of an inhomogeneous illumination is illustrated in Fig. 15.4. Two types of circles are contained in the image; both are circles but of different color. The radiance of the brighter circles comes close to the background. Indeed, a histogram (Fig. 7.7b) shows that the gray values of these brighter circles no longer form a distinct maximum. It overlaps with the wide distribution of the background.

Consequently, the global thresholding fails (Fig. 15.4c). Even with an optimal threshold, some of the background in the right upper and lower corners are segmented as objects and the brighter circles are still segmented only partly. If we first correct for the inhomogeneous illumination as illustrated in Fig. 7.7, the segmentation is perfect (Fig. 15.4b and d).

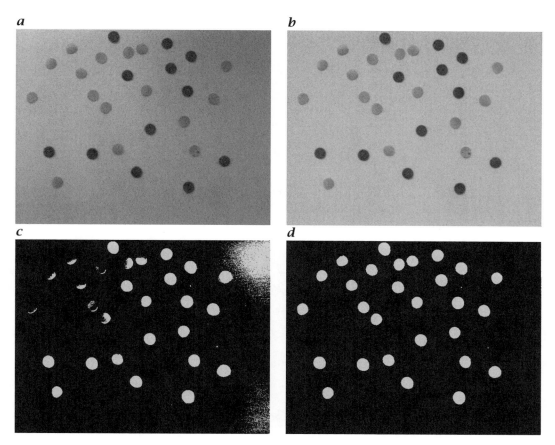

Figure 15.4: *Correction of uneven illumination with an inhomogeneous point operation:* ***a*** *original image with inhomogeneous background illumination (for histogram, see Fig. 7.7b);* ***b*** *same as* ***a***, *but corrected for the inhomogeneous background (Section 7.4.2, for histogram, see Fig. 7.7f);* ***c*** *and* ***d*** *show optimal segmentation with a global threshold of the images in* ***a*** *and* ***b***, *respectively.*

15.4.2 Pyramid Linking

Burt [12] suggested a *pyramid-linking* algorithm as an effective implementation of a combined segmentation feature computation algorithm. We will demonstrate it using the illustrative example of a noisy *step edge* (Fig. 15.5). In this case, the computed feature is simply the mean gray value. The algorithm includes the following steps:

1. *Computation of the Gaussian pyramid.* As shown in Fig. 15.5a, the gray values of four neighboring pixels are averaged to form a pixel on the next higher level of the pyramid. This corresponds to a smoothing operation with a box filter.

2. *Segmentation by pyramid-linking.* Since each pixel contributes to two pixels on the higher level, we can now decide to which it most likely belongs. The decision is simply made by comparing the gray values and choosing the pixel. The link is pictured in Fig. 15.5b by an edge connecting the two pixels. This procedure is repeated through all the levels of the pyramid. As a result, the links on the pyramid constitute a new data structure. Starting from the top of the pyramid one pixel is connected with several pixels on the next lower level. Such a data structure is called a *tree* in computer science. The links are called *edges*; the data points are the gray

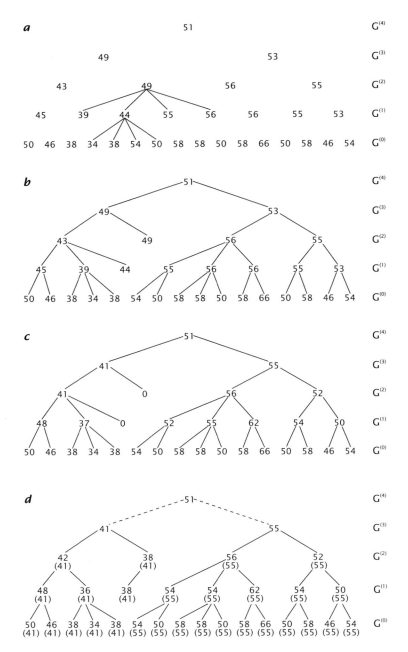

Figure 15.5: *Demonstration of the pyramid-linking segmentation procedure with a one-dimensional noisy edge: **a** first step: computation of the Gaussian pyramid; **b** second step: node-linking: each node is linked with the father node at the next higher level whose value is closest to the value of the node; **c** third step: re-computation of the mean gray values at the father nodes, now only using the linked son nodes; **d** fourth step: final result after several iterations of steps **b** and **c** . The braced values indicate the regional means of the sub-trees with their roots in the third level of the pyramid. At the lowest level, these values represent the estimate of the noisy edge by a step-edge [12].*

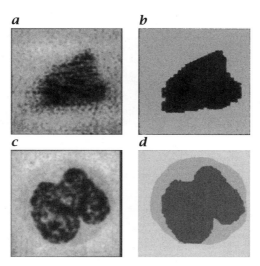

Figure 15.6: *Noisy images of a tank (a) and a blood cell (c) segmented with the pyramid-linking algorithm in two and three regions (b) and (d), respectively; after Burt [12].*

values of the pixels, and are denoted as *nodes* or *vertices*. The node at the highest level is called the *root* of the tree; the nodes which have no further links are called the *leaves* of the tree. A node linked to a node at a lower level is denoted as the *father node* of this node. Correspondingly, each node linked to a node at a higher level is defined as the *son node* of this node.

3. *Averaging of linked pixels.* Next, the resulting link structure is used to recompute the mean gray values, now using only the linked pixels (Fig. 15.5c), i.e., the new gray value of each father node is computed as the average gray value of all the son nodes. This procedure starts at the lowest level and is continued through all the levels of the pyramid.

The last two steps are repeated iteratively until we reach a stable result which is shown in Fig. 15.5d. An analysis of the link-tree shows the result of the segmentation procedure. In Fig. 15.5d we recognize two *subtrees*, which have their roots in the third level of the pyramid. At the next lower level, four subtrees originate. But the differences in the gray values at this level are significantly smaller. Thus we conclude that the gray value structure is obviously parted into two regions. Then we obtain the final result of the segmentation procedure by transferring the gray values at the roots of the two subtrees to the linked nodes at the lowest level. These values are shown as braced numbers in Fig. 15.5d.

The application of the pyramid-linking segmentation algorithm to two-dimensional images is shown in Fig. 15.6. Both examples point out that even very noisy images can be successfully segmented with this procedure. There is no restriction to the form of the segmented area.

The pyramid-linking procedure merges the segmentation and the computation of mean features for the objects extracted in an efficient way by building a tree on a pyramid. It is also advantageous that we do not need to know the number of segmentation levels beforehand. They are contained in the structure of the tree. The tree also includes the correctly averaged gray or feature values for the segmented areas. Further details of pyramid-linking segmentation are discussed in Burt et al. [14] and Pietikäinen and Rosenfeld [113].

a   *b*

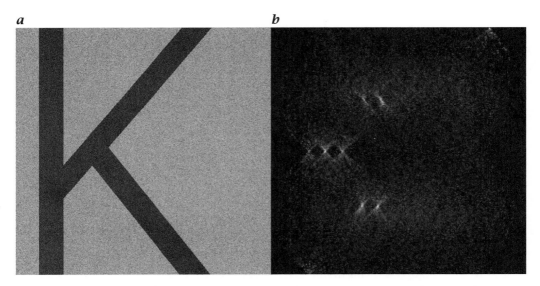

Figure 15.7: *Orientation-based fast Hough transform:* **a** *noisy letter "K";* **b** *Hough parameter space with the distance d (horizontal axis) and the angle θ (vertical axis) of the line according to Eq. (15.3).*

15.4.3 Orientation-Based Fast Hough Transformation

The basic idea of the orientation-based Hough transform is simple. If we do not simply detect edges but also estimate the orientation of the edges, we have two pieces of information with each point: a point through which the line passes and its orientation. This already completely describes the line. Consequently, each point on a line in the image space corresponds no longer to a line — as discussed in Section 15.3.4 — but to a single point in the parameter space.

Given a point on a line and the slope of the line through this point results in the following equation for the line

$$\bar{n}x = d \quad \text{or} \quad x \cos \theta + y \sin \theta = d. \tag{15.3}$$

The angle θ is given by the analysis of the local orientation and the distance d to the center of the coordinate system is computed from the coordinates $[x, y]^T$ using Eq. (15.3). Thus, the Hough parameter space is spanned by the distance d and the angle θ.

The one-to-one correspondence considerably speeds up the computation of the Hough transform. An application of the orientation-based Hough transform is demonstrated in Fig. 15.7. Figure 15.7a shows a noisy image of the letter "K". From this letter, the structure tensor is computed and used as the basis of the Hough transform. From each point in the image, θ and d are computed according to Eq. (15.3). Then the corresponding point in the Hough parameter space is incremented by the length of the orientation vector. In this way, points are weighted according to the certainty measure for the local orientation and thus edge strength. No thresholding is applied; all points are considered. In the Hough parameter space Fig. 15.7b, six clusters show up, corresponding to the six different lines of the letter "K". The clusters occur in pairs since two lines are parallel to each other and differ only by the distance to the center of the image. These pairs are thus located on a row of the parameter space.

15.5   Advanced Reference Material

References to advanced topics

15.1

L. G. Shapiro and G. C. Stockman, 2001. Computer Vision, Prentice Hall, Upper Saddle River, NJ. *A modern textbook on computer vision that deals in detail with image segmentation.*

D. A. Forsyth and J. Ponce, 2003. Computer Vision, A Modern Approach, Prentice Hall, Upper Saddle River, NJ. *Another modern textbook on computer vision with two extended chapters on image segmentation.*

A. Blake and M. Isard, 1998. Active Contours, Springer, London. *A monograph on snakes and their use as active contours to track dynamic and deformable objects in image sequences.*

C. Schnörr, 1999. Variational methods for adaptive image smoothing and segmentation, in Handbook of Computer Vision and Applications, Volume 2, edited by B. Jähne and H. Haußecker and P. Geißler, Academic Press, San Diego, pp. 451-484. *Review paper on variational methods for image segmentation.*

16 Size and Shape

16.1 Highlights

In this section, the transition is made from pixels to objects. While all image processing up to this point took place on spatial data (matrices), now new data structures are required to describe the geometrical shape and other properties of objects. Of course, it is still possible to treat object shape and also to manipulate it with spatial data on regular grids. The versatile class of morphological operators (Section 16.3.1) on binary and gray scale images can be used to correct for distortions of the segmentation process such as holes in objects, to smoothen object boundaries, to extract object boundaries, and to filter out objects of certain size ranges (Sections 16.4.1 and 16.4.2).

Significantly more compact data structures and faster algorithms are achieved when objects are represented by the boundary. This representation is complete as it entirely describes the shape of a connected region. We treat run-length code (Section 16.3.2), chain code (Section 16.3.3), Fourier descriptors (Section 16.3.4), and moments (Section 16.3.5) for the description of the shape of objects.

Fourier descriptors are the most versatile data structures for object shape description. It may come as a surprise that convolution operators can be used on these data structures in the very same way as on images to smooth boundaries (Section 16.4.1), and to compute the slope and curvature of object boundaries. Moreover, it is very easy to extract translation, rotation, and scale-invariant shape parameters from Fourier descriptors which still allow a complete description of the shape (Section 16.4.4). The only disadvantage of Fourier descriptors is that it is difficult to extract an accurate boundary line from the image from which the Fourier descriptors can be extracted.

16.2 Task

Once objects have been segmented from the background (Chapter 15), the geometrical form of them, the shape, can be processed and analyzed.

The tasks related to shape analysis are summarized in task list 12. In the first place, it might be useful and of interest to correct the segmented shape (Section 16.4.1). This task can be performed by a class of neighborhood operators on binary images, known as *morphological operators* (Section 16.3.1). These operators can also extract object boundaries, sort objects according to their size, etc. In short, they offer a versatile tool to manipulate the shape of objects. They play the same important role for object shape extraction and manipulation as convolution operators do for feature extraction.

Morphological operations work on binary images represented on a matrix. This means that we still store each pixel of the object and all the background pixels. The question is whether all information contained in binary images can be stored in a much more compact form. It is, for example, obvious that all information on the object shape

Task List 12: Shape

Task	Procedures
Manipulate object shape parameters, sort objects according to size	Binary and gray-scale morphology
Extract object boundary	Binary or gray-scale morphology, edge
Find appropriate data structures to represent shape in a compact form	Run-length code, chain code, Fourier descriptors, moments
Determine basic shape parameters	Centroid, center of gravity, area, eccentricity, etc.
Determine set of rotation and scale invariant shape parameters	Derive these parameters from moments or Fourier descriptors

is contained in its boundary pixels. It is, therefore, sufficient to store only the boundary pixels of an object. This is a much more compact representation of a binary image. It is worthwhile studying whether the extraction of shape parameters or operations modifying the shape of the object can also be performed using this data structure. As alternative data structures to represent the shape of objects, we will study run-length code (Section 16.3.2), chain code (Section 16.3.3), Fourier descriptors (Section 16.3.4), and moments (Section 16.3.5).

Object shape manipulation is discussed in Section 16.4.1, while Section 16.4.2 handles the extraction of object boundaries. The determination of shape parameters is split into two sections. Section 16.4.3 handles elementary shape parameters such as area, radius, and eccentricity. Rotation and scale-invariant shape parameters are discussed in Section 16.4.4. This is a central issue of shape analysis. In most practical applications, neither the distance to the object is constant nor the orientation of the objects is known. Then, only scale and rotation-invariant parameters are applicable. These parameters are even not sufficient when an image of an object is generated by a perspective projection as with any optical system (Section 4.3.2a). However, shape features that are invariant under perspective projection are still subject to intensive research and no fast solution can be provided yet. Thus, we focus in this handbook on rotation and scale-invariant shape features. They can be used with perspective projection for flat objects that are oriented parallel to the image plane or are observed under a small field of view. This is the case for most inspection tasks.

16.3 Concepts

16.3.1 Morphological Operators

16.3.1a Neighborhood Operations on Binary Images. Operators which relate pixels in a small neighborhood have emerged as a versatile and powerful tool for scalar and vectorial images. In Sections 10.3.3 and 10.3.9, we discussed the two basic operations to combine neighboring pixels of gray value images: convolution ("weight and sum up") and rank value filtering ("sort and select"). With binary images, we do not have much choice as to which kind of operations to perform. We can combine pixels only with the logical operations of Boolean algebra. With *binary convolution*, the multiplication of the image and mask pixels are replaced by an *and operation* and the summation by an *or operation*

$$G'_{mn} = \bigvee_{k=-r}^{r} \bigvee_{l=-r}^{r} M_{k,l} \wedge G_{m-k,n-l}. \tag{16.1}$$

a

b

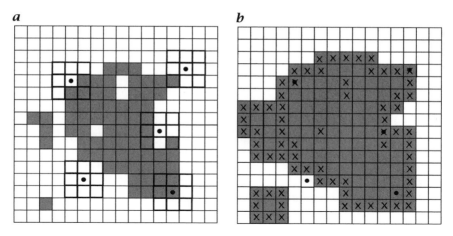

Figure 16.1: *Dilation of a binary object with the binary convolution operation as defined in Eq. (16.1). Shown is the application of a 3 × 3 mask. Crosses mark pixels added to the object by the dilation operation and black dots some positions of the 3 × 3 mask.*

a

b

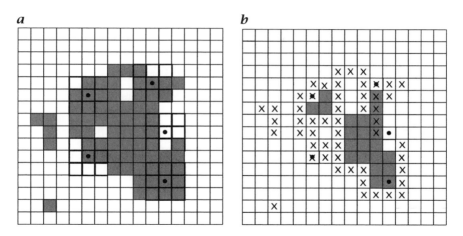

Figure 16.2: *Erosion of a binary object with a 3 × 3 mask. Crosses mark pixels deleted from the object by the erosion operation and black dots some positions of the 3 × 3 mask.*

The ∧ and ∨ denote the logical *and* and *or* operations, respectively. The binary image *G* is convolved with a symmetric $(2r + 1) \times (2r + 1)$ mask **M**.

What does this operation achieve? Let us assume that all the coefficients of the mask are set to 'one'. If one or more object pixels, i. e., 'ones', are within the mask, the result of the operation will be one; otherwise it is zero (Fig. 16.1). Hence, the object will be dilated. Small holes or cracks will be filled and the contour line will become smoother, as shown in Fig. 16.1b. The operator defined by Eq. (16.1) is known as the *dilation operator*. Interestingly, we can end up with an identical operator if we apply *rank-value filter* operations (see Section 10.3.9) to binary images. Let us take the *maximum operator*. The maximum will then be one if one or more 'ones' are within the mask, just as with the binary convolution operation in Eq. (16.1).

The *minimum operator* has the opposite effect. Now, the result is only one if the mask is completely within the object (Fig. 16.2). In this way, the object is eroded. Object pixels at which the mask is not completely covered by the object disappear (Fig. 16.2b).

The *erosion* of an object can also be performed using binary convolution. In order to erode the object, we dilate the background:

$$G'_{mn} = \overline{\bigvee_{k=-R}^{R} \bigvee_{l=-R}^{R} M_{k,l} \wedge \overline{G_{m-k,n-l}}}.$$ (16.2)

In this equation, the image is negated to convert the background to the object and vice versa. The result must be negated to reverse this negation.

By transferring the concepts of neighborhood operations from gray value images to binary images, we have gained an important tool to operate on the form of objects. We have already seen in Figs. 16.1 and 16.2 that these operations can be used to fill small holes and cracks or to eliminate small objects. The size of the mask governs the effect of the operators; therefore, the mask is often called the *structure element*. For example, an erosion operation works like a net which has holes in the shape of the mask. All object pixels for which the object does not completely fill the hole will slip through and disappear from the image. The operations that work on the form of objects are called *morphological operators*. The name originates from the research area of morphology which describes the form of objects in biology and geosciences.

We used a rather unconventional way to introduce morphological operations. Normally, these operations are defined as operations on sets of pixels. We regard G as the set of all the pixels of the matrix which are not zero. M is the set of the nonzero mask pixels. With M_p we denote the mask shifted with its reference point (generally but not necessarily its center) to the pixel p. Erosion is then defined as

$$G \ominus M = \{p : M_p \subseteq G\}$$ (16.3)

and dilation as

$$G \oplus M = \{p : M_p \cap G \neq \varnothing\}$$ (16.4)

where $\varnothing$ is the empty set. The operator $\cap$ denotes the union of two sets, while $\subseteq$ indicates the condition that M_p must be completely contained in G_1, i. e., M_p is a subset of G.

These definitions are equivalent to Eqs. (16.1) and (16.2), respectively, except for the fact that the mask in the convolution operation is rotated by 180° (see Section 10.3.4). We can now express the erosion of the set of pixels G by the set of pixels M as the set of all the pixels p for which M_p is completely contained in G. In contrast, the dilation of G by M is the set of all the pixels for which the intersection between G and M_p is not an empty set. Since the approach of set theory leads to more compact and illustrative formulas, we will use it now. Equations Eqs. (16.1) and (16.2) still constitute the basis from which to implement morphological operations. The erosion and dilation operator can be regarded as elementary morphological operators from which other more complex operators can be built. Their properties are studied in detail in the next section.

16.3.1b General Properties of Morphological Operations. Morphological operators share many of the properties of convolution operators (Section 10.3.6) but some properties are different. Morphological operators are in general nonlinear operators.

Shift Invariance. Shift invariance results directly from the definition of the erosion and dilation operator as convolutions with binary data in Eqs. (16.1) and (16.2). Using the *shift operator* $\circlearrowleft$ as defined in Eq. (10.21) and the operator notation, we can write the shift invariance of any morphological operator $\mathcal{M}$ as

$$\mathcal{M}\left(^{kl}\circlearrowleft G\right) = \,^{kl}\circlearrowleft (\mathcal{M}G).$$ (16.5)

Principle of Superposition. What does the *superposition principle* mean for binary data? For gray value images it is defined as

$$\mathcal{H}(a\boldsymbol{G} + b\boldsymbol{G}') = a\mathcal{H}\boldsymbol{G} + b\mathcal{H}\boldsymbol{G}'. \tag{16.6}$$

The factors a and b make no sense for binary images; the addition of images corresponds to the *union* or *logical or* of images. The superposition principle for binary images is given as

$$\mathcal{M}(G \cup G') = (\mathcal{M}G) \cup (\mathcal{M}G') \ \text{ or } \ \mathcal{M}(\boldsymbol{G} \vee \boldsymbol{G}') = (\mathcal{M}\boldsymbol{G}) \vee (\mathcal{M}\boldsymbol{G}'). \tag{16.7}$$

The operation $\boldsymbol{G} \vee \boldsymbol{G}'$ means a point-wise *logical or* of the elements of the matrices $\boldsymbol{G}$ and $\boldsymbol{G}'$. Generally, morphological operators are not additive in the sense of Eq. (16.7). While the dilation operation meets the superposition principle, the erosion does not. The erosion of the union of two objects is generally a superset of the union of two eroded objects:

$$\begin{aligned}
(G \cup G') \ominus M &\supseteq (G \ominus M) \cup (G' \ominus M) \\
(G \cup G') \oplus M &= (G \oplus M) \cup (G' \oplus M).
\end{aligned} \tag{16.8}$$

In conclusion, only the dilation operator is a linear shift-invariant operator; erosion is, in general, not linear.

Commutativity and Associativity. Also, morphological operators are not generally commutative:

$$M_1 \oplus M_2 = M_2 \oplus M_1, \text{ but } M_1 \ominus M_2 \neq M_2 \ominus M_1. \tag{16.9}$$

We can see that the erosion is not commutative if we take the special case that $M_1 \supset M_2$. Then, the erosion of M_2 by M_1 yields the empty set. However, both erosion and dilation masks consecutively applied in a cascade to the same image G are commutative:

$$\begin{aligned}
(G \ominus M_1) \ominus M_2 &= G \ominus (M_1 \oplus M_2) = (G \ominus M_2) \ominus M_1 \\
(G \oplus M_1) \oplus M_2 &= G \oplus (M_1 \oplus M_2) = (G \oplus M_2) \oplus M_1.
\end{aligned} \tag{16.10}$$

These equations are important for the implementation of morphological operations. Generally, the cascade operation with k structure elements $M_1, M_2, \ldots, M_k$ is equivalent to the operation with the structure element $M = M_1 \oplus M_2 \oplus \ldots \oplus M_k$. In conclusion, we can decompose large structure elements in the very same way as we decomposed linear shift-invariant operators. An important example is the composition of separable structure elements by the horizontal and vertical element $M = M_x \oplus M_y$. Another less trivial example is the build-up of large one-dimensional structure elements by structure elements including many zeros:

$$[1\,1\,1\,1\,1\,1\,1\,1\,1] = [1\,1\,1] \oplus [1\,0\,0\,1\,0\,0\,1]. \tag{16.11}$$

In this way, we can build up large structure elements with a minimum number of logical operations just as we built up large smoothing masks by cascaded convolution (Section 10.4.4). It is more difficult to obtain isotropic, i.e., circular-shaped, structure elements. The problem is that the dilation of horizontal and vertical structure elements always results in a rectangular-shaped structure element, but not in a circular mask. A circular mask can, however, be approximated with one-dimensional structure elements running in more directions than only along the axes.

Monotony. Erosion and dilation are monotonous operations

$$\begin{aligned}
G_1 \subseteq G_2 &\rightsquigarrow G_1 \oplus M \subseteq G_2 \oplus M \\
G_1 \subseteq G_2 &\rightsquigarrow G_1 \ominus M \subseteq G_2 \ominus M.
\end{aligned} \tag{16.12}$$

The monotony property means that the subset relations are invariant with respect to erosion and dilation.

Distributivity. Linear shift-invariant operators are distributive with regard to addition. The corresponding distributivities for erosion and dilation with respect to the union and intersection of two images G_1 and G_2 are more complex:

$$
\begin{aligned}
(G_1 \cap G_2) \oplus M &\subseteq (G_1 \oplus M) \cap (G_2 \oplus M) \\
(G_1 \cap G_2) \ominus M &= (G_1 \ominus M) \cap (G_2 \ominus M)
\end{aligned}
\tag{16.13}
$$

and

$$
\begin{aligned}
(G_1 \cup G_2) \oplus M &= (G_1 \oplus M) \cup (G_2 \oplus M) \\
(G_1 \cup G_2) \ominus M &\supseteq (G_1 \ominus M) \cup (G_2 \ominus M).
\end{aligned}
\tag{16.14}
$$

Erosion is distributive over the intersection operation, while dilation is distributive over the union operation.

Duality. Erosion and dilation are *dual operations*. By negating, the binary image erosion converts to dilation and vice versa:

$$
\begin{aligned}
\overline{G \ominus M} &= \overline{G} \oplus M \\
\overline{G \oplus M} &= \overline{G} \ominus M.
\end{aligned}
\tag{16.15}
$$

16.3.1c Hit-Miss Operator. Although all other morphological operators can be composed from the elementary morphological *erosion* and *dilation* operators, one additional operator deserves to be added here because of its central importance for advanced morphological operations. The *hit-miss operator* originates from the question of whether it is possible to filter out objects of a given shape. The following example shows that this is not possible with a single morphological operation but requires the combination of two operators. We want to extract from a binary image exactly all objects which are a square of 3×3 pixels.

If we erode the image with a 3×3 mask

$$
M_1 = \begin{bmatrix} 1 & 1 & 1 \\ 1 & 1 & 1 \\ 1 & 1 & 1 \end{bmatrix}
\tag{16.16}
$$

we will certainly remove all objects that are smaller but still retain all objects which are larger than the mask, i. e., where the shifted mask is a subset of the object $G(M_p \le G)$ (Fig. 16.3b). A 3×3 object, however, is characterized by the fact that it is surrounded by the background. Thus, we can use as a second step an erosion of the background with a 5×5 mask M_2 in which all coefficients are zero except for the border pixels:

$$
M_2 = \begin{bmatrix} 1 & 1 & 1 & 1 & 1 \\ 1 & 0 & 0 & 0 & 1 \\ 1 & 0 & 0 & 0 & 1 \\ 1 & 0 & 0 & 0 & 1 \\ 1 & 1 & 1 & 1 & 1 \end{bmatrix}.
\tag{16.17}
$$

The eroded background then contains all pixels in which the background has the shape of M_2 or larger ($M_2 \subseteq \overline{G}$, Fig. 16.3c). This leaves only pixels in the image of objects that fit into the 3×3 hole of the mask M_2. The intersection of the image eroded with M_1 and the background eroded with M_2 gives then all center pixels where 3×3 objects are located. In our example only one pixel remains (Fig. 16.3d). In general, the *hit-miss operator* is defined as

$$
\begin{aligned}
G \ominus (M_1 M_2) &= (G \ominus M_1) \cap (\overline{G} \ominus M_2) \\
&= (G \ominus M_1) \cap \overline{(G \oplus M_2)}
\end{aligned}
\tag{16.18}
$$

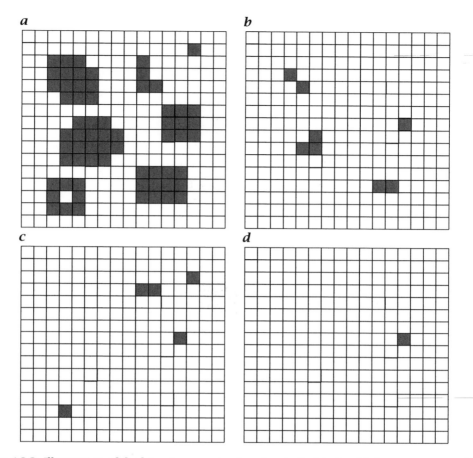

Figure 16.3: *Illustration of the hit-miss operator to extract all objects which are a square of 3×3 pixels: **a** original image; **b** object eroded by a 3×3 mask; **c** background eroded by a 5×5 mask in which only the border pixels are one; **d** intersection of **b** and **c** extracts all 3×3 objects.*

with $M_1 \cap M_2 = \emptyset$. The condition $M_1 \cap M_2 = \emptyset$ is required because otherwise the hit-miss operator would result in an empty set. With the hit-miss operator, a flexible tool is given with which it is possible to detect objects of a given shape. If a class of objects is to be detected, hit-miss operators are applied to detect any element of the class. Then, the union of the individual hit-miss images includes all elements of the object class.

16.3.2 Run-Length Code

A compact representation of a binary image is the *run-length code*. In order to extract the run-length code, a binary image is scanned line by line. If a line contains a sequence of p equal pixels, we do not store p times the same figure, but store the value of the pixel and indicate that it occurs p times. In this way, large uniform line segments can be stored in a very efficient way.

For binary images, the code can be especially efficient since we have only the two pixel values zero and one. Since a sequence of zeros is always followed by a sequence of ones, there is no need to store the pixel value. We only need to store the number of times a pixel value occurs (see Example 16.1).

We must be careful, however, at the beginning of a line since it may begin with a one or a zero. This problem can be resolved if we assume a line to begin with zero. If a line starts with a sequence of ones, we start the run-length code with a zero to indicate that the line begins with a sequence of zero zeros. Two examples of run-length code are given in Example 16.1.

Example 16.1: Run-length code for gray scale and binary images

The line of a gray scale image

$$12\ 12\ 12\ 20\ 20\ 20\ 20\ 25\ 27\ 25\ 20\ 20\ 20\ 20\ 20\ 20$$

written in run-length code in hexadecimal numbers is

$$82\ 12\ 83\ 20\ 2\ 25\ 27\ 25\ 85\ 20.$$

In this code, a sequence where the most significant bit is set (hexadecimal 80) indicates that the following number is to be repeated $n - 80 + 1$ times.
The following line of a binary image

$$1111110001110010000011111111$$

results in the following run-length code:

$$06332158.$$

Run-length code is suitable for compact storage of images. It has become an integral part of several standard image formats, for example, the TGA or the *TIFF* formats. Run-length code is normally applied to a whole image for image compression. In this form, it is, however, not useful for object representation. Run-length code can also be used object-oriented. Then, the stripes which belong to one object are packed together. In Section 2.2.4 and Fig. 2.6, it is demonstrated that the area of the object can be computed much faster in this representation.

16.3.3   Chain Code

The *chain code* or *contour code* is a data structure to represent the boundary of a binary image on a discrete grid in an efficient way. Instead of storing the positions of all the boundary pixels, we select a starting pixel and store its coordinates. If we use an algorithm which scans the image line by line, this will be the uppermost left pixel of the object (Fig. 16.4a and b). Then, we follow the boundary in clockwise direction. In a 4-neighborhood, there are 4; in an 8-neighborhood there are 8 possible directions to go which we can decode with a 3-bit or 2-bit code as indicated in Fig. 16.4c and d. If the object is not connected or if it has holes, we need more than one chain code to represent it. We must also include the information whether the boundary surrounds an object or a hole.

The chain code shows a number of obvious advantages over the matrix representation of a binary object:

Compact representation. The chain code is a compact representation of a binary object. Let us assume a disk-like object with a diameter of R pixels. In a direct matrix representation, we need to store the *bounding box* of the object, i. e., about R^2 pixels which are stored in R^2 bits. The bounding box is the smallest rectangle enclosing the object. If we use an 8-connected boundary, the disk shows about πR boundary points. The chain code of the πR points can be stored in about $3\pi R$ bit. From these figures, it follows that for objects with a diameter larger than 10 the chain code is a more compact representation.

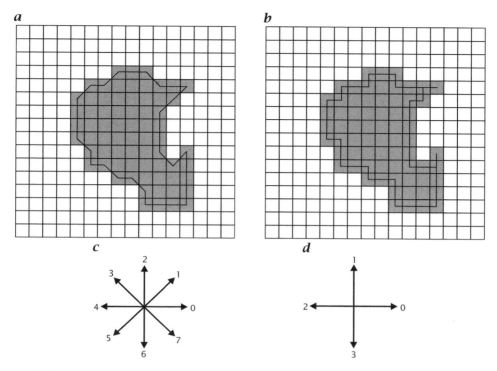

Figure 16.4: *Boundary representation with the chain code: **a** 8-connected boundary; **b** 4-connected boundary; **c** direction coding in 8-neighborhood; **d** direction coding in 4-neighborhood.*

Translation invariance. The chain code is a *translation invariant* representation of a binary object. This property makes the comparison of objects easier.

Fast algorithms. Since the chain code is a complete representation of an object or curve, we can principally compute any shape feature from the chain code. We can compute a number of shape parameters — including the perimeter and area — more efficiently using the chain-code representation than in the matrix representation of the binary image.

Fast object reconstruction. Reconstruction of the binary image from a chain code is an easy procedure. First we might draw the outline of the object and then use a *fill operation* to paint it.

The chain code has also a number of significant disadvantages.

No rotation invariance. This is the most significant disadvantage. The chain code cannot directly be used as a base for scale and rotation-invariant object recognition. It is only possible to extract scale and rotation invariant parameters from it. The simplest rotation-invariant parameter is, of course, the area of the object.

No scale invariance. The same remarks apply as for rotation invariance.

No subpixel accuracy. As a digital curve on a discrete grid, the curve goes only in 4 or 8 directions.

Noise sensitivity. The row chain code is sensitive to noise. Similar objects which have almost the same shape can have a quite different chain code.

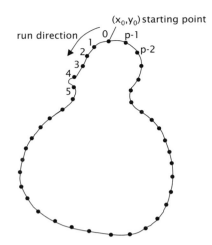

Figure 16.5: *Illustration of a parametric representation of a closed curve. The parameter p is the path length from the starting point (x_0, y_0). The equidistant sampling of the curve is also shown.*

16.3.4 Fourier Descriptors

The Fourier descriptors, as the chain code, use only the boundary of the object. In contrast to the chain code, the Fourier descriptors describe no curve on a discrete grid. They can be formulated for continuous or sampled curves. Consider the closed boundary curve sketched in Fig. 16.5. We can describe the boundary curve in a parametric description by taking the path length p from a starting point (x_0, y_0) as a parameter. Then, we get a curve of the form $x(p)$ and $y(p)$. We can combine these two curves to one complex curve $z(p) = x(p) + iy(p)$. This curve is cyclic. If P is the perimeter of the curve, then

$$z(p + nP) = z(p) \quad n \in \mathbb{Z}. \tag{16.19}$$

A cyclic or periodic curve can be expanded in a *Fourier series*. The coefficients of the Fourier series are given by

$$\hat{z}_u = \frac{1}{P} \int_0^P z(p) \exp\left(\frac{-2\pi i u p}{P}\right) dp, \quad u \in \mathbb{Z}. \tag{16.20}$$

The periodic curve can be reconstructed from the Fourier coefficients by

$$z(p) = \sum_{u=-\infty}^{\infty} \hat{z}_u \exp\left(\frac{2\pi i u p}{P}\right). \tag{16.21}$$

The coefficients $\hat{z}_u$ are known as the *Fourier descriptors* of the boundary. Their meaning is straightforward. The first coefficient

$$\hat{z}_0 = \frac{1}{P} \int_0^P z(p) dp = \frac{1}{P} \int_0^P x(p) dp + \frac{i}{P} \int_0^P y(p) dp \tag{16.22}$$

Figure 16.6: *Illustration of the reconstruction of shape by an increasing number of Fourier descriptors as indicated for* **a** *the letter "F" (with all, 2, 3, 4, and 8 descriptor pairs), and* **b** *a square (with all, 2, 4, 6, and 8 descriptor pairs). The corresponding Fourier descriptors are listed in Table 16.1.*

gives the mean vortex or *centroid* of the boundary. The second coefficient describes a circle

$$z_1(p) = \hat{z}_1 \exp\left(\frac{2\pi i p}{P}\right)$$
$$= r_1 \exp\left(\frac{2\pi i(\varphi_1 + p)}{P}\right) \tag{16.23}$$

with the radius r_1 and a starting point at an angle of φ_1 around the centroid ($\hat{z}_1 = r_1 \exp(i\varphi_1)$). The coefficient $\hat{z}_{-1}$ also results in a circle

$$z_{-1}(p) = r_{-1} \exp\left(\frac{2\pi i(\varphi_{-1} - p)}{P}\right) \tag{16.24}$$

but this circle is traced, in contrast to Eq. (16.23), in a clockwise direction. With both complex coefficients together — in total four parameters — an ellipse can be formed with arbitrary half axes a and b, orientation ϑ of the main axis a, and starting angle φ_0 on the ellipsis. As an example, we take $r_1 = a^2$, $\varphi_1 = 0$, $r_{-1} = b^2$, $\varphi_{-1} = \pi/2$. Then,

$$\hat{z}_1(p) + \hat{z}_{-1}(p) = a^2 \cdot \cos\left(\frac{2\pi p}{P}\right) + ib^2 \sin\left(-\frac{2\pi p}{P}\right). \tag{16.25}$$

This curve is an ellipse where the axes are oriented with the coordinate system and the starting point lies on the x axis.

From this discussion it is obvious that the Fourier descriptors $\hat{z}_u$ and $\hat{z}_{-u}$ must always be paired. The Fourier descriptor pair $\hat{z}_u$ and $\hat{z}_{-u}$ also form an ellipse. The only difference to $\hat{z}_1$ and $\hat{z}_{-1}$ is that the ellipse is cycled u times. Since according to Eq. (16.21) the ellipses are added up to form the curve, the higher Fourier descriptors add more and more details to the curve. For further illustration, the reconstruction of the letter F and a square is shown with an increasing number of Fourier descriptors (Fig. 16.6). Table 16.1 lists the first Fourier descriptors of these two boundary lines. The examples show that only a few coefficients are required to describe even complex shapes.

Fourier descriptors can also easily be computed from sampled boundaries z_n. If the perimeter of the closed curve is P, N samples must be taken at equal distances of

Table 16.1: *The first few Fourier descriptors of the objects shown in Fig. 16.6, the letter "F", a square, and a diagonal line. The complex-valued descriptors $\hat{h}_u$ are given in pairs, $\hat{h}_u, \hat{h}_{-u}$. The first descriptor, $\hat{h}_0$, is omitted.*

$|u|$	"F"	Square	Diagonal line
1	8.372 -0.830i, -13.33 -11.41i	-25.93 -25.93I, 0.000 0.000i	-25.93 -25.93i, -25.93 -25.93i
2	2.144 0.789i, -0.806 -8.613i	0.000 0.000i, 0.000 0.000i	0.000 0.000i, 0.000 0.000i
3	-6.515 -1.501i, -2.604 -1.002i	0.000 0.000I, -2.883 -2.883I	-2.883 -2.883i, -2.883 -2.883i
4	1.283 -0.514i, 0.241 1.687i	0.000 0.000i, 0.000 0.000i	0.000 0.000i, 0.000 0.000i
5	-0.157 -0.232i, -1.529 0.597i	-1.038 -1.038I, 0.000 0.000i	-1.038 -1.038i, -1.038 -1.038i
6	-0.501 -0.297i, 1.365 -1.288i	0.000 0.000i, 0.000 0.000i	0.000 0.000i, 0.000 0.000i
7	0.213 -0.151i, -0.316 -0.124i	0.000 0.000I, -0.530 -0.530i	-0.530 -0.530i, -0.530 -0.530i
8	-0.245 -0.101i, -0.149 -0.005i	0.000 0.000i, 0.000 0.000i	0.000 0.000i, 0.000 0.000i
9	-0.076 0.030i, -0.476 -0.207i	-0.321 -0.321I, 0.000 0.000i	-0.321 -0.321i, -0.321 -0.321i

P/N (Fig. 16.5). Then,

$$\hat{z}_u = \frac{1}{N} \sum_{n=0}^{N-1} z_n \exp\left(\frac{-2\pi i u n}{N}\right). \tag{16.26}$$

All other equations are valid also for sampled boundaries. Only the range of the wave number coefficient u has to be restricted from all integers to the interval $[0, N-1]$ or $[-N/2, N/2-1]$.

The Fourier descriptors have a number of significant advantages for shape representation and description.

Area. The *area* is directly given by the Fourier descriptors as

$$A = \pi \sum_{u=-N/2}^{N/2-1} u|\hat{z}_u|^2. \tag{16.27}$$

Translation invariance. The position of the object is confined into one single coefficient $\hat{z}_0$. All other coefficients are translation-invariant.

Scale invariance. If the contour is scaled by a factor of α, all Fourier descriptors are also scaled by α. If the contour is traced counterclockwise, the first coefficient is always unequal to zero. Thus, we can simply scale all Fourier descriptors by $|\hat{z}_1|$ to obtain scale-invariant shape descriptors. Note that these scaled descriptors are still complete.

Rotation invariance. If a contour is rotated by the angle φ_0 in counterclockwise direction, the Fourier descriptor $\hat{z}_u$ is multiplied by the phase factor $\exp(u\varphi_0)$ according to the shift theorem (Appendix B.4). This simple shift property makes the construction of rotation invariant Fourier descriptors easy. We can, for example, relate the phases of all Fourier descriptors to the phase of $\hat{z}_1$, φ_1, and subtract the phase shift $u\varphi_1$ from all coefficients. Then, all remaining Fourier descriptors are rotation invariant.

Symmetries. Symmetries can easily be detected with Fourier descriptors. If a contour has m-rotational symmetry, then only $z_{1\pm um}$ are unequal to zero (see Fourier descriptors of the square in Table 16.1). If the contour is the mirror contour of another, the Fourier descriptors are conjugate complex to each other.

The Fourier descriptors can also be used for nonclosed curves. To make it closed, we simply trace the curve backward and forward. It is easy to recognize such curves, since their area is zero. From Eq. (16.27), we can then conclude that $|\hat{z}_{-u}| = |\hat{z}_u|$. If the trace begins at one of the endpoints, even $\hat{z}_{-u} = \hat{z}_u$.

Given all these significant advantages of Fourier descriptors, there is unfortunately also one big disadvantage. It is not easy to generate a boundary curve with equidistant samples from digital images. In an 8-neighborhood, the points connected by the chain code are not equidistant. In a 4-neighborhood, the samples are equidistant but the boundary is jagged because the pieces of the boundary curve can only go in horizontal and vertical directions. Therefore, the perimeter tends to be too long. Consequently, it appears not a good idea to form a continuous boundary curve from points on a regular grid. The only alternative is to extract subpixel accurate object boundary curves directly from the gray scale images. This is, however, not an easy task. Thus, the accurate determination of Fourier descriptors from contours in images still remains a challenging research problem.

16.3.5 Moments

16.3.5a Definitions. Object description by *moments* has the significant advantage that it does not only use the boundary of an object but all pixels of an object. It can also be applied to gray scale images and as such is sensitive to the distribution of the gray values in the object.

We have used moments in Section 7.3.1b to describe the probability density function for gray values. Here we extend this description to two dimensions and define the moments of the gray value function $g(\mathbf{x})$ of an object as

$$m_{p,q} = \int (x_1 - \overline{x_1})^q (x_2 - \overline{x_2})^p \, g(\mathbf{x}) \, \mathrm{d}^2 x, \qquad (16.28)$$

where

$$\overline{x_i} = \int x_i g(\mathbf{x}) \, \mathrm{d}^2 x \bigg/ \int g(\mathbf{x}) \, \mathrm{d}^2 x. \qquad (16.29)$$

The integration includes the area of the object. Instead of the gray value, we may use more generally any pixel-based feature to compute object moments. The vector $\overline{\mathbf{x}} = (\overline{x_1}, \overline{x_2})$ is called the *center of mass* or *center of gravity* of the object in analogy to classical mechanics. Think of $g(\mathbf{x})$ as the density $\rho(\mathbf{x})$ of the object; then, the zero-order moment $m_{0,0}$ becomes the total mass of the object.

All the moments defined in Eq. (16.28) are related to the center of mass. Therefore they are often denoted as *central moments*. Central moments are invariants under a translation of the coordinates and thus are useful features to describe the shape of objects.

For discrete binary images, the moment calculation reduces to

$$m_{p,q} = \sum (x_1 - \overline{x_1})^q (x_2 - \overline{x_2})^p. \qquad (16.30)$$

The summation includes all pixels x_i belonging to the object. The zero-order moment $m_{0,0}$ gives the number of pixels and, thus, the area A of the object.

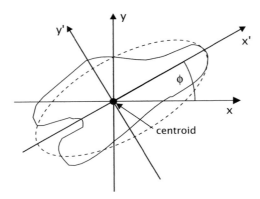

Figure 16.7: *Principal axes of the inertia tensor of an object for rotation around the center of mass.*

16.3.5b   Normalized Moments.   Often it is necessary to use shape parameters which do not depend on the size of the object. This is always required if objects must be compared which are observed from different distances. Moments can be normalized in the following way to obtain scale-invariant shape parameters. If we scale an object $g(x)$ by a factor of α, $g'(x) = g(x/\alpha)$, its moments are scaled according to

$$m'_{p,q} = \alpha^{p+q+2}\, m_{p,q}.$$

We can then normalize the moments with the zero-order moment, $m_{0,0}$, to gain scale-invariant moments

$$\overline{m}_{p,q} = \frac{m_{p,q}}{m_{0,0}^{(p+q+2)/2}}.$$

Since the zero-order moment of a binary object gives the area of the object Eq. (16.30), the *normalized moments* are scaled by the area of the object; for a gray scale object, they are scaled by the total mass.

16.3.5c   Second-Order Moments; the Inertia Tensor.   Shape analysis starts with the second-order moments. The zero order moment just gives the area or "total mass" of a binary or gray value object, respectively. The first-order central moments are, by definition, zero.

The analogy to mechanics is again helpful to understand the meaning of the second-order moments $m_{2,0}$, $m_{0,2}$, and $m_{1,1}$. They contain terms in which the gray value function, i.e., the density of the object, is multiplied by squared distances from the center of mass. Such expressions are known from the inertia in mechanics. The three second-order moments form the components of the *inertia tensor* for rotation of the object around its center of mass:

$$J = \begin{bmatrix} m_{0,2} & -m_{1,1} \\ -m_{1,1} & m_{2,0} \end{bmatrix}. \tag{16.31}$$

The properties of symmetric tensors are discussed in detail in Section 13.3.3. Therefore we can transfer all the results from there to shape description with second-order moments. The *orientation* of the object is defined as the angle between the x axis and the axis around which the object can be rotated with minimum inertia which is given by the eigenvector to the minimal eigenvalue. The object is most elongated in this direction (Fig. 16.7). According to Eq. (13.19), this angle is given by

$$\phi = \frac{1}{2} \arctan \frac{2m_{1,1}}{m_{2,0} - m_{0,2}}. \tag{16.32}$$

As a measure for the *eccentricity* ε, we can use what we have defined as a coherency measure for local orientation Eq. (13.22):

$$\varepsilon = \frac{(m_{2,0} - m_{0,2})^2 + 4m_{1,1}^2}{(m_{2,0} + m_{0,2})^2}. \tag{16.33}$$

The eccentricity ranges from 0 to 1. It is zero for a circular object and one for a line-shaped object. Shape description by second-order moments essentially models the object as an *ellipse* as illustrated in Fig. 16.7.

Once the orientation of an object is known, we can draw a box around it which is aligned with the principal axes and just large enough to contain all object pixels. This box is known as the *bounding box* of an object. The width and height of the bounding rectangle are two other orientation-independent parameters to describe the shape of an object.

16.4 Procedures

In the procedure section, we discuss four elementary tasks related to the shape of objects.

Object shape manipulation. Operators of this type are required to correct segmentation errors, to improve the "noisy" object shapes, or to detect objects of a certain shape (Section 16.4.1) .

Boundary extraction. This operation is required to generate all data structures that are based on the boundary of an object such as the chain code (Section 16.3.3) and Fourier descriptors (Section 16.3.4).

Basic shape parameters. Here, we discuss all elementary shape parameters such as area and perimeter and global measures for the form of an object (Section 16.4.3).

Scale and rotation invariance. Here, we discuss the generation of scale and rotation invariant parameters.

In all sections, the different concepts for shape description including the *label image*, *run-length code*, *chain code*, *Fourier descriptors*, and *moments* are compared with each other with respect to accuracy, robustness, and speed of computation in the extraction of the parameters.

16.4.1 Object Shape Manipulation

16.4.1a Removal of Small Objects. A common task for shape manipulation is the removal of small objects, for instance, erroneously segmented pixels of the background. The elementary erosion operation performs this task as discussed in Section 16.3.1a. According to Eq. (16.3) all objects O disappear which meet the condition

$$O \ominus M = \varnothing \equiv M_p \nsubseteq O \ \forall p. \tag{16.34}$$

By a proper choice of the structure element, we can eliminate objects with a certain form which meet the criterion given in Eq. (16.34). However, the erosion operation

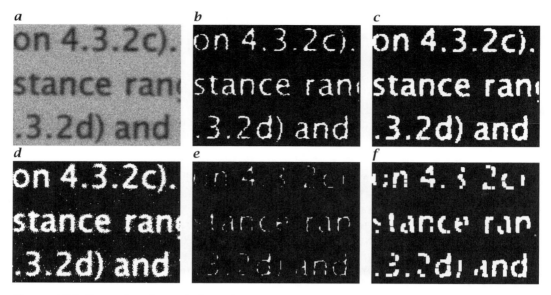

Figure 16.8: *Examples of morphological operations: **a** original image; **b** binary erosion of **d** with a 3 × 3 mask; **c** binary opening of **d** with a 3 × 3 mask; **d** binary image obtained by optimum global thresholding from **a**; **e** binary erosion of **d** with a 5 × 5 mask; **f** binary opening of **d** with a 5 × 5 mask.*

shows the disadvantage that all the remaining objects shrink in size. We can avoid this effect by dilation of the image after erosion with the same structure element. This combination of operations is called an *opening* operation

$$G \circ M = (G \ominus M) \oplus M. \tag{16.35}$$

The opening sieves out objects which are smaller than the structure element, but avoids a general shrinking of the size (Fig. 16.8c and d). It is also an ideal operation to remove lines with a diameter that is smaller than the diameter of the structure element.

16.4.1b Filling Holes and Cracks. In contrast, dilation enlarges objects and closes small holes and cracks. General enlargement of the object by the size of the structure element can be reversed by a subsequent erosion (Fig. 16.9c and d). This combination of operations is called a *closing* operation

$$G \bullet M = (G \oplus M) \ominus M. \tag{16.36}$$

The relation between the sizes of objects that are modified with the morphological operations erosion, opening, closing, and dilation using the *same* structure element is summarized by the following relations:

$$G \ominus M \subseteq G \circ M \subseteq G \subseteq G \bullet M \subseteq G \oplus M. \tag{16.37}$$

Opening and closing are idempotent operations

$$\begin{aligned} G \bullet M &= (G \bullet M) \bullet M \\ G \circ M &= (G \circ M) \circ M, \end{aligned} \tag{16.38}$$

i.e., a second application of a closing and opening with the same structure element does not show any further effects.

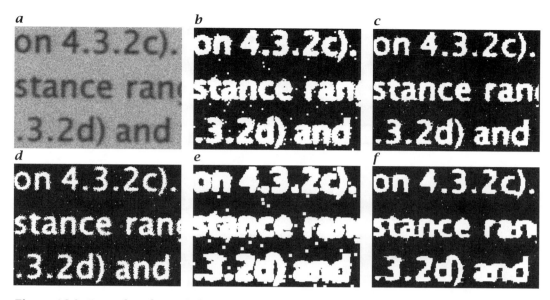

Figure 16.9: *Examples of morphological operations: a original image; b binary dilation of d with a 3 × 3 mask; c binary closing of d with a 3 × 3 mask; d binary image obtained by optimum global thresholding from a; e binary dilation of d with a 5 × 5 mask; f binary closing of d with a 5 × 5 mask.*

16.4.1c   Shape Detection and Selection.   This task is performed by the hit-miss operator discussed in Section 16.3.1c.

16.4.2   Extraction of Object Boundaries

Boundary points can be extracted by morphological operators by analyzing what characterizes a border point. A border point misses at least one of its neighbors. As we discussed in Section 6.3.1b, we can define 4- and 8-neighborhoods on a rectangular grid. We can remove the boundary points by eroding the object with a structure element that contains all the possible neighbors of the central pixel:

$$M = \underbrace{\begin{bmatrix} 1 & 1 & 1 \\ 1 & 1 & 1 \\ 1 & 1 & 1 \end{bmatrix}}_{\text{8-neighborhood}} \quad \text{and} \quad M = \underbrace{\begin{bmatrix} 0 & 1 & 0 \\ 1 & 1 & 1 \\ 0 & 1 & 0 \end{bmatrix}}_{\text{4-neighborhood}}. \tag{16.39}$$

The boundary is then gained by the set difference (/ operator) between the object and the eroded object

$$B = G/(G \ominus M) = G \cap \overline{(G \ominus M)} = G \cap (\overline{G} \oplus M). \tag{16.40}$$

As shown in the formula, we can also understand the set difference as the intersection of the object with the dilated background. It is important to note that the boundary line shows the dual connectivity to the connectivity of the eroded object. If we erode the object with the 8-neighbor structure element, the boundary is 4-connected, and vice versa. An example for boundary extraction is shown in Fig. 16.10.

a

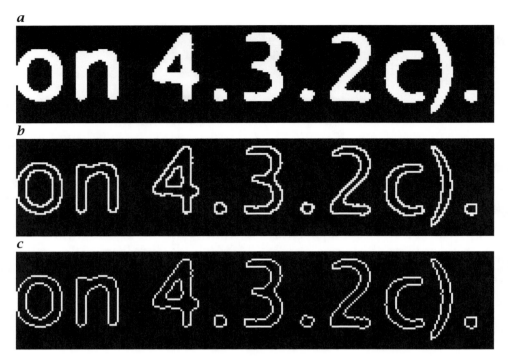

b

c

Figure 16.10: *Boundary extraction:* ***a*** *original binary image;* ***b*** *4-connected boundary;* ***c*** *8-connected boundary.*

16.4.3   Basic Shape Parameters

16.4.3a   Area. The most trivial shape parameter is the *area A* of an object. Here we discuss how it can be computed from run-length code, chain code, and Fourier descriptors.

Run-length Code. The area of an object can be computed by simply summing up the lengths of the streaks belonging to the object. This is a simple and fast operation requiring generally less operations than the number of pixels of the object.

Chain code. In a digital binary image the area is given by the number of pixels that belong to the image. In the matrix or pixel list representation of the object, area computing simply means counting the number of pixels. At first glance, area computation of an object which is described by its chain-code seems to be a complex operation. However, the contrary is true. Computation of the area from the chain code is much faster than counting pixels since the boundary of the object contains only a small fraction of the object's pixels and requires only two additions per boundary pixel.

The algorithm works in a similar way as numerical integration. We assume a horizontal base line drawn at an arbitrary vertical position in the image. Then we start the integration of the area at the uppermost pixel of the object. The distance of this point to the base line is B.

We follow the boundary of the object and increment the area of the object according to the figures in Table 16.2. If we, for example, move to the right (chain code 0), the area increases by B. If we move upwards to the right (chain code 1), the area also increases by B, but B must be incremented, because the distance between the boundary pixel and the base line has increased. For all movements to the left, the area is decreased

Table 16.2: *Computation of the area of an object from the contour code. Initially, the area is set to zero. With each step, the area and the parameter B are incremented corresponding to the value of the contour code; after Zamperoni [156] with corrections.*

Contour code	Area increment	Increment of B
0	+B	0
1	+B	1
2	0	1
3	-B+1	1
4	-B+1	0
5	-B+1	-1
6	0	-1
7	+B	-1

by $B - 1$. The value $B - 1$ has to be used, because an object that is just one pixel high would otherwise get an area of zero.

In this way, we subtract the area between the lower boundary line of the object and the base line, which was included in the area computation when moving to the right. The algorithm terminates, when we come to the start pixel again. The area A is set initially to 1.

Fourier descriptors. The area is directly given by the Fourier descriptors as

$$A = \pi \sum_{u=-N/2}^{N/2-1} u|\hat{z}_u|^2. \tag{16.41}$$

Again, this is a fast algorithm, which requires at most as many operations as points at the boundary line of the curve. The Fourier descriptors show the additional advantage that we can compute the area for a certain degree of smoothness by taking only a certain number of Fourier descriptors. The more Fourier descriptors we take, the more detailed the boundary curve can be, as demonstrated in Fig. 16.6. From the Fourier descriptors listed in Table 16.1, you can see that the contribution of a Fourier descriptor to the area rapidly decreases with u.

16.4.3b Perimeter.

Chain code. The *perimeter* is another geometrical parameter, which can easily be obtained from the chain code of the object boundary. We just need to count the length of the chain code and take into consideration that steps in diagonal directions are by a factor of $\sqrt{2}$ longer. The perimeter P is then given by an 8-neighborhood chain code:

$$P = n_e + \sqrt{2}n_o, \tag{16.42}$$

where n_e and n_o are the number of even and odd chain code steps, respectively. In contrast to the area, the perimeter is a parameter which is sensitive to the noise level in the image. The more noisy the image, the more rugged and thus longer the boundary of an object will become in the segmentation procedure. This means that care must be taken in comparing perimeters which have been extracted from different images. We must be sure that the smoothness of the boundaries in both images is comparable.

Figure 16.11: *Illustration of the importance of the phase for the description of shape with Fourier descriptors. The Fourier descriptors of the letter "F" are taken and the phase of the descriptors is changed randomly. Besides the original "F", three random phase modifications are shown.*

16.4.3c Form Parameters.

Circularity. Area and perimeter are two parameters which describe the size of an object in one or the other way. In order to compare objects which are observed from different distances, it is important to use shape parameters which do not depend on the size of the object on the image plane. The *circularity* c is one of the simplest parameters of this kind. It is defined as

$$c = \frac{P^2}{A}. \tag{16.43}$$

The circularity is a dimensionless number with a minimum value of $4\pi \approx 12.57$ for circles. The circularity is 16 for a square and $12\sqrt{3} \approx 20.8$ for an equilateral triangle. Generally, it shows large values for elongated objects.

Eccentricity. This is a measure similar to the circularity but with a better defined range. The parameter is extracted from the second-order moments as:

$$\varepsilon = \frac{(m_{2,0} - m_{0,2})^2 + 4m_{1,1}^2}{(m_{2,0} + m_{0,2})^2}. \tag{16.44}$$

The eccentricity ranges from 0 to 1. It is zero for a circular object and one for a line-shaped object. Shape description by second-order moments essentially models the object as an *ellipse* as illustrated in Fig. 16.7.

16.4.3d Object Orientation.

Moments. Second-order moments model an object as an ellipse. Therefore the orientation of an object can be defined as the direction of the main axis. This angle is given by

$$\phi = \frac{1}{2} \arctan \frac{2m_{1,1}}{m_{2,0} - m_{0,2}} \tag{16.45}$$

in analogy to the orientation extracted from the structure tensor (Section 13.4.2b). The moment tensor (Section 16.3.5) and the structure tensor (Section 13.3.3) are both symmetrical tensors.

16.4.4 Scale and Rotation Invariant Shape Parameters

Both Fourier descriptors (Section 16.3.4) and moments (Section 16.3.5) provide a framework for scale and rotation invariant shape parameters as already discussed in the corresponding sections. The Fourier descriptors are the more versatile instrument. They restrict, however, the object description to the boundary line while moments of gray scale objects are sensitive to the spatial distribution of the gray values in the object.

The most challenging task is to find a complete and unique description. This means that different shapes must not be mapped onto the same set of features. Here we just want to demonstrate how dangerous it is to form scale and rotation invariant shape features from Fourier descriptors by taking their magnitude. Figure 16.11 shows the several boundaries obtained by taking the Fourier descriptors of the letter "F" and changing the phase randomly. Once the Fourier descriptors are made rotation invariant with the technique discussed in Section 16.3.4, the Fourier descriptors constitute a complete rotation and scale invariant description. By leaving out higher-order Fourier descriptors, we can gradually relax fine details from the shape description.

16.5   Advanced Reference Material

References to advanced topics   16.1

T. H. Reiss, 1993. Recognizing Planar Objects using Invariant Image Features, Lecture Notes in Computer Science, Vol. 676, Springer, Berlin. *Covers moment-based shape description.*

J. Serra, 1982. Image Analysis and Mathematical Morphology, Academic Press, London. *The mathematical foundation of morphological operations including complete proofs for all the properties stated in this chapter can be found in this classic book.*

P. Soille, 2003. Morphological Image Analysis, Principles and Applications, 2nd ed., Springer, Berlin. *Integral presentation of morphological image processing including theory, methods, and applications.*

17 Classification

17.1 Highlights

Recognizing objects in images as members of a certain class is – as many other aspects of image processing and analysis – a truly interdisciplinary problem. Firstly, it requires solid knowledge concerning the application area. The reason is that images are very high-dimensional objects (an $N \times M$ image can be considered a point in an $N \cdot M$-dimensional feature space) and do not lend themselves to direct classification. Instead, preprocessing or feature extraction is required to reduce noise and dimensionality. Secondly, predicting class membership from measured features is a problem that has surfaced in many different disciplines, ranging from speech recognition (which word was uttered?) to industrial quality control (part intact or defect?).

This chapter presents algorithms which "learn" automatically how to predict the true class membership from previously extracted features. This process is known as *pattern recognition*, *discriminant analysis*, *classification*, or *supervised learning*. Another family of techniques which seek to detect inherent structure or "natural groupings" in a data set with no class labels goes by the name of *unsupervised learning* or *cluster analysis* (or, unfortunately, also by the name of classification). These methods are not treated here.

17.2 Task

In this chapter, we assume that the goal is to predict, as well as possible, the type, or the class membership of an object in an image. We further assume that the user has access to a historical database or *training set* of objects and their features, along with their true type, or class membership. Previous chapters have dealt with the extraction of appropriate features describing, for instance, shape, geometrical or chromatic properties of an object.

We give a few examples to show the range of practical applications. In recent years, a central topic in environmental science has been "*Waldsterben*", i. e., large-scale forest damage by acid rain and other environmental pollution. Here, the task is to map and classify the extent of the damage in forests from remote sensing techniques such as *aerial* and *satellite imagery*. In this setting, the training set would consist of remotely sensed images along with "ground truth" data from human forestry experts.

In medicine, much effort has been devoted to the automatic analysis of mammographic images (X-ray images of the female breast), with the aim of detecting the mamma carcinoma. A more recent development is the automatic analysis of spectroscopic images from magnetic resonance tomography, with the aim of detecting brain or prostate tumors. In these cases, the training set consists of a large number of images from both patients and healthy volunteers, along with the medical diagnosis.

In all of the above examples, we can expect to need more than one feature to identify the appropriate class, and we cannot expect a simple relationship between features and classes. The algorithms introduced in this chapter should help find such a relationship.

The following sections discuss a quality criterion for classification, and how a classifier is validated. In Section 17.4, we introduce a very stiff classifier (Section 17.4.1) and a generalization thereof (Section 17.4.2), as well as an extremely flexible classifier (Section 17.4.3), and we discuss their respective merits. Section 17.4.4 gives guidance on the informed choice of features or classification methods.

17.3 Concepts

17.3.1 Statistical Decision Theory

Objects can be characterized in many ways; in particular, their features can be measured on the nominal scale, the ordinal scale, or the interval scale [18]. For simplicity, we will assume that a total of P features have been measured on the continuous scale. Assuming that our training set contains a total of N objects, we can interpret these as N points in a P-dimensional feature space. If P is two or three, it is easy to visualize the data in a scatter plot. The class membership could then be illustrated using, for instance, different colors; in a quality control problem, one might choose to color those parts that have passed a test green and the others red.

We now seek to parameterize an algorithm such that it takes the measured features of an unknown object as input and gives an estimate or prediction of the true class membership as output. Such an algorithm is a *classifier*. The more modest term "estimate" already indicates that this may not always be possible without error. The frequency of false predictions depends on the degree of overlap of the two classes in feature space: if the two classes from the training set are clearly separated, and if the examples in the training set are representative for the process under investigation, a prediction with no or little error can be expected.

If, on the other hand, the two classes overlap severely, and if the algorithm does not have access to information other than the objects' position in feature space, it cannot be expected to make accurate predictions.

In a region of feature space in which two or more classes overlap, the algorithm's estimate should depend on the relative seriousness of the consequences of a wrong prediction. Let us assume that you need to set up an automatic quality control for a small but safety-relevant part delivered to the aeronautical industry; assume, in addition, that the costs incurred by delivering a faulty part are very high. In this case, the algorithm should take a "conservative" stance: if there is only a trace of doubt, that is, if an object lies in a part of feature space in which the faulty class has some density (i.e., the training set contained a few faulty objects lying in that region of feature space), it is safer to assume the part is faulty and either discard it or subject it to closer scrutiny.

The relative costs incurred by the decision process can be coded in a *loss matrix* (Table 17.1). Typically, correct predictions (true positives, true negatives) incur zero loss, whereas false predictions (false positives, false negatives) are undesirable; as was illustrated above, the seriousness of wrong decisions can be unequal, and the off-diagonal elements in the loss matrix can be chosen to reflect this imbalance.

We stated above that the aim of classification was to "predict as well as possible" the class of an object. We can now sharpen this concept, by choosing to minimize the *expected loss* of a classifier. The idea is to choose the parameters in an algorithm such that, when an infinite number of training samples are drawn randomly and classified

Table 17.1: *Loss matrix: diagonal elements are typically set to zero, off-diagonal elements to positive constants which can differ. An optimal classifier minimizes the expected loss.*

	pos. prediction	neg. prediction
truth: pos.	true positive	false negative
truth: neg.	false positive	true negative

and the loss of each single prediction is looked up in the loss matrix and added up, the sum of all these losses becomes minimal.

The resultant classifier will be the *best possible* for this choice of problem and loss matrix. It is also called the Bayes classifier. Details can be found, e. g., in Ripley [120]. In the following and for simplicity, we will assume that false positive and false negative predictions are equally bad, and that correct predictions incur no loss; in short, we choose the loss matrix to be of the form

$$\begin{bmatrix} 0 & 1 \\ 1 & 0 \end{bmatrix}. \tag{17.1}$$

We state here without proof that for this specific choice of loss function, the best possible or Bayes classifier is the one which predicts, at each point in feature space, that class which has the highest density at that point [120]. Unfortunately, this statement shifts our problem without solving it – for we do not know the true class densities. We will, in the procedures section Section 17.4, introduce two methods: linear discriminant analysis, discussed in Section 17.4.1, seeks to estimate the densities and relative frequencies of the classes; whereas k-nearest neighbors, discussed in Section 17.4.3, aim to directly predict the locally dominant class.

17.3.2 Model Optimization and Validation

The solution of a classification problem requires a number of choices. The first of these is the proper choice of features, and this is of paramount importance! For what information is lost in the feature extraction can never be compensated for later on, no matter how sophisticated the machinery. We stated above that classification results are best if two classes have little or no overlap in feature space. Accordingly, it makes sense to include (combinations of) features that separate the classes as well as possible. Note that this separation need not be linear, all that is required is that a scatter plot would reveal, say, the "green" and the "red" areas to be as disjunct as possible. "Then why", you may ask, "not include all possible features, just to be on the safe side?" The reason is that the effective number of parameters that need be determined in an algorithm grows with the dimensionality of the feature space.

When the training set is small, there may not be enough data to reliably determine all these extra parameters. As a consequence, it is best to choose a small subset of all conceivable features, that subset which allows for the best discrimination. If subsets with only two members are sought, it is possible to produce scatter plots of the training set in this representation. While this may become tedious if the set of candidate features is large (for P candidate features, you would need to consider $P(P-1)/2$ scatter plots), it becomes outright impossible if the subset sought should comprise more than three features: straightforward visualization in a scatter plot is no longer feasible. In this case, an automatic selection is desirable, which will be discussed below.

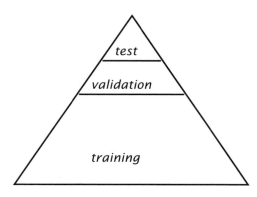

Figure 17.1: *If the database is sufficiently large, it can be split into training, validation, and test sets. The latter should be used only once, for a final evaluation before the algorithm is deployed.*

Another choice required is that of the capacity, complexity, or flexibility of the classifier. Some classifiers allow for an explicit tuning of their complexity, e.g., neural nets with their variable number of units in the hidden layer. Finally, there is the choice between different classifiers; modern software packages for the statistical analysis of data (such as [62]) offer a wide range of classification methods and it is, unfortunately, impossible to ascertain the general superiority of a specific method.

In summary, there is a number of choices which we would like to automate and render more objective. If no additional constraints such as easy interpretability of the algorithm or meager computational resources exist, one possibility is to take the expected loss (that is the overall classification performance, see above) as the only criterion.

Now, in practice, we do not have an infinite training set. One way forward is to split the available training set as indicated in Fig. 17.1. The performance on the validation set guides the optimization of the parameters in the algorithm, whereas the classification performance on the test set gives an estimate for the method's performance on unknown data.

While honest, this method makes poor use of the data; frequently, experiments are so expensive or tedious that the historical dataset is small, and it would be wasteful not to use all of the data to improve the classification algorithm. On the other hand, it is a very bad idea to use all of the data for training with no external check; the 1-nearest neighbor method discussed below will make zero errors no matter how large or complex the training set, i.e., it is a perfect fit to the data. However, due to *overfitting*[1], generalization performance on unseen data can be poor. Generally speaking, the performance on the training set is better than the true performance, because the method is optimized to do as well as possible on the training set. The difference between the empirical error on the training set and the true error can be called optimism, and it grows with the flexibility of a classifier.

Besides certain analytical schemes (such as Akaike's Information Criterion, AIC [120]) which seek to avoid overfitting, there are a number of *resampling techniques*, such as *cross-validation* or bootstrapping [51]. In these schemes, the data is repeatedly partitioned into training and test sets; see Section 17.4.4.

---

[1]The phenomenon of overfitting may be more familiar in the setting of regression: if there is noise on the observations or an element of randomness in the observed process, a high-order polynomial or a spline with many nodes that exactly interpolates all observations will generally not make good predictions.

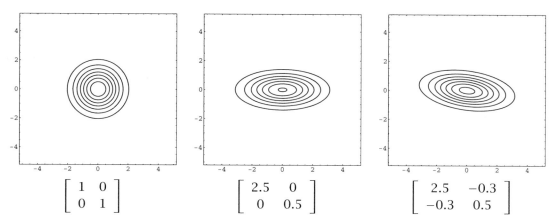

$$\begin{bmatrix} 1 & 0 \\ 0 & 1 \end{bmatrix} \qquad \begin{bmatrix} 2.5 & 0 \\ 0 & 0.5 \end{bmatrix} \qquad \begin{bmatrix} 2.5 & -0.3 \\ -0.3 & 0.5 \end{bmatrix}$$

Figure 17.2: Contour plots of the densities of two-dimensional normal distributions along with their covariance matrices. The first density is spherical or isotropic, the second has a diagonal covariance matrix, and the third shows some anti-correlation: a positive value of the first component tends to go together with a negative value of the second component and vice versa.

17.4   Procedures

17.4.1   Linear Discriminant Analysis (LDA)

In this approach, the distribution of each class in feature space is modeled using a multivariate normal (i.e., P-dimensional Gaussian) distribution. Formally, each class k is assumed to have density p at position x in feature space:

$$p(x|k) = \frac{1}{((2\pi)^P |\Sigma_k|)^{\frac{1}{2}}} \cdot \exp\left[-\frac{1}{2}(x - \mu_k)^T \Sigma_k^{-1}(Vx - \mu_k) \right]. \qquad (17.2)$$

In this expression, Σ is the *covariance matrix* describing the shape of a normal distribution, and μ is its center of mass. The diagonal elements of the covariance matrix (see Section 7.3.3b) are the variances. These indicate the spread of the distribution along the various coordinate axes. The off-diagonal elements indicate the amount of correlation: a spherically symmetric distribution has zero covariances, whereas one which is extended and "points" in a direction other than one of the coordinate axes is characterized by positive or negative correlations; see Fig. 17.2.

   We stated that the best possible classifier is the one predicting the locally most dominant class. Assuming that we seek to solve a two-class problem ($k \in \{1, 2\}$), the discrimination surface separating these two classes must then be the set of all points at which the classes have the same probability, that is, the set of all points where neither class dominates. In other words, the discrimination surface is the solution x of

$$p(1|x) \quad = \quad p(2|x) \qquad (17.3)$$

$$\frac{p(x|1)\pi(1)}{p(x)} \quad = \quad \frac{p(x|2)\pi(2)}{p(x)}, \qquad (17.4)$$

where the transition from the first equation to the second is by Bayes theorem[2], $p(k|x)$ is the *posterior* density of class k at location x, $p(x|k)$ is the density of class k at x

---

[2]Bayes theorem $p(k|x) = p(x|k)\pi(k)/p(x)$ [23] is illustrated visually in Figs. 17.3–17.5; $p(a|b)$ is the *conditional probability density* of a given b.

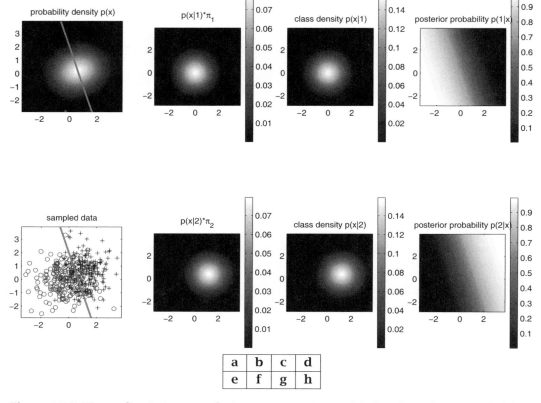

Figure 17.3: *Linear discriminant analysis as a generative model: data have been sampled from the model. See text for a detailed description of the panels.*

and $\pi(k)$ is the *prior probability*, or relative frequency of occurrence of class k, with $\sum_k \pi(k) = 1$.

If the two classes are described by the same covariance matrix $\Sigma_1 = \Sigma_2$, then after taking the logarithm of Eq. (17.4) the quadratic terms in x cancel and what remains is a (hyper-) plane, that is, a straight line in two dimensions, a plane in three dimensions, etc. This property has inspired the name of *linear* discriminant analysis.

The estimation of the parameters involved is simple: the vectors μ_k are given by the arithmetic mean or center of mass of all the samples from class k

$$\mu_k = \frac{1}{N_k} \sum_{i=1}^{N_k} x_i^k, \tag{17.5}$$

where x_i^k is the ith sample out of a total of N_k examples of class k in the training set. The relative frequencies are simply given by $\pi_k = N_k / (N_1 + N_2)$.

The pooled covariance matrix is computed by

$$\Sigma = \frac{1}{N_1 + N_2} \left(\sum_{i=1}^{N_1} \left(x_i^1 - \mu_1 \right) \left(x_i^1 - \mu_1 \right)^T + \sum_{i=1}^{N_2} \left(x_i^2 - \mu_2 \right) \left(x_i^2 - \mu_2 \right)^T \right) \tag{17.6}$$

where the superscripts on x again indicate class membership.

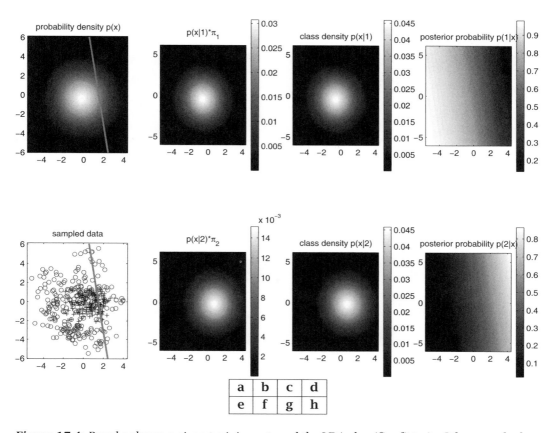

*Figure 17.4: Panel **e** shows a given training set, and the LDA classifier fit to it. Other panels show the estimated class densities, total density, and posterior densities, see description of Fig. 17.3 in text for details. Clearly, the linear classifier does not do the data justice in this example.*

We now illustrate linear discriminant analysis on two training sets: Fig. 17.3 shows what kind of data LDA is suitable for (in fact, the data has in this case been sampled from the model), while Fig. 17.4 illustrates a case for which LDA is not appropriate.

Starting with Fig. 17.3, panels **c** and **g** give two isotropic class densities; these are identical except for their location. Panels **b** and **f** show these densities multiplied by their prior probabilities which were chosen as $\pi(1) = \pi(2) = 0.5$. Panel **a** illustrates the total density

$$p(\boldsymbol{x}) = p(\boldsymbol{x}|1)\pi(1) + p(\boldsymbol{x}|2)\pi(2). \tag{17.7}$$

Panels **d** and **h** give the posterior probabilities of the classes at each point in this two-dimensional feature space; they can be computed from the information in the other panels by application of Bayes' theorem, Eqs. (17.3) and (17.4). The set of all points at which the two posterior probabilities are equally large is indicated by the straight line in panels **a** and **e**. This is the linear discriminant function. Since it was computed from the exact densities, it is in this case the best possible or Bayes classifier. Finally, panel **e** shows a set of points which have been sampled from this model. Note that there is significant overlap of the two classes and thus a significant number of prediction errors even for the best possible classifier. On an intuitive level, the discriminant function seems appropriate for the data.

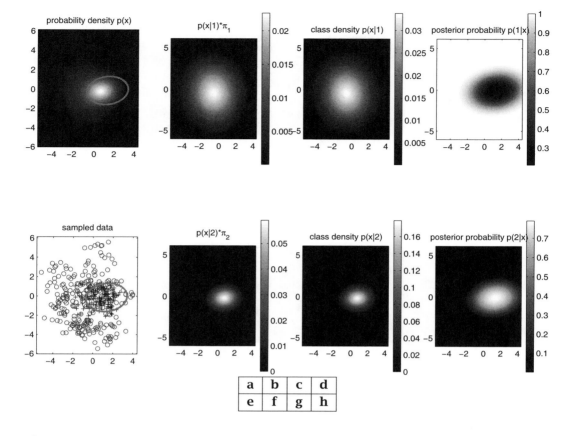

Figure 17.5: *Panel **e** shows the same training set as in Fig. 17.4, and the QDA classifier fit to it. Other panels show the estimated class densities, total density, and posterior densities. This nonlinear classifier offers a much improved fit to the data, and promises superior classification performance.*

Figure 17.4 shows a training set in panel **d**, along with the linear discriminant model obtained by estimating parameters according to Eqs. (17.5) and (17.6). In this example, the assumptions inherent in linear discriminant analysis do not match the data: one of the classes is much more compact, such that one covariance matrix Σ cannot adequately describe both classes. The discriminant function obtained is non-sensical. Apparently, we need to relax some of our assumptions in order to give the model more flexibility and allow for nonlinear classifiers.

17.4.2 Quadratic Discriminant Analysis (QDA)

As in linear discriminant analysis, each class is assumed to follow a multivariate normal distribution. The important difference is that the classes are now allowed to have different covariance matrices, and these are estimated separately for each class. As a consequence, the term quadratic in x does not cancel anymore in Eq. (17.4) and the resultant discriminant functions are quadrics, i.e. parabolas, hyperbolas, etc. Figure 17.5 shows an example.

Since *QDA* offers added flexibility and contains LDA as a special case, why not use it always? We had asked a similar question in Section 17.3.2, and our answer here is similar: QDA requires determination of more parameters, and this endeavor becomes difficult if the training set is small compared to the dimensionality P of the feature space. Which of these two should then be used? As always, cross-validation (see Section 17.4.4) offers a convenient answer. Let us add that it is possible to "compromise" [51] between LDA and QDA by using a covariance matrix of the form

$$\Sigma_k^{\text{mod}}(\alpha) = \alpha\Sigma_k + (1 - \alpha)\Sigma, \quad 0 \le \alpha \le 1, \tag{17.8}$$

where Σ_k are estimates of the class-covariance matrices and Σ is an estimate of the pooled covariance matrix, cf. Eq. (17.6).

If even linear discriminant analysis has too many parameters compared to the size of the training set, further simplifications are possible: all off-diagonal elements of the covariance matrix may be set to zero, or the entire covariance matrix may be biased towards sphericity using

$$\Sigma^{\text{mod}}(\beta) = \beta\Sigma + \sigma^2(1 - \beta)\boldsymbol{I}, \quad 0 \le \beta \le 1 \tag{17.9}$$

with $\boldsymbol{I}$ the unit matrix and σ^2 the variance of the data [51]. Again, an optimal value of β can be found using cross-validation. This modification of the estimated covariance may be necessary in the case of very high-dimensional data. For the sake of argument, assume that you compute a large number P of features which are thought to be relevant. N examples in feature space can only span an $(N-1)$-dimensional subspace. If $N-1 < P$, then the covariance matrix becomes singular and a stabilization is required.

17.4.3 *k*-Nearest Neighbors (*k*-NN)

The methods discussed above make strong assumptions concerning the distribution of the data, and yield succinctly formulated models. *k-nearest neighbor methods* lie at the other extreme in that they make very little assumptions, but are difficult to formulate in a concise manner: these memory-based methods "grow" with the size of the training set.

The basic idea is simple: for each new sample which should be classified, find that example from the training set which lies closest in feature space. Give the label of that closest sample as prediction.

An obvious generalization is to search not only for the closest sample, but for the closest k^3 , k uneven, samples from the training set. A "democratic vote" then reveals the locally dominant class which is given as prediction. Taking this vote with a larger number of neighbors gives the result with greater certainty (the change in the classifier when going from one training set to another will be smaller), but it also involves averaging over a larger region, making the discrimination surface smoother, and maybe overly so. A suitable value of k can again be found by cross-validation; see Section 17.4.4.

It may appear as if the method had only one parameter, k. This is far from the truth: the effective number of parameters is much larger, up to N/k [51]. One justification of this view is to consider 1-NN: the training set partitions the feature space into influence regions of each sample (these are the Voronoi regions). Each sample transmits its class membership to that region, thus each individual region has one parameter.

Since Euclidean distance (or its square, which is computationally cheaper) is used for identifying the k nearest neighbors, proper scaling of the feature axes is vital: if the units on one axis are converted, say, from meters to millimeters, the classification

---

^{3}In previous sections, k was a class index; now it has become an integer denoting the number of nearest neighbors that are used for prediction.

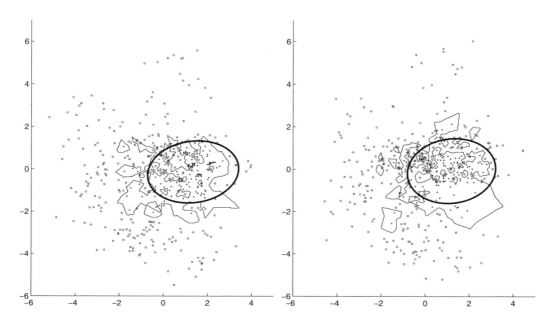

Figure 17.6: *The left panel shows the same training set as in Figs. 17.4 and 17.5. The right panel shows another training set which was sampled from the same underlying distribution. The ellipses are from a quadratic discriminant analysis on these data sets; the contours come from a 1-NN analysis. The two ellipses are quite similar, whereas the 1-NN classifier varies a lot from one training set to the next. This is what statisticians call the* variance, *and the severity of this undesirable feature generally increases with the flexibility of a classifier. In turn, overly stiff classifiers as illustrated in Fig. 17.4 cannot always do the data justice, leading to large* bias *[51].*

results may change drastically. This is in contrast to LDA which can correct affine distortions of the entire training set, such as scaling, "by itself" through the estimation of the covariance structure.

Among the advantages of k-NN are its flexibility and ease of use; disadvantages are the computational burden in routine use if the available training set is large, and the difficult interpretation of this "model".

Overall, the classification schemes presented here, though simple, are among the most performant [24]. If results prove unsatisfactory, advanced methods such as support vector machines [11, 124] may be applied, but generally speaking, no great leap in performance should be expected. In many practical situations, improper selection, evaluation or transformation (taking the logarithm, the inverse, etc.) of features is the weakest link in the chain and should be amended first.

17.4.4 Cross-Validation

Cross-validation can be used to estimate the predictive power of different methods when there is not enough data to split it in the way indicated in Fig. 17.1. In this context, LDA, QDA, 5-NN, and 7-NN would all qualify as "different methods", and so would LDA based on two different sets of features (one of which may be a subset of the other). The complete available database is split randomly[4] into K fractions of equal size. A specific method is then parameterized using all but one of these fractions as

---

[4]It is important to really choose samples randomly to avoid biasing by temporal trends.

the training set. The classifier is then used to predict the labels of the samples in the fraction that had been left out, and the predicted labels are compared to the true ones. The error rate is stored. This procedure of taking one subset out while training on all others is repeated for each of the K fractions, and the mean error rate is computed. The method with the smallest mean error rate can be expected to yield the greatest prediction performance.

There is a trade-off concerning the size of the fractions: on one hand, the smallest possible amount of data should be left out in order not to "waste" it; on the other hand, leaving out a small number of samples corresponds to a small perturbation of the complete training set only. This in turn means that results may differ if a totally different training set from the same process is used. A good compromise seems to be to choose K of the order $5, \ldots, 10$. Another frequent choice is $K = N$, the number of samples, which is also called "leave-one-out-cross-validation".

There are many variations on this theme, summarized under the term *resampling methods*. One enjoying much popularity recently is the *bootstrap* [51] in which the fractions are chosen by sampling from the complete training set with replacement.

In summary, cross-validation has a number of beneficial properties which were not discussed here, and is easy to implement and use. While often computationally prohibitive in the past, it is a method that merits its popularity nowadays.

17.5 Advanced Reference Material

References to advanced topics

17.1

R. O. Duda and P. E. Hart and D. G. Stork, 2000. Pattern Classification, John Wiley & Sons, New York, USA. *Verbose and highly recommendable for absolute beginners; also features a very well-structured introduction to clustering methods.*

T. Hastie and R. Tibshirani and J. Friedman, 2001. The Elements of Statistical Learning, Springer, New York, USA. *Puts more emphasis on recent developments. A fine book, even though its derivations are sometimes not very detailed.*

B. D. Ripley, 1997. Pattern Recognition and Neural Networks, Cambridge University Press, Cambridge, UK. *A fine introduction to the statistical foundations of pattern recognition, requires more background knowledge than the previous two.*

Part V

Appendices

A Notation

A.1 General

Symbols	Description
$a, b, \ldots$	Italic: *scalars*
$\boldsymbol{k}, \boldsymbol{u}, \boldsymbol{x}, \ldots$	Lowercase italic bold: *vectors*
$G, H, J, \ldots$	Uppercase italic bold: matrices, tensors, i.e., discrete images (there are some exceptions from this rule, see below)
$\mathbf{G}$	Uppercase bold: vectorial (multichannel) discrete image, *matrix, tensor*
$\bar{k}, \bar{\boldsymbol{n}}, \ldots$	A bar indicates a *unit vector*
$\tilde{k}, \tilde{\boldsymbol{k}}, \tilde{\boldsymbol{x}}, \ldots$	A tilde indicates a *dimensionless* or *normalized* quantity
$\hat{G}_{u,v}, \hat{\boldsymbol{G}}, \ldots$	A hat indicates a quantity in the *Fourier domain*
$\mathcal{B}, \mathcal{R}, \ldots$	Calligraphic letters indicate representation-independent *operators*
$\boldsymbol{x} = [x, y]^T = [x_1, x_2]^T$	Image coordinates in the spatial domain
$\boldsymbol{k} = \left[k_x, k_y\right]^T = [k_1, k_2]^T$	Image coordinates in the Fourier domain
$\boldsymbol{X} = [X, Y, Z]^T = [X_1, X_2, X_3]^T$	World coordinates
K, L, M, N	Extension of images in t, z, y, and x directions
k, l, m, n	Indices of images in t, z, y, and x directions
k', l', m', n'	Indices in sums of images in t, z, y, and x directions
r, s, u, v	Indices of images in Fourier domain in t, z, y, and x directions
r	Size of masks for neighborhood operators
p, q	Index for component in multichannel image or a quantization level
P	Number of components in a multichannel image
Q	Number of quantization levels
w	Index for a coordinate in a W-dimensional image
W	Dimension of an image

A.2 Image Operators

Symbols	Description
·	Pointwise multiplication of two images
$*$	Convolution
$\star$	Correlation
$\ominus, \oplus$	Morphological erosion and dilation operators
$\circ, \bullet$	Morphological opening and closing operators
$\otimes$	Morphological hit-miss operator
$\vee, \wedge$	Boolean *or* and *and* operators
$\cup, \cap$	Union and intersection of sets
$\subset, \subseteq$	Set is subset, subset or equal
$\circlearrowright$	Shift operator
$\Downarrow_s$	Sample or reduction operator: take only every sth pixel, row, etc.
$\Uparrow_s$	Expansion or interpolation operator: increase resolution in every coordinate direction by a factor of s, the new points are interpolated from the available points
$\boldsymbol{B}, \hat{b}(\tilde{\boldsymbol{k}}), \mathcal{B}$	Binomial smoothing operator (mask, transfer function, operator)
$\boldsymbol{C}, \hat{c}(\tilde{\boldsymbol{k}}), \mathcal{C}$	Curvature operator, second-order difference (derivative) operator
$\boldsymbol{D}, \hat{d}(\tilde{\boldsymbol{k}}), \mathcal{D}$	First-order difference (derivative) operator
$\mathcal{F}$	Fourier transform
$\mathcal{F}^{-1}$	Inverse Fourier transform
$\boldsymbol{H}, \hat{h}(\tilde{\boldsymbol{k}}), \mathcal{H}$	Hilbert operator
$\mathcal{I}$	Identity operator
$\boldsymbol{L}, \hat{l}(\tilde{\boldsymbol{k}}), \mathcal{L}$	Laplacian operator
$\boldsymbol{R}, \hat{r}(\tilde{\boldsymbol{k}}), \mathcal{R}$	Box operator
$\boldsymbol{U}, \hat{u}(\tilde{\boldsymbol{k}}), \mathcal{U}$	Elementary recursive lowpass (relaxation) operator
$\boldsymbol{V}, \hat{v}(\tilde{\boldsymbol{k}}), \mathcal{V}$	Elementary recursive highpass operator
$\boldsymbol{Q}, \hat{q}(\tilde{\boldsymbol{k}}), \mathcal{Q}$	Elementary recursive bandpass (resonance) operator
$\mathcal{H}_x$	1-D operator is applied in x direction
$\mathcal{H}_{2x}$	Operator is applied in x direction with double step width
$\mathcal{H}_y \mathcal{H}_x$	1-D operator is applied first in x direction and then in y direction
$\mathcal{H}$	If operator is a 1-D separable operator, it is applied in all directions
$\mathcal{H}^r \underbrace{\mathcal{H}\mathcal{H}\ldots\mathcal{H}}_{r\ \text{times}}$	Operator is applied r times
$^r\mathcal{H}$	A leading superscript indicates a variant of an operator, e. g., the number of coefficients, etc.
$\mathcal{H}^{(p)}$	A superscript in parenthesis specifies the level of a multi-grid data structure (pyramid) at which the operator $\mathcal{H}$ is applied

A.3 Alphabetical List of Symbols and Constants

Symbol	Definition, Value, [Units]	Meaning	Chapter
$\cdot$		Pointwise multiplication of two images	10
$*$		Convolution	10
$\star$		Correlation	10
$\ominus, \oplus$		Morphological erosion and dilation operators	16
$\circ, \bullet$		Morphological opening and closing operators	16
$\otimes$		Morphological hit-miss operator	16
$\vee, \wedge$		Boolean *or* and *and* operators	16
$\cup, \cap$		Union and intersection of sets	16
$\subset, \subseteq$		Set is subset, subset or equal	16
$\circlearrowleft$		Shift operator	10
$\Downarrow_s$		Sample or reduction operator: take only every sth pixel, row, etc.	11
$\Uparrow_s$		Expansion or interpolation operator: increase resolution in every coordinate direction by a factor of s, the new points are interpolated from the available points	11
α	$[\text{m}^{-1}]$	Absorption coefficient	3
$[\alpha]$	$[\text{m}^2/\text{mol}], [\text{cm}^2/\text{g}]$	Specific rotation for an optically active material	3
β	$[\text{m}^{-1}]$	Scattering coefficient	3
β_{ad}		Adiabatic compressibility	3
δ		δ distribution	B
Δ	$\displaystyle\sum_{w=1}^{W} \frac{\partial^2}{\partial x_w^2}$	Laplacian operator	3
ϵ	$[1]$	Specific emissivity	3
ϵ	$[\text{m}]$	Radius of blur disk	3
κ	$[\text{m}^{-1}]$	Extinction coefficient, sum of absorption and scattering coefficient	3
∇	$\left[\dfrac{\partial}{\partial x_1}, \ldots, \dfrac{\partial}{\partial x_W}\right]^T$	Gradient operator	3
λ	$[\text{m}]$	Wavelength	3
ν	$[1/\text{s}], [\text{Hz}]$	Frequency	3
$\nabla\times$		Rotation operator	3
η	$n + \mathrm{i}\xi, [1]$	Complex index of refraction	3
η	$[1]$	Quantum efficiency	5
ϕ	$[\text{rad}], [°]$	Phase shift, phase difference	3
ϕ_e	$[\text{rad}], [°]$	Azimuth angle	3
Φ	$[\text{J/s}], [\text{W}], [\text{s}^{-1}], [\text{lm}]$	Radiant or luminous flux	3
Φ_e	$[\text{J/s}], [\text{W}]$	Energy-based radiant flux	3
Φ_p	$[\text{s}^{-1}]$	Photon-based radiant flux	3
Φ_v	$[\text{lm}]$ (lumen)	Luminous flux	3
ρ	$[1]$	Reflectivity for unpolarized light	3

continued on next page

Symbol	Definition, Value, [Units]	Meaning	Chapter
continued from previous page			
$\rho_\parallel$	[1]	Reflectivity for parallel polarized light	3
$\rho_\perp$	[1]	Reflectivity for perpendicularly polarized light	3
ρ	[kg/m^3]	Density	3
σ		Standard deviation, σ^2 variance	7
σ	$5.6696 \cdot 10^{-8}$ Wm^{-2}K^{-4}	Stefan-Boltzmann constant	3
σ_s	[m^2]	Scattering cross section	3
Σ		Covariance matrix	17
τ	[1]	Optical depth (thickness)	3
τ	[1]	Transmissivity	3
τ	[s]	Time constant	10
θ	[rad], [°]	Angle of incidence	3
θ_b	[rad], [°]	Brewster angle (polarizing angle)	3
θ_c	[rad], [°]	Critical angle (for total reflection)	3
θ_e	[rad], [°]	Polar angle	3
θ_i	[rad], [°]	Angle of incidence	3
Ω	[sr] (steradian)	Solid angle	3
ω	[s^{-1}], [Hz]	Circular frequency	4
$\boldsymbol{B}$	[Vs/m^2]	Magnetic field	3
$\boldsymbol{B}$		Binomial filter mask	11
$\hat{\boldsymbol{B}}, \hat{B}(\tilde{\boldsymbol{k}})$		Transfer function of binomial mask	11
$\mathcal{B}$		Binomial convolution operator	11
c	$2.9979 \cdot 10^8$ ms^{-1}	Speed of light	3
d	[m]	Diameter (aperture) of optics	3
d_i	[m]	Image distance	4
D	[m^2/s]	Diffusion coefficient	11
$\boldsymbol{D}$		First-order difference filter mask	11
$\hat{\boldsymbol{D}}, \hat{D}$		Transfer function of $\boldsymbol{D}$	11
$\mathcal{D}$		First-order difference operator	11
E	[W/m^2], [lm/m^2], [lx]	Radiant (irradiance) or luminous (illuminance) incident energy flux density	3
$\boldsymbol{E}$	[V/m]	Electric field	3
e	$1.6022 \cdot 10^{-19}$ As	Elementary electric charge	3
$\bar{\boldsymbol{e}}$	[1]	Unit eigenvector of a matrix	14
f, f_e	[m]	(Effective) focal length of an optical system	3
f_b	[m]	Back focal length of an optical system	3
f_f	[m]	Front focal length of an optical system	3
$\boldsymbol{f}$	[m/s]	Optical flow	13
h	$6.6262 \cdot 10^{-34}$ Js	Planck's constant (action quantum)	3
$\hbar$	$h/(2\pi)$ [Js]		3
i	$\sqrt{-1}$	Imaginary unit	3
I	[W/sr], [1/(s sr)], [lm/sr]	Radiant or luminous intensity	3
I	[A]	Electric current	5
$\boldsymbol{I}$		Identity matrix	4
$\mathcal{I}$		Identity operator	10

continued on next page

Symbol	Definition, Value, [Units]	Meaning	Chapter
	continued from previous page		
$\boldsymbol{J}$		Structure tensor	10
k_B	$1.3806 \cdot 10^{-23}$ J/K	Boltzmann constant	3
k	$1/\lambda$, $[\mathrm{m}^{-1}]$	Magnitude of wave number	3
$\boldsymbol{k}$	$[\mathrm{m}^{-1}]$	Wave number	3
$\tilde{k}$	$k\Delta x/\pi$	Wave number normalized to the maximum wave number that can be sampled (Nyquist wave number)	3
$\bar{k}$	$1/\lambda$, $[\mathrm{m}^{-1}]$	Wave number divided by 2π	3
K_q	[l/mol]	Quenching constant	6
K_r	Φ_v/Φ_e, [lm/W]	Radiation luminous efficiency	3
K_s	Φ_v/P [lm/W]	Lighting system luminous efficiency	3
K_I	[1]	Indicator equilibrium constant	3
L	$[\mathrm{W/(m^2 sr)}]$, $[1/(\mathrm{m}^2\mathrm{sr})]$, $[\mathrm{lm/(m^2 sr)}]$, $[\mathrm{cd/m^2}]$	Radiant (radiance) or luminous (luminance) flux density per solid angle	3, 4
M	$[\mathrm{W/m^2}]$, $[1/(\mathrm{s\,m^2})]$	Excitant radiant energy flux density (excitance, emittance)	3
M_e	$[\mathrm{W/m^2}]$	Energy-based excitance	3
M_p	$[1/(\mathrm{s\,m^2})]$	Photon-based excitance	3
m	[1]	Magnification of an optical system	4
n	[1]	Index of refraction	3, 4
n_a	[1]	Numerical aperture of an optical system	4
n_f	f/d, [1]	Aperture of an optical system	4
$\bar{\boldsymbol{n}}$	[1]	Unit vector normal to a surface	
p	[kg m/s], [W m]	Momentum	3
p	$[\mathrm{N/m^2}]$	Pressure	3
p		Probability density function	7
P		Distribution function	7
pH	[1]	pH value, negative logarithm of proton concentration	3
Q	[Ws] (Joule), [lm s] number of photons	Radiant or luminous energy	3
Q_s	[1]	Scattering efficiency factor	3
r	[m]	Radius	3
R	Φ/s, [A/W]	Responsivity of a radiation detector	5
$\boldsymbol{S}$	$[\mathrm{W/m^2}]$	Poynting vector	3
s	[A]	Sensor signal	3
T	[K]	Absolute temperature	3
t	[s]	Time	4
t	[1]	Transmittance	4
u	[m/s]	Velocity	3, 13
$\boldsymbol{u}$	[m/s]	Velocity vector	3, 13
$V(\lambda)$	[lm/W]	Spectral luminous efficacy for photopic human vision	3
$V'(\lambda)$	[lm/W]	Spectral luminous efficacy for scotopic human vision	3

B Mathematical Toolbox

B.1 Matrix Algebra

B.1.1 Vectors and Matrices

An ordered set of M elements such as a time series of some data is known as a *vector* and written as

$$\boldsymbol{g} = \begin{bmatrix} g_0 \\ g_1 \\ \vdots \\ g_{M-1} \end{bmatrix} = [g_0 \ g_1 \ \cdots \ g_{M-1}]^T . \tag{B.1}$$

The nth element of the vector $\boldsymbol{g}$ is denoted by g_n. In image processing, we require vectors with real-valued and complex-valued elements. A *matrix* $\boldsymbol{G}$ of size $M \times N$ is a double-indexed ordered set with MN elements

$$\boldsymbol{G} = \begin{bmatrix} g_{0,0} & g_{0,1} & \cdots & g_{0,N-1} \\ g_{1,0} & g_{1,1} & \cdots & g_{1,N-1} \\ \vdots & \vdots & \ddots & \vdots \\ g_{M-1,0} & g_{M-1,1} & \cdots & g_{M-1,N-1} \end{bmatrix} . \tag{B.2}$$

The matrix consists of M *rows* and N *columns*. The element in row m and column n of the matrix $\boldsymbol{G}$ is denoted by $g_{m,n}$. The first and second indices of a matrix element denote the row and column numbers, i. e., the y and x coordinates, respectively. Thus the matrix notation differs from the standard Cartesian coordinate representation by a clockwise 90° rotation.

Discrete images will be represented in this book by matrices with the starting indices 0 as shown above. All vectors and matrices which are related to coordinates in the two- and three-dimensional space will start with the index of 1. For a diagonal element of a matrix, the first and second index are equal.

In image processing, we require vectors and matrices with integers, real numbers, and complex numbers as elements. An $M \times 1$ matrix is also denoted as a *column vector*, $\boldsymbol{g}$, and a $1 \times N$ matrix as a *row vector*, $\boldsymbol{g}^T$. Note that generally vectors Eq. (B.1) are regarded as column vectors. A row vector is written as

$$\boldsymbol{g} = [g_0 \ g_1 \ \cdots \ g_{M-1}] . \tag{B.3}$$

B.1.2 Operations with Vectors and Matrices

Some important operations with vectors and matrices are summarized in the following table:

Operation	Definition
Transposition	$(G^T)_{mn} = g_{nm}$
Matrix multiplication $\underbrace{G'}_{M \times N} = \underbrace{G}_{M \times K} \underbrace{H}_{K \times N}$	$G'_{mn} = \sum_{k=0}^{K-1} g_{mk} h_{kn}$
Inner vector or scalar product gh	$\sum_{m=0}^{M-1} g_m h_m$
Outer vector product $g \otimes h$	$g_{mn} = g_m h_n$
Trace[1] trace(G)	$\sum_{m=0}^{M-1} g_{mm}$
Inverse matrix[1] G^{-1}	$G^{-1}G = I$
Eigenvalues[1] λ	$Gx = \lambda x$
Eigenvectors[1] to eigenvalue λ	all $x \neq 0$ with $(G - \lambda I)x = 0$

[1]Only defined for square matrices (see table below)

B.1.3   Types of Matrices

Here some often used special types of matrices are summarized. The superscript $*$ denotes the complex conjugate for a scalar or a matrix.

Name	Definition
Square matrix	$M = N$
Diagonal square matrix	$g_{mn} = 0 \ \forall \ m \neq n$
Identity matrix I	$i_{m,n} = \delta_{m-n} = \begin{cases} 1 & m = n \\ 0 & \text{else} \end{cases}$
Symmetric matrix	$G = G^T, g_{mn} = g_{nm}$
Hermitian matrix	$G = G^*, g_{mn} = g_{nm}^*$
Orthogonal matrix	$G^{-1} = G^T$
Unitary matrix	$G^{-1} = G^*$

B.2   Least-Squares Solution of Linear Equation Systems

Overdetermined linear equation systems occur at many places in image processing: the optimum design of filters (Sections 8.3.2d and 12.4.1d), reconstruction from projections, determination of coordinate transforms from point correspondences (Sections 8.3.1b and 8.3.1c), and with various techniques to determine the optical flow [73]. Here we derive the general solution. Given is a set of measurements collected in an Q-dimensional *data vector* d and an $P \times Q$ *design matrix* G which relates the measured data to the P unknown *model parameters* in the *parameter vector* m:

$$d = Gm. \tag{B.4}$$

Generally for $Q > P$, no exact solution is possible but only an estimated solution which minimizes the norm of the *error vector*, e, which is the difference between the measured data, d, and the data predicted by the model, d_{pre},

$$e = d - d_{\text{pre}} = d - Gm_{\text{est}}, \tag{B.5}$$

$$||e||_2^2 = \sum_{q'=1}^{Q} \left(d_{q'} - \sum_{p'=1}^{P} g_{q'p'} m_{p'} \right)^2 .$$

Factorizing the sum and interchanging of the two summations yields

$$||e||_2^2 = \underbrace{\sum_{p'=1}^{P} \sum_{p''=1}^{P} m_{p'} m_{p''} \sum_{q'=1}^{Q} g_{q'p'} g_{q'p''}}_{A}$$

$$- \underbrace{2 \sum_{p'=1}^{M} m_{p'} \sum_{q'=1}^{N} g_{q'p'} d_{q'}}_{B}$$

$$+ \sum_{q'=1}^{Q} d_{q'} d_{q'} .$$

We find a minimum for this expression by computing the partial derivatives with respect to the parameters m_p to be optimized. Only the expressions A and B depend on m_p:

$$\frac{\partial A}{\partial m_p} = \sum_{p'=1}^{P} \sum_{p''=1}^{P} \left(\delta_{p'p} m_p' + \delta_{p''p} m_p'' \right) \sum_{q'=1}^{Q} g_{q'p'} g_{q'p''}$$

$$= \sum_{p'=1}^{P} m_p' \sum_{q'=1}^{Q} g_{q'p'} g_{q'p} + \sum_{p''=1}^{P} m_p'' \sum_{q'=1}^{Q} g_{q'p''} g_{q'p''}$$

$$= 2 \sum_{p'=1}^{P} m_p' \sum_{q'=1}^{Q} g_{q'p'} g_{q'p}$$

$$\frac{\partial B}{\partial m_p} = 2 \sum_{q'=1}^{Q} g_{q'p} d_{q'} .$$

We add both derivatives and set them equal to zero:

$$\frac{\partial ||e||_2^2}{\partial m_p} = \sum_{p'=1}^{P} m_p' \sum_{q'=1}^{Q} g_{q'p} g_{q'p'} - \sum_{q'=1}^{Q} g_{q'p} d_q' = 0.$$

In order to obtain matrix-matrix and matrix-vector multiplications, we substitute the matrix G at two places by its transpose G^T:

$$\sum_{p'=1}^{P} m_p' \sum_{q'=1}^{Q} g_{pq'}^T g_{q'p'} - \sum_{q'=1}^{Q} g_{pq'}^T d_q' = 0$$

and finally obtain the matrix equation

$$\underbrace{\underbrace{G^T}_{P \times Q} \underbrace{G}_{Q \times P} \underbrace{m_{\text{est}}}_{P}}_{P \times P} = \underbrace{\underbrace{G^T}_{P \times Q} \underbrace{d}_{Q}}_{P} . \tag{B.6}$$

This equation can be solved if the quadratic and symmetric $P \times P$ matrix $G^T G$ is invertible. Then

$$m_{\text{est}} = \left(G^T G\right)^{-1} G^T d. \tag{B.7}$$

The matrix $(G^T G)^{-1} G^T$ is known as the *pseudo inverse* or *generalized inverse* of G.

B.3 Fourier Transform

B.3.1 Definition

In one dimension, the *Fourier transform* of a complex-valued function $f(x)$ is defined as

$$\hat{g}(k) = \int_{-\infty}^{\infty} g(x) \exp(-2\pi \mathrm{i} kx) \mathrm{d}x, \tag{B.8}$$

where $k = 1/\lambda$ is the *wave number* of the complex exponential $\exp(-2\pi \mathrm{i} kx)$ with the wavelength λ. The back transformation is given by

$$g(x) = \int_{-\infty}^{\infty} \hat{g}(k) \exp(2\pi \mathrm{i} kx) \mathrm{d}k. \tag{B.9}$$

A function $g(x)$ and its Fourier transform $\hat{g}(k)$ form a Fourier transform pair denoted by

$$g(x) \circ\!\!-\!\!\bullet \hat{g}(k). \tag{B.10}$$

The complex exponentials, the *kernel* of the Fourier transform, constitute an orthonormal basis

$$\int_{-\infty}^{\infty} \exp(-2\pi \mathrm{i} k'x) \exp(2\pi \mathrm{i} kx) \mathrm{d}x = \delta(k' - k). \tag{B.11}$$

The W-dimensional Fourier transform is defined by

$$\hat{g}(\boldsymbol{k}) = \int_{-\infty}^{\infty} g(\boldsymbol{x}) \exp(-2\pi \mathrm{i} \boldsymbol{k}\boldsymbol{x}) \mathrm{d}^W x. \tag{B.12}$$

The kernel of the multidimensional Fourier transform is separable:

$$\exp(-2\pi \mathrm{i} \boldsymbol{k}\boldsymbol{x}) = \prod_{w=1}^{W} \exp(-2\pi \mathrm{i} k_w x_w). \tag{B.13}$$

Therefore the W-dimensional Fourier transform can be separated in W one-dimensional Fourier transforms. For example, the two-dimensional Fourier transform can be written as

$$\hat{g}(\boldsymbol{k}) = \int_{-\infty}^{\infty} \left[\int_{-\infty}^{\infty} g(\boldsymbol{x}) \exp(-2\pi \mathrm{i} k_1 x_1) \mathrm{d}x_1 \right] \exp(-2\pi \mathrm{i} k_2 x_2) \mathrm{d}x_2. \tag{B.14}$$

The W-dimensional inverse Fourier transform is defined by

$$g(\boldsymbol{x}) = \int_{-\infty}^{\infty} \hat{g}(\boldsymbol{k}) \exp(\mathrm{i} \boldsymbol{k}\boldsymbol{x}) \mathrm{d}^W k. \tag{B.15}$$

B.3.2 Properties of the Fourier Transform

The theorems summarized here are valid for the Fourier transform in W dimensions. Let $g(x)$ and $h(x)$ be complex-valued functions, the Fourier transforms of which $\hat{g}(k)$ and $\hat{h}(k)$ do exist. Let a and b be complex-valued constants, s a real-valued constant, and R an orthogonal rotation matrix.

Property	Space domain	Fourier domain
Linearity	$ag(x) + bh(x)$	$a\hat{g}(k) + b\hat{h}(k)$
Scaling	$g(sx)$	$\hat{g}(k/s)/\lvert s\rvert$
Rotation	$g(Rx)$	$\hat{g}(Rk/s)/\lvert s\rvert$
Separability	$\displaystyle\prod_{i=1}^{W} g(x_w)$	$\displaystyle\prod_{i=1}^{W} \hat{g}(k_w)$
Shifting	$g(x - x_0)$	$\exp(-2\pi i k x_0)\hat{g}(k)$
Modulation	$\exp(2\pi i k_0 x)g(x)$	$\hat{g}(k - k_0)$
Derivation	$\dfrac{\partial^p g(x)}{\partial x_w^p}$	$(2\pi i k_w)^p \hat{g}(k)$
Derivation	$(-2\pi i x_w)^p g(x)$	$\dfrac{\partial^p \hat{g}(k)}{\partial k_w^p}$
Moments	$\displaystyle\int_{-\infty}^{\infty} x_p^m x_q^n g(x')\mathrm{d}^W x'$	$\left(\dfrac{i}{2\pi}\right)^{m+n} \dfrac{\partial^{m+n} \hat{g}(\tilde{k})}{\partial k_p^m \partial k_q^n}\Big\rvert_0$
Convolution	$\displaystyle\int_{-\infty}^{\infty} g(x')h(x - x')\mathrm{d}^W x'$	$\hat{h}(k)\hat{g}(k)$
Multiplication	$g(x)h(x)$	$\displaystyle\int_{-\infty}^{\infty} \hat{g}(k')\hat{h}(k - k')\mathrm{d}^W k'$
Correlation	$\displaystyle\int_{-\infty}^{\infty} g(x')h(x' + x)\mathrm{d}^W x'$	$\hat{g}(k)\hat{h}^*(k)$
Inner product	$\displaystyle\int_{-\infty}^{\infty} g(x)h^*(x)\mathrm{d}^W x$	$\displaystyle\int_{-\infty}^{\infty} \hat{g}(k)\hat{h}^*(k)\mathrm{d}^W k$
Norm, energy conservation	$\displaystyle\int_{-\infty}^{\infty} \lVert g(x)\rVert_2^2\, \mathrm{d}^W x$	$\displaystyle\int_{-\infty}^{\infty} \lVert \hat{g}(k)\rVert_2^2\, \mathrm{d}^W k$

The following table lists important symmetry properties of the Fourier transform:

Space domain	Fourier domain
Even, odd $g(-x) = \pm g(x)$	Even, odd $\hat{g}(-k) = \pm\hat{g}(k)$
Real $g(x) = g^*(x)$	Hermitian $\hat{g}(-k) = \hat{g}^*(k)$
Imaginary $g(x) = -g^*(x)$	Antihermitian $\hat{g}(-k) = -\hat{g}^*(k)$
Rotational symmetric $g(\lvert x\rvert)$	Rotational symmetric $\hat{g}(\lvert k\rvert)$

B.3.3 Important Fourier Transform Pairs

Space domain	Fourier domain
δ function $\delta(x)$	const., 1
Cosine function $\cos(2\pi k_0 x)$	$\frac{1}{2}\left(\delta(k-k_0)+\delta(k+k_0)\right)$
Sine function $\sin(2\pi k_0 x)$	$\frac{i}{2}\left(\delta(k-k_0)-\delta(k+k_0)\right)$
δ comb $\sum_{n=-\infty}^{\infty}\delta(x-n\Delta x)$	$\sum_{u=-\infty}^{\infty}\delta(k-u/\Delta x)$
Box function $\Pi(x) = \begin{cases} 1 & \|x\| \le 1/2 \\ 0 & \|x\| > 1/2 \end{cases}$	$\operatorname{sinc}(k) = \dfrac{\sin(\pi k)}{\pi k}$
2-D disk $\frac{1}{\pi}\Pi(\boldsymbol{x}) = \begin{cases} 1 & \|\boldsymbol{x}\| \le 1/2 \\ 0 & \|\boldsymbol{x}\| > 1/2 \end{cases}$	$\dfrac{J_1(\pi\|\boldsymbol{k}\|)}{\pi\|\boldsymbol{k}\|/2}$
Signum $\operatorname{sgn}(x) = \begin{cases} 1 & x \ge 0 \\ -1 & x < 0 \end{cases}$	$\dfrac{-i}{\pi k}$
Gaussian function $\exp\left(-\pi x^2\right)$	$\exp\left(-\pi k^2\right)$
$x\exp\left(-\pi x^2\right)$	$-ik\exp\left(-\pi k^2\right)$
$x\exp\left(-\|x\|\right)$	$\dfrac{2}{1+(2\pi k)^2}$

B.4 Discrete Fourier Transform (DFT)

B.4.1 Definition

The one-dimensional DFT maps a *complex-valued vector* $\boldsymbol{g}$ onto another vector $\hat{\boldsymbol{g}}$ of a vector space with the same dimension M:

$$\hat{g}_u = \frac{1}{M}\sum_{m=0}^{M-1} g_m \exp\left(-\frac{2\pi i m u}{M}\right) = \frac{1}{M}\sum_{m=0}^{M-1} g_m W_M^{-mu}, \tag{B.16}$$

where

$$W_M = \exp\left(\frac{2\pi i}{M}\right). \tag{B.17}$$

The back transformation is given by

$$g_m = \sum_{u=0}^{M-1} \hat{g}_u W_M^{mu}. \tag{B.18}$$

In two dimensions, the DFT maps a complex-valued $M \times N$ matrix onto another matrix of the same size:

$$\begin{aligned} \hat{g}_{uv} &= \frac{1}{MN}\sum_{m=0}^{M-1}\sum_{n=0}^{N-1} g_{mn} \exp\left(-\frac{2\pi i m u}{M}\right)\exp\left(-\frac{2\pi i n v}{N}\right) \\ &= \frac{1}{MN}\sum_{m=0}^{M-1}\left(\sum_{n=0}^{N-1} g_{mn} W_N^{-nv}\right) W_M^{-mu}. \end{aligned} \tag{B.19}$$

The inverse 2-D DFT is given by

$$g_{mn} = \sum_{u=0}^{M-1} \sum_{v=0}^{N-1} \hat{g}_{uv} W_M^{mu} W_N^{nv}. \tag{B.20}$$

B.4.2 Important Properties

The following theorems apply to the 2-D DFT. Let G and H be complex-valued $M \times N$ matrices, $\hat{G}$ and $\hat{H}$ their Fourier transforms, and a and b complex-valued constants.

Property	Space domain	Wave number domain
Mean	$\dfrac{1}{MN} \sum\limits_{m=0}^{M-1} \sum\limits_{n=0}^{N-1} g_{mn}$	$\hat{g}_{0,0}$
Linearity	$aG + bH$	$a\hat{G} + b\hat{H}$
Shifting	$g_{m-k,n-l}$	$W_M^{-ku} W_N^{-lv} \hat{g}_{uv}$
Modulation	$W_M^{-kp} W_N^{-lq} g_{m-k,n-l}$	$\hat{g}_{u-p,v-q}$
Finite difference	$(g_{m+1,n} - g_{m-1,n})/2$	$i\sin(2\pi u/M)\hat{g}_{uv}$
Finite difference	$(g_{m,n+1} - g_{m,n-1})/2$	$i\sin(2\pi v/N)\hat{g}_{uv}$
Convolution	$(G * H)_{mn} = \sum\limits_{k=0}^{M-1} \sum\limits_{l=0}^{N-1} g_{kl} h_{m-k,n-l}$	$MN\hat{g}_{uv}\hat{h}_{uv}$
Spatial correlation	$(G \star H)_{mn} = \sum\limits_{k=0}^{M-1} \sum\limits_{l=0}^{N-1} g_{kl} h_{m+k,n+l}$	$MN\hat{g}_{uv}\hat{g}_{uv}^{*}$
Multiplication	$g_{mn} h_{mn}$	$(\hat{G} * \hat{H})_{uv} = \sum\limits_{p=0}^{M-1} \sum\limits_{q=0}^{N-1} \hat{g}_{pq} \hat{h}_{u-p,v-q}$
Inner product	$\sum\limits_{m=0}^{M-1} \sum\limits_{n=0}^{N-1} g_{mn} h_{mn}^{*}$	$\sum\limits_{u=0}^{M-1} \sum\limits_{v=0}^{N-1} \hat{g}_{uv} \hat{h}_{uv}^{*}$
Norm, energy conservation	$\sum\limits_{m=0}^{M-1} \sum\limits_{n=0}^{N-1} |g_{mn}|^2$	$\sum\limits_{u=0}^{M-1} \sum\limits_{v=0}^{N-1} |\hat{g}_{uv}|^2$

B.4.3 Important Transform Pairs

Space domain	Fourier domain
δ function $\delta_{mn} = \begin{cases} 1 & m=0, n=0 \\ 0 & \text{else} \end{cases}$	$\dfrac{1}{MN}$
Constant function $c_{mn} = 1$	δ_{uv}
Cosine function $\cos\left(\dfrac{2\pi pm}{M} + \dfrac{2\pi qn}{N}\right)$	$\dfrac{1}{2}\left(\delta_{u-p,v-q} + \delta_{u+p,v+q}\right)$

B.5 Suggested Further Readings

B.1   **Matrix algebra**

Hoffman, K. and R. Kunze, 1971. Linear Algebra, 2nd ed., Prentice-Hall, Englewood Cliffs, NJ. *Classical textbook on linear algebra.*

Menke, W., 1989. Geophysical Data Analysis: Discrete Inverse Theory, Academic Press, San Diego. *An intuitive introduction to discrete inverse problems.*

Golub, G. H. and C. F. van Loan, 1989. Matrix Computations. Johns Hopkins Series in the Mathematical Sciences, No. 3, The Johns Hopkins University Press, Baltimore. *Detailed treatment of algorithms for matrix computations.*

Press, W. H., B. P. Flannery, S. A. Teukolsky, and W. T. Vetterling, 1992. Numeral Recipes in C: The Art of Scientific Computing, 2nd ed., Cambridge University Press, New York. *A very valuable and remarkably complete general resource to numerical algorithms.*

B.2   **Fourier transform**

Blahut, R., 1985. Fast Algorithms for Digital Signal Processing, Addison-Wesley, Reading, Mass. *Includes a detailed discussion of fast algorithms for the discrete Fourier transform.*

Bracewell, R., 1986. The Fourier Transform and Its Applications, 2nd edition, revised, McGraw-Hill, New York. *The classical reference to the continuous and discrete Fourier transform.*

Gonzalez, R. C. and P. Wintz, 1987. Digital Image Processing, 2nd ed., Addison-Wesley, Reading, MA. *Contains a good introduction to the Fourier transform with respect to image processing.*

Parker, J. A., 1990. Image Reconstruction in Radiology, CRC Press, Boca Raton, FL. *The title of the book is somewhat misleading. It is actually an excellent and easy to understand treatment of all basic mathematical topics for image processing.*

C Glossary

Abbe constant [V-value, ν-value]; *g.* **Abbesche Zahl:** constant that describes the ratio of the refractivity to the dispersion of an optical medium:

$$V_d = (n_d - 1)/(n_f - n_c).$$

n is the *index of refraction*. The indices indicate wavelengths. d: Helium line *d* (587.6 nm), *f* and *c*: hydrogen lines *C* and *F* (656.3 nm and 486.1 nm). Sometimes, slightly different wavelengths are used.

aberration; *g.* **Aberration, Bildfehler:** defects of an optical system which cause its image to deviate from the rules of *paraxial imagery*. ≻ *Astigmatism, chromatic aberration, coma, curvature of field, distortion, and spherical aberration.*

absolute calibration; *g.* **absolute Kalibrierung:** the task of determining the relationship of a sensor's output to a known input with traceable reference to known standards. An absolute calibration requires the traceability trail to an accepted standard and the repeatability accuracy for the entire test setup (including the instrument to be calibrated). Typically, radiometric calibration is only accurate to approximately 1 and 5 percent in the visible and the infrared, respectively.

absolute orientation; *g.* **absolute Orientierung:** *photogrammetry* term. The rotation and translation transform(s) by which one or more camera coordinate system(s) coincide(s) with the world coordinate system.

absolute temperature; *g.* **absolute Temperatur:** Temperature measured on the Kelvin scale, whose base is absolute zero temperature (-273.15° C) that can only be approached but never be reached. The ∼ is often used in *radiometry* and *thermography*.

absorptance; *g.* **Absorptionsvermögen:** ratio of the absorbed radiant or luminous flux to the incident flux under a given condition. The ∼ is not only a property of the material but depends also on a given setup. ≻ *absorption coefficient.*

absorption band; *g.* **Absorptionsband:** a range of wavelengths in the EM spectrum where a material absorbs EM energy incident upon it and, thus, attenuates the radiant flux.

absorption coefficient [absorptivity]; *g.* **Absorptionskoeffizient:** property of an optical medium indicating the internal absorptance α. Equal thicknesses absorb equal fractions of the radiant flux: $d\Phi/\Phi = -\alpha(\lambda)dx$ resulting in Beer's exponential law of absorption: $\Phi = \Phi_0 \exp(-\alpha x)$. Units: [1/m]. Compare with *absorptance.*

accuracy; *g.* **Genauigkeit:** related to the deviation of the measured value of a quantity from the (mostly unknown) true value. A high accuracy means a low deviation. The deviation is also known as the systematic error and caused by calibration errors and the influence of parameters that are not controlled by the experimental setup.

achromatic lens [achromat]; *g.* **Achromat:** a lens consisting of two elements, usually of crown and flint glass, that has been corrected for *chromatic aberration* with respect to two selected wavelengths.

active imaging; *g.* **aktives Sehsystem:** an imaging system based on the illumination of a scene by artificial radiation and the collection of the reflected energy returned to the system. Examples are *radar* and *lidar* systems and techniques using *structured light*.

active vision; *g.* **aktives Sehen:** a paradigm of *computer vision* that emphasizes vision as an active participant in the world. This includes tracking of objects of interest and any other adaptations of the vision system useful to solve a certain task in an optimum way. Active vision refers not to sensing technology (≻ *active imaging*) but to strategies for observation.

adaptive coding; *g.* **adaptive Kodierung:** the application of two or more image compression techniques to a single image, based on properties of different parts of the image.

adaptive filter; *g.* **adaptives Filter:** filter techniques where the filter applied depends in one or the other way on the neighborhood of a pixel. As such, a ∼ is a nonlinear operation.

~ techniques include, among others, normalized convolution and steerable filters.

ADC: *analog to digital converter*

affine transform; *g.* **affine Transformation:** a linear coordinate transformation which maps a triangle into a triangle by ($x' = Ax$), where A is an arbitrary 2×2 matrix. An ~ includes rotation, scaling, and shearing. It maps straight lines onto straight lines, parallel lines onto parallel lines, and circles onto ellipses.

AGC: *automatic gain control*

AI; *g.* **KI:** *artificial intelligence*

airy disk; *g.* **Airy Scheibe:** the central spot on a focal plane produced by a diffraction-limited optical system when viewing a point source.

albedo; *g.* **Albedo:** the ratio of the sun's total radiant energy reflected from a rough surface to that incident on it.

algebraic reconstruction: ≻ *image reconstruction*

aliasing; *g.* **Aliasing:** artifact in an image (or more generally any signal) caused by a sampling distance that is too coarse to preserve the wave number (frequency). ≻ *sampling theorem*

alpha radiation; *g.* **Alphastrahlung:** particulate radiation emitted by many radioactive elements and consisting of nuclei of helium (nucleus with two neutrons and two protons).

ALU: *arithmetic-logical unit*

ambient light; *g.* **Hintergrundbeleuchtung:** light which is persistent in the environment around a vision system and which is generated from sources outside the system. Might adversely affect the image acquisition. Care is usually taken to minimize its effect.

analog-to-digital converter [ADC]; *g.* **Analog-Digital-Wandler:** electronic device that converts an electric signal (voltage, current) from analog form to digital representation for further digital processing.

angle of incidence; *g.* **Einfallswinkel:** the angle formed between a ray of light striking a surface and the *surface normal* at the point of incidence.

angle of reflection; *g.* **Reflexionswinkel:** the angle formed between the *surface normal* and the reflected ray. This angle lies in a common plane with the *angle of incidence* and is equal to it.

angle of refraction; *g.* **Brechungswinkel:** the angle formed between a refracted ray and the *surface normal*. This angle lies in a common plane with the *angle of incidence*.

angstrom; *g.* **Angström:** outdated unit of length measure equal to 10^{-10} m; frequently used to measure wavelength in the visible spectrum and surface errors of optical surfaces. Preferred replacement unit 1 nm = 10^{-9} m.

antireflection coating; *g.* **Antireflexbeschichtung:** a coating applied to the surfaces of a lens that reduces reflection and thereby increases transmission. A single layer with the thickness of a quarter wavelength and with a refractive index of a value that is the square root of the index of the lens material is the simplest type of an ~.

aperture; *g.* **Blende:** an opening to pass light. The effective diameter of an optical system controlling the amount of light passing it.

aperture problem; *g.* **Blendenproblem:** refers to the problem that the component of the *optical flow* tangential to the edge of an object cannot be recovered by neighborhood operations.

apogee; *g.* **Apogäum, Erdferne:** the furthermost point from the earth in the orbit of a satellite.

application-specific integrated circuit [ASIC]; *g.* **applikationsspezifischer integrierter Schaltkreis:** a custom-made single integrated circuit (IC) chip to perform specific user-required functions.

arcsecond; *g.* **Bogensekunde:** a one-dimensional measurement of angle. A degree is divided into sixty arcminutes and an arcminute is divided into 60 arcseconds. Therefore, an arcsecond equals 1/3600 of a degree or 4.85 μrad.

area of interest: a rectangular region of an image that is selected for further processing since it contains potentially interesting patterns.

arithmetic-logical unit [ALU]; *g.* **arithmetisch-logische Einheit:** the processing unit which performs elementary arithmetic (addition, subtraction) and logical (and, or, xor, not, etc) operations. Part of the *CPU*.

artificial intelligence [AI]; *g.* **künstliche Intelligenz [KI]:** an interdisciplinary research area that aims at "intelligent" in the sense that a human being can solve a given task in an intelligent way. Since vision is the most complex recognition system, it is an important issue and field of application in ~. An important aspect of ~ is the representation, retrieval, and expansion (learning) of knowledge.

aspect ratio; *g.* **Seiten-Höhen-Verhältnis**: the ratio of the width to the height of an image. The ~ is sometimes also referred to as the ratio of the width to the height of an individual pixel.

aspherics; *g.* **Asphärische Linse/Spiegel**: an optical element (lens or mirror) that does not have a spherical surface.

astigmatism; *g.* **Astigmatismus**: lens *aberration* which results in the tangential and sagittal image planes being separated axially.

asynchronous; *g.* **asynchron**: computer operations, especially communications, triggered successively by external, untimed events and not by a fixed frequency clock.

ATM: *asynchronous* transfer mode, a switch-based network standard.

atmospheric window; *g.* **Atmospärisches Fenster**: a range of electromagnetic wavelengths where radiation can pass through the Earth's atmosphere with relatively little attenuation.

autocorrelation; *g.* **Autokorrelation**: correlation of a signal with itself. The autocorrelation function $R_{gg}(s)$ computes the dot product between a signal $g(x)$ and a shifted copy $g(x+x)$: $R_{gg}(s) = \int_{-\infty}^{\infty} g(x)g(xs)\mathrm{d}^W x$. The Fourier transform of the autocorrelation function is the *power spectrum*.

autocovariance; *g.* **Auto-Kovarianz**: same as *autocorrelation* after subtracting the mean value from a signal.

automatic gain control [AGC]; *g.* **automatische Verstärkungsregelung**: a circuit by which the electronic gain is automatically adjusted as a function of input intensity or other specified parameter, in order to retain the output at constant level. Common feature for CCD cameras.

automatic visual inspection; *g.* **automatische visuelle Inspektion**: an inspection process which uses an imaging sensor, image processing, pattern recognition, or computer vision techniques to measure and/or interpret the imaged objects in order to determine whether they have been manufac-tured within permitted tolerances. ~ systems usually integrate the technologies of material handling, illumination, image acquisition and — if required —special-purpose computer hardware along with the appropriate image analysis algorithms into a system.

avalanche photodiode; *g.* **Avalanche-Photodiode**: a solid-state detector that produces a response by generating numerous hole-electron pairs for each absorbed photon. Solid state equivalent to *photomultiplier*.

AVHRR: advanced very high resolution radiometer, a multispectral imaging system carried by the TIROS-NOAA series of meteorological satellites.

azimuth; *g.* **Azimut**: angular dimension measured from a local horizontal. Azimuth is perpendicular to elevation. The horizontal may be determined by the earth, the body of an aircraft, ship, or spacecraft. Depending on application, the reference direction for the azimuth is the south, the north, or a symmetry axis of the reference object/system.

B-spline; *g.* **B-Spline**: a special class of piecewise polynomial curves with smoothness constraints between the polynomial segments. An n-order B-spline function is given by convolving the box function n+1 times with itself and continuous in the first n-1 derivatives. The continuous B-spline curves do not go through the discrete grid points from which they are computed.

back focal length; *g.* **bildseitige Brennweite**: the distance from the last surface of a lens to its image plane.

back lighting; *g.* **Durchlichtbeleuchtung**: illumination arrangement in which the light source is on the opposite side of the object from the camera. ~ tends to produce images with silhouettes of the imaged objects. Often used in machine vision systems for precise geometric measurements.

back projection; *g.* **Rückprojektion**: the recreation of the "bundle of rays" that produced a given image in *tomography*.

background; *g.* **Hintergrund**: a connected component in an image, completely surrounding connected regions identified as objects of interest.

baffle; *g.* **Strahlenfalle**: structure(s) that obstruct(s) stray light from irradiating the image plane.

bandpass; *g.* **Bandpaß**: (a) spectroscopy: a spectral range of transmission; (b) signal processing: a range of frequential wavelengths remaining after a linear convolution operator is applied.

bandwidth; *g.* **Bandbreite**: the difference between the lower and upper limiting frequencies of a frequency band of a *bandpass filter* or an electronic circuit. Usually, the limiting frequencies are defined by an amplitude of 50 % of the peak amplitude.

BDRF: *bidirectional reflectance distribution function*

beamsplitter; *g.* **Strahlteiler**: a device for dividing a light beam into two or more separate beams.

beta radiation; *g.* **Betastrahlung**: particulate radiation emitted by many radioactive elements and consisting of *electrons*.

Betacam: a SMPTE video standard for professional quality using 1/2" cassette recorders. The color content of the video signal (*chrominance*) is stored with a lower bandwidth (1.5 MHz) than the luminance signal (4.1 MHz). Thus, a Betacam signal does not contain the full color resolution available with a 3-chip color CCD camera.

bidirectional reflectance distribution function [BDRF, bidirectional reflectivity]; *g.* **Bidirektionale Reflexionsverteilungsfunktion**: specifies the reflectance of a surface in terms of both incident and reflected beam geometry, i.e., the ratio of reflected radiance towards the viewer to the irradiance in the direction towards the light source.

bifrigent; *g.* **doppelbrechend**: property of an anisotropic optical medium. Light polarized in different directions has different *indices of refraction*.

bin: one of a series of equal intervals in a range of data employed to describe the divisions in a *histogram*.

binary; *g.* **binär**: a numerical system using the base 2. Examples are $0 = 0$, $1 = 2^0 = 1$, $10 = 2^1 = 2$, $101 = 2^2 + 2^0 = 5$.

binary image; *g.* **Binärbild**: a digital image in which a value of either zero or one is assigned to each pixel.

binning: a method in which on-chip addition of pixels is performed to decrease resolution, but to increase sensitivity.

binomial distribution; *g.* **Binomialverteilung**: a statistical probability density function; the discrete analog to the *normal distribution*. The coefficients of the ~ can be computed by cascaded convolution of the $1/2[1\ 1]$ mask with itself (equivalent approach to the computation scheme known as Pascal's triangle).

binomial filter; *g.* **Binomialfilter**: a separable smoothing filter based on the coefficients of the *binomial distribution*, which quickly converges to a *Gaussian filter* with increasing filter masks. Examples are $1/4[1\ 2\ 1]$, $1/16[1\ 4\ 6\ 4\ 1]$.

bit; *g.* **Bit**: a contraction of binary digit; the fundamental unit of digital computing. A bit is either 1 or 0, expressing the binary state of on or off, true or false, yes or no.

bit reversal; *g.* **Bitumkehr**: an operation reversing the order of the bits in a digital word. The least-significant bit becomes the most-significant bit, and vice versa. Important part of a radix-2 fast Fourier transform (*FFT*) algorithm.

black (material or coating): a material or coating that has a high *emissivity* close to one for radiation through the spectral *bandpass* of interest.

black box; *g.* **schwarzer Kasten**: technical slang for a component or subassembly that performs functions without requiring the user to have detailed knowledge of its internal design. This implies easily definable and accurate interface requirements and specifications. Frequently, these are self-contained modules that can easily be replaced with an identical one. Also used in linear system theory, where the knowledge of the input/output relation is sufficient to describe a system completely without knowing of which components it actually consists.

blackbody; *g.* **Schwarzkörper**: an ideal radiator and absorber that completely absorbs all radiant energy striking it and emits radiation with an *emissivity* of one. The radiant excitance of a ~ is only determined by its temperature and described exactly by Planck's law. An industrial ~ usually approaches the ideal by using a cone or spherically shaped cavity with *black coatings* achieving an effective emissivity exceeding 0.99.

blind spot; *g.* **blinde Fleck**: the point of entry of the optic nerve to the retina where no radiation is detected by the eye.

block distance [city distance]; *g.* **Blockdistanz**: distance measure in a digital image or n-dimensional feature space. Length of shortest part between the two points when walking only in directions along the coordinate axis is allowed.

blocks world; *g.* **Blockwelt**: a simple world consisting of planar-faced solids, such as cubes and pyramids, that is often used as the experimental domain for image analysis problems.

blooming: a phenomenon of excessive bleeding and crosstalk of the signal from one sensor element (*sel*) to neighboring sensor elements in a sensor array. Blooming can occur in both the time and spatial domains. It is common with *CCD*s when the sels are heavily saturated or an *electronic shutter* with a short illumination time is used.

blur; *g.* **Unschärfe**: pertaining to elements in an image that are indistinct or not readily discernible. Contrast with: sharp.

blur circle; *g.* **Unschärfekreis**: the image formed by a lens system, on its focal surface, of a point source object. The size of the ~ will be dictated by the precision of the lens and the state of focus; the blur can be caused by aberrations, defocusing and manufacturing defects.

BMP: bitmap data. The standard format for raster images in Microsoft's Windows operation system. Includes binary, gray scale, and color (RGB) images, limited to 8 bit per channel.

border [boundary]; *g.* **Rand**: the set of pixels in a region of a digital image that are adjacent to the *background*. Contrast with: interior. ≻ *edge; perimeter*.

bottom-up: a data-driven control strategy to problem solving. It employs no object models in its early stages and uses only general knowledge about the world being sensed. The features extracted from the observed image data are interpreted and aggregated to generate a sufficiently high level of description of the scene.

boundary: ≻ *border*

boundary detection [boundary delineation]; *g.* **Randdetektion**: any process which determines a chain of pixels separating one image region from a neighboring image region.

boundary following; *g.* **Konturverfolgung**: the sequential procedure by which the chain of the boundary pixels of a region can be determined.

bounding box: ≻ *bounding rectangle*

bounding rectangle [bounding box]: the ~ of a region *R* is a rectangle which circumscribes *R*. It is just as large that the region just fits into the ~. Sometimes its sides are aligned with the row and column directions, sometimes with axes of the coordinate system aligned with the *inertia tensor* computed from the *moments* of the region.

box classification; *g.* **Kastenklassifikation**: a classification technique in which the decision boundary in the *feature space* is a rectangular box.

box filter; *g.* **Rechteckfilter**: a linear spatial smoothing filter in which each pixel in the filtered image is the equally weighted average of the surrounding pixels in a rectangular window.

box function; *g.* **Kastenfunktion [Rechteckfunktion]**: discontinuous function which is zero except for a finite interval in which the ~ is one.

BRDF: *bidirectional reflectance distribution function*

bremsstrahlung; *g.* **Bremsstrahlung**: continuous *EM* radiation emitted by a decelerated electron while moving through matter.

brightness; *g.* **Helligkeit**: either the amount of light received or emitted by a surface. Because of the ambivalence and common misuse of this term it is better to use more precisely defined terms such as *radiance, irradiance, luminance*, and *illuminance*.

buffer; *g.* **Pufferspeicher**: device used to temporarily store data. It is often located between two devices of differing speeds, e.g., a computer output is faster than the output of a printer.

bus; *g.* **Bus**: in computer hardware, a circuit or group of circuits providing communication paths between two or more devices, such as CPU, peripherals, and memory, or between functions on a single PC board (e.g., VMEbus, ISA, EISA, PCI).

byte; *g.* **Byte**: in digital data transmission, the number of bits (usually eight) used to represent a character. Abbreviated by B.

C-mount: threaded lens mount developed as the 16 mm movie standard; used extensively for imaging sensor arrays such as CCD cameras. The threads have a major diameter of 1"

(25.4 mm) and a pitch of 32 threads per inch. The *flange focal distance* is 0.69" (17.526 mm).

camera constant; *g.* **Kamerakonstante:** ≻ *principal distance*

camera coordinates; *g.* **Kamerakoordinaten:** 3-D coordinate system that is attached to the camera. The Z axis is aligned with the optical axis. The origin of the camera coordinate system either coincides with the focal plane (image plane) or with one of the principal planes. Denoted by $X = [X, Y, Z]^T = [X_1, X_2, X_3]^T$.

camera link: interface standard for digital cameras and frame grabbers using the serial channel link transmission standard with low voltage differential signaling (LVDS) and standard MDR-26 pin connectors. The standard comes in three configurations (base, medium, and full with up to 24, 48, and 64 bit parallel) and image data transmission rates up to 500 MB/s. Two extra LVDS cable pairs are provided for asynchronous and bidirectional communication between frame grabber and camera. Unfortunately, the communication protocol is not standardized.

camera model; *g.* **Kameramodell:** a mathematical specification for mapping the 3-D world to a 2-D image, expressed as a 4×3 matrix when the mapping is described in terms of *homogeneous coordinates*.

candela: units for the *luminous flux*. The candela is one of the basic units of the international metric system of units.

Cartesian product; *g.* **kartesisches Produkt:** the ~ of two sets A and B, denoted by $A \times B$, is the set of all ordered pairs where the first component is an element from the first set and the second component is an element from the second set.

CCD: *charge-coupled device*

CCIR: Comité Consultatif International dés Radiocommunication. Name for the basic standard for European video signal with 25 frames/sec and 625 lines/frame. ≻ *RS-170*.

CD: compact disk, ≻ *CD-ROM*.

CD-ROM [CD]: compact disk, the most widespread form of optical storage. Used for storing audio, or for holding massive amounts of fixed, unchangeable, data (about 600 MB). Ideal storage media for image data, compatible between different computer systems and operation systems. With a CD-writer, data can be written once to a CD.

center of perspectivity; *g.* **Projektionszentrum:** the common point in a *perspective projection* where all rays meet.

central moments; *g.* **zentrale Momente:** *moments* referred to the mean values.

centroid; *g.* **Zentroid:** the ~ of a region is the center of mass of a region. It is the mean (row, column) position for all pixels in the region.

certainty measure; *g.* **Zuverlässigkeitsmaß:** measure how accurate or reliable a measurement is performed in a statistical sense.

CGM: computer graphics metafile. Device and operating system-independent file format for picture information with many graphic primitives. ANSI standard. Mostly used in the CAD community.

CGS-units: centimeter-gram-second. Metric unit system using the units centimeter, gram, and second for length, mass, and time, respectively.

chain code [contour code]; *g.* **Kettencode:** a representation of a digital curve using line segments with a fixed set of orientations. A curve can be compactly expressed as the sequence of integers representing the orientation of each line segment.

change detection; *g.* **Änderungsdetektion:** an image processing technique in which the pixels of two registered images (especially consecutive images of a sequence) are compared, pixel by pixel, to detect differences. A binary one value is given to the output pixel whenever corresponding pixels on the input images have significantly different gray levels.

channel; *g.* **Kanal:** the range of wavelengths recorded by a single detector to form an image. A multispectral image is recorded in several channels simultaneously. The term can also be used to refer to a synthetic channel (i. e., not a simple spectral band as originally recorded), such as one created by rationing or principal-components transformation.

charge transfer efficiency [CTE]; *g.* **Ladungstransporteffizienz:** the percentage of charge that is transferred from one CCD shift register element to the adjacent register.

charge-coupled device [CCD]: a semiconductor device in which finite isolated charge-packets are transported from one position in the semiconductor to an adjacent position by sequential clocking of an array of gates. Dominant type of solid-state image sensor.

charge-injection device [CID]: specific fabrication scheme for solid-state image sensors. The photo-generated charge is sensed by injecting it from the sensor into the substrate.

chromatic aberration; *g.* chromatische Aberration: lens aberration resulting from the increase in the refractive index of all common optical materials toward shorter wavelengths (the blue end of the spectrum). The change in image size from one color to another is known as lateral color or chromatic difference of magnification, the change in focal length as axial color.

chromatic vision; *g.* Farbsehen: the perception of the human eye of changes in *hue*.

chromaticity diagram; *g.* Farbenkarte: a 2-D diagram with one of the three chromaticity coordinates against another. The chromaticity coordinates are 3-D color coordinates normalized by their sum. With two chromaticity coordinates, all possible colors can be generated.

chrominance; *g.* Farbwert: the color content of the color tristimulus. *Luminance* and chrominance together make up a full color signal.

CID: *charge-injection device*

CIE: Commission Internationale de l'Èclairage, the international commission on illumination; sets the standards related to *photometry* including color coordinate systems.

circular frequency; *g.* Kreisfrequenz: frequency multiplied by 2π. Can be thought as the speed of an oscillatory motion represented as a rotation on the unit circle.

CISC: complex instruction set computer

city distance [city block distance]; *g.* Blockdistanz: > *block distance*

class; *g.* Klasse: > *pattern class*

classification; *g.* Klassifizierung: > *pattern classification*

classifier; *g.* Klassifikator: a device or process that sorts patterns into categories or classes.

closing; *g.* Schließen: an increasing and *idempotent* morphological operation. It is the dual operation to *opening*. Closing an image with a disk shaped *structuring element* smoothes the contours, fuses narrow breaks and long thin gulfs, eliminates holes smaller in size than the structuring element and fills gaps on the contour.

CLSM: *confocal laser scanning microscopy*

cluster; *g.* Cluster: a set of points in a *feature space* that are close to each other and well separated from other sets of points. A cluster is a natural candidate for a *class* of objects.

cluster analysis; *g.* Clusteranalyse: the detection and description of *clusters* in a *feature space*.

CMOS: complementary metal-oxide-semiconductor. The most important standard production process for digital and analog integrated circuits including memory chip: (DRAM, SRAM) and central processor units (CPUs). Advantages: low power dissipation and small size for elementary devices such as transistors.

co spectrum: real part of the complex-valued *cross-correlation spectrum*.

coarse/fine; *g.* Grob/Fein-Strategie: a general strategy in which computations are first carried out on a "coarse" version of an image. Subsequent computations are carried out on finer and finer versions, with each additional pass more tightly constrained by the preceding ones.

Coastal Zone Color Scanner (CZCS): a multispectral imaging system carried by the Nimbus series of meteorological satellites.

coherent radiation; *g.* kohärente Strahlung: electromagnetic radiation in which the phase relationship between any two points in the radiation field has a constant difference.

color burst; *g.* Burst: a burst of the reference sub-carrier frequency required for color decoding of *PAL*, *NTSC*, and *SECAM* video signals.

color space; *g.* Farbraum: color can be represented in a computer by a triple of values in a number of ways: (1) the intensity of the red, green, and blue components (*RGB*); (2) the values of the hue, saturation, and intensity (*HIS*); and (3) intensity and a set of color differences (red-green), (blue-yellow) (*YUV*). Each of these representational systems defines a ~ with different distance relationships existing between a given pair of color vectors.

coma; *g.* Koma: a lens aberration, resulting from different magnifications in the various lens zones, that occurs in that part of the image field which is at some distance from the principal axis of the system. Object points appear as short comet-like images with the brighter small head toward the center of the field (positive coma) or away from the center (negative coma).

composite video signal; *g.* **FBAS-Signal**: the information required to produce a video image (synchronization, *luminance* and *chrominance* combined into a single signal. Three compatible ~s are in wide use for broadcasting: *NTSC*, *PAL*, *SECAM*.

compression: > *image compression*

computer graphics; *g.* **Computergrafik**: research area in computer science dealing with the generation and representation of images from two- and higher-dimensional scenes. Much effort is devoted to the generation of natural looking images (photorealism). Image analysis can be regarded as the (much more complex) inverse procedure to compute graphics.

computer tomography [CT]: > *tomography*

computer vision [image understanding, scene analysis]; *g.* **Maschinelles Sehen**: computer analysis of one or more images or an image sequence combing *image processing*, *pattern recognition*, and *artificial intelligence* technologies. The analysis recognizes, locates the position and orientation, and provides a sufficiently detailed symbolic description or recognition of those imaged objects deemed to be of interest in the three-dimensional environment.

concave; *g.* **konkav**: a region is ~ for which at least one straight line segment between two points of the region is not entirely contained within the region. Contrast with *convex*.

cone; *g.* **Zapfen**: receptor in the retina which is sensitive to color. There are three types of cones sensitive in the red, green, and blue wavelength range of *light*. > *rods*.

confocal laser scanning microscopy [CLSM]; *g.* **konfokale Laserabtastmikroskopie**: a microscopy technique where a laser raster scanning through the imaging optics is applied. Only one point at the focal plane is illuminated at a time. A small aperture (pinhole) is used in front of the detector obstructing most of the light reflected from out-of-focus objects. In contrast to conventional microscopy, high-resolution 3-D images can be acquired. ~ has become the central tool for biological and medical research at the cell level.

contour map; *g.* **Konturgrafik**: a two-dimensional representation of the topography of a (terrain) surface using a set of closed curves, each of which represents a constant surface elevation.

contour tracing; *g.* **Konturverfolgung**: a searching or traversing process by which the bounding contour of a region can be identified.

contrast; *g.* **Kontrast**: the difference in the radiance between an object and its background or the difference in the minimum and maximum radiance that a sensor can measure. Usually expressed as a *contrast ratio*, i.e., $10,000 : 1$.

contrast ratio; *g.* **Kontrastverhältnis**: > *contrast*

contrast stretching; *g.* **Kontrastspreizung**: monotonically increasing point operator expanding a measured range of digital numbers in an image to a larger range, to increase or enhance the visibility of an image's detail in a small gray value interval.

convex; *g.* **konvex**: a region is ~ for which a straight line segment between any two points of the region is entirely contained within the region. Contrast with: *concave*.

convolution; *g.* **Faltung**: the most important linear operator in image processing. The ~ takes a weighted linear combination of the pixels in a neighborhood around a pixel to compute a new value for the pixel: $G'_{mn} = \sum_{m',n'} H_{m',n'} G_{m-m',n-n'}$. The weights H are known as the convolution mask or *point spread function* of the convolution operator. In the *Fourier domain*, ~ reduces to a complex-valued multiplication. Depending on the wave number, each sinusoidal structure is changed in its amplitude and phase. Averaging (smoothing) and derivation are simple examples of ~ operations.

correlation; *g.* **Korrelation**: A numerical measure of similarity between two vector quantities that is computed as the dot (scalar) product of the vectors, i.e., the projection of one vector onto the other. The definition of ~ as used in statistics, sometimes called "normalized ~", involves adjusting each measurement (vector component) by its population mean and dividing by its population standard deviation. Under these adjustments, ~, now a value between $+1$ and -1, measures the cosine of the angle between two vectors.

corresponding point; *g.* **korrespondierender Punkt:** A point p on one image and a point q on a second image are said to form a corresponding point pair (p, q) if p and q are each a different sensor projection of the same 3-D point. The visual correspondence problem consists of matching all pairs of ~s from two images of the same scene.

cosine transform; *g.* **Kosinustransformation:** unitary transform using the cosine function as a kernel: $\hat{g}_u = \sum_{n=0}^{N-1} g_n \cos(\pi u n)$. In contrast to the discrete Fourier transform used for image compression ($\succ$ *JPEG*).

covariance matrix; *g.* **Kovarianzmatrix:** $P \times P$ matrix of the *cross covariances* of P signals with each other.

CPU: central processing unit. A computer's main processing chip.

cross correlation; *g.* **Kreuzkorrelation:** correlation of two signals computed as the dot product of the vector or matrix representing the signals.

cross covariance; *g.* **Kreuzkovarianz:** same as *cross correlation* with zero-mean signals.

cross section; *g.* **Wirkungsquerschnitt:** measure for the absorption or scattering of radiation. Can be thought as the effective area of the scattering media that entirely absorbs or scatters radiation.

cross-correlation spectrum; *g.* **Kreuzkorrelationsspektrum:** *Fourier transform* of the *cross correlation* function.

CT: *computer tomography*

curvature; *g.* **Krümmung:** rate of change of the direction of a curve given by the second spatial derivative. Description of the ~ of a surface at a point requires three parameters. At each point of the surface, the maximum and minimum ~s (called the principal ~) correspond to curves at right angles to each other formed by intersecting the surface with two perpendicular planes containing the normal to the surface at the given point. The product of the principal ~s at the point is called the Gaussian ~.

curvature of field [field curvature]; *g.* **Feldkrümmung:** a lens aberration that causes a flat object surface to be imaged onto a curved surface rather than a plane.

DAC: *digital to analog converter*

dark current [dark signal]; *g.* **Dunkelstrom [Dunkelsignal]:** the current flowing or signal amplitude generated in a photosensor placed in total darkness. This is caused by thermally generated electrons and increases linearly with integration time. The ~ generally increases by a factor of two with each 7° C increase. $\succ$ *fixed pattern noise*

DCT: discrete cosine transform

DDB: device dependent bitmap

deblurring: $\succ$ *sharpening*

decision rule [classifier]; *g.* **Entscheidungsregeln:** a rule or algorithm used in pattern classification to assign an image pixel or an object to a pattern class based on features extracted from the image.

deconvolution [inverse filtering]; *g.* **Entfaltung:** operation to cancel the effect of a convolution operation. Important practical reconstruction task, since the blurring caused by optical systems can be described by convolution. ~ requires the application of the inverse convolution operator that only exists if the transfer function of the operator has no zeros.

depth map [range image]; *g.* **Tiefenbild, Tiefenkarte:** a digital range image in which the value in each pixel's position is the distance between an image plane and the surface patch corresponding to the pixel.

depth of field; *g.* **Schärfentiefe (im Gegenstandsraum):** the in-focus range of an imaging system. Measured from the distance behind an object to the distance in front of the object for which all objects are appearing in focus with a limit set by a specified maximum *blur circle*.

depth of focus; *g.* **Schärfentiefe (im Bildraum):** the range of lens to image plane distance having the image formed by the lens appearing in focus within a specified maximum *blur circle*.

DFT: *discrete Fourier transform*

diaphragm; *g.* **Blende:** a flanged or plain ring with a restricted aperture, located in an optical system at any of several points, that cuts off marginal light rays not essential to image formation. ~s are used as field stops, to limit the field of view to that portion which is fully illuminated, as aperture stops, to limit the light gathering power of the instrument, and as antiglare ~s, to eliminate reflections from the sides of the tube and consequent glare in the field of view. Lens cells or the sides of the tube may act as ~s.

DIB: device-independent bitmap

difference of Gaussian [DoG]: linear convolution operator obtained by subtracting two images filtered by Gaussian filters with different variance.

diffraction; *g.* **Beugung:** as a wave front of light passes by an opaque edge or through an opening, secondary weaker wave fronts are generated, apparently originating at that edge. These secondary wave fronts will interfere with the primary wave front as well as with each other to form various diffraction patterns.

diffraction-limited; *g.* **beugungsbegrenzt:** an optical system where the aberrations have been corrected to the extent that only the effects of diffraction limit the resolution.

diffuse reflection; *g.* **diffuse Reflexion:** non-specular reflection from a rough surface.

diffusion; *g.* **Diffusion:** a process that tends to level temperature and concentration of chemical species by molecular processes, i. e., the motion of the molecules and atoms and exchange processes between them. The flux density j is proportional to the concentration difference and directed in the opposite direction of the concentration gradient: $j = -D\nabla c$. The instationary diffusion equation is given by $\partial c \partial t = \nabla(D\nabla c)$. ~ is an important model for spatial averaging in image processing. It is equivalent to smoothing with a *Gaussian filter*.

digital image; *g.* **digitales Bild:** an image in digital format consisting of an array of pixels. It is obtained by partitioning the area of the image into a finite two-dimensional array of small uniformly shaped mutually exclusive regions called pixels or resolution cells and assigning a representative image value to each such spatial region.

digital photogrammetry; *g.* **digitale Photogrammetrie:** computer processing of digital images for the automatic interpretation of scene or image content, especially the measurement of positions and distances in aerial and satellite images.

digital terrain model [DTM]; *g.* **digitales Geländemodell:** a numerical array representing terrain elevation as a function of geographic or spatial location.

digital-to-analog converter [DAC, D/A Converter]; *g.* **Digital-Analog-Wandler:** in image processing, a device that transforms the digital data into a time-sequential voltage sequence, such as an image that can be viewed on a monitor, stored on recorders, etc.

digitization; *g.* **Digitalisierung:** the process of converting an image into a digital image. ≻ *quantization; sampling.*

digitized image; *g.* **digitalisiertes Bild:** ≻ *digital image*

dilation; *g.* **Dilatation:** morphological operator increasing the size of objects and filtering small holes and gaps. Dilating is the dual operation to ≻ *erosion.* ≻ *morphology.*

direct imaging; *g.* **direkte Abbildung:** an imaging system that results in an image at a certain (not necessary planar) surface. All classical optical systems are ~ devices. Contrast with *indirect imaging.*

direct memory access [DMA]: a technique where a peripheral device gains direct access to the memory of a computer without using the *CPU.*

directional filter; *g.* **Richtungsfilter:** a spatial-frequency filter which enhances features in an image in selected directions.

discrete Fourier transform [DFT]; *g.* **diskrete Fouriertransformation:** the ~ of a digital image represents the image in terms of a linear combination of periodic functions (complex exponentials).

disparity; *g.* **Disparität:** the relative displacement of corresponding points in the two images of a stereo pair; ~ is inversely related to the distance of the object from the camera.

displacement vector; *g.* **Verschiebungsvektor:** vectorial displacement of an image feature between two consecutive frames of an image sequence caused by the motion projected by the imaging system onto the image plane.

displacement vector field; *g.* **Verschiebungsvektorfeld:** continuous field of *displacement vector.*

distance transform; *g.* **Abstandstransformation:** the ~ of a binary image is an image having in each pixel's position its distance from the nearest binary zero pixel of the input image. Distance can be ≻ *city block distance, Euclidian distance.*

distortion; *g.* **geometrische Verzerrung:** aberration of an optical system related to deviations of the position at the image plane from those given by an ideal optical system with perspective projection. For accurate position and size measurements, it is required to correct for ~s.

DLL: dynamic link library

DMA: *direct memory access*

DoG: *difference-of-Gaussian*

Doppler shift; *g.* **Dopplerverschiebung**: a change in the observed frequency of *EM* or other waves caused by the relative motion between source and detector.

DRAM: dynamic random access memory

DTM: *digital terrain model*

DXF: drawing exchange file. Most widely used interchange format for 2-D and 3-D geometric data; originates from Autodesk Inc., supported by virtually every CAD program and many other programs.

dynamic range; *g.* **Dynamik**: ~ of an image sensor array can be specified with respect to optical or electrical quantities. Optical ~ is the ratio of maximum light intensity required to saturate a pixel to light intensity which produces an output equal to noise output. The electrical ~ is the ratio of the saturation output voltage to the noise output voltage. For an image sensor array with a linear response, the two are equivalent.

early vision; *g.* **frühes Sehen**: the first stages of processing in the human visual system. ≻ *low-level vision*.

edge; *g.* **Kante**: a discontinuity or other transition in the spatial gray value function. Strong candidate for a border between an object and the background or another object.

edge detection; *g.* **Kantendetektion**: image processing techniques in which edge pixels are identified by examining their neighborhoods.

edge enhancement; *g.* **Kantenverschärfung**: image enhancement techniques in which the steepness and contrast of edges are increased.

edge operator; *g.* **Kantenoperator**: a neighborhood operation which determines the extent to which pixels in an image are edge pixels, i.e., contain the boundary between two regions. Some ~s can also produce the tangent direction of the boundary that passes through the pixel. ~s include: *gradient operators, Laplacean operators*, and *morphologic edge operators*.

edgel; *g.* **Kantenelement**: short for edge element; a triplet whose first component is the (row, column) location of a pixel, whose second component is the position and orientation of an edge running through the pixel, and whose third component is the strength of the edge.

effective focal length: ≻ *focal length*

electromagnetic radiation [EM]; *g.* **elektromagnetische Strahlung**: propagating evolved electric and magnetic fields characterized by the frequency of oscillation ν and the wavelength λ and propagation speed. In vacuum, ~ propagates with the speed of light $c \approx 3 \cdot 10^8$ m/s, one of the most fundamental physical constants. The wavelength of ~ covers an enormous 24 decades including gamma, X, ultraviolet, visible (light), infrared, microwave, and radiowave radiation.

electron; *g.* **Elektron**: elementary particle with a mass of $m_e \approx 9.1 \cdot 10^{-31}$ kg carrying the smallest negative unit of electrical charge $e \approx 1.6 \cdot 10^{-19}$ As.

electron microscopy; *g.* **Elektronenmikroskopie**: an imaging device using an electronic beam to generate an image. Lenses for ~ consist of properly formed electrical and magnetic fields. The magnification of an ~ can be much higher than with a light microscopy because of the much lower wavelength $(h/(m_e u))$ of the particulate radiation.

electronic shutter; *g.* **elektronischer Verschluß**: limitation of the charge accumulation time and, thus, exposure time on a CCD imager shorter than the frame time by draining the charges for most of the frame time. Exposure times can generally be between 1/15 000 s and 1/60 s.

EM: *electromagnetic radiation*

emissivity; *g.* **Emissionskoeffizient**: a measure of how well a surface emits, or radiates, EM radiation thermally, defined as the ratio of the thermal excitance from the surface to the thermal excitance from a *blackbody* (perfect emitter) at the same temperature. A blackbody therefore has an emissivity of 1 and the emissivity of natural materials ranges from 0 to 1.

emittance: ≻ *excitance*

EOS: earth observing system, a proposed multinational series of remote-sensing satellites to be deployed in the decade beginning in the mid-1990s.

epipolar lines; *g.* **Epipolarlinien**: the corresponding pairs of lines in the two images of a stereo pair that are the intersections of the two image planes and the set of planes passing through the two lens centers. Corresponding points in the two images fall on corresponding ~. Use of the epipolar constraint reduces stereo matching from a 2-D to a 1-D search problem.

EPS: encapsulated postscript. *Postscript* Interchange format to incorporate single images into application-specific programs (drawing programs, type-set programs); widely supported, but inefficient with respect to storage space and plagued by many incompatibilities due to incompatible or erroneous implementations.

equatorial orbit; *g.* **äquitoriale Umlaufbahn**: an orbit of a satellite around the Earth in which the orbital plane makes an angle of less than 45° with the Equator.

erosion; *g.* **Erosion**: morphological operator decreasing the size of objects and removing small objects. It is the dual operation to *dilation*. ≻ *morphology*.

ERS-1, ERS-2: earth remote sensing satellites, European (*ESA*) satellites launched in 1992 and 1995, targeted principally on marine applications by the use of active and passive microwave techniques.

ESA: European Space Agency, based in Paris. A consortium between several European states for the development of space science, including the launch of remote-sensing satellites.

Euclidian distance; *g.* **Euklidsche Distanz**: distance between two points given by the magnitude of the vector connecting the two points.

Euler number of a region; *g.* **Eulerzahl**: the number of its connected components minus the number of its holes.

excitance [**emittance**]; *g.* **Strahlungsflußdichte**: the *radiant flux density* leaving a surface.

exterior orientation [**outer orientation**]; *g.* **externe Orientierung**: position and orientation of a camera reference frame with respect to a world reference frame.

extinction coefficient; *g.* **Extinktionskoeffizient**: wavelength-dependent material property describing the extinction of EM radiation by absorption and scattering. The ~

κ is the relative decrease of the radiant flux per unit length: $d\Phi/\Phi = -\kappa(\lambda)dx$.

f-number [**f-stop**]; *g.* **Blendenzahl, Blende**: ratio of the focal length of a system to the diameter of the entrance pupil.

f-stop: ≻ *f-number*

false identification [**false alarm, type II error**]; *g.* **Fehlidentifikation**: in pattern classification, the assignment of a pattern to a pattern class other than its true pattern class. Contrast with: misidentification.

false-color image; *g.* **Falschfarbenbild**: ≻ *pseudo-color image*

fan beam projection; *g.* **Fächerprojektion**: ≻ *image reconstruction*

Faraday effect; *g.* **Faradayeffekt**: induction of optical activity, i. e., the rotation of the polarization plane by magnetic fields.

fast algorithm; *g.* **schneller Algorithmus**: computation of a certain procedure with a minimum number of arithmetic operations (addition/multiplication) and/or minimum number of data access. Usually much less computations are required than with the straightforward implementation of the procedure. ≻ *FFT*

feature; *g.* **Merkmal**: an attribute that may contribute to pattern classification; for example, size, texture, or shape. A ~ can be related either to an entire region (extracted object) or a pixel.

feature extraction; *g.* **Merkmalsextraktion**: a step in image processing, in which measurements or observations are processed to find attributes that can be used to assign patterns to pattern classes.

feature image; *g.* **Merkmalsbild**: an image containing a feature computed by low-level image processing operators and used to recognize and classify objects in images.

feature space; *g.* **Merkmalsraum**: a set of all possible n-tuples that can be used to represent n features of a pattern.

feature vector; *g.* **Merkmalsvektor**: an n-tuple (vector) containing n features either of a region (extracted object) or a pixel.

FFT: *fast Fourier transform*; an algorithm to perform the discrete Fourier transform (*DFT*) with maximal speed, i. e., a minimum number of computations.

field; *g.* **Halbbild**: one of the partial scans that are interlaced to make up a frame. Standard video signals contain two fields by scanning the odd and even rows after each other.

field curvature: > *curvature of field*

field of view; *g.* **Gesichtsfeld**: the area of object space imaged at the focal plane of a camera.

filter; *g.* **Filter**: an operator that replaces the value of a pixel by a function of the gray values of the pixels and its neighbors. Filtering is used to smooth, enhance, or detect specific features of an image.

filtered back projection; *g.* **gefilterte Rückprojektion**: > *image reconstruction*

finite impulse response [FIR]; *g.* **endliche Impulsantwort**: a class of linear filter operations with a finite impulse response or *point spread function*. All linear convolution operations are of ~. Contrast with *infinite impulse response.*

FIR: *finite impulse response*

FireWire [IEEE-1394]: serial bus system with data rates up to 400 Mb/s connecting up to 63 peripherals. Can also be used to connect digital cameras to a personal computer and to transmit up to 40MB/s of image data isosynchronously. In contrast to the *camera link* standard, communication with the camera and the image formats are standardized.

fish eye lens; *g.* **Fischauge**: a type of wide-angle lens that has an angular field above 140° and exhibits barrel distortion. Available with fields of view up to 200°.

FITS: Flexible Image Transport Service: data format for *N*-dimensional images established by the international Astronomical Union (1982) and later *NASA* (1990) to transfer astronomical images. Pixels are encoded in 8-bit, 16-bit, or 32-bit integers, or 32-bit or 64-bit IEEE floating point.

flange focal distance [FFD]: the distance between the locating surface of the lens mount and the image plane.

FlashPIX: new standard format for easy handling of digital images proposed by Kodak in cooperation with other companies in 1996. URL: http://www.Kodak.com.

fluorescence; *g.* **Fluoreszenz**: the emission of EM radiation of longer wavelengths by a substance as a result of the absorption of some other radiation of shorter wavelengths. In contrast to phosphorescence, emission occurs with only short delay after the absorption.

fluorescence quenching; *g.* **Fluoreszenzlöschung**: suppression of fluorescence by radiationless decays of the excited state (due to collisions, for example).

focal length; *g.* **Brennweite**: the effective focal length (EFL) is the distance from the principal point to the focal point. The *back focal length* (BFL) is the distance from the vertex of the last lens to the second focal point. The front focal length (FFL) is the distance from the first lens surface to the first focal point.

focal plane; *g.* **Brennebene, Fokalebene**: a plane through the focal point at right angles to the principal axis of a lens.

focal plane assembly [focal plane array]: an infrared imaging device comprising a detector array and read-out electronics. It may include a cryogenic cooling system.

focal point; *g.* **Brennpunkt, Fokus**: the point of focus for parallel incident light through an optical system.

focus of contraction; *g.* **Kontraktionspunkt**: the ~ of a motion field arises from the relative motion of the camera away from a stationary scene. It is defined as the point on the image at which the projected motion field is zero. The motion field of the neighboring points are directed toward it.

focus of expansion [FOE]; *g.* **Expansionspunkt**: same as *focus of contraction* with the exception that the camera is moving toward the static scene and that the motion sectors are all directed away from the ~.

focus series; *g.* **Fokusserie**: a series of microscopic images taken by changing the focal plane step by step using a micropositioning device. Because of the low *depth of field* of high-resolution microscopy, a three-dimensional image is acquired. With conventional microscopy, the depth resolution is limited and the ~ is significantly disturbed by blurred structures at other depths. *Confocal laser scanning microscopy* avoids these problems.

FOE: *focus of expansion*

foot candle: outdated units of *illuminance.*

Fourier descriptors; *g.* **Fourierdeskriptoren**: the ~ of a closed planar curve (boundary of a region) are the coefficients of the Fourier series of the spatial positions of the curve as a function of arc length. Versatile data structure for object shape recognition that can easily be made scale- and rotation invariant.

Fourier domain [Fourier space]; *g.* **Fourierraum:** Another representation of temporal (time series) and spatial (images) signals by decomposing them into sinusoidal signals. In the ~ the signal is a complex-valued function of the *frequency* or *wave number* including the amplitude and the phase of the sinusoidal signal at the corresponding frequency or wave number. The ~ representation of a signal is equivalent to its spatial representation. The conversion between the two representations is provided by the *Fourier transform.*

Fourier slice theorem; *g.* **Fourierscheibentheorem:** theorem used in *tomography.* It states that each *parallel projection* of an object gives one slice of the *Fourier transform* of the object on a plane through the origin of the *Fourier domain* and parallel to the projection plane.

Fourier transform; *g.* **Fouriertransformation:** a technique to convert data from the space or time domain to the wave number or frequency domain. It represents data as the sum of a series of sinusoidal waves, with varying amplitude, frequency, and phase.

fovea: the region around that point on the retina intersected by the eye's optic axis, where receptors are most densely packed. It is the most sensitive part of the retina.

FPN: *fixed pattern noise*

frame grabber; *g.* **Bildspeicher:** a device which interfaces with a camera and is capable of storing sampled video converted to digital signals in memory.

frame rate; *g.* **Bildrate:** number of times per second that the frame is scanned. The RS-170 and European CCIR standards are 30 and 25 frames per second, respectively.

frame transfer: a method to transfer the accumulated charges in a *CCD* sensor. The whole frame of the imaging sensor is transferred to a second nonilluminated storage array where it is subsequently read out row by row and converted into a sequential video signal while the charges for the next frame are already accumulated.

frequency; *g.* **Frequenz:** the number of times an event occurs per unit of time (temporal ~) or space (spatial ~). More commonly, spatial ~ies are referred to as wave numbers (number of wavelengths per unit of space).

frequency domain; *g.* **Frequenzraum:** an explicit representation of a time series in terms of frequency-parameterized basis functions, such as produced by the *Fourier transform.* ≻ *Fourier domain.*

frequency response; *g.* **Frequenzantwort:** the range or band of frequencies at which the system has sensitivity or detectivity.

Fresnel lens; *g.* **Fresnellinse:** a compact and lightweight lens resembling a plano-convex or plano-concave lens that is cut into narrow rings and flattened out. If the steps are narrow, the surface of each step is generally made conical and not spheric.

front lighting; *g.* **Auflichtbeleuchtung:** an illumination arrangement in which the light source is on the same side of the object as the camera.

Gabor filter; *g.* **Gabor-Filter:** a *quadrature filter* pair. The ~ is a bandpass filter with the shape of the Gaussian function $\exp(-|\boldsymbol{k}|^2/(2\sigma_k^2)$ centered at a certain wave number $\boldsymbol{k}_0$.

gamma radiation; *g.* **Gammastrahlen:** high-energy *EM* radiation emitted by radioactive isotopes.

gauging: ≻ *optical gauging*

Gaussian filter; *g.* **Gauß-Filter:** a linear spatial smoothing filter using the Gaussian function as a convolution kernel.

Gaussian focal length: ≻ *principal distance; focal length*

Gaussian pyramid; *g.* **Gauß-Pyramide:** in the ~, the resolution is decreased by successive convolutions of the image at the previous level of the pyramid with a Gaussian-like kernel. After the lowpass filtering, the sample density is typically decreased by sampling every other pixel in every other row.

GB: gigabyte

genlock: the ability to synchronize two separate video signals — essential when mixing video images.

geographic information system (GIS); *g.* **geographisches Informationssystem:** a datahandling and analysis system based on sets of data distributed spatially in two dimensions. The data sets may be map oriented, when they comprise qualitative attributes of any area recorded as lines, points, and areas in vector format, or image oriented, when the data are quantitative attributes referring to cells in a rectangular grid, usually in raster format.

geometric correction; *g.* **geometrische Korrektur:** an image restoration technique in which a geometric transformation is performed on an image to compensate for geometric *distortions*.

geometric optics; *g.* **geometrische Optik:** ~ approximates the propagation of EM radiation as if it were composed of rays diverging in various directions from the source and abruptly bent by refraction or turned back by reflection into paths. The concept that light travels in a straight line is the basis of ~, which neglects *diffraction*. ≻ *refraction*.

geostationary orbit; *g.* **geostationäre Umlaufbahn:** an orbit at 4100 km height in the direction of the earth's rotation, which matches speed so that a satellite remains over a fixed point on the earth's surface.

GIF: graphics interchange format. Established by CompuServe for efficient transmission of image data over networks. Widely used, limited to monochrome and color images with up to 256 colors.

gradient operator; *g.* **Gradientenoperator:** vectorial operator containing the partial derivatives in all directions. In digital images, the partial derivatives are approximated by discrete differences. The magnitude and the angle of the ~ are a measure for the edge strength and direction, respectively. ≻ *edge operator*.

gradient space; *g.* **Gradientenraum:** a two-dimensional space whose axes represent the first order partial derivatives of a surface of the form $z = f(x, y)$. Each point in ~ corresponds to the orientation of a possible surface normal.

gray level [gray value, image intensity, image density, image value]; *g.* **Grauwert:** A number or value assigned to a position on an image. The ~ is proportional to the integrated output, *reflectance*, or *transmittance* of a small area, usually called a resolution cell or pixel.

gray scale; *g.* **Grauwertbereich:** the range of gray levels that occur in an image. Most commonly, a gray value image contains 256 gray values stored in one byte. The digital values 0 and 255 are assigned to black and white, respectively. Alternatively in a binary offset representation, the values -128 (the most negative number) and 127 (the largest positive number) can be assigned to black and white, respectively. In this representation, a gray value of 0 means a mean irradiance.

HDF: Hierarchical Data Format. A versatile multiobject file format for graphical and scientific data, maintained by the National Center for Supercomputing Applications (NCSA) at the University of Illinois at Urbana-Champaign. Used in the remote sensing and medical community.
`http://hdf.ncsa.uiuc.edu`.

HDTV: high definition television. Systems of equipment to generate, broadcast, and display television images of high quality. Ideally, the images have 1125 lines with a 16:9 aspect ratio and a luminance bandwidth of 60 MHz. Unfortunately, a host of different standards exists in the US, Europe, and Japan.

hierarchical; *g.* **hierarchisch:** an approach to problem solving in which the given problem is solved by dividing it up into a set of subproblems, each of which encapsulates an important or major aspect of the original problem. Then, each subproblem is successively divided into more detailed subproblems. The refinement continues until the most refined subproblems can be solved directly.

high-level vision: the part of the image analysis that derives a goal-oriented description and understanding of a scene from previously extracted features.

highlight; *g.* **Glanzlicht:** the image of a light source reflected from a (generally curved) but smooth surface to the camera. ~s are not fixed to the object surface but come and go and wander around on the object's surface to the point where the *reflection* condition is met. They disturb the computation of object features significantly and make object recognition difficult.

highpass filter; *g.* **Hochpaßfilter:** a linear spatial filter which attenuates the low spatial frequencies of an image and accentuates the high spatial frequencies of an image. Used to enhance small details, edges, and lines.

Hilbert transform [Hilbert filter]; *g.* **Hilbert-Transformation:** transforms a signal in such a way that all sinusoidal components into which it can be decomposed are shifted in phase by $\pi/2$ (90°) without changing their amplitude. The ~ is useful to compute the *phase* and *local frequency* (*local wave number*) of a signal.

HIS: *hue, intensity, saturation* color coordinate system.

histogram; *g.* **Histogramm**: discrete representation of a distribution function. A range of a quantity is parted into (normally equidistant) intervals, called bins, and the distribution function gives the number of observations that occur in each interval.

hit-miss operator; *g.* **Hit–Miss-Operator**: a versatile morphological operator that can, e. g., be used to detect binary objects of a certain shape. ≻ *morphology*.

homogeneous coordinates; *g.* **homogene Koordinaten**: a point in Cartesian n-space is represented as a line in homogeneous $(n+1)$-space by adding a scaling parameter. ~ allow many important geometrical transformations including translation and perspective projection to be represented uniformly and elegantly as matrix multiplications rather.

Hough transform; *g.* **Hough-Transformation**: a multidimensional histogram used to estimate model parameters; each point in the parameter space corresponds to a complete specification of the model parameters. For example, for each edge point detected in an image, the ~ will increment the counters for each point in Hough space that corresponds to the parameters of a line that could pass through the given point. All the points in the image on the same line contribute a weight to the same point, allowing this line to be detected as a local maximum in the Hough space. The ~ can be generalized to detect arbitrary shapes.

HP-GL: *Hewlett Packard Graphics Language*. A simple vector graphics language for sending graphical commands to graphical output devices. HP-GL/2 is a refinement of HP-GL, which is only partly compatible with ~.

HP-PCL: *Hewlett Packard Printer Command Language*. A simple language to control printers.

hue; *g.* **Farbton**: that aspect of color described by words such as red, yellow or blue. Achromatic colors, such as white, gray and black, do not exhibit hue.

IC: *integrated circuit*

idempotent: An operation that does not cause any further changes when applied a second time. The morphological operators *opening* and *closing* are examples for ~ operators.

IEEE-1394 [i.Link]: ≻ *FireWire*

IIR: *infinite impulse response*

illuminance; *g.* **Beleuchtungsstärke**: *luminous flux* incident per unit area of a surface. Units: Lux, lx = lm m^{-2}; symbol: E$_v$.

image; *g.* **Bild**: a reproduction of an object produced by light rays. An ~-forming optical system gathers beams of light diverging from an object point and transforms them into beams that converges toward another point, thus producing an ~.

image analysis; *g.* **Bildanalyse**: the process of extracting objects from images and measuring their geometrical and radiometric properties to gain an understanding of the presented scene. ≻ *high-level* vision.

image compression; *g.* **Bildkompression**: an operation which preserves all or most of the information in the image and which reduces the amount of memory needed to store an image or the time needed to transmit an image.

image coordinates; *g.* **Bildkoordinaten**: A 2-D coordinate system fixed to the focal plane or image plane of a camera. The origin of the ~is given by the intersection of the image plane with the *optical axis* of the camera.

image enhancement; *g.* **Bildverbesserung**: any type of operations which improve the detectability of objects or the appearance of an image.

image intensity; *g.* **Bildintensität**: ≻ *gray level*

image matching; *g.* **Bildvergleich**: the process of determining the correspondence between two images taken of the same scene but with different sensors, different lighting, or a different viewing angle. ~ can be used in the spectral/temporal pattern classification of remote sensing, or in determining corresponding points for stereo, tracking, *change detection*, and motion analysis. In the pixel-based approach, subimages of one image are translated over a second image and the translation with maximum similarity and minimum derivation is determined.

image processing [picture processing]; *g.* **Bildverarbeitung**: the manipulation of images by computer. Encompasses a wide variation of operations which can be applied to an image in order to restore degradations, to enhance its appearance, to extract features for object recognition and classification, or to compress it for storage and transmission.

image reconstruction; *g.* **Bildrekonstruktion:** the process of reconstructing an object from a set of its projections. The most commonly employed reconstruction techniques are the *filtered back projection* and the *algebraic reconstruction* techniques. ~ techniques are important in computer tomography, nuclear medicine, and ultrasonic imaging.

image registration; *g.* **Bildausrichtung:** the process of positioning two images of the same scene with respect to one another so that corresponding points in the images represent the same point in the scene.

image restoration; *g.* **Bildrestaurierung:** ➢ *restoration*

image segmentation; *g.* **Bildsegmentierung:** ➢ *segmentation*

imaging radar; *g.* **bildaufnehmendes Radar:** a radar that constructs an image of the terrain or sea surface by complex processing of the echoes reflected back to the antenna.

imaging spectrometer; *g.* **bildaufnehmendes Spektrometer:** nonscanning spectrometer in which the spectrum generated by a dispersive optical element (prism or grid) is directly imaged onto a linear sensor array.

impedance; *g.* **Impedanz:** the effective resistance of a circuit's input, measured in Ohms.

impulse response; *g.* **Impulsantwort:** output signal of a system to an impulse (δ distribution). A *linear shift-invariant system* is completely described by the ~.

incoherent; *g.* **inkohärent:** lack of a fixed phase relationship between two waves. If two incoherent waves are superimposed, interference effects cannot last longer than the individual coherent times of the waves.

index of refraction [refractive index]; *g.* **Brechungsindex:** the ratio of the velocity of light in vacuum to the velocity of light in a refractive material for a given wavelength.

indirect imaging; *g.* **indirekte Abbildung:** any imaging technique that does not directly result in an image of an object but delivers a spatial or temporal signal that can be used to reconstruct the object. The most common ~ techniques are *tomography* and holography.

infinite impulse response [IIR]; *g.* **unendliche Impulsantwort:** A linear operator has an ~ if its *impulse response* or *point spread function* is not limited. All *recursive filters* have an ~.

infrared; *g.* **Infrarot:** the invisible part of the *EM* radiation with a wavelength between about 0.75 and 1000 μm. Radiation in the near ~ produces a sensation of heat.

inner orientation; *g.* **innere Orientierung:** description of the relation between the 2-D *image coordinates* and the 3-D *camera coordinates*. This includes the position of the principal point (optical axis) in the image coordinates, the focal length, and eventually parameters which describe the lens distortion.

integrated circuit [IC]; *g.* **integrierter Schaltkreis:** thousands to millions of digital circuits integrated into a single chip of silicon, and covered in plastic or ceramic.

integration time [exposure time]; *g.* **Belichtungszeit:** the time interval that the photoelements are allowed to collect charge.

interlaced format; *g.* **Zwischenzeilenformat:** video image format. The video signal is divided into two *fields*. The first field contains all the odd numbered lines and the second field all the even numbered lines. Rows are separated by horizontal sync pulses and successive fields by vertical sync pulses. All standard TV videosignals (RS-170, CCIR) are in the interlaced format.

interline transfer: a method to transfer the charges accumulated at the photodetectors of a CCD array. The charges are first transferred from the photodetectors to storage sites between the lines of sensors and then row by row into a secondary storage area where they are clocked out to form a sequential video signal.

internal orientation; *g.* **interne Orientierung:** ➢ *inner orientation*

invariance; *g.* **Invarianz:** An important concept to classify image processing operators and object features. An operator or feature is called invariant if it does not change under a certain transform. For image features that are invariant under scale, rotation, affine, and perspective transforms are of importance.

inverse filtering; *g.* **inverses Filter:** a technique used in *image restoration*.

inverse operator; *g.* **inverser Operator:** An ~ restores the image to which an operator is applied. For many image processing operators, especially those reducing the information contents, no ~ exists. ➢ *image restoration*

IR: *infrared*

irradiance; *g.* **Bestrahlungsstärke:** Radiant flux incident per unit area of a surface. Also called *radiant flux density*. Units: Wm^{-2}, symbol: E.

IT: *information technology*

JFIF: JPEG interchange format. $\succ$ *JPEG*.

JPEG: *Joint Photographic Experts Group*; widely standard for compressing still pictures into smaller amounts of data. `ftp://ftp.uu.unet/graphics/jpeg/`

JPEG2000: lossy and lossless new standard for compressing still pictures based on wavelet transforms to replace the *JPEG* standard. Achieves better compression rates and less artifacts than the JPEG standard.

kb: kilobit. One thousand bits (actually 2^{10} = 1024).

kB: kilobyte. One thousand bytes (actually 2^{10} = 1024).

kernel; *g.* **Kern:** the $\sim$ of a linear spatial filter is equivalent to the filter's *point spread function*.

kHz: kilohertz. Frequency units, thousand oscillations per second.

knowledge-based vision; *g.* **wissensbasiertes Sehen:** strategy which includes a knowledge data base and a reasoning or inference component. $\sim$ is typically rule based and integrates the information produced by image processing with the information in the knowledge data base and reasons about what hypothesis should be generated and validated, what new information can be inferred from what has already been established, and what new primitives should be extracted next.

kurtosis: forth-order moment describing the shape of the symmetrical part of a distribution function.

Lambertian surface [Lambertian reflector]; *g.* **Lambertscher Strahler:** a perfect diffusion surface that has a reflectance function that is a constant times the cosine of the angle between the incident radiation and the surface normal. The radiance of a Lambertian surface is isotropic, i. e., constant regardless of the angle from which it is viewed.

Landsat: a series of remote-sensing satellites in sun-synchronous, polar orbit that began in 1972. Initially administered by *NASA*, then *NOAA*, and since 1985 by *EOSAT*.

Laplace of Gaussian [LoG]: a Laplace operator regularized by smoothing with a *Gaussian filter*. Presumably playing a significant role in human vision.

Laplacian (operator); *g.* **Laplace-Operator:** a mathematical operator for computing the sum of the second partial derivatives of a multidimensional function. The $\sim$ is isotropic. It does not provide information about the direction in which the function is changing. Widely used for *edge detection*.

Laplacian pyramid; *g.* **Laplace-Pyramide:** in the Laplacian image pyramid, each successive layer is obtained by taking the difference between two corresponding levels on the *Gaussian pyramid*. This procedure leads to a bandpass decomposition of the image with bandpasses of one octave (factor two) distance. The original image can be reconstructed from the $\sim$ by iteratively expanding one level and adding it to the next finer level.

laser; *g.* **Laser:** an acronym of light amplification by stimulated emission of radiation. A $\sim$ is an optical cavity filled with lasable material and with mirrors at the ends. This is any material, the atoms of which are capable of being excited to a semi-stable state by radiation absorption or an electric discharge. The light emitted by an atom as it drops back to the ground state releases other nearby, excited atoms. Thus the light oscillates back and forth between the mirrors and is continuously increased in intensity. If one mirror transmits a few percent of the light, a beam of highly monochromatic, coherent radiation is emitted through the mirror.

laser diode; *g.* **Laserdiode:** a laser with a forward-biased semiconductor junction as the active medium.

learning classifier; *g.* **lernender Klassifikator:** a classifier with an iterative training procedure that increases the *classification* performance accuracy of the *classifier* after each few iterations.

least squares [LS]; *g.* **Methode der kleinsten Quadrate:** selecting the parameters of a model to describe a set of data points so that the sum of the squared differences between the points and the model is minimized.

LDA: *linear discriminant analysis*, a *pattern classification* method.

LED: $\succ$ *light-emitting diode*

light-emitting diode [LED]; *g.* **Leuchtdiode:** a semiconductor device that gives off radiation when biased in the forward direction.

limiting resolution [resolution limit]; *g.* **Auflösungsgrenze**: the smallest dimension of the target or object that can just be discriminated or observed. ≻ *resolution*.

line detection; *g.* **Liniendetektion**: an image *segmentation* technique in which line pixels are identified by examining their *neighborhoods*.

linear and shift invariant [LSI]; *g.* **linear und verschiebungsinvariant**: a system or an operator that is (a) linear and (b) does not explicitly depend on the position. A *convolution* operator is ∼ while morphological operators are only shift invariant.

linear symmetry [local orientation]; *g.* **lineare Symmetrie**: a property of a *neighborhood* in which the gray values change only in one direction. It can therefore be described by a function of the type $g(x\bar{k})$, where $\bar{k}$ is a unit vector in the direction of the changing gray values.

local frequency, local wave number; *g.* **lokale Frequenz, lokale Wellenzahl**: the instantaneous frequency or wave number of the periodical signal. Can be computed from the *phase* of the signal.

local orientation; *g.* **lokale Orientierung**: ≻ *linear symmetry*

look-up table (LUT); *g.* **Nachschautabelle**: a table that stores the values for *point operations*. The gray value of the input pixel is taken as the address to the ∼. The output of the point operator is then the value stored at that location in the ∼.

lossless encoding; *g.* **verlustfreie Kodierung**: any *image compression* technique that represents *gray levels* compactly but permits exact reconstruction of the image. Example: *run length encoding*; contrast with *lossy encoding*.

lossy encoding; *g.* **verlustbehaftete Codierung**: any image compression technique which only approximates the signal. An exact reconstruction is not possible. Example: *JPEG*; contrast with *lossless encoding*.

low-level vision: The first phase of image analysis. In ∼ the processing is local and is independent of purpose or content. The output is a set of local features that are used to detect, characterize, and classify objects.

LS (least squares): *least squares*

lumen: Units for the *luminous flux*.

luminance; *g.* **Leuchtdichte**: *luminous flux* emitted from a surface per unit solid angle per unit area, projected onto a plane normal to the direction of propagation. Units: $\mathrm{lm\,sr^{-1}\,m^{-2}}$ or $\mathrm{cd\,m^{-2}}$.

luminous efficacy [luminous efficiency]; *g.* **Lichtausbeute**: (1) radiant ∼: quotient of total luminous flux divided by total radiant flux; measure of the effectiveness of radiation in stimulating the perception in the human eye. (2) lighting system ∼: quotient of total luminous flux divided by lamp power input. Units: lm/W.

luminous energy; *g.* **Lichtmenge**: a measure of the time-integrated amount of *luminous flux*. It might be used to describe such things as the radiant energy that the eye would receive from a photographic flash. Units: lm s or talbot; symbol: Q_v.

luminous excitance [luminous emittance]; *g.* **spezifische Lichtausstrahlung**: luminous flux emitted per unit area of a source. Units: $\mathrm{lm\,m^{-2}}$; symbol: M_v.

luminous flux; *g.* **Lichtstrom**: radiant power of visible light modified by the eye response. It is defined as the amount of flux radiated by a source of one candela into a solid angle of one steradian. Units: lm (*lumen*); symbol: Φ_v.

luminous intensity; *g.* **Lichtstärke**: luminous flux emitted by a source in a given range of directions. Units: lm/sr or cd (*candela*); symbol: I_v.

LUT: *look-up table*

lux: units of *illuminance*, $\mathrm{lm/m^2}$.

machine vision system; *g.* **Bildauswertungssystem**: a system that acquires one or more images of an object, processes, analyzes, and measures various characteristics of the acquired images, and interprets the results of the measurements in such a way that a decision can be made about the object. Functions of machine vision systems include locating, inspecting, gauging, identifying, recognizing, counting, and motion estimating.

mathematical morphology; *g.* **mathematische Morphologie**: ≻ *morphology*

matte reflectivity; *g.* **diffuse Reflexion**: a surface with a dull (diffuse) reflectivity. ≻ *Lambertian surface*.

MB: megabyte. One million bytes (actually 2^{20} = 1,048,567).

median filter; *g.* **Median-Filter**: a nonlinear *neighborhood* operator in which the value of an output pixel is the median value of all the input pixels in the supporting neighborhood of the filter about the given pixel's position. ~s are used to smooth and remove *noise* from images.

MFLOPS: million floating point operations per second

microwave; *g.* **Mikrowellen**: *EM* radiation with frequencies between about 1 GHz (30 cm wavelength) and 100 GHz (3 mm wavelength)

mid-infrared (MIR); *g.* **mittleres Infrarot**: *EM* radiation with wavelengths between 8 and 14 μm dominated by emission of thermally generated radiation from materials. Also known as thermal infrared.

Mie scattering; *g.* **Mie-Streuung**: the scattering of *EM* energy by particles in the atmosphere with similar dimensions to the wavelength involved.

MIMD: multiple instruction stream, multiple data stream

minimum distance classification: a classification technique in which a class is modeled by a vector in the *feature space*. The distance between the object's feature vector and the class vector decides to which class the object is associated. The class with the minimum distance is taken.

MIPS: million instructions per second

misdetection: ≻ *misidentification*

misidentification [type I error]: In *pattern classification*, the failure to assign a *pattern* to its true *pattern class*. Contrast with *false identification*.

MKSA-units: meter, kilogram, seconds, and ampere. Basic units of the standard international (*SI*) metric system of units.

MMX: multimedia extension, *SIMD* instruction set extension for pixel processing. Introduced by Intel in January 1997 with a new generation of the Pentium processor family.

model fitting; *g.* **Modellfit**: selecting the parameters of a model to describe a set of data points. ≻ *least squares*.

model-based image analysis; *g.* **modellbasierte Bildanalyse**: a strategy for image analysis which employs an explicit model of the object to be recognized. Recognition proceeds in a *top-down* control by matching the object data structure inferred from the features extracted from the image to the model data structure.

modulation transfer function [MTF]; *g.* **Modulationstransferfunktion**: contrast of the image of a pattern related to its original contrast as a function of the wave number. The wave number is normalized to the limit given by the *sampling theorem*, which is half of the numbers of photo detectors per unit length.

moment; *g.* **Moment**: ~s can be composed from binary and gray value regions to describe their shape. The zero-order ~ is the area and gray value sum, the first-order ~ the centroid and center-of-gravity for binary and gray value regions, respectively. Second-order ~s model the region as an ellipse.

morphological edge operator; *g.* **morphologischer Kantenoperator**: *edge operator* based on morphological operations.

morphology; *g.* **Morphologie**: an area of image processing concerned with the analysis of shape. The basic morphologic operations consist of dilating, eroding, opening, and closing an image with a structuring element.

motion field; *g.* **Geschwindigkeitsfeld**: the projection of the 3-D ~ of the scene onto the image plane. Each pixel of the ~ image contains a 2-D velocity vector.

motion-JPEG: compression method for motion video. It enables video to be recorded and played back from the computer's hard disk by saving the data as a series of compressed still *JPEG* frames.

MPEG: MPEG 1: compression standard optimized for storing compressed audio and video onto a CD. MPEG 2: same, but optimized and improved for broadcast use.

MTF: *modulation transfer function*

multiband image: ≻ *multispectral image*

multichannel image [vectorial image]; *g.* **Mehrkanalbild, Vektorbild**: any image with more than one channel. In a ~ with P channels, a pixel thus is a P-dimensional vector. Examples for ~s are color images, stereo images, and multispectral images.

multigrid [multiresolutional]; *g.* **Mehrgitter**: refers to image processing techniques that use images at different resolution levels. Simple implementations of ~ data structures for images are the *Gaussian* and *Laplacian pyramids*. ~ data structures result in a significant speed up of the processing at coarser scales.

multispectral image; *g.* **Multispektralbild**: a *multichannel image* in which each channel is an image taken at the same time, but sensitive in a different part of the electromagnetic spectrum.

nadir: the point on the ground vertically beneath the center of a remote-sensing system.

NASA: National Aeronautics and Space Administration, USA

near-infrared (NIR); *g.* **nahes Infrarot**: the shorter wavelength range of the infrared region of the *EM* spectrum, from 0.7 to 2.5 μm. It is often divided into the very-near infrared (VNIR) covering the remainder of the NIR atmospheric window from 1.0 to 2.5 μm.

NEDT [NEΔT]: noise equivalent temperature difference

NEE: noise equivalent exposure

neighborhood; *g.* **Nachbarschaft**: a set of pixels located in a region around a given pixel.

neighborhood operator; *g.* **Nachbarschaftsoperator**: An *image operator* that assigns a *gray level* to each output pixel based on the gray levels in a neighborhood of the corresponding input pixel. The three basic classes of ~s are *convolution* operators, *rank-value filters*, and binary *morphological operators*; contrast with *point operator*.

neural net; *g.* **neuronales Netz**: an interconnected network of nonlinear processing elements for object recognition and classification capable of learning and self organizing. The response of a unit or a processing element is a nonlinear monotonic function of a weighted sum of the inputs to the processing elements. The weights, called synaptic weights, are modified by a learning or reinforcement algorithm. In a simple ~ each processing element contributes one component to the output response vector and its inputs are selected from the components of the input pattern vector. Processing units whose outputs do not directly influence the components of the output response vector are called hidden units.

neutron; *g.* **Neutron**: an elementary particle without electrical charge. The nucleus of atoms consists of ~ and *protons*.

Nimbus-7: an experimental polar orbiting satellite launched in 1987, carried the Coastal Zone color Scanner (CZCS), the Scanning Multichannel Radiometer (SMMR),

and the Total Ozone Mapping Spectrometer (TOMS).

NOAA: National Oceanic and Atmospheric Administration, USA

noise; *g.* **Rauschen**: random variations in a signal or an image, caused by the processes and transformations that are employed in sensing the signal or scene and converting it to a computer-usable form but that are irrelevant or meaningless and not related to the sensed object.

noise equivalent exposure [NEE]: the radiative (luminous) energy per unit area required to generate an output signal equal to the output noise level of a sensor. The ~ describes the lower limit on detectable *irradiance* (*illuminance*).

normal distribution; *g.* **Normalverteilung**: the standard statistical distribution for which the probability density function is the Gaussian function $\exp(-(x - \overline{x})^2/(2\sigma^2))$. The ~ is completely described by the mean $\overline{x}$ and the variance σ^2. Most statistical data are well described by a ~. The discrete analog to the ~ is the *binomial distribution*.

normalized convolution; *g.* **normalisierte Faltung**: a versatile nonlinear *neighborhood operator* based on *convolution*. Each pixel has not only a value but also a weight. ~ performs two convolutions: one with the product of the value with the weight and one with the weight using the same convolution mask. The result of the ~ is the ratio of both convolution operators. If the weight is the same for all pixels, ~ reduces to linear convolution.

NTSC: National Television Standards Committee. The original color television standard developed in the USA.

object extraction; *g.* **Objektextraktion**: ≻ *segmentation*

occluding edge; *g.* **Okklusionskante**: a boundary appearing on an image due to a discontinuity in range or depth of an object in the observed scene. Step edges in depth maps are always occluding edges.

occlusion; *g.* **Okklusion**: the hiding from view of part of one object by another.

OCR: optical character recognition

opening; *g.* **Öffnen**: a *idempotent* morphological operator. It is the dual operator to *closing*. Opening an image with a disk shaped structuring element smoothes the contour, breaks narrow isthmuses, and eliminates all objects or part of objects smaller in size or width than the structuring element.

operator; *g.* **Operator:** transforms an image into another image by applying certain computations. The three basic classes of operators are *point operators, neighborhood operators,* and global transforms such as the *Fourier transform.*

optical axis [principal axis]; *g.* **optische Achse:** the straight line which passes through the symmetry center of an optical system.

optical depth [optical density]; *g.* **optische Dichte:** a logarithmic measure of the *transmittance* along a path through an optical medium. The ~ is equal to the base 10 logarithm of the reciprocal of the transmittance. In a homogeneous medium, the optical depth is directly proportional to the path length and the *extinction coefficient.*

optical flow [optical flux]; *g.* **optischer Fluß:** apparent motion at the image plane based on the visual perception. Ideally but not necessarily identical to the *motion field.*

optical gauging [visual gauging]; *g.* **optische Vermessung:** measuring specific positions or dimensions of a manufactured object by using imaging sensors to compare these measurements to preselected tolerance limits for quality inspection and sorting decision. Gauging has wide application in manufacturing since it can determine the diameters of holes, openings, or cutouts, the widths of shafts, components, gaps, wires, or rods, and the relative locations of holes, folds, features, components, openings, or breaks.

optical image; *g.* **optisches Bild:** the result of projecting a scene onto a surface using an optical system.

optical transfer function [OTF]; *g.* **optische Transferfunktion:** the amplitude damping and spatial phase shift of the image of a sinusoidal object as a function of its wave number.

orbital period; *g.* **Umlaufzeit:** the time taken by a satellite to complete an orbit.

orthographic projection; *g.* **orthographische Projektion:** ≻ *parallel projection*

OTF: *optical transfer function*

outer orientation; *g.* **äußere Orientierung:** ≻ *exterior orientation*

PAL: phase alternate line. The television standard currently in use in most parts of the Commonwealth and Europe.

parallax; *g.* **Parallaxe:** the observed positional difference of a projected 3D point on a pair of 2D perspective images. The difference in position is caused by a shift in the position of the perspective centers and optical axis orientation and is used to estimate the depth of the point.

parallel projection [orthographic projection]; *g.* **Parallelprojektion:** a ~ of a point $[X, Y, Z]^T$ onto a point $[x, y]^T$ on a plane perpendicular to the Z-axis is defined by $x = X, y = Y$.

paraxial; *g.* **paraxial:** characteristic of optical system that is limited to infinitesimally small apertures. Also called first-order or Gaussian optics. Practically, the range in which the angles between the rays and the optical axis are sufficiently small to set $\alpha = \tan \alpha = \sin \alpha$.

pattern [feature]; *g.* **Muster [Merkmal]:** a meaningful regularity or a collection of features (≻ *feature vector*) that can be used to classify objects or other items of interest.

pattern class [pattern category]; *g.* **Musterklasse:** one of a set of mutually exclusive categories into which a pattern can be classified.

pattern classification [pattern identification]; *g.* **Musterklassifikation:** the process of assigning patterns to *pattern classes.*

pattern recognition; *g.* **Mustererkennung:** identification of objects on the basis of their *feature.* In statistical pattern recognition, the features are P-dimensional feature vectors. In syntactic pattern recognition, they have the form of sentences from the language of a phrase structure grammar. In structural pattern recognition, the object being measured is encoded in terms of its parts and the relationships as well as properties of the parts.

PCI bus; *g.* **PCI-Bus:** peripheral component interconnect bus. A local bus standard offering fast data transfer within a computer, especially real-time image data transfer from image input devices, e. g., frame grabbers, to the *RAM* of the computer.

PCX: PC Paintbrush file format

pel: ≻ *pixel*

perimeter; *g.* **Umfang:** the ~ of a connected region R is the length of the bounding contour of R.

perspective projection; *g.* **perspektivische Projektion, Zentralprojektion**: a projection in which points in 3D space are mapped onto an image by lines or "rays" passing through a single point in space called the *center of perspectivity*. The ideal model of these systems is the pinhole camera. Optical systems, the human eye, and X-ray imaging imply ~.

phase; *g.* **Phase**: The ~ of a signal is an amplitude independent robust feature of a signal. The phase ϕ of a sinusoidal spatial signal $\exp(i\mathbf{kx})$ is given by $\phi = \mathbf{kx}$. For an arbitrary signal, the phase can be computed by applying a *quadrature filter* or a *Hilbert filter* to get a signal pair $^+q, ^-q$ that consists of two signals with the same local amplitude but a phase shift of $\pi/2$ (90°). Then the local phase of the signal is given by $\phi = \arctan(^-q/^+q)$.

photogrammetry; *g.* **Photogrammetrie**: engineering discipline dealing with measurements from photographs. Main applications are mapmaking and surveying using aerial and satellite photography.

photographic infrared; *g.* **Photographisches Infrarot**: that part of the near-infrared spectrum to which photographic films respond, wavelengths between 0.7 μm and about 1.0 μm. Virtually equivalent with *very near-infrared* (VNIR).

photometric stereo; *g.* **photometrisches Stereo**: two images obtained from a stationary camera, but using a differently positioned light source for each image, provide sufficient information for depth recovery. ~ eliminates the matching problem by the fixed positioning of the camera relative to the scene.

photometry; *g.* **Photometrie**: the science of the measurement of light intensity, where "light" refers to the range of radiation to which the eye is sensitive. ~ relates *radiometry* to the human eye's response.

photon; *g.* **Photon**: a quantum of electromagnetic energy. The energy of a photon equals $h\nu$, h being Planck's constant and ν the frequency of the propagating electromagnetic wave. The momentum of the photon in the direction of propagation is $h\nu/c$, c being the velocity of light.

photonics; *g.* **Optoelektronik**: The technology of generating and detecting light and other radiation including radiation emission, transmission, deflection, amplification, and detection; lasers and other light sources, fiber optics, and electro-optical instrumentation.

photopic vision; *g.* **Hellsehen**: vision under conditions of bright illumination, when both *rods* and *cones* are employed.

photoresponse non-uniformity [PRNU]: variation in response of the sensor elements in a sensor array under uniform illumination.

PICT: PICTure data format. Data format on Apple's Macintosh computers for 2-D graphics and 2-D raster images.

picture: ≻ *image*

picture element: ≻ *pixel*

pixel [pel, resolution cell]; *g.* **Bildpunkt**: contraction of picture element. A small, usually square or rectangular area at the image plane, in which an average *irradiance* value is determined. A complete *digital image* is formed by an array of ~s.

Planck's law: an expression for the variation of emittance of a *blackbody* at a particular temperature as a function of wavelength.

PLL: phase locked loop

point operator; *g.* **Punktoperator**: an *image operator* that assigns a *gray level* to each output pixel based on the gray level of the corresponding input pixel. If the assignment depends on the position in the image the ~ is inhomogeneous; otherwise the ~ is homogeneous. Contrast with: *neighborhood operator*.

point spread function [PSF]; *g.* **Punktantwort**: response of a linear system or operator to a point (impulse) input. Gives a complete description of the properties of a linear system, i. e., its output for any input can be computed by convolving the ~ with the input signal. Syn: *impulse response*, Green's function.

polar orbit; *g.* **polare Umlaufbahn**: an orbit that passes close to the poles, thereby enabling a satellite to pass over most of the earth's surface, except the immediate vicinity of the poles themselves.

polarization; *g.* **Polarisation**: the restriction of the vibrations of the magnetic or electric field of *EM* radiation to a single plane. In a beam of *EM* radiation, the ~ direction is the direction of the electric field vector. The ~ vector is always in the plane at right angles to the beam direction. The ~ direction in the beam can vary at random (unpolarized), can remain constant (linear ~), or can have two coherent plane-polarized elements whose ~

directions make a right angle. In the latter case, depending on the amplitude of the two waves and their relative phase, the combined electric vector traces out an ellipse (elliptical ~). Elliptical and plane ~s can be converted into each other by means of birefringent optical systems.

Postscript: a high-level, device-independent, stack-based page description language developed by Adobe Systems Inc. for output to raster-based graphics devices.

power spectrum; *g.* **Leistungsspektrum**: squared magnitude of the Fourier transform of a signal (time series or spatial data). If the amplitude of the signal is related to a physical oscillation, the ~ directly gives the energy distribution in the *Fourier domain*.

precision; *g.* **Präzison**: refers to degree of deviation between repeated measurements taken with the same experimental setup. A ~ measurement results in a low scatter of the measurement. A ~ measurement is not necessarily accurate, i. e., close to the true value. ≻ *accuracy*.

preprocessing; *g.* **Vorverarbeitung**: an operation performed before a primary process; for example, in *pattern recognition*, processing in which patterns are simplified to make *classification* easier.

principal axis: ≻ *optical axis*

principal axis transform: ≻ *principal components*

principal components; *g.* **Hauptachsen**: directions, or axes, in multichannel image, defined by approximating the data from a whole image, or selected area in an image, to a multidimensional ellipsoid. The first ~ direction is parallel to the longest axis of this ellipsoid, all other ~s are orthogonal to it. The ~s are computed by the principal axes transform based on the *covariance matrix* of the data. The ~s are an optimal representation of multichannel data since they are uncorrelated. ~s with low variance can be omitted without affecting the data quality, thus reducing the number of required channels in a multichannel image.

principal distance [camera constant, Gaussian or effective focal length]; *g.* **Brennweite**: the distance between the *center of perspectivity* and the image projection plane.

principal point; *g.* **Hauptpunkt**: that point on the image which is the intersection of the image plane with the *optic axis*.

PRNU: *photoresponse non-uniformity*

probability density function (PDF); *g.* **Wahrscheinlichkeitsdichte**: a function indicating the relative frequency with which any measurement may be expected to occur. Can be estimated by *histograms*.

proton; *g.* **Proton**: an elementary particle carrying one positive elementary charge. The nucleus of atoms consists of ~s and *neutrons*.

PS: *postscript*

pseudo-color image [false color image]; *g.* **Falschfarbenbild**: A color image in which a gray-scale image or parts of the nonvisible *EM* spectrum are expressed as one or more of the red, green, and blue components, so that the colors produced do not correspond to normal visual experience. The most commonly seen false-color images display the very near-infrared as red, red as green, and green as blue.

PSF: *point spread function*

pyramid [image pyramid]; *g.* **Pyramide**: a sequence of copies of an image in which both sample density and *resolution* are decreased in regular steps. The bottom level of the ~ is the original image. Each successive level is obtained from the previous level by a smoothing operator followed by a sampling operator.

QDA: *quadratic discriminant analysis*, a *pattern classification* method.

quad spectrum: imaginary part of the complex-valued *cross-correlation spectrum*.

quadrature filter; *g.* **Quadraturfilter**: a pair of filters with the same magnitude of the transfer function. One filter is of even symmetry and has a real transfer function; the other is of odd symmetry and has a purely imaginary transfer function. Thus the response from the two filters differs by a phase shift of $\pi/2$ (90°).

quantization; *g.* **Quantisierung**: in *image processing*, a process in which each pixel in an image is assigned one of a finite set of *gray levels*.

quantum; *g.* **Quant**: the elementary quantity of *EM* energy. According to quantum mechanics, *EM* radiation is emitted, transmitted, and absorbed as numbers of quanta. ≻ *photon.*

quantum efficiency; *g.* **Quantenausbeute**: (1) For a radiating source, the ratio of photons emitted per second to electrons flowing per second. (2) For detectors, the ratio of the number of electron/hole pairs generated to the number of incident photons.

radar; *g.* **Radar**: the acronym for radio detection and ranging, which uses pulses of micro- or radio waves that are emitted to locate objects which reflect the radiation. The position of the object is given by the time that a pulse takes to reach it and to return to the antenna.

radar altimeter: a nonimaging device that records the time of radar returns from vertically beneath a platform to estimate the distance to and hence the elevation of the surface. Carried by Seasat and the *ERS* satellites.

radar scatterometer: a nonimaging device that records radar energy backscattered from terrain as a function of the *angle of incidence.* Used over the oceans to derive global maps of the wind speed and directions.

radiance; *g.* **Strahlungsdichte**: radiant power per unit source area per unit solid angle. Units: $W\,m^{-2}\,sr^{-1}$; symbol: L.

radiant; *g.* **Strahlungs–**: pertaining to electromagnetic radiation, with the contributions at all wavelengths of interest weighted equally.

radiant efficiency; *g.* **Strahlungsausbeute**: the ratio of the radiant flux emitted by a source to the power supplied; dimensionless; symbol: η.

radiant energy; *g.* **Strahlungsenergie**: the energy passed on as *EM* radiation. Units: Joule [J]; symbol: Q.

radiant excitance [emittance]; *g.* **(emittierte) Strahlungsflußdichte**: the radiant flux per unit area emitted from a surface. Units: $W\,m^{-2}$; symbol M.

radiant flux [radiant power]; *g.* **Strahlungsfluß**: the power incident on or leaving a body in the form of EM radiation. Units: W; symbol: Φ.

radiant flux density; *g.* **Strahlungsflußdichte**: the power incident on (*irradiance*) or leaving (*radiant excitance*) a body in the form of *EM* radiation per unit area of surface. Units: W/m^2; symbol: E or M.

radiant heat; *g.* **Wärmestrahlung**: *infrared* radiation emitted from a source that is not heated sufficiently to give off visible radiation.

radiant intensity; *g.* **Strahlstärke**: the radiant flux emitted per unit solid angle. Unit: $W\,sr^{-1}$; symbol: I.

radio waves; *g.* **Radiowellen**: *EM* radiation with frequencies between 30 kHz and 1 GHz used to broadcast radio and television signals.

radiometry; *g.* **Radiometrie**: the science of radiation measurement: detection and measurement of *radiant energy*, either as separate wavelengths or integrated over a broad wavelength band, and the interaction of radiation with matter in such ways as absorption, reflection, scattering, and emission.

radiosity; *g.* **Radiosität**: total radiance of a surface, i. e., emitted *and* reflected radiance.

radix; *g.* **Radix**: Total number of characters available to each position of a digital numeric system.

RAM: random access memory

range image; *g.* **Tiefenbild**: an image in which the value on each *pixel* value is a function of the distance between the pixel and the object surface patch imaged on the pixel. Depending on the sensor and *preprocessing* used to create the ~ either the absolute or only a relative distance is measured.

rangel: the range data element produced by a range sensor. ≻ *range image.*

rank value filter; *g.* **Rangordnungsfilter**: a class of shift-invariant nonlinear filter operators based on sorting the pixels in a neighborhood and selecting one of the sorted pixels. The best known ~ is the *median filter.* Others include the minimum and maximum filters for gray scale morphology.

raster; *g.* **Raster**: the scanned and illuminated area of a video display, produced by a modulated beam of electrons sweeping the phosphorescent screen line by line from top to bottom at a regular rate of repetition.

raster scan order; *g.* **Abtastrichtung**: refers to the sequence of *pixel* locations obtained by scanning the spatial domain of an image in a left to right scan of each image row with the rows taken in a top to bottom ordering. Frame format video images are images scanned in raster scan order.

ray tracing; *g.* **Strahlverfolgung**: an approach to creating synthetic images of modeled 3-D scenes in which the intensity at a particular image *pixel* is determined by tracing all rays from light sources to the 3-D scene surfaces that contribute to the intensity at that pixel.

Rayleigh criterion; *g.* **Rayleigh-Kriterium**: a way of quantifying surface roughness with respect to wavelength λ of *EM* radiation, to determine whether the surface will act as a specular reflector or as a diffuse reflector. A surface can be considered rough if the root mean square height of surface irregularities is greater than $\lambda/(8 \cos \theta)$, where θ is the *angle of incidence*.

Rayleigh scattering; *g.* **Rayleigh-Streuung**: scattering by particles very small compared to the wavelength of the radiation being considered. The scattered flux is inversely proportional to the fourth power of the wavelength. Thus blue light is scattered more strongly by the molecules of the air than longer wavelengths, accounting for the blue color of the sky.

real-aperture radar: an imaging radar system where the azimuth resolution is determined by the physical length of the antenna, the wavelength, and the range.

real-time; *g.* **Echtzeit**: processing of an image at the same or faster rate than the frame rate of the sensor.

recursive filter; *g.* **rekursives Filter**: a linear shift-invariant filter that works similar to convolution but adds also previous results of the filter in the convolution sum. This gives the filter a run direction which is natural for time series but not for spatial data.

reflectance; *g.* **Reflektivität**: the ratio of radiant flux reflected from a surface to that which falls upon it. The suffix -ance implies a property of a specific piece of material and may include multiple reflections such as from a thin plate where reflections from the front and back face and multiple reflections from rays bouncing back and force between the reflecting surfaces contribute to the total fraction of radiant flux being reflected and thus contributing to the reflectance. Units: 1; symbol R. Contrast with *reflectivity*.

reflectivity; *g.* **Reflexionskoeffizient**: the ratio of the *EM* energy reflected by a surface to that which falls upon it. In contrast to the *reflectance*, the ~ includes only a *single* reflection from the surface and as such is an *intrinsic* property of the optical properties of the media on both sides of the surface. Units: 1; symbol ρ.

refraction; *g.* **Brechung**: the bending of an oblique incident ray as it passes from one optical medium to another optical medium with a different *index of refraction*.

refractive index: $\succ$ *index of refraction*

region; *g.* **Region**: a connected subset of an image, usually representing an object of interest.

region growing; *g.* **Regionenwachstum**: a sequential image *segmentation* procedure in which pixels are successively added to existing regions or initiate new regions when it is not appropriate to make them part of any of the existing regions.

registration; *g.* **Ausrichtung**: the geometrical transforms (translation, rotation, affine and perspective transforms) required by which two images of the same set of objects are positioned coincident with one another so that corresponding points of the imaged scene appear in the same position on the registered images. Required if a scene is observed with multiple imaging systems, e. g., with a CCD camera in the visible and an thermal imager in the infrared.

relative orientation; *g.* **relative Orientierung**: in analytic *photogrammetry* the relative position and orientation of multiple camera systems to each other.

remote sensing; *g.* **Fernerkundung**: techniques that utilize *EM* or other radiation to detect and quantify physical, chemical, and biological properties about objects that are not in contact with the sensing apparatus. Term mostly used for observation of the earth's surface and atmosphere from satellites and airplanes.

resolution; *g.* **Auflösung**: describes how well an imaging system reproduces closely spaced objects or lines. ~ may depend on object contrast and spatial position as well as its shape (point, line etc.). $\succ$ *limiting resolution*.

resolution cell: $\succ$ *pixel*

responsivity [sensitivity]; *g.* **Empfindlichkeit**: the ratio of the electrical signal generated by a photosensor to the irradiance. Units: A/W or V/W; symbol: R. In the linear range of the sensor, the ~ is constant.

restoration; *g.* **Restaurierung**: the process of returning an image to its original condition by reversing the effects of known or estimated distortions and degradations such as blurring by lens aberrations, velocity smearing, defocusing, vibration, and geometrical distortions.

RGB: color coordinate system for a color image, where R, G, and B are the luminance of the red, green, and blue channel.

RISC: reduced instruction set computer

robust; *g.* **robust**: a vision procedure is said to be robust if small changes in the assumed model on which the procedure or technique was developed produce only small changes in the result. Small fractions of the data which do not fit the assumed model and which in fact are very far from fitting the assumed model, constitute a small change in the assumed model. Data not fitting an assumed model may be due to rounding or quantizing errors, gross errors, or because the model itself is only an idealized approximation to reality.

rod; *g.* **Stäbchen**: the receptors in the retina which are sensitive to the *luminance* and not to the color of the light. ≻ *cone.*

RS-170: Electronic Industries Association (EIA) standard governing monochrome television signals. Basic standard of US video signal with 30 frames/sec and 525 lines/frame. ≻ *CCIR.*

RS-232-C: standard electrical interface for connecting peripheral devices to computers employing serial binary data interchange.

RS-422: specifies the electrical aspects for wideband communication over balanced lines at data rates up to 20 Mbits per second. Widely used for cameras with digital output.

run-length encoding; *g.* **Lauflängencodierung**: compact representation for binary (and also gray scale) images. ~ makes use of the fact that longer pieces of zeros and ones are contained in binary images. Thus it is only necessary to store the lengths of these pieces and not each pixel separately. There are a variety of run length encoding formats.

sampling; *g.* **Abtastung**: conversion of a continuous image (as projected by an imaging system onto the image plane) into a digital image, selecting an array of points, *pixels*, to represent the whole image.

sampling theorem; *g.* **Abtasttheorem**: general theorem of signal theory stating the conditions under which an array of discrete points, a digital image, can completely represent a continuous image. Expressed in words, the ~ simply states that each periodic component contained in the continuous image must be sampled at least twice per wavelength.

SAR: synthetic aperture radar

saturation; *g.* **Sättigung**: (1) The decrease in the *absorption coefficient* of a medium when the power of the incident radiation exceeds a certain value. (2) With respect to color, attribute how intense a color is perceived. The saturation of a pure (monochromatic) color is one. By mixing a pure color with white, any saturation of the color between 0 and 1 is obtained. ≻ *hue.*

saturation exposure; *g.* **Sättigungsbelichtung**: the exposure (irradiance × integration time) level that produces a saturation level in the pixel charge of a photosensor. For a given exposure time, the ~ sets the upper limit for the irradiance that can be measured.

scale space; *g.* **Skalenraum**: a multiscale representation of a signal. It is obtained by convolving the signal by a Gaussian filter with continuously increasing standard deviation. This variable constitutes the additional coordinate of the ~. The physical analog to the ~ is a *diffusion* process. The time coordinate is equivalent to the scale parameter.

SCART: Syndicat des Constructeurs d'Appareils Radio Récepteurs et Téléviseur. A standardized 21-pin multifunction connector found on TVs and videos.

scattering; *g.* **Streuung**: an atmospheric effect where *EM* radiation, usually of short visible wavelength, is propagated in all directions by the effects of gas molecules and aerosols. See *Rayleigh* and *Mie scattering.*

scattering coefficient; *g.* **Streukoeffizient**: property of an optical medium that part of the radiation is directed, e.g., by inhomogeneities in the index of refraction, into other directions. Equal thicknesses scatter equal fractions of the radiant flux: $d\Phi/\Phi = -\beta(\lambda)dx$. Units: [1/m]. ≻ *absorption coefficient, extinction coefficient.*

scotopic vision; *g.* **Dunkelsehen**: vision under conditions of low illumination, when only the rods are sensitive to light. Visual acuity under these conditions is highest in the blue part of the spectrum.

Seasat: polar-orbiting satellite launched in 1978 by NASA to monitor the oceans, using imaging radar and a radar altimeter. It survived for only a few months.

SECAM: *séquentielle colour a mémoire*. A color television standard used mostly in France and the Commonwealth of Independent States.

segmentation; *g.* **Segmentierung**: the process of dividing an image into regions. ≻ *edge detection; line detection; region growing*.

sel: sensor element

sensitivity: ≻ *responsivity*

separable filter; *g.* **separierbares Filter**: a multdimensional filtering operation is called separable if the *convolution* can be decomposed into successive one-dimensional convolutions in all coordinate directions. A ∼ can be computed much faster than a nonseparable filter.

shape from contour; *g.* **Gestalt aus Konturen**: inferring the 3-D shape of an object from a 2-D perspective projection view of the non-hidden edges of the objects.

shape from shading; *g.* **Gestalt aus Schattierung**: inferring the 3-D shape of an object, i. e., the *surface normals*, from the *gray value* shading resulting from the relations between the incidence angle(s) of the light source(s), the viewing angle of the camera, and the *bidirectional reflectivity* of the surface.

shape from texture; *g.* **Gestalt aus Textur**: inferring the 3-D shape of a homogeneously textured surface from the *texture* density variations manifested by the surface on a *perspective projection* or *orthographic projection* image.

sharpening; *g.* **Verschärfung**: Any image enhancement technique which increases the contrast at edges in an image.

short-wavelength infrared (SWIR); *g.* **kurzwelliges Infrarot**: the part of the near or reflected infrared with wavelengths between 1.0 μm and 3.0 μm. These wavelengths are too long to affect infrared photographic film.

SI-units; *g.* **Si-Einheiten**: units of the standard international (SI) metric system of units. ≻ *MKSA-units*.

signal-to-noise-ratio [S/N]; *g.* **Signal-zu-Rausch-Verhältnis**: the ratio of the maximum value of an output signal to the amplitude of the *noise* on the signal.

SIMD: single instruction stream, multiple data stream

simple neighborhood [linear symmetry]; *g.* **einfache Nachbarschaft**: a neighborhood in a multidimensional image in which the gray values change only in one direction. ≻ *linear symmetry*.

SIR: shuttle imaging radar; synthetic-aperture radar experiments carried aboard the NASA Space Shuttle in 1981 and 1984.

SISD: single instruction stream, single data stream

skewness; *g.* **Schiefheit**: third-order moment describing the elementary asymmetry of a distribution. ≻ *moments, kurtosis*.

smoothing [noise cleaning; noise suppression]; *g.* **Glättung**: any image enhancement technique in which the effect of noise in the original image is reduced or fine structures are diminished or removed.

space domain; *g.* **Ortsraum**: representation of a continuous (discrete) image by a function (matrix) explicitly describing the gray values as a function of the spatial coordinates. ≻ *Fourier domain*.

space-time image [image sequence, spatiotemporal image]; *g.* **Orts-Zeit-Bild [Bildsequenz]**: an image with both a spatial and a temporal coordinate.

spatiotemporal image: ≻ *space-time image*

spectroradiometer; *g.* **Spektralradiometer**: a device which measures the radiance reflected, transmitted, or emitted by an object as a function of the wavelength of the *EM* radiation.

specular reflection; *g.* **spiegelnde Reflexion**: reflection of *EM* radiation from a smooth surface, which behaves like a mirror. A specular reflector has a surface that is smooth on the scale of the wavelength of *EM* radiation concerned. If the surface is rough on the scale of the wavelength of the *EM* radiation, it will behave as a diffuse reflector.

spherical aberration; *g.* **sphärische Aberration**: that basic aberration which leads to the failure of a lens to form a perfect image of a monochromatic, on-axis point source object. When rays from a point on the axis passing through the outer lens zones are focused closer to the lens than rays passing the central zones, the lens is said to have negative ∼; if the outer zones have a longer focal length than the inner zones, the lens is said to have positive ∼. In the first case, the lens is said to be uncorrected or undercorrected; in the second it is overcorrected.

spline; *g.* **Spline**: an interpolation technique using piecewise polynomial curves with smoothness constraints between the polynomial segments.

SPOT: satellite probatoire pour l'observation de la terre, a French satellite carrying two pushbroom imaging systems, one for three wavebands in the visible and *VNIR* with 20-m resolution, the other producing panchromatic images with 10-m resolution. Each system comprises two devices which are pointable so that offnadir images are possible, so that stereo image pairs can be taken. Launched in February 1986.

SRAM: static random access memory. Fast but expensive memory, typically used as cache memory to decrease the access time of the *CPU* to memory.

statistical pattern recognition; *g.* **statistische Mustererkennung**: an approach to *pattern recognition* that uses probability and statistical methods to assign patterns to *pattern classes*.

steerable filter; *g.* **einstellbares Filter**: a suitable filter set with a small number of basis filters that can continuously be steered in one or more of its properties by suitable interpolation between the basis filters. A simple example of a steerable filter is a smoothing filter that can be steered to smooth in any direction.

Stefan-Boltzmann law: a radiation law stating that the energy radiated by a blackbody is proportional to the fourth power of its absolute temperature T.

stereo vision; *g.* **Stereosehen**: the use of two different views of the same scene to recover 3D scene geometry.

structure from motion; *g.* **Struktur durch Bewegungsanalyse**: (1) determining a moving object's shape characteristics — and its position and velocity as well — from an image sequence taken with a stationary camera. (2) Determining an object's shape characteristics, and its position and the camera's velocity, from a sequence taken by the moving camera.

structured light; *g.* **strukturiertes Licht**: Projection of a carefully designed light pattern on a scene and viewing the scene from a different direction. Usually the pattern consists of successive planes of light at different positions and orientations.

structuring element; *g.* **Strukturelement**: the ~ of a morphologic operator corresponds to the mask of a *convolution* operation. It determines in which way a morphological operator changes the shape of a binary or gray-scale object.

subtractive primary colors; *g.* **subtraktive Grundfarben**: the colors cyan, magenta, and yellow, the subtraction of which from white light in different proportions allows all colors to be created.

sun-synchronous orbit; *g.* **sonnensynchrone Umlaufbahn**: a polar orbit where the satellite always crosses the Equator at the same local solar time.

supervised classification; *g.* **überwachte Klassifizierung**: a classification technique whereby a human operator identifies training areas on the image that are intended to be representative examples of each class.

surface normal; *g.* **Oberflächennormale**: a unit vector that is normal to a surface.

surface reconstruction; *g.* **Oberflächenrekonstruktion**: procedures by which a 3-D surface is analytically described on the basis of processing *stereo*, *range*, or *spatiotemporal* images of the observed surface.

symbolic image; *g.* **symbolisches Bild**: a *digital image* in which the value associated with each pixel is a symbol, rather than a gray level.

syntactic pattern recognition; *g.* **syntaktische Mustererkennung**: a type of structural *pattern recognition* that identifies primitives and relationships in natural or artificial language patterns.

synthetic-aperture radar [SAR]: a radar imaging system in which high resolution in the azimuth direction is achieved by using the *Doppler shift* of backscattered waves to identify waves from ahead of and behind the platform, thereby simulating a very long antenna.

TCP/IP: transmission control protocol/Internet protocol, the standard network protocol used on the Internet.

texel; *g.* **Texturelement**: short for texture element, a pixel that contains a vector of texture properties.

texture; *g.* **Textur**: texture describes the way the gray values change in a local *neighbor-*

hood. ~ is characterized by the mean and variance of the gray values, the orientation of the gray values and a characteristic scale.

thermal infrared (TIR); *g.* **thermisches Infrarot**: that part of the spectrum with wavelengths between 3 and 100 μm. These are the wavelengths at which thermal emission is greatest for surfaces at normal environmental temperatures. However, hot objects emit thermally at wavelengths shorter than the thermal infrared.

thresholding; *g.* **Schwellwertsegmentierung**: an image point operation which produces a *binary image* from a *gray scale image*. A binary one is produced on the output image whenever a pixel value on the input image is in a certain gray scale range. A binary zero is produced otherwise.

TIFF: Tag Image File Format. A portable, versatile, and widely used data format for image data. It is operation system and platform independent and suitable to store images with more than 8-bit depth, color images, multi-channel images, image sequences, and higher dimensional images. `ftp://sgi.com/graphics/tiff`.

timecode: unique number assigned to each frame in a video film. Used for frame accurate video editing.

tomography; *g.* **Tomographie**: an *indirect imaging* technique using penetrating radiation. The 3-D shape of an object is reconstructed from multiple projections taken from different directions. Each (parallel) projection of the object gives one slice of the Fourier transform of the object in the corresponding direction (*Fourier slice theorem*).

top-down: a control strategy to problem solving that is goal-directed or expectation directed. A form of solution is hypothesized. Assuming the hypothesis is true and using the information in the knowledge data base, the inference mechanism then infers, if possible, some consistent set of values for the unknown variables or parameters. If a consistent set can be inferred, then the problem has been solved. Otherwise, a new form of solution is generated.

transmissivity; *g.* **Transmissionskoeffizient**: the ratio of the radiant flux passing through the surface of an optical medium to that incident to it. The ~ is a material property and considers the processes at a single surface. ≻ *reflectivity, transmittance*.

transmittance; *g.* **Transmissivität**: the ratio of the radiant flux to the incident flux under specified conditions. The ~ is not a material property but depends on the setup under which the ~ is measured. For the ~ of a thin plate, for example, the transmission of the radiative flux at the front and back face must be considered. Units: 1. ≻ *reflectance, transmissivity*.

triangulation; *g.* **Triangulation**: procedure of determining the coordinates of a 3-D point from the observed position of two perspective projections of the point. The *centers of perspectivity* and the *perspective projection* planes are assumed to be known.

tristimulus: color perception requires three signals, called the ~. In the human eye, three types of *cones* exist that are sensitive to different wavelength ranges.

TTL: transistor-transistor logic. Logic design familiy, where binary 0 is represented as less than 0.8 V, and 1 is represented as more than 2 V.

TV: television

type I error; *g.* **Fehler erster Art**: ≻ *misidentification*

type II error; *g.* **Fehler zweiter Art**: ≻ *false identification*

UHF: ultra high frequency. *Radio waves* with frequencies between 300 and 3 GHz used to broadcast television signals.

ultraviolet; *g.* **Ultraviolett**: that invisible region of the spectrum just beyond the violet end of the visible region. Wavelengths range from 1 to 400 nm.

uncertainty relation; *g.* **Unschäferelation**: the ~ states that the product of the resolutions with which the wave number and position of a signal can be measured is limited. The ~ originates from the basic relations between the *Fourier domain* and the *space domain*.

unsharp masking: an image enhancement technique in which an intentionally blurred version of the image is subtracted from the image to diminish contrast differences in images.

unsupervised classification: a classification technique in which classes are identified by a computer driven search for clusters in multispectral data space without any preassignment of classes by using a training set.

UV; *g.* **ultraviolet**: *ultraviolet*

variance; *g.* **Varianz**: a measure of the dispersion of the actual values of a variable about its mean. It is the mean of the squares of all the deviations from the mean value of a range of data.

variational calculus; *g.* **Variationsrechnung**: many problems in image analysis can be phrased in terms of an optimization problem where some objective function is to be maximized or minimized.

velocity flow field: ≻ *motion field*

very near-infrared (VNIR): the shortest wavelength part of the near-infrared, with wavelengths between 0.7 and 1.0 *μm*. Virtually equivalent with the *photographic infrared*.

VGA: video graphics array. Graphics standard set by IBM.

VHF: very high frequency. *Radio waves* with frequencies between 30 and 300 MHz used to broadcast television signals.

VHS: most common consumer standard for recording of video images on video tape recorders.

video image; *g.* **Videobild**: an image represented as a time serial electronic signal for transmission, storage, or display. A video signal includes the image signal scanned row by row and additional synchronization to indicate the start of frames and rows. Video images have two common formats, the frame format (progressive scanning) and the *interlaced format*.

vidicon: an imaging device based on a sheet of transparent material whose electrical conductivity increases with the irradiance of *EM* radiation. The variation in conductivity across the plate is measured by a sweeping electron beam and converted into a video signal. Now largely replaced by *charge-coupled devices* (CCDs).

vignetting; *g.* **Vignettierung**: the gradual reduction of image *irradiance* as the off-axis angle increases, resulting from limitations of the clear *apertures* of elements within an optical system.

VIS: Visual Instruction Set. *SIMD* instruction set extension for fast pixel processing added by Sun to the UltraSparc processor family.

visible radiation [light]; *g.* **sichtbare Strahlung**: the part of *EM* radiation to which the human eye responds. It lies between the *ultraviolet* and the *infrared*, with wavelengths from 400 to 700 nm.

voxel; *g.* **Volumenelement**: short for volume element; a small volume element representing the properties of a rectangular parallelepiped volume in a volumetric image. ≻ *pixel*.

wave; *g.* **Welle**: a signal traveling through space. If the travel speed does not depend on the *wavelength*, the shape of the signal does not change (nondispersive wave). An *EM* wave is an oscillating electrical and magnetic field where the two field vectors are perpendicular to each other and the propagation direction traveling with the speed of light through the vacuum.

wave number; *g.* **Wellenzahl**: the number of wavelength λ per unit lengths. Often used with an additional factor 2π as with the *circular frequency*. This handbook uses $\bar{k} = 1/\lambda$ and $k = 2\pi/\lambda$.

wave optics; *g.* **Wellenoptik**: the part of optics concerned with optical phenomena such as *diffraction* and interference related to the wave characteristics of *EM* radiation.

wavelength; *g.* **Wellenlänge**: distance between two successive points of a periodic wave in the direction of propagation in which the oscillation has the same phase. Units: m; symbol: λ. ≻ *frequency*.

white noise; *g.* **weißes Rauschen**: noise which is equally distributed over all frequencies. The *power spectrum* of white noise is flat.

white point; *g.* **Weißpunkt, Unbuntpunkt**: point in the *chromaticity diagram* where the color *saturation* is zero.

Wien's displacement law; *g.* **Wiensches Verschiebungsgesetz**: states that the peak of radiant heat emitted by a material shifts to shorter wavelengths as the absolute temperature increases.

wire frame; *g.* **Drahtgittermodell**: a representation of a solid in which an object is shown as if it were constructed of wires for all its edges. This representation is often used to store models of objects to be matched or recognized by a vision system.

working distance; *g.* **Arbeitsabstand**: distance of the focal plane from the front lens of an optical system.

world coordinates; *g.* **Weltkoordinaten**: 3-D coordinate system $[X, Y, Z]^T$ attached to the scene observed by a camera. Contrast with *camera coordinates*.

WORM: write once read many times. Optical storage method that allows data to be stored as well as retrieved on a CD-like disk.

X-ray; *g.* **Röntgenstrahlen**: high-energy penetrating *EM* radiation emitted from the electronic transition in the inner shells of atoms.

YC: two-channel color video signal. One channel carries the *luminance* signal with high bandwidth, the other the *chrominance* signals where the hue and saturation of the color are encoded as the phase and amplitude of the signal.

YUV: color system with a luminance signal (Y) and two color difference signals (U, V) used to encode broadcast-quality color television signals.

zero crossing; *g.* **Nulldurchgang**: The locations in an image where the Laplace-filtered image undergoes a sign change. The ~s are good candidates for *edge* pixels. ≻ *Laplace operator, edge detection.*

zero-mean homogeneous noise; *g.* **mittelwertfreies homogenes Rauschen**: ideal type of noise in a sensor array. It is equal for each sensor without an offset.

Bibliography

[1] M. L. Banner, I. S. F. Jones, and J. C. Trinder. Wave number spectra of short gravity waves. *J. Fluid Mech.*, 198:321–344, 1989.

[2] M. Bass, E. W. V. Stryland, D. R. Williams, and W. L. Wolfe, eds. *Handbook of Optics*. McGraw-Hill, 2nd edn., 1995.

[3] P. R. Bevington. *Data Reduction and Error Analysis*. McGraw-Hill, New York, 3rd edn., 2002.

[4] L. M. Biberman, ed. *Electro Optical Imaging: System Performance and Modeling*. SPIE, Bellingham, WA, 2001.

[5] J. Bigün and J. M. H. du Buf. N-folded symmetries by complex moments in Gabor spaces and their application to unsupervised texture analysis. *IEEE Trans. PAMI*, 16:80–87, 1994.

[6] J. Bigün and G. H. Granlund. Optimal orientation detection of linear symmetry. In *Proc. ICCV'87*, pp. 433–438. IEEE Computer Society, Washington, DC, 1987.

[7] R. Blahut. *Fast Algorithms for Digital Signal Processing*. Addison-Wesley, Reading, MA, 1985.

[8] A. Blake and M. Isard. *Active Contours*. Springer, London, 1998.

[9] R. Bracewell. *The Fourier Transform and its Applications*. McGraw-Hill, New York, 2nd edn., 1986.

[10] C. W. Brown and B. J. Shepherd. *Graphics File Formats, Reference and Guide*. Manning, Greenwich, CT, 1995.

[11] C. J. C. Burges. A Tutorial on Support Vector Machines for Pattern Recognition. *Knowledge Discovery and Data Mining*, 2(2), 1998.

[12] P. J. Burt. The pyramid as a structure for efficient computation. In A. Rosenfeld, ed., *Multiresolutional Image Processing and Analysis*, vol. 12 of *Springer Series in Information Sciences*, pp. 6–35. Springer, New York, 1984.

[13] P. J. Burt and E. H. Adelson. The Laplacian pyramid as a compact image code. *IEEE Trans. COMM*, 31:532–540, 1983.

[14] P. J. Burt, T. H. Hong, and A. Rosenfeld. Segmentation and estimation of image region properties through cooperative hierarchical computation. *IEEE Trans. SMC*, 11:802–809, 1981.

[15] J. F. Canny. A computational approach to edge detection. *PAMI*, 8:679–698, 1986.

[16] E. Catmull and A. R. Smith. 3-D transformations of images in scanline order. *Computer Graphics (SIGGRAPH '80 Proc.)*, 14:279–285, 1980.

[17] G. Cloud. *Optical Methods of Engineering Analysis*. Cambridge Univ. Press, Cambridge, UK, 1995.

[18] T. F. Cox and M. A. A. Cox. *Multidimensional Scaling*. Monographs on Statistics and Applied Probability. Chapman & Hall, 1995.

[19] C. DeCusatis, ed. *Handbook of Applied Photometry*. Springer, New York, 1997.

[20] P. DeMarco, J. Pokorny, and V. C. Smith. Full-spectrum cone sensitivity functions for X-chromosome-linked anomalous trichromats. *J. Optical Soc.*, A9:1465–1476, 1992.

[21] E. B. Dobson. Measurements of the fine-scale structure of the sea. *J. Gephys. Res.*, 75: 2853–2856, 1970.

[22] S. A. Drury. *Image Interpretation in Geology.* Chapman and Hall, London, 2nd edn., 1993.

[23] R. O. Duda, P. E. Hart, and D. G. Stork. *Pattern Classification.* Wiley, New York, 2000.

[24] S. Dudoit, J. Fridlyand, and T. P. Speed. Comparison of discrimination methods for the classification of tumors using gene expression data. *Journal of the American Statistical Association*, 97(457):77–87, 2002.

[25] R. Eils, E. Bertin, K. Saracoglu, B. Rinke, E. Schröck, and F. Parazza. Application of confocal laser microscopy and three-dimensional Voronoi diagrams for volume and surface estimates of interphase chromosomes. *J. Microscopy*, 117(2):150–161, 1995.

[26] W. C. Elmore and M. A. Heald. *Physics of Waves.* Dover Publications, New York, 1985.

[27] A. Erhardt, G. Zinser, D. Komitowski, and J. Bille. Reconstructing 3D light microscopic images by digital image processing. *Applied Optics*, 24:194–200, 1985.

[28] L. J. C. et al. The directional spectrum of a wind generated sea as determined from data obtained by the Stereo Wave Observation Project. *Meteorological Papers*, 2(6):1–88, 1960.

[29] O. Faugeras. *Three-dimensional Computer Vision. A Geometric Viewpoint.* MIT Press, Cambridge, MA, 1993.

[30] M. Felsberg and G. Sommer. A new extension of linear signal processing for estimating local properties and detecting features. In G. Sommer, N. Krüger, and C. Perwass, eds., *Mustererkennung 2000, 22. DAGM Symposium, Kiel*, Informatik aktuell, pp. 195–202. Springer, Berlin, 2000.

[31] D. G. Fink and D. Christiansen. *Electronics Engineers' Handbook.* McGraw-Hill, New York, 1989.

[32] D. J. Fleet. *Measurement of Image Velocity.* Kluwer Academic Publisher, Dordrecht, 1992.

[33] D. J. Fleet and A. D. Jepson. Computation of component image velocity from local phase information. *Int. J. Comp. Vision*, 5:77–104, 1990.

[34] J. D. Foley, A. van Dam, S. K. Feiner, and J. F. Hughes. *Computer Graphics, Principles and Practice.* Addison-Wesley, Reading, MA, 1990.

[35] W. Förstner. Image preprocessing for feature extraction in digital intensity, color and range images. In A. Dermanis, A. Grün, and F. Sanso, eds., *Geomatic Methods for the Analysis of Data in the Earth Sciences*, vol. 95 of *Lecture Notes in Earth Sciences*. Springer, Berlin, 2000.

[36] D. A. Forsyth and J. Ponce. *Computer Vision, A Modern Approach.* Prentice-Hall, Upper Saddle River, NJ, 2003.

[37] W. T. Freeman and E. H. Adelson. The design and use of steerable filters. *IEEE Trans. PAMI*, 13:891–906, 1991.

[38] L.-L. Fu and A. Cazenave, eds. *Satellite Altimetry and Earth Sciences.* Academic Press, San Diego, 2001.

[39] G. Gaussorgues. *Infrared Thermography.* Chapman & Hall, London, 1994.

[40] R. Godding. Gormetric calibration of digital imaging systems. In B. Jähne and H. Haußecker, eds., *Computer Vision and Applications*, pp. 153–175. Academic Press, San Diego, 2000.

[41] H. Goldstein. *Classical Mechanics.* Addison-Wesley, Reading, MA, 1980.

[42] G. H. Golub and C. F. van Loan. *Matrix Computations.* The Johns Hopkins University Press, Baltimore, 1989.

[43] R. C. Gonzalez and P. Wintz. *Digital Image Processing.* Addison-Wesley, Reading, MA, 2nd edn., 1987.

[44] G. H. Granlund. In search of a general picture processing operator. *Comp. Graph. Imag. Process.*, 8:155–173, 1978.

[45] G. H. Granlund and H. Knutsson. *Signal Processing for Computer Vision.* Kluwer, Dordrecht, 1995.

[46] F. J. Green. *The Sigma-Aldrich Handbook of Stains, Dyes and Indicators.* Aldrich Chemical Company, Milwaukee, WI, 1990.

[47] A. Gruen and H. A. Beyer. System calibration through self-calibration. In A. Gruen and T. S. Huang, eds., *Calibration and Orientation of Cameras in Computer Vision*, pp. 163–193. Springer, Berlin, 2001.

[48] R. J. Gurney, J. L. Foster, and C. L. Parkinson, eds. *Atlas of Satellite Observations Related to Global Change*. Cambridge University Press, Cambridge, 1993.

[49] A. Haase, J. Ruff, and M. Rokitta. Nuclear Magnetic Resonance Microscopy. In B. Jähne, H. Haußecker, and P. Geißler, eds., *Handbook of Computer Vision and Applications*, vol. 1, pp. 601–612. Academic Press, San Diego, 1999.

[50] P. Hariharan. *Optical Holography, Principles, Techniques and Applications*. Cambridge Univ. Press, Cambridge, UK, 1994.

[51] T. Hastie, R. Tibshirani, and J. Friedman. *The Elements of Statistical Learning*. Springer Series in Statistics. Springer, New York, 2001.

[52] R. P. Haugland. Handbook of Fluorescent Probes and Research Chemicals. Molecular Probes (http://www.probes.com), 1996.

[53] H. Haußecker and H. Spies. Motion. In B. Jähne and H. Haußecker, eds., *Computer Vision and Applications*, pp. 347–395. Academic Press, San Diego, 2000.

[54] H. Haußecker. *Messung und Simulation kleinskaliger Austauschvorgänge an der Ozeanoberfläche mittels Thermographie*. Diss., Univ. Heidelberg, 1996.

[55] F. Hering, D. Wierzimok, and B. Jähne. Measurements of enhanced turbulence in short wind-induced water waves. In B. Jähne and E. Monahan, eds., *Air-Water Gas Transfer, Selected Papers, 3rd Intern. Symp. on Air-Water Gas Transfer*, pp. 125–134. AEON, Hanau, 1995.

[56] K. Hoffmann and R. Kunze. *Linear Algebra*. Prentice-Hall, Englewood Cliffs, NJ, 2nd edn., 1971.

[57] G. Holst, I. Klimant, M. Kühl, and O. Kohls. Optical microsensors and microprobes. In M. S. Varney, ed., *Chemical Sensors in Oceanography*, pp. 143–188. Gordon and Breach, 2000.

[58] G. C. Holst. *CCD Arrays, Cameras, and Displays*. SPIE, Bellingham, WA, 2nd edn., 1998.

[59] G. C. Holst. *Common Sense Approach to Thermal Imaging*. SPIE, Bellingham, WA, 2000.

[60] L. H. Holthuijsen. Observations of the directional distribution of ocean-wave energy in fetch-limited conditions. *J. Phys. Oceanogr.*, 13:191–207, 1983.

[61] S. B. Howell. *Handbook of CCD Astronomy*. Cambridge University Press, Cambdridge, UK, 2000.

[62] R. Ihaka and R. Gentleman. R: A Language for Data Analysis and Graphics. *J. Comput. Graph. Stat.*, 5(3):299–314, 1996. http://www.r-project.org/.

[63] K. Iizuka. *Engineering Optics*. Springer, Berlin, 1987.

[64] M. Ikeda and F. W. Dobson, eds. *Oceanographic Applications of Remote Sensing*. CRC Press, Boca Raton, FL, 1995.

[65] A. F. Inglis. *Video Engineering*. McGraw-Hill, New York, 1993.

[66] B. Jähne. Image sequence analysis of complex physical objects: nonlinear small scale water surface waves. In *Proceedings ICCV'87, London*, pp. 191–200. IEEE Computer Society, Washington, DC, 1987.

[67] B. Jähne. Imaging of gas transfer across gas/liquid interfaces. In S. Sideman and K. Hijikata, eds., *Imaging in Transport Processes*, pp. 247–256. Begell House Publishers, New York, 1993.

[68] B. Jähne, ed. *Image Sequence Analysis to Investigate Dynamic Processes*. Lecture Notes in Computer Science. Springer, Berlin, 2004. in preparation.

[69] B. Jähne and P. Geißler. Depth from focus with one image. In *Proc. Conference on Computer Vision and Pattern Recognition (CVPR '94), Seattle, 20.-23. June 1994*, pp. 713–717, 1994.

[70] B. Jähne and P. Geißler. An imaging optical technique for bubble measurements. In M. J. Buckingham and J. R. Potter, eds., *Proc. Sea Surface Sound '94*, pp. 290–303. World

Scientific, Singapore, 1995.

[71] B. Jähne, H. Scharr, and S. Körgel. Principles of filter design. In B. Jähne, H. Haußecker, and P. Geißler, eds., *Computer Vision and Applications, volume 2, Signal Processing and Pattern Recognition*, chapter 6, pp. 125–151. Academic Press, San Diego, 1999.

[72] B. Jähne. *Spatio-temporal Image Processing*, vol. 751 of *Lecture Notes in Computer Science*. Springer, Berlin, 1993.

[73] B. Jähne. *Digital Image Processing - Concepts, Algorithms, and Scientific Applications.* Springer, Berlin, 4th edn., 1997. Includes CD-ROM.

[74] B. Jähne. *Digital Image Processing.* Springer, Berlin, 5th edn., 2002.

[75] A. K. Jain. *Fundamentals od Digital Image Processing.* Prentice-Hall, Englewood Cliffs, NJ, 1989.

[76] J. R. Janesick. *Scientific Charge-Coupled Devices.* SPIE, Bellingham, WA, 2001.

[77] J. Karlholm. *Efficient Spatiotemporal Filtering and Modelling.* Diss., Linköping Univ., Linköping, Sweden, 1996. Diss. No. 562.

[78] M. Kass, A. Witkin, and D. Terzopoulos. Snakes: active contour models. In *Proc. ICCV'87*, pp. 259–268. IEEE Computer Society, Washington, DC, 1987.

[79] B. Klinke and B. Jähne. Measurement of short ocean wind waves during the MBL-ARI west coast experiment. In B. Jähne and E. Monahan, eds., *Air-Water Gas Transfer, Selected Papers, 3rd Intern. Symp. on Air-Water Gas Transfer*, pp. 165–173. AEON, Hanau, 1995.

[80] H. Knutsson. *Filtering and Reconstruction in Image Processing.* Diss., Linköping Univ., Linköping, Sweden, 1982.

[81] H. E. Knutsson, R. Wilson, and G. H. Granlund. Anisotropic nonstationary image estimation and its applications: part I - restoration of noisy images. *IEEE Trans. COMM*, 31(3): 388–397, 1983.

[82] E. Kohlschütter. Die Forschungsreise S.M.S. Planet, Band 2, Stereophotogrammetrische Aufnahmen. *Ann. Hydrographie*, 34:219, 1906.

[83] B. M. Krasovitskii and B. M. Bolotin. *Organic Luminescent Materials.* VCH, New York, 1988.

[84] S. Kraus, K. Degreif, N. Smoljar, M. Korniyenko, T. Wagner, M. Wenig, B. Jähne, and U. Platt. Spectroscopic imaging. In B. Jähne, ed., *Image Sequence Analysis to Investigate Dynamic Processes*, Lecture Notes in Computer Science. Springer, Berlin, 2003.

[85] P. W. Kruse. *Uncooled Thermal Imaging, Arrays, Systems, and Applications.* SPIE, Bellingham, WA, 2001.

[86] Labsphere. A Guide to Integrating Sphere Theory and Applications. Techguide, see http://www.labsphere.com/, 2003.

[87] Labsphere. Integrating Sphere Uniform Light Source Applications. Techguide, see http://www.labsphere.com/, 2003.

[88] Laurin Publishing. Photonics Handbook. 49th international edition, Laurin Publishing Company, Pittsfield, MA, 2003.

[89] M. Levine. *Vision in Man and Machine.* McGraw-Hill, New York, 1985.

[90] J. S. Lim. *Two-dimensional Signal and Image Processing.* Prentice-Hall, Englewood Cliffs, NJ, 1990.

[91] T. Lindeberg. *Scale-Space Theory in Computer Vision.* Kluwer, Dordrecht, 1994.

[92] M. Loose, K. Meier, and J. Schemmel. A self-calibrating single-chip CMOS camera with logarithmic response. *IEEE J. Solid-State Circuits*, p. 36, 2001.

[93] T. Luhmann. *Nahbereichsphotogrammetrie.* Wichmann, Heidelberg, 2000.

[94] H.-G. Maas. A highspeed camera system for the acquisition of flow tomography sequences for 3-D least squares matching. In *IAPRS*, vol. 29 of *B5*, pp. 241–249, 1994.

[95] V. K. Madisetti and D. B. Williams, eds. *The Digital Signal Processing Handbook.* CRC Press, Boca Raton, FL, 1998.

[96] D. Malacara. *Color Vision and Colorimetry.* SPIE, Bellingham, WA, 2002.

[97] S. L. Marple Jr. *Digital Spectral Analysis with Applications*. Prentice-Hall, Englewood Cliffs, NJ, 1987.

[98] D. Marr. *Vision*. W. H. Freeman and Company, New York, 1982.

[99] D. Marr and E. Hildreth. Theory of edge detection. *Proc. Royal Society, London, Ser. B*, 270:187–217, 1980.

[100] G. F. Marshall, ed. *Optical Scanning*. Marcel Dekker, New York, 1991.

[101] E. A. Maxwell. *General Homogeneous Coordinates in Place of Three Dimensions*. University Press, Cambridge, 1951.

[102] R. McCluney. *Introduction to Radiometry and Photometry*. Artech, Boston, 1994.

[103] W. Menke. *Geophysical Data Analysis: Discrete Inverse Theory*. Academic Press, San Diego, 2nd edn., 1989.

[104] J. L. Miller. *Principles of Infrared Technology. A Practical Guide to the State of the Art*. Van Nostrand Reinhold, New York, 1994.

[105] T. Münsterer, H. J. Mayer, and B. Jähne. Dual-tracer measurements of concentration profiles in the aqueous mass boundary layer. In B. Jähne and E. Monahan, eds., *Air-Water Gas Transfer, Selected Papers, 3rd Intern. Symp. on Air-Water Gas Transfer*, pp. 637–648. AEON, Hanau, 1995.

[106] T. Münsterer. Messung von Konzentrationsprofilen gelöster Gase in der wasserseitigen Grenzschicht. Dipl., Univ. Heidelberg, 1993. D-334 Institut für Umweltphysik.

[107] J. D. Murray and W. vanRyper. *Encyclopedia of Graphics File Formats*. O'Reilly & Associates, Sebastopol, CA, 2nd edn., 1996.

[108] A. V. Oppenheim and R. W. Schafer. *Discrete-Time Signal Processing*. Prentice-Hall, Englewood Cliffs, NJ, 1989.

[109] A. W. Paeth. A fast algorithm for general raster rotation. In *Graphics Interface '86*, pp. 77–81, 1986.

[110] A. Papoulis. *Probability, Random Variables, and Stochastic Processes*. McGraw-Hill, New York, 1991.

[111] J. A. Parker. *Image Reconstruction in Radiology*. CRC Press, Boca Raton, FL, 1990.

[112] P. Perona and J. Malik. Scale space and edge detection using anisotropic diffusion. In *Proc. Workshop on Computer Vision*, pp. 16–20. IEEE Computer Society Press, Washington, 1987.

[113] M. Pietikäinen and A. Rosenfeld. Image segmentation by texture using pyramid node linking. *IEEE Trans. SMC*, 11:822–825, 1981.

[114] W. H. Press, B. P. Flannery, S. A. Teukolsky, and W. T. Vetterling. *Numerical Recipes in C: The Art of Scientific Computing*. Cambridge University Press, New York, 1992.

[115] J. G. Proakis and D. G. Manolakis. *Digital Signal Processing. Principles, Algorithms, and Applications*. McMillan, New York, 1992.

[116] A. R. Rao. *A Taxonomy for Texture Description and Identification*. Springer, New York, 1990.

[117] T. H. Reiss. *Recognizing Planar Objects using Invariant Image Features*, vol. 676 of *Lecture Notes in Computer Science*. Springer, Berlin, 1993.

[118] J. A. Rice. *Mathematical Statistics and Data Analysis*. Duxbury Press, Belmont, CA, 1995.

[119] A. Richards. *Alien Vision: Exploring the Electromagnetic Spectrum with Imaging Technology*. SPIE, Bellingham, WA, 2001.

[120] B. D. Ripley. *Pattern Recognition and Neural Networks*. Cambridge Univ. Press, 1997.

[121] A. Rosenfeld and A. C. Kak. *Digital Picture Processing, Volume I and II*. Academic Press, Orlando, 1982.

[122] T. Scheuermann, G. Pfundt, P. Eyerer, and B. Jähne. Oberflächenkonturvermessung mikroskopischer Objekte durch Projektion statistischer Rauschmuster. In G. Sagerer, S. Posch, and F. Kummert, eds., *Mustererkennung 1995, 17. DAGM Symposium Bielefeld*, pp. 319–326. Springer, Berlin, 1995.

[123] C. Schnörr. Variational methods for adaptive image smoothing and segmentaton. In B. Jähne, H. Haußecker, and P. Geißler, eds., *Handbook of Computer Vision and Applications*, vol. 2, pp. 451–484. Academic Press, San Diego, 1999.

[124] B. Schölkopf and A. J. Smola. *Learning with Kernels*. MIT Press, Boston, 2002.

[125] T. Scholz. *Ein Depth from Focus-Verfahren zur On Line-Bestimung der Zellkonzentration bei Fermentationsprozessen*. Diss., Univ. Heidelberg, 1995.

[126] W. G. Schreiber and G. Brix. Magnetic resonance imaging in medicine. In B. Jähne, H. Haußecker, and P. Geißler, eds., *Handbook of Computer Vision and Applications*, vol. 1, pp. 563–599. Academic Press, San Diego, 1999.

[127] A. Schuhmacher. Stereophotogrammetrische Wellenaufnahmen, wissenschaftliche Ergebnisse der Deutschen Atlanischen Expedition auf dem Forschungs- und Vermessungsschiff "Meteor" 1925–1927, 1939.

[128] P. Seitz. Solid-State Image Sensing. In B. Jähne, H. Haußecker, and P. Geißler, eds., *Handbook of Computer Vision and Applications*, vol. 1, pp. 165–222. Academic Press, San Diego, 1999.

[129] P. Seitz. Solid-State Image Sensing. In B. Jähne and H. Haußecker, eds., *Computer Vision and Applications*, pp. 111–151. Academic Press, San Diego, 2000.

[130] J. Serra. *Image Analysis and Mathematical Morphology*. Academic Press, London, 1982.

[131] L. G. Shapiro and G. C. Stockman. *Computer Vision*. Prentice-Hall, Upper Saddle River, NJ, 2001.

[132] O. H. Shemdin and H. M. Tran. Measuring short surface waves with stereophotography. *Photogrammetric Eng. and Remote Sens.*, 93:311–316, 1992.

[133] O. H. Shemdin, H. M. Tran, and S. C. Wu. Directional measurements of short ocean waves with stereophotography. *J. Geophys. Res.*, 93:13891–13901, 1988.

[134] A. Singh. *Optic Flow Computation: A Unified Perspective*. IEEE Computer Society Press, Los Alamitos, CA, 1991.

[135] W. J. Smith. *Modern Optical Engineering. The Design of Optical Systems*. McGraw-Hill, New York, 2nd edn., 1990.

[136] P. Soille. *Morphological Image Analysis, Principles and Applications*. Springer, Berlin, 2nd edn., 2003.

[137] G. Sommer, ed. *Geometric Computing with Clifford Algebras*. Springer, Berlin, 2001.

[138] J. Steurer, H. Giebel, and W. Altner. Ein lichtmikroskopisches Verfahren zur zweieinhalbdimensionalen Auswertung von Oberflächen. In G. Hartmann, ed., *Mustererkennung 1986*, vol. 125 of *Informatik Fachberichte*, pp. 66–70. Springer, Berlin, 2000.

[139] R. H. Stewart. *Methods of Satellite Oceanography*. University of California Press, Berkeley, 1985.

[140] D. J. Stilwell. Directional energy spectra of the sea from photographs. *J. Geophys. Res.*, 74:1974–1986, 1969.

[141] H. Suhr, G. Wehnert, K. Schneider, C. Bittner, T. Scholz, P. Geißler, B. Jähne, and T. Scheper. In-situ microscopy for on-line characterization of cell-populations in bioreactors, including concentration measurements by depth from focus. *Biotechnology and Bioengineering*, 47:106–116, 1995.

[142] A. Tanaka, M. Kameyama, S. Kazama, and O. Watanabe. A rotation method for raster image using skew transformation. In *Proc. IEEE Conf. Computer Vision and Pattern Recognition (CVPR'86)*, pp. 272–277, 1986.

[143] P. Thévenaz, T. Blu, and M. Unser. Interpolation revisited. *IEEE Medical Imaging*, 19:739–758, 2000.

[144] M. Unser, A. Aldroubi, and M. Eden. Fast B-spline transforms for continuous image representation and interpolation. *IEEE Trans. PAMI*, 13:277–285, 1991.

[145] M. Unser, A. Aldroubi, and M. Eden. B-spline signal processing: part I — theory. *IEEE Trans. Signal Proc.*, 41:821–832, 1993.

[146] M. Unser, A. Aldroubi, and M. Eden. B-spline signal processing: part II— efficient design and applications. *IEEE Trans. Signal Proc.*, 41:834–848, 1993.

[147] D. Uttenweiler, M. Vogel, T. Ober, H. Haußecker, H. Scharr, F. Raisch, C. Veigel, and R. H. A. Fink. The dynamics of motor proteins. In B. Jähne, ed., *Image Sequence Analysis to Investigate Dynamic Processes*, Lecture Notes in Computer Science. Springer, Berlin, 2003.

[148] D. Uttenweiler, C. Weber, B. Jähne, R. Fink, and H. Scharr. Spatiotemporal anisotropic diffusion filtering to improve signal-to-noise ratios and object restoration in fluorescence microscopic image sequences. *J. Biomed. Opt.*, 8:40–47, 2003.

[149] H. C. van de Hulst. *Light Scattering by Small Particles*. Dover Publications, New York, 1981.

[150] W. M. Vaughan and G. Weber. Oxygen quenching of pyrenebutyric acid fluorescence in water. *Biochemistry*, 9:464, 1970.

[151] J. Weickert. Multiscale texture enhancement. In V. Hlaváč and R. Sára, eds., *Proc. Comp. Analysis of Images and Patterns (CIAP'95)*, Lecture Notes in Computer Science, pp. 230–237. Springer, Berlin, 1995.

[152] J. Weickert. *Anisotropic Diffusion in Image Processing*. Diss., Univ. Kaiserslautern, 1996.

[153] J. Weickert. *Anisotropic Diffusion in Image Processing*. Teubner, Stuttgart, 1998.

[154] M. Wenig, S. Beirle, J. Hollwedel, S. Kraus, C. Leue, T. Wagner, U. Platt, and B. Jähne. Atmospheric emission, transport, trends and fate of tropospheric trace gases. In B. Jähne, ed., *Image Sequence Analysis to Investigate Dynamic Processes*, Lecture Notes in Computer Science. Springer, Berlin, 2003.

[155] G. Wolberg. *Digital Image Warping*. IEEE Computer Socierty Press, Los Alamitos, 1990.

[156] P. Zamperoni. *Methoden der Digitalen Bildsignalverarbeitung*. Vieweg, Braunschweig, 1989.

Index

D   Color Plates

Plate 1: *Optical measurement of oxygen in corals: **a** Sample place for the coral, the lagoon of Heron Island, Capricorn Islands, Great Barrier Reef, Australia. **b** Set-up with the coral in the glass container placed on top of the LED frame while the measuring camera was looking at the cut coral surface from below. **c** Close up of the set-up with a cut coral porite placed on top of a transparent planar oxygen optode, which was fixed at the bottom of a glass container, filled with lagoon sea water. The blue light is the excitation light, coming from blue LEDs arranged in a frame below the glass;, the orange-red light corresponds to the excited luminescence of the optode. The fixed white tube (upper right part) served for flushing the water surface with a constant air stream to aerate the water. **d** Measured 2D oxygen distribution (view to cut coral surface) given in % air saturation. The oxygen image is blended into a grayscale image from the coral structure. The high oxygen values were generated by endolithic cells, which live within the coral skeleton. The production was triggered by a weak illumination through the oxygen sensor, which simulated the normal light situation of these cells at daylight in the reef. Images courtesy of Dr. Holst, PCO AG, Kelheim, Germany. (See also Fig. 1.10, p. 10.)*

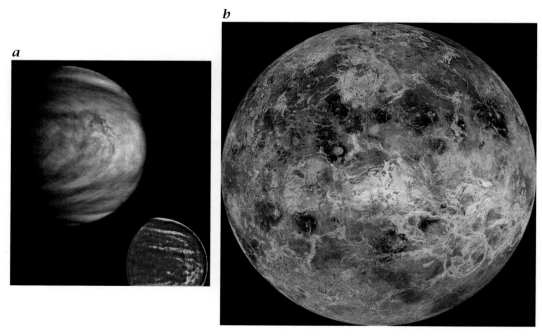

*Plate 2: Images of the planet Venus: **a** Images in the ultraviolet (top left) show patterns at the very top of Venus' main sulfuric acid haze layer while images in the near infrared (bottom right) show the cloud patterns several km below the visible cloud tops. **b** Topographic map of the planet Venus shows the elevation in a color scheme as it is usually applied for maps (from blue over green to brownish colors). This image was computed from a mosaic of Magellan radar images that have been taken in the years 1990 to 1994. Source: http//www.jpl.nasa.gov. (See also Fig. 1.15, p. 15.)*

Plate 3: *Synthetic color image of a* 30.2 × 21.3 *km sector of the tropical rain forest in west Brazil. Three SAR images taken with different wave lengths (lower three images: Left: X-band, Middle: C-band, Right: L-band) have been composed into a color image. Pristine rain forest appears in pink colors while clear areas for agricultural usage are greenish and bluish. A heavy rain storm appears in red and yellow colors since it scatters the shorter wavelength micro waves. Image taken with the imaging radar-C/X-band aperture radar (SIR-C/X-SAR) on April 10, 1994 on board the space shuttle Endeavour (source: image p-46575 published in http://www.jplinfo.jpl.nasa.gov). (See also Fig. 1.18, p. 18.)*

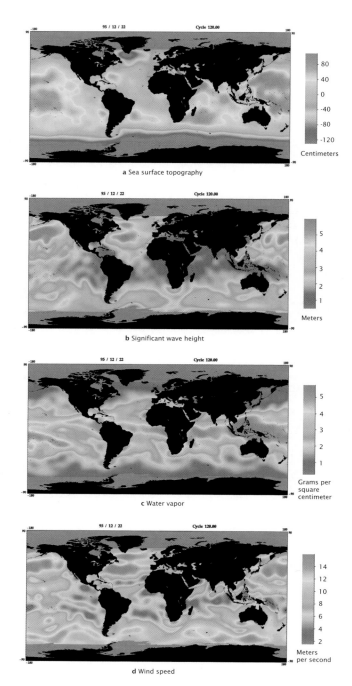

Plate 4: *Images of the Topex/Poseidon mission derived from radar altimetry and passive microwave radiometry. **a** dynamical topography (highs and lows of the ocean currents) shown as a deviation from an area of constant gravitational energy. **b** significant wave height, **c** water vapor content of the atmosphere in g/cm² measured by passive micro wave radiometry, **d** wind speed determined from the strength of the backscatter. All images have been averaged over a period of 10 days around the 22 December 1995 (source: http://www.jplinfo.jpl.nasa.gov). (See also Fig. 1.17, p. 17.)*

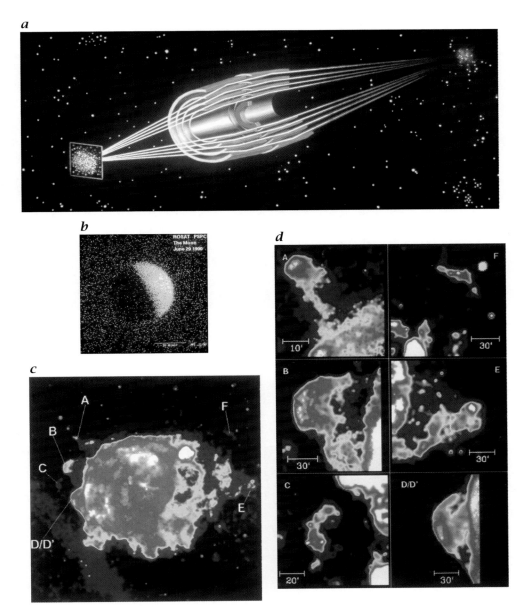

Plate 5: **a** *Artist's impression of an X-ray telescope, the so-called Wolters telescope, the primary X-ray instrument on the German ROSAT satellite. Because of the low index of refraction, only telescopes with grazing incident rays can be built. Four double-reflecting pairs of concentric mirrors are used to increase the aperture and thus brightness of the image. **b** X-ray image of the moon, with 15' (1/4 °) diameter a small object in the sky as compared to **c** Explosion cloud of the supernova Vela with a diameter of about 4 °as observed with the German X-ray satellite ROSAT. Scientists of the Max Planck Institute for Extraterrestrial Physics (MPE) discovered six fragments of the explosion marked from A to F. The intensity of the X-ray radiation in the range from 0.1 to 2.4 keV is shown in pseudo colors from light blue over yellow, red to white and covers an intensity range of 500. The bright white circular object at the upper right edge of the explosion cloud is the remains of another supernova explosion which lies behind the VELA explosion cloud and has no connection to it. **d** Enlarged parts of figure **c** of the fragments A through F at the edge of the explosion cloud. (See also Fig. 1.19, p. 20.)*

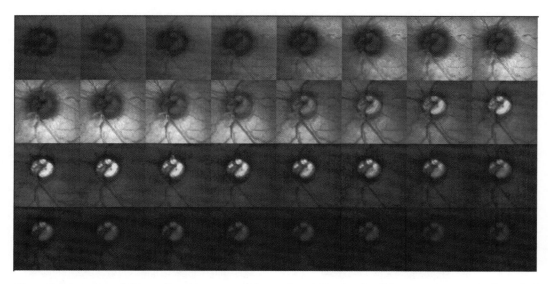

Plate 6: *A series of 32 confocal images of the retina. The depth of the scan increases from left to right and top to bottom. From Zinser, 1995. (See also Fig. 1.23, p. 24.)*

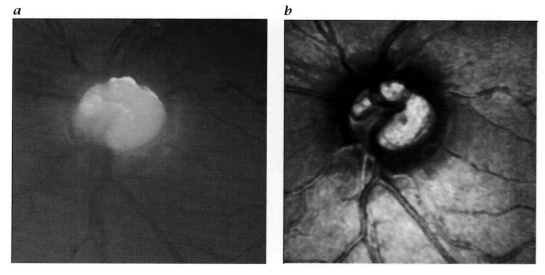

a   *b*

Plate 7: *Reconstruction of the topography of the retina from the focus series in Fig. 1.23. **a** Depth map: deeper lying structures as the exit of the optical nerve are coded brighter. **b** Reconstructed reflection image showing all parts of the image sharp despite significant depth changes. From Zinser, 1995. (See also Fig. 1.24, p. 24.)*

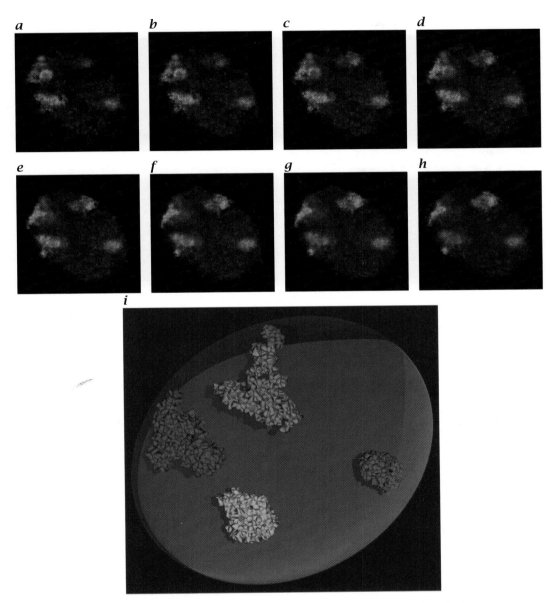

a *b* *c* *d*

e *f* *g* *h*

i

Plate 8: *a - h Part of a depth scan using confocal microscopy across a female human nucleus. Visible are chromosomes X and 7 (green). For differentiation between chromosomes X and 7, a substructure of chromosome 7 has been additionally colored with a red dye. The depth increment between the individual 2-D images is 0.3 μm, the image sector is 30 × 30 μm. **i** 3-D reconstruction. The image shows the inactive chromosome X in red, the active chromosome X in yellow, chromosome 7 in blue, and its centromeres in magenta. The shape of the whole nucleus has been modeled as an ellipsoid. From Eils, 1995. (See also Fig. 1.25, p. 26.)*

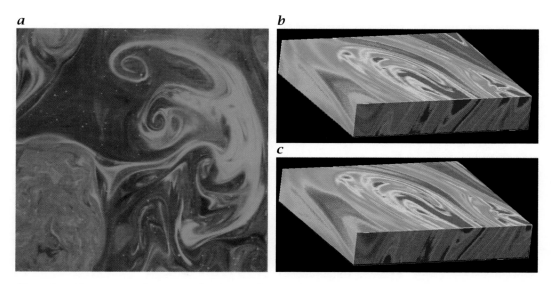

Plate 9: *Fluorescent dyes can also be used to make flows visible:* **a** *visualization of a mixing process;* **b** *and* **c** *two consecutive* $15 \times 15 \times 3\ mm^3$ *volumetric images consisting of 50 layers of* 256×256 *pixels taken 100 ms apart. The voxel size is* $60 \times 60 \times 60\ \mu m^3$*. From Maas [94]. (See also Fig. 1.29, p. 29.)*

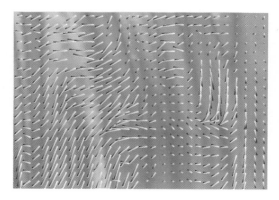

Plate 10: *Flow field determined by the least squares matching technique from volumetric image sequences. The flow vectors are superimposed to the concentration field. From Maas, 1995. (See also Fig. 1.30, p. 29.)*

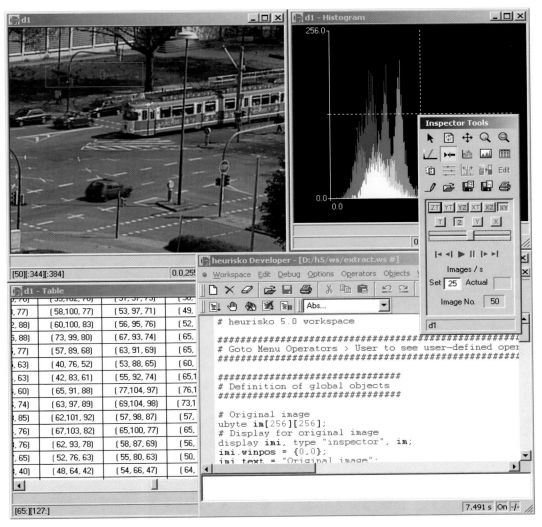

Plate 11: *Screenshot of the graphical interface of a modern image processing software* *(**heurisko**®) featuring multi-window image display, flexible image "inspectors" for interactive graphical evaluation of images, and a powerful set of image operators, which can easily be combined to user-defined operators adapted to specific image processing tasks. This software package is not limited to 2-D image processing with 8 bit depth (256 gray scales) but can also handle multi-channel images of various data types, multi-scale images (pyramids), volumetric images, and image sequences. (See also Fig. 2.11, p. 47.)*

a *b*

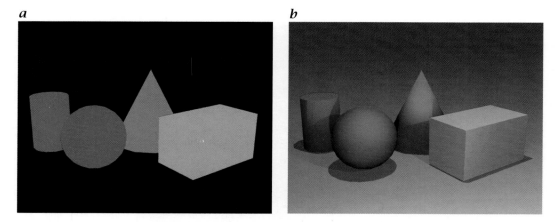

Plate 12: *A computer generated (rendered) 3-D scene illustrating the complexity of the inter-actions between illumination and objects. **a** flat shading (uniformly radiating objects), the ideal world for image processing. **b** Gouraud shading (matte Lambertian surface) with multiple point light sources. Now the radiance of the objects is no longer uniform. It depends on the angle between the surface normal and the direction to the light sources. Furthermore, other objects that are between a light source and the object obstruct light of this light source to reach the object resulting in shadows. (See also Fig. 3.18, p. 86.)*

a *b* *c* *d*

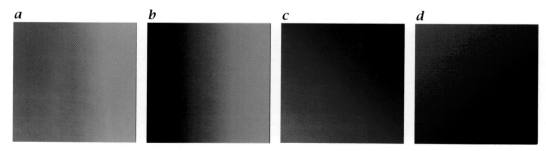

Plate 13: *Color wedge (**a**) produced by additive color mixing from the **b** green, **c** red, and **d** blue color wedges according to Eq. (3.78). (See also Fig. 3.25, p. 94.)*

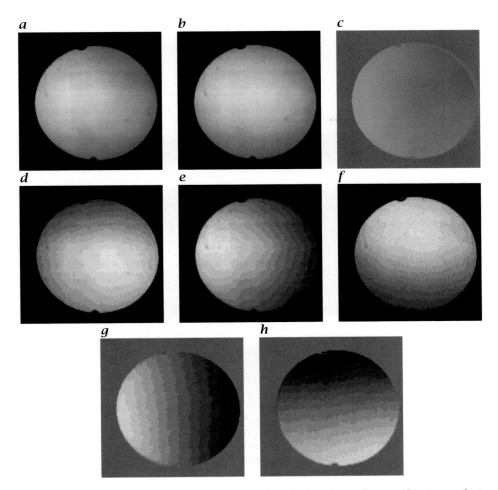

Plate 14: *Illustration of color ratio imaging used with the shape from reflection technique to measure the slope of water surfaces. It is demonstrated here with a calibration target, a spherical lens. **a** original color image; **b** intensity of the original image; watch the intensity fall off towards the edge of the lens (higher slopes); **c** color image normalized by **b**; **d** - **f** : Red, green, and blue channels of the original color image shown in discrete intensity steps; **g** and **h** : x, y positions in the telecentric illumination system computed after Eq. (3.80). These quantities are according to Eq. (3.77) directly proportional to the x and y components of the surface slope. (See also Fig. 3.26, p. 95.)*

a

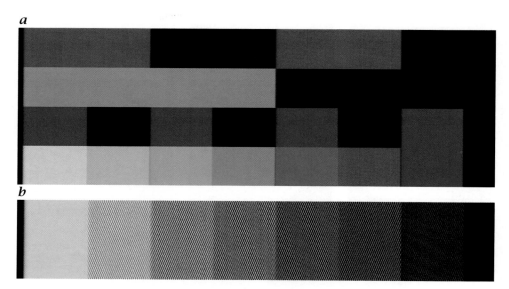

b

Plate 15: *Color bar used as a standard video signal: **a** The four horizontal stripes contain the following components of the signal from top to bottom: red channel only, green channel only, blue channel only, red, green, and blue channel together. **b** Color bar signal shown as a PAL composite video signal digitized with a gray scale frame grabber that does not suppress the color carrier frequency and thus shows the colors as high frequency patterns. Notice that the white stripe is free of the high frequency patterns since it contains no color (the U and V components are zero). Compare also with Fig. 5.26.*

599

Plate 16: *a Red, b green, and c blue channels of a color image (d) taken with a 3-CCD RGB color camera. (See also Fig. 5.27, p. 202.)*

Plate 17: *Illustration of pseudocolor display of grayscale images: **a** Original gray scale image, pseudo color displays **b** rainbow colors, **c** glow colors (from red via yellow to white) **d** under/overflow marking: blue underflow, green low, yellow high, red overflow. (See also Fig. 7.4, p. 258.)*

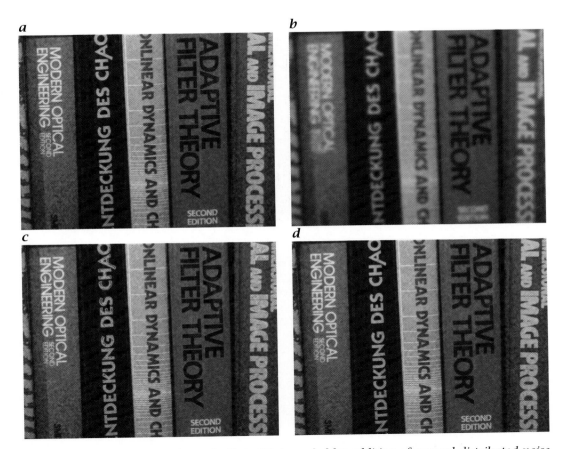

Plate 18: *a Color image shown in Plate 16 degraded by addition of normal distributed noise with a standard deviation of 50 smoothed by b Image shown in a after smoothing the red, green, blue channels by a binomial mask with a standard deviation (pixel radius of smoothing) of 3.6, c Smoothing of only the hue channel in the HSI color model with the same filter as in b. d Smoothing of only the U and V channels in the YUV color model with the same filter as in b. (See also Fig. 11.3, p. 367.)*

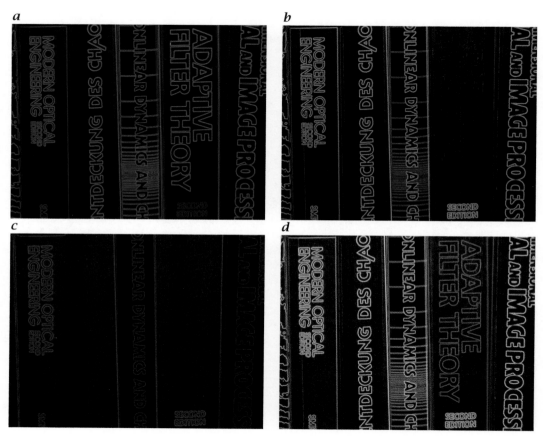

Plate 19: *Edges in color images* **a** *,* **b** *, and* **c** *edge coefficient computed from the red, green, and blue channnels, respectively.* **d** *Edge coefficients of the red, green, and blue channels combined. (See also Fig. 12.3, p. 401.)*

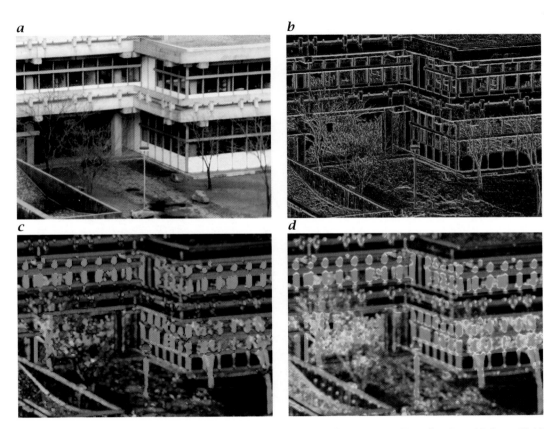

Plate 20: *Different possibilities for color coding of edge directions.* *(See also Fig. 13.5, p. 424.)*

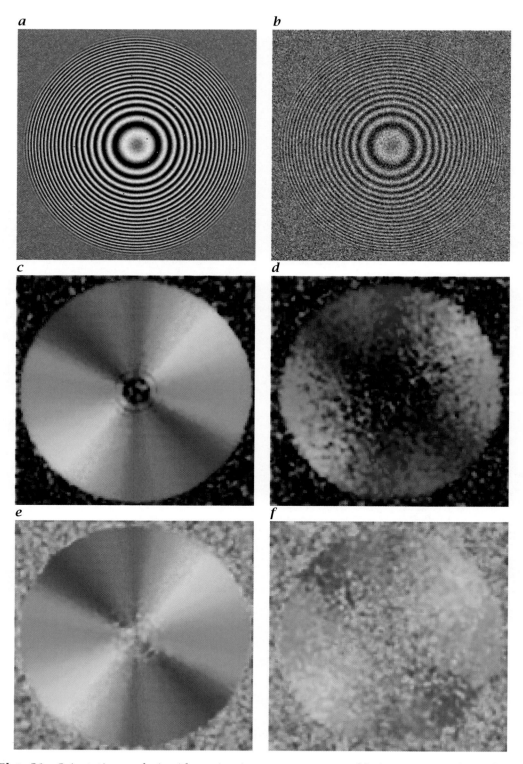

Plate 21: *Orientation analysis with a noisy ring test pattern.* ***a*** *and* ***b*** *ring pattern with amplitude 100 and 50, standard deviation of normal distributed noise 10 and 50, SNR 10 and 2, respectively;* ***c*** *and* ***d*** *color representation of the orientation vector;* ***e*** *and* ***f*** *color representation of the structure tensor.* *(See also Fig. 13.10, p. 438.)*

Plate 22: Plant leaf growth studies: images of young ricinus leaves taken in the visible (**a** and **c**, left) and near infrared wavelength range (800 – 900 nm) (**b** and **c**, right); **d** example image of a sequence taken to measure leaf growth; color representation of **e** the orientation vector and **f** the structure tensor of the image shown in **d**. (See also Fig. 13.12, p. 440.)

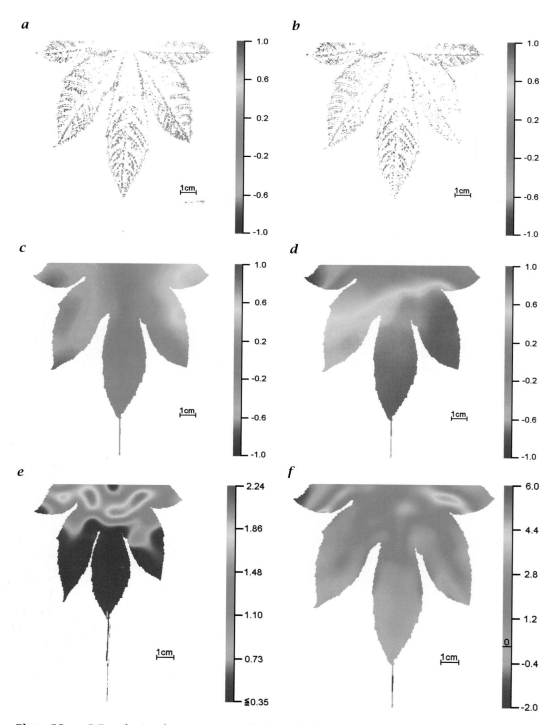

Plate 23: *2-D velocity determination of plant leaf growth:* **a** *Masked horizontal velocity,* **b** *masked vertical velocity,* **c** *interpolated horizontal velocity,* **d** *interpolated vertical velocity,* **e** *and* **f** *area dilation (divergence of the velocity vector field).* *(See also Fig. 13.13, p. 441.)*

Plate 24: *Examples of natural textures.* *(See also Fig. 14.1, p. 445.)*

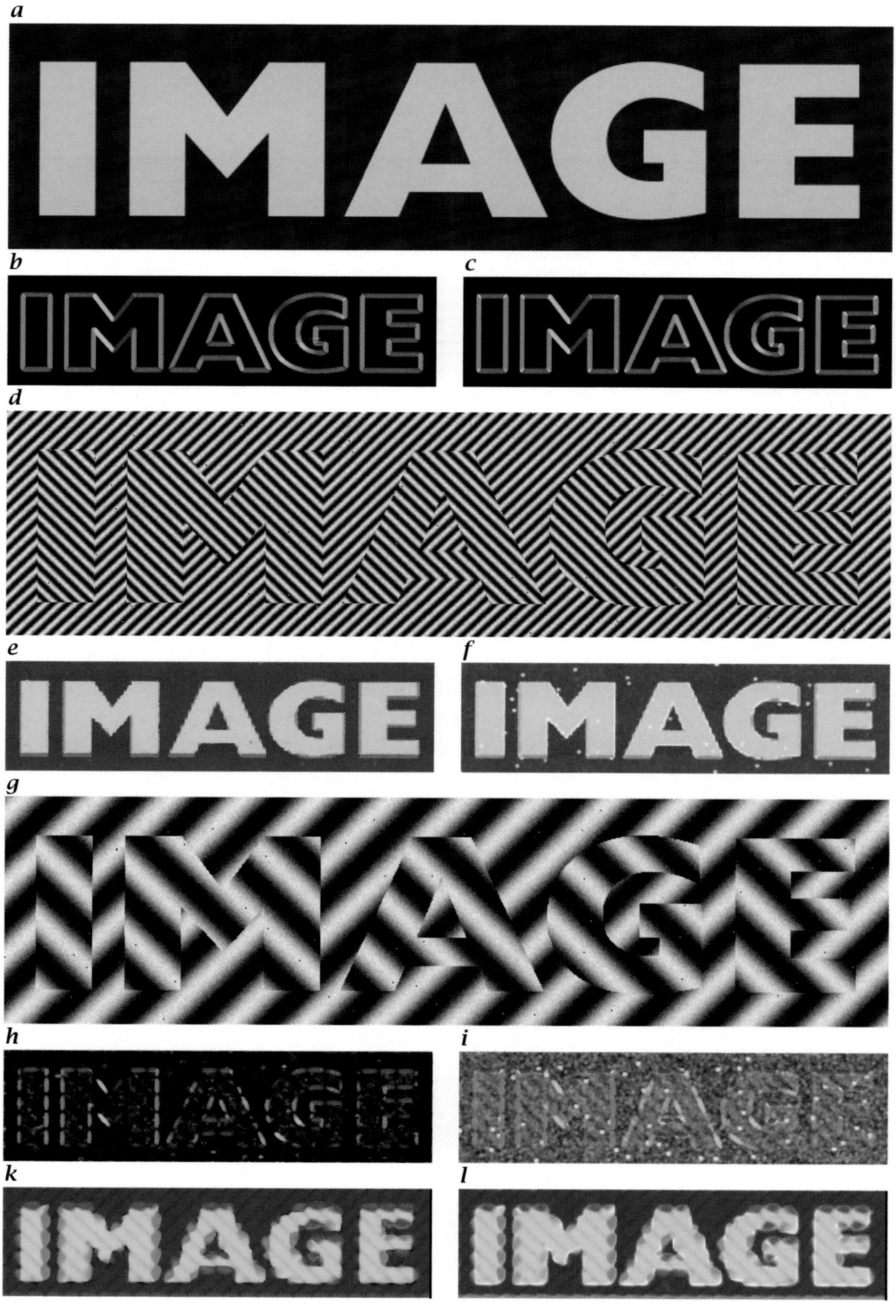

Plate 25: *Analysis of textures with the local structure tensor technique.* *(See also Fig. 14.2, p. 447.)*

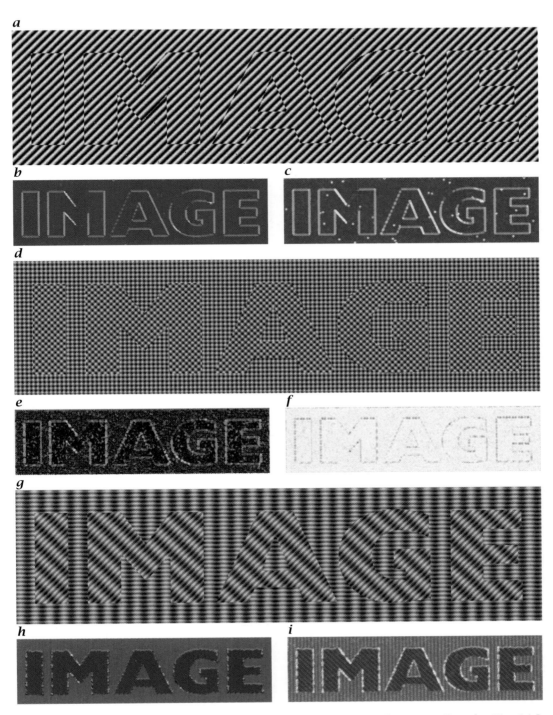

Plate 26: *Analysis of textures with the local structure tensor technique. (See also Fig. 14.3, p. 448.)*

Plate 27:   *Analysis of textures with the local structure tensor technique.   (See also Fig. 14.4, p. 449.)*